Gasdynamics: Theory and Applications

George Emanuel
The University of Oklahoma
Norman, Oklahoma

AIAA EDUCATION SERIES
J. S. Przemieniecki
Series Editor-in-Chief
Air Force Institute of Technology
Wright-Patterson Air Force Base, Ohio

Published by
American Institute of Aeronautics and Astronautics, Inc.
1633 Broadway, New York, N.Y. 10019

Texts Published in the AIAA Education Series

Re-Entry Vehicle Dynamics
  Frank J. Regan, 1984
Aerothermodynamics of Gas Turbine and Rocket Propulsion
  Gordon C. Oates, 1984
Aerothermodynamics of Aircraft Engine Components
  Gordon C. Oates, Editor, 1985
Aircraft Combat Survivability Analysis and Design
  Robert E. Ball, 1985
Intake Aerodynamics
  J. Seddon and E.L. Goldsmith, 1985
Composite Materials for Aircraft Structures
  Brian C. Hoskin and Alan A. Baker, Editors, 1986

American Institute of Aeronautics and Astronautics, Inc.
New York, New York

Library of Congress Cataloging in Publication Data

Emanuel, George.
  Gasdynamics, theory and applications.

(AIAA education series)
Includes index.
1. Gasdynamics. 2. Fluid dynamics. 3. Aerodynamics, Supersonic. I. Title. II. Series.
QC168.E6   1986   620.1'074   86-10771
ISBN 0-930403-12-6

Copyright © 1986 by the American Institute of Aeronautics and Astronautics, Inc. All rights reserved. Printed in the United States of America. No part of this publication may be reproduced, distributed, or transmitted, in any form or by any means, or stored in a data base or retrieval system, without the prior written permission of the publisher.

Foreword

Gasdynamics: Theory and Applications by George Emanuel is the seventh in the Education Series of textbooks and monographs published by the American Institute of Aeronautics and Astronautics (AIAA). Embracing a broad spectrum of theory and applications of many different disciplines of aerospace, including aerospace design practice, the Education Series develops both teaching texts and reference materials for practicing engineers and scientists. George Emanuel's text clearly serves both aims.

Emanuel covers all the important aspects of gasdynamics, with particular emphasis on compressible-flow theory and applications. The book is divided into two parts. Part I covers thermodynamic laws, conservation equations, steady streamtube flow, normal and oblique shock waves, Prandtl-Meyer flow and shock expansion theory, nozzle and diffuser flow, heat transfer, and unsteady one-dimensional flow. Part II deals with more advanced topics, such as shock waves, two-dimensional flow, theory of characteristics, aerodynamic windows, flows with shock waves, and waverider aerodynamics. The text should prove to be of great value for senior-level undergraduate and graduate courses and as a reference for the practicing aerospace engineer.

J. S. PRZEMIENIECKI
Editor-in-Chief
AIAA Education Series

Table of Contents

ix   **Preface**

PART I

1   **Chapter 1.   Introduction**

3   **Chapter 2.   Thermodynamics**
- 2.1   First Law of Thermodynamics
- 2.2   State Variables
- 2.3   Entropy
- 2.4   Reciprocity
- 2.5   Equations of State
- 2.6   Specific Heats
- 2.7   Isentropic Process
- 2.8   Second Law of Thermodynamics
- 2.9   Speed of Sound
-       Problems

23   **Chapter 3.   One-Dimensional Conservation Equations**
- 3.1   Streamtube Flow
- 3.2   Substantial Derivative
- 3.3   Conservation Equations
- 3.4   Change in Entropy
- 3.5   Acoustic Approximation
- 3.6   Summary
-       Problems

39   **Chapter 4.   Steady Streamtube Flow**
- 4.1   Integrated Form of the Governing Equations
- 4.2   Isentropic Relations
- 4.3   Area and Mass Flow Rate
-       Problems

49   **Chapter 5.   Normal and Oblique Shock Waves**
- 5.1   Steady, Normal Shock Waves
- 5.2   Oblique Shock Waves
-       Problems

69   **Chapter 6.   Prandtl-Meyer Flow and Shock-Expansion Theory**
- 6.1   Prandtl-Meyer Flow
- 6.2   Shock-Expansion Theory
-       Problems

89   **Chapter 7.   Nozzle and Diffuser Flow**
         7.1   Nozzle Flow
         7.2   Diffuser Flow
         Problems

113   **Chapter 8.   Ducts with Area Change, Heat Transfer, and Friction**
         8.1   Influence Coefficient Method
         8.2   Rayleigh Flow
         8.3   Fanno Flow
         Problems

147   **Chapter 9.   Unsteady, One-Dimensional Flow**
         9.1   Normal Shock Waves
         9.2   Reflected Normal Shock Waves
         9.3   Method of Characteristics
         9.4   Unsteady Expansion Waves
         Problems

181   **Chapter 10.   Applications of Unsteady, One-Dimensional Flow**
         10.1   Shock-Tube Flow
         10.2   Piston Expansion Tube
         Problems

PART II

Theory

197   **Chapter 11.   Governing Equations**
         11.1   Volume Dilatation
         11.2   Conservation Equations
         11.3   Conservative Form
         11.4   Boundary and Initial Conditions
         Problems

203   **Chapter 12.   Shock Waves**
         12.1   One-Dimensional Flow
         12.2   Oblique Shock Waves
         Problems

213   **Chapter 13.   Transformation of the Conservation Equations**
         13.1   General Theory
         13.2   Steady Two-Dimensional or Axisymmetric Flow—I
         13.3   Steady Two-Dimensional or Axisymmetric Flow—II
         13.4   Natural Coordinates
         13.5   Hodograph Transformation
         Problems

249 Chapter 14. Definitions and Theorems
    14.1 Basic Concepts
    14.2 Stream and Potential Functions
    14.3 Homogeneity of the Conservation Equations
    Problems

273 Chapter 15. Exact Solutions of Steady Homentropic Flow of a Perfect Gas
    15.1 Preliminary Remarks
    15.2 Spiral Flow
    15.3 Supersonic Flow Past a Cone
    Problems

283 Chapter 16. Theory of Characteristics
    16.1 Steady Two-Dimensional or Axisymmetric Flow
    16.2 Compatibility and Characteristic Equations
    Problems

Applications

307 Chapter 17. Minimum-Length Nozzles
    17.1 Preliminary Remarks
    17.2 Curved Sonic Line MLN
    17.3 Straight Sonic Line MLN
    17.4 MLN Comparisons
    Problems

329 Chapter 18. Aerodynamic Window
    18.1 Preliminary Remarks
    18.2 Theory for a Free-Vortex AW
    18.3 Design Procedure
    Problems

347 Chapter 19. Flows with Shock Waves
    19.1 Preliminary Remarks
    19.2 Stability of Shock Waves
    19.3 Flow Over a Compressive Ramp
    19.4 Formation of Shock Waves in Jets
    19.5 Shock Wave Reflection from a Wall in Steady Flow
    19.6 Pseudo-Steady Flow Over a Planar Compressive Ramp
    19.7 Shock Wave Interference
    Problems

399 Chapter 20. Waverider Aerodynamics
    20.1 Preliminary Remarks
    20.2 Caret-Shaped Waveriders
    20.3 C_L and C_D for an Arbitrary Waverider
    20.4 Waveriders Derived from a Conical Flowfield
    Problems

415	Appendix A.	SI Units and Nomenclature
419	Appendix B.	Thermodynamic Summary
421	Appendix C.	Streamtube Flow Equations
423	Appendix D.	Normal and Oblique Shock Summary
425	Appendix E.	Shock Wave Angle β vs Flow Deflection Angle θ
427	Appendix F.	Prandtl-Meyer Flow Summary
429	Appendix G.	Shock-Expansion Summary
431	Appendix H.	Nozzle Flow Summary
433	Appendix I.	Summary of Equations for Ducts with Area Change, Heat Transfer, and Friction
435	Appendix J.	Rayleigh Flow Summary
437	Appendix K.	Fanno Flow Summary
439	Appendix L.	Unsteady, Normal Shock Summary
441	Appendix M.	Reflected Shock Wave Summary
443	Appendix N.	Jacobian Theory
447	Subject Index	

Preface

Over the past four decades, a number of texts on compressible flow have appeared. *Gasdynamics: Theory and Applications* presents traditional material in a new way, as, for example, on unsteady flow in Chaps. 9 and 10 and conical flow in Chap. 15. The material in Part II is largely unavailable in other books.

Part I contains introductory material suitable for a one-semester senior-level course. To assist the student, the chapters in Part I contain many fully worked examples; and the appendices give concise summaries of the material.

Part II contains ample material for a one-semester first-year course at the graduate level. The subject matter builds directly on Part I, but is more analytically oriented. Part II is subdivided into theory, Chaps. 11–16, and applications, Chaps. 17–20. In both parts of the text, but especially in Part II, unresolved issues or difficulties are exposed; gasdynamics is an evolving subject, a point that should be brought home to students.

The material in Part II should provide a useful background for other advanced courses, such as transonic or hypersonic flow, and particularly for computational fluid dynamics (CFD). That is, Part II —largely unavailable in other books—bridges the gap between traditional gasdynamics and CFD, making this text do double service as a useful reference book.

Although not designed for a course in CFD, *Gasdynamics: Theory and Applications* includes many topics important to CFD, such as Jacobian theory, homogeneity, and the conservative form, which is given special treatment and considerable emphasis. Other topics of more general significance, such as shock waves, conical flow, and transformation theory, are treated in a manner consistent with CFD.

Gasdynamics: Theory and Applications emphasizes supersonic gasdynamics, in which the continuum flow is inviscid and adiabatic and body forces are negligible. Hence, the book does not consider viscous layers, magnetohydrodynamics, and rarefied gas flows. The book attempts to strike a balance between theory and applications. The applications often justify the assumptions of a nonreactive, nonradiating, nondiffusing perfect gas and a steady, two-dimensional or axisymmetric flow. The theory is tailored somewhat to these conditions.

Homework problems represent an integral and essential element of both parts of the text. By working through difficult problems and those requiring a theoretical derivation, students should develop the ability to handle unusual situations in future courses and in their professional careers.

A number of problems require a programmable calculator for solution. These are listed separately. I have not included the otherwise standard gas tables. Students are expected to produce their own sets. Problems for generating these tables appear at the start of the computational problem sections of Chaps. 4–9. The assignment is made early so that the tables can be utilized for the usual homework problems. Appendices C–K summarize the material in these chapters. Students can consult these appendices for the equations to be tabulated. Little time is required for this effort, and most of the students have enjoyed using one of the computer systems available for the task.

It is a pleasure to acknowledge the encouragement of my friend and colleague, Professor Maurice L. Rasmussen. I also wish to thank Farid Moslehi for his assistance in editing the text. Most of all, I wish to thank the University of Oklahoma for its generous support over the years that made this book possible. As author, I take full responsibility for any inaccuracies or oversights, and welcome comments or corrections.

GEORGE EMANUEL
The University of Oklahoma
Norman, Oklahoma

PART I

1. INTRODUCTION

The principles of gasdynamics are largely a product of the first half of this century. Interest in this topic was spurred by the early development of the supersonic wind tunnel, which preceded applications by many decades. In fact, most applications, such as supersonic flight or large thrust rocket nozzles, did not reach maturity until after 1950. Many excellent textbooks also first appeared in the 1950s, when gasdynamic courses were introduced into the mechanical and aerospace engineering curricula.

Our subject is the compressible motion of a gas. Of course, a liquid or a solid is compressible, but extreme pressure changes are required to effect minute changes in their density. In a gas, this is not the case because pressure and density changes are of comparable magnitude. Furthermore, only a factor of 2 pressure change is required for the speed of the gas to become appreciable.

The nondimensional parameter that measures compressibility in a fluid flow is the Mach number M, the ratio of the flow speed to the speed of sound. Generally, the demarcation between incompressible flow, as studied in earlier courses, and compressible flow is a Mach number value of about 0.4. Thus, natural phenomena such as hurricanes and tornadoes, which fortunately have a peak Mach number well below 0.4, belong to the incompressible flow regime. On the other hand, commercial aircraft typically cruise at, or slightly above, $M = 0.8$, where aerodynamic compressibility effects are significant.

The effects of compressibility, however, do not really become dominant until after the Mach number exceeds unity, when the flow is referred to as supersonic. We note that the flow at the exit of a thrust rocket nozzle is supersonic. Indeed, density and pressure in the combustion chamber can exceed by several orders of magnitude the density and pressure in the nozzle's exit plane. These enormous changes can occur over a distance as short as a few millimeters.

Applications of gasdynamic principles abound. These include turbine flow, gas lasers, aerodynamic windows, waverider missile aerodynamics, jet engines, and flow around a body entering the atmosphere (such as a meteorite). As evident from this list, most applications are man-made. In nature, thunder is the most common phenomenon in which compressibility is important; it is caused by a rapid and large energy release, i.e., lightning. In turn, this energy release creates a shock wave that propagates outward at a supersonic speed. If we are close to the lightning source, the pressure disturbance caused by the shock wave will be heard as a sharp crack rather than the more familiar rumble of thunder.

Thus, a sufficiently large change in density or pressure is associated with supersonic flow and with shock waves. Roughly speaking, a shock wave is an abrupt discontinuity that occurs in a supersonic flow in which the flow makes a sudden transition to a slower speed. Shock waves are a central feature of gasdynamics and will therefore receive considerable attention in our discussion.

The state of any substance is governed by thermodynamics. Often, this fact can be overlooked if the motion is incompressible. In such a case, we merely set the density equal to a constant. However, in a compressible flow the state can change appreciably and thermodynamics cannot be ignored. Thermodynamics enters in two principal ways: through the first and second laws of thermodynamics and through the equations of state. Compressible flow thus requires a merging of thermodynamics and mechanics. The two primary contributions of mechanics are the principles of conservation of matter and of momentum.

In earlier fluid mechanics courses, incompressible flow problems are often analyzed using just the momentum principle or conservation of mass. In general, this is not possible in a compressible flow, where all conservation laws plus thermodynamic state equations are necessary to obtain a solution. As we will discover, new and interesting phenomena can occur as a consequence of this complexity.

Fortunately, a number of analytical simplifications that are consistent with most gasdynamic applications can be made. For example, we assume a continuum flow with no body or viscous forces. The continuum assumption means that gas molecules are not treated as individual particles. Exclusion of body forces means that gravitational effects are negligible and that the gas is not hot enough to ionize appreciably. The disregard of viscous forces is the most restrictive of our assumptions. In an inviscid flow, the realistic no-slip wall condition cannot be imposed. Our approach is applicable to the bulk of the gaseous medium; it is invalid only in a thin boundary layer adjacent to the walls. For the forces on a vehicle or body moving at high speed in a gas, the inviscid assumption is often reasonable. In this circumstance, the pressure forces often dominate the viscous forces.

Our principal thermodynamic assumption is that the fluid is a thermally and calorically perfect gas. This is an excellent approximation for monatomic and diatomic gases, especially near room temperature. In particular, air is well represented by the perfect gas assumption for most applications.

The subject of gasdynamics consists of three components: theoretical, computational, and experimental gasdynamics. We will be concerned primarily with the first component, although some consideration is given to the others. For instance, equations are often specially formulated in a computationally suitable manner, and some of the homework problems require a programmable calculator.

The use of the Système International d'Unités (SI units) came into vogue in the 1970s. In this text we only use SI units, with occasional use of other common units for pressure. Appendix A contains a brief SI units table. Most of the time we will use dimensionless quantities so that choice of units is not of great importance. This appendix also contains the nomenclature.

2. THERMODYNAMICS

The science of thermodynamics is concerned primarily with two topics: (1) the states of matter and (2) the transformation of energy from one form to another. These topics are closely connected, inasmuch as a transfer of energy results in a change of state.

In this chapter, we will discuss thermodynamics largely to the extent needed for understanding material in subsequent chapters. A summary of important results is contained in Appendix B.

2.1 FIRST LAW OF THERMODYNAMICS

A thermodynamic system is a region containing energy and matter that is separated from its surroundings. This definition, however, is too general for our purposes. We will be concerned only with a simple closed system. In a simple system, the substance is homogeneous and isotropic. Furthermore, chemical reactions, electromagnetic phenomena, or a gravitational field are not allowed. In a closed system, no mass crosses the boundary of the system in either direction.

When we refer to a unit mass of gas, we mean a simple, closed thermodynamic system that contains a unit mass of material in the gaseous state. Thus, no material crosses the system's boundary either by diffusion or by convection. However, transport of energy—for example, by heat conduction, radiative energy transfer, or a rotating paddle wheel—into or out of the system is allowed. In addition, work may be done by the surroundings on the system, or vice versa. This work is performed by the boundary of the system either pushing, or being pushed by, the surroundings. Thus, the system of interest can interact with its surroundings in only two ways: by energy transfer or by work.

We would like to connect the above transfer mechanisms with the properties of the system's material. Clearly, heat transfer into the system will alter it in some manner. The connection between the transfer processes and the state of the system is provided by the first law of thermodynamics. In its most elementary form, this law connects infinitesimal changes in the heat transfer and work with the internal energy of a simple, closed thermodynamic system:

$$\mathrm{d}e = \mathrm{d}q + \mathrm{d}w \tag{2.1}$$

In Eq. (2.1), we have:

e = specific internal energy,

q = heat transfer into the gas, per unit mass of gas, and

w = work performed on the gas, per unit mass of gas.

The term specific means the property is for a unit mass. Lowercase symbols, such as q and w, also refer to a unit mass system. When dq is positive, heat is transferred into the system, and similarly when dw is positive, work is performed on the system. Positive values for dq and dw result in an increase in internal energy.

Equation (2.1) is a statement of conservation of energy. This terminology is ambiguous, since it is the energy of the system in combination with its surroundings that is conserved.

We now derive an expression for the mechanical work term by considering a frictionless piston/cylinder arrangement, shown in Fig. 2.1. The volume V contains a mass m of gas and constitutes a closed thermodynamic system. By definition, the specific volume is

$$v = V/m$$

or in differential form

$$dV = m\, dv \tag{2.2}$$

The volume also is given by

$$V = A(x_0 - x)$$

where A is the piston's surface area and x_0 is a constant. Again, by differentiation

$$dV = -A\, dx \tag{2.3}$$

We eliminate dV from Eqs. (2.2) and (2.3) to obtain

$$dx = -\frac{m}{A}\, dv \tag{2.4}$$

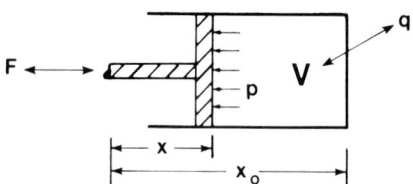

Fig. 2.1 Frictionless piston/cylinder arrangement.

Not only is friction ignored but also the mass of the piston and the external, or ambient, pressure. The pressure force on the piston then is

$$F = pA \tag{2.5}$$

where p is the gas pressure in volume V. From mechanics, the differential of the work, per unit mass of gas, done on the system is given by

$$dw = \frac{F}{m} dx$$

We use Eqs. (2.4) and (2.5) to eliminate F and dx to obtain

$$dw = \frac{1}{m}(pA)\left(-\frac{m}{A} dv\right) = -p\, dv \tag{2.6}$$

Hence, the first law becomes

$$de = dq - p\, dv \tag{2.7}$$

The replacement of dw with $-p\, dv$ requires a reversible process, which is defined shortly. At this time we simply require a slow piston motion so that the pressure is very nearly uniform throughout the gaseous volume.*

2.2 STATE VARIABLES

Consider a simple thermodynamic system consisting of a uniform gas. The state of the system is described by variables, such as specific volume v, pressure p, and internal energy e. By the expression state variable, we mean a quantity whose value does not depend on the past history of the system. Thus, a state variable, or property, informs us about the current status of the system, not about the processes that previously occurred.

Of particular importance are state variables that are intensive and thus do not depend on the size, weight, or configuration of the system. The quantities v, p, and e are all intensive state variables.

A more precise definition of a state variable is that its integral between two end states depends only on the end states. In other words, pressure and specific volume are state variables, since

$$\int_1^2 dp = p_2 - p_1, \qquad \int_1^2 dv = v_2 - v_1$$

where indices 1 and 2 denote the two end states. A knowledge of the process between states 1 and 2 is not required. On the other hand, w and q

---

*If the piston has a rapid compressive motion, then a shock wave is generated. The pressure is different on each side of the shock wave and the system is no longer homogeneous and, therefore, not simple; nor is it close to an equilibrium state.

are not state variables, since

$$\int_1^2 dw \neq w_2 - w_1, \qquad \int_1^2 dq \neq q_2 - q_1$$

A path that represents the actual process would have to be used for these integrals.

In a simple thermodynamic system, the state of a uniform gas depends on only two intensive state variables. Consequently, the internal energy is determined by the pressure and specific volume, i.e.,

$$e = e(p, v)$$

Since we are free to choose any two state variables, we can also write

$$p = p(e, v)$$

or

$$v = v(p, e)$$

A fourth state variable of importance is the temperature T. In addition, we shall need the density ρ and enthalpy h, which are defined by

$$\rho = 1/v \qquad (2.8)$$

$$h = e + pv = e + (p/\rho) \qquad (2.9)$$

Since neither w nor q are state variables, it is desirable to replace the differentials, dw and dq, in Eq. (2.1) with expressions that involve only state variables. This is the reason behind the piston/cylinder derivation of Sec. 2.1. The result of the derivation, Eq. (2.6), replaces dw with such an expression. In the next section, dq is replaced by a state variable expression.

2.3 ENTROPY

Any real process is irreversible. The unique aspect about an irreversible process is that energy dissipation occurs. For example, this dissipation is due to frictional heating, or heat transfer in which the temperature difference is finite.

In many processes, or parts of processes, the dissipation is negligible. Except in a very thin boundary layer along the wall, gas flow in a nozzle is one such process. When energy dissipation can be ignored, we have a reversible process. The piston/cylinder process described in Sec. 2.1 is such a process, since the piston's motion is frictionless. Equally important is the requirement of a slow motion for the piston. With this condition, viscous dissipation within the gas can be ignored.

Our interest in reversible processes is motivated by the need to eliminate dq from the first law. This goal is accomplished by defining a new state variable, called entropy, by means of

$$ds = \frac{(dq)_{rev}}{T} \tag{2.10}$$

where the heat transfer must be for a reversible process. We now choose a reversible heat-transfer process

$$(dq)_{rev} = T ds$$

to use in Eq. (2.7) for the first law

$$de = T ds - p dv \tag{2.11}$$

This relation is of greater value than either Eq. (2.1) or (2.7) since it contains only state variables.

In deriving Eq. (2.11), we twice assumed a reversible process, namely,

$$(dw)_{rev} = -p dv, \quad (dq)_{rev} = T ds$$

However, Eq. (2.11) is *not* restricted to a reversible process. The reason for this is that now only state variables occur in it. The change in any state variable, such as

$$\int_1^2 de = e_2 - e_1, \quad \int_1^2 ds = s_2 - s_1$$

is independent of whether the actual path from state 1 to state 2 is reversible or irreversible. Any convenient reversible path from 1 to 2 can be chosen for the integration of Eq. (2.11), even though the actual process is irreversible.

In general, the integrals of dw and dq depend on the path of a given process. Often there is a way around this restriction. Consider an irreversible but adiabatic (i.e., $dq = 0$) process, such as occurs in Fanno flow, which is analyzed in Chap. 8. Then, from Eq. (2.1), we have

$$dw = de$$

or

$$w = \int_1^2 de = e_2 - e_1$$

where the integral holds for any convenient path between states 1 and 2.

Thermodynamics is well defined only for equilibrium states. Although the process between states 1 and 2 may be reversible or irreversible, the end states themselves must be equilibrium states. Equilibrium thermodynamics is then used to connect end states by assuming a reversible path between them. Thus, an adiabatic (reversible or irreversible) process between equilibrium states 1 and 2 allows the calculation of, say, $e_2 - e_1$, by using a

$ds = 0$ process. Of course, we must also know something else about states 1 and 2, say, T_1 and T_2, in order to obtain a numerical value for $e_2 - e_1$.

2.4 RECIPROCITY

For a simple system, only two types of state equations occur, and both are needed. In a general form, they can be written as

$$p = p(v, T), \text{ thermal state equation} \qquad (2.12a)$$

and

$$e = e(v, T), \text{ caloric state equation} \qquad (2.12b)$$

These relations are not independent of each other. The derivation of a unique constraint between them is the objective of this section.

We start by differentiating the caloric state equation to obtain

$$de = \left.\frac{\partial e}{\partial T}\right|_v dT + \left.\frac{\partial e}{\partial v}\right|_T dv$$

This is substituted into Eq. (2.11)

$$\left.\frac{\partial e}{\partial T}\right|_v dT + \left.\frac{\partial e}{\partial v}\right|_T dv = T\,ds - p\,dv$$

which rearranges to

$$ds = \frac{1}{T}\left.\frac{\partial e}{\partial T}\right|_v dT + \left(\frac{1}{T}\left.\frac{\partial e}{\partial v}\right|_T + \frac{p}{T}\right) dv$$

Since entropy is a state variable, we can write

$$s = s(T, v)$$

$$ds = \left.\frac{\partial s}{\partial T}\right|_v dT + \left.\frac{\partial s}{\partial v}\right|_T dv$$

Upon comparison with the first ds equation, we have

$$\left.\frac{\partial s}{\partial T}\right|_v = \frac{1}{T}\left.\frac{\partial e}{\partial T}\right|_v$$

$$\left.\frac{\partial s}{\partial v}\right|_T = \frac{1}{T}\left.\frac{\partial e}{\partial v}\right|_T + \frac{p}{T} \qquad (2.13)$$

However, mathematics requires that the second partial derivatives

$$\frac{\partial}{\partial v}\left[\left.\frac{\partial s}{\partial T}\right|_v\right]_T = \frac{\partial}{\partial T}\left[\left.\frac{\partial s}{\partial v}\right|_T\right]_v$$

THERMODYNAMICS

be equal. Applying this to Eqs. (2.13) yields

$$\frac{\partial}{\partial v}\left[\frac{\partial s}{\partial T}\bigg|_v\right]_T = \frac{\partial}{\partial v}\left[\frac{1}{T}\frac{\partial e}{\partial T}\bigg|_v\right]_T = \frac{1}{T}\frac{\partial}{\partial v}\left[\frac{\partial e}{\partial T}\bigg|_v\right]_T = \frac{1}{T}\frac{\partial}{\partial T}\left[\frac{\partial e}{\partial v}\bigg|_T\right]_v$$

and

$$\frac{\partial}{\partial T}\left[\frac{\partial s}{\partial v}\bigg|_T\right]_v = \frac{\partial}{\partial T}\left[\frac{1}{T}\frac{\partial e}{\partial v}\bigg|_T + \frac{p}{T}\right]_v$$

$$= -\frac{1}{T^2}\frac{\partial e}{\partial v}\bigg|_T + \frac{1}{T}\frac{\partial}{\partial T}\left[\frac{\partial e}{\partial v}\bigg|_T\right]_v - \frac{p}{T^2} + \frac{1}{T}\frac{\partial p}{\partial T}\bigg|_v$$

By equating the above relations for the second partial derivatives of s, we obtain

$$\frac{1}{T}\frac{\partial}{\partial T}\left[\frac{\partial e}{\partial v}\bigg|_T\right]_v = -\frac{1}{T^2}\frac{\partial e}{\partial v}\bigg|_T + \frac{1}{T}\frac{\partial}{\partial v}\left[\frac{\partial e}{\partial T}\bigg|_v\right]_T - \frac{p}{T^2} + \frac{1}{T}\frac{\partial p}{\partial T}\bigg|_v$$

or

$$\frac{\partial e}{\partial v}\bigg|_T = T\frac{\partial p}{\partial T}\bigg|_v - p \qquad (2.14)$$

This equation is called a reciprocity or a Maxwell relation. It represents a constraint on the form of either of Eqs. (2.12), given the other. In a simple system, this is the only constraint on Eq. (2.12a) or (2.12b).

2.5 EQUATIONS OF STATE

Instead of the general thermal state equation, Eq. (2.12a), our discussion often will be limited to a thermally perfect gas

$$p = RT/v = \rho RT \qquad (2.15)$$

where

$$R = \frac{\tilde{R}}{W} = \frac{8314}{W} \qquad (2.16)$$

The gas constant R is in J/kg-K, W is the molecular weight, and ρ is the density, which is defined by Eq. (2.8). Since this approximation is more than adequate in most applications, the use of a thermally perfect gas is not a severe limitation.

In SI units, the thermally perfect gas equation for room temperature air at 1-atm pressure is

$$\underbrace{1.013 \times 10^5}_{p} = \underbrace{\frac{8314}{28.97}}_{\tilde{R}/W} \times \underbrace{1.226}_{\rho} \times \underbrace{288}_{T}$$

where p is in N/m² (Pascals), W is in kg/kmole, $\tilde{R}$ (the universal gas constant) is in J/kmole-K, ρ is in kg/m³, and T is in K (see Appendix A).

We now substitute Eq. (2.15) into the right side of the reciprocity equation, to obtain

$$\left.\frac{\partial e}{\partial v}\right|_T = T\left.\frac{\partial p}{\partial T}\right|_v - p = T\frac{R}{v} - \frac{RT}{v} = 0$$

Consequently, for a thermally perfect gas

$$e = e(T) \tag{2.17}$$

Such a gas has no intermolecular forces and cannot condense to a liquid or a solid. (The critical point for a thermally perfect gas does not occur at a finite temperature.) Conversely, if the caloric equation of state is $e = e(T)$, then the thermal equation of state must satisfy reciprocity

$$T\left.\frac{\partial p}{\partial T}\right|_v - p = 0$$

This equation is integrated as follows:

$$\frac{1}{p}\left.\frac{\partial p}{\partial T}\right|_v = \frac{1}{T}$$

$$\ln p = \ln T + \ln\left[R\frac{g(v)}{v}\right]$$

$$p = (RT/v)g(v) \tag{2.18}$$

where $g(v)$ is an arbitrary function of integration. In summary, we have from reciprocity

$$p = RT/v \Rightarrow e = e(T)$$

and

$$e = e(T) \Rightarrow p = (RT/v)g(v)$$

It is convenient, at this point, to introduce enthalpy, which is defined by Eq. (2.9). If the gas is thermally perfect, then

$$h = e(T) + RT = h(T) \tag{2.19}$$

and both e and h only depend on T. Furthermore, their derivatives then differ by the gas constant

$$\frac{dh}{dT} = \frac{de}{dT} + R \tag{2.20}$$

THERMODYNAMICS

If we are to use a thermally perfect gas, then the caloric state equation is Eq. (2.17). A further realistic simplification occurs if we assume

$$e = (\text{const}) \cdot T$$

which defines a calorically perfect gas. In this case, we see from Eq. (2.20) that

$$h = (\text{const} + R) \cdot T$$

Note that a calorically perfect gas does not necessarily imply a thermally perfect gas because of the $g(v)$ factor in Eq. (2.18).

When a gas is both thermally and calorically perfect, we will refer to it hereafter as a perfect gas.

Most of the time, we will assume a perfect gas for the analysis. This assumption greatly simplifies the analysis and is often warranted by the nature of the flow. There are, however, important situations when the perfect gas assumption is not valid. For instance, in the test section of a cryogenic wind tunnel the gas can be close to liquefaction. In this circumstance, real gas effects are important, and both the thermal and caloric state equations are nonperfect. The reciprocity equation must be satisfied if the two state equations are to be thermodynamically consistent with each other.

2.6 SPECIFIC HEATS

The heat transfer dq is associated with a temperature difference dT. This motivates us to determine

$$c = \frac{dq}{dT}$$

Since q is not a state variable, it is necessary to specify a heat-transfer process. As evidenced by alternate forms of the first law,

$$dq = de + p\,dv$$

and

$$dq = dh - v\,dp$$

two processes prove convenient. Dividing both sides by dT, yields

$$c = \frac{dq}{dT} = \frac{de}{dT} + p\frac{dv}{dT}$$

and

$$c = \frac{dq}{dT} = \frac{dh}{dT} - v\frac{dp}{dT}$$

If the heat transfer is for a constant volume process, $dv = 0$, then from the first equation, with c redefined as c_v,

$$c_v = \left.\frac{\partial e}{\partial T}\right|_v \tag{2.21a}$$

If the heat transfer is for a constant pressure process, $dp = 0$, then from the second equation, with c redefined as c_p,

$$c_p = \left.\frac{\partial h}{\partial T}\right|_p \tag{2.21b}$$

Both c_v and c_p are state variables and hold even when the process is neither at constant volume nor at constant pressure.

If the gas is thermally perfect, Eqs. (2.21) become

$$c_v = \frac{de}{dT}, \quad c_p = \frac{dh}{dT}$$

and c_v and c_p can depend only on the temperature. In fact, we have from Eq. (2.20)

$$c_p(T) = c_v(T) + R \tag{2.22}$$

It is convenient to introduce the dimensionless ratio of specific heats

$$\gamma = c_p/c_v$$

which only depends on T for a thermally perfect gas. With Eq. (2.22), it is easy to show that

$$c_v = \frac{1}{\gamma - 1}R, \quad c_p = \frac{\gamma}{\gamma - 1}R$$

If the gas is perfect, then

$$e = c_v T, \quad c_v = \text{const}$$
$$h = c_p T, \quad c_p = \text{const}$$
$$\gamma = c_p/c_v = \text{const}$$

and one can show that

$$\begin{aligned}\gamma &= 5/3, \text{ any monatomic gas} \\ &= 7/5, \text{ any diatomic gas}\end{aligned} \tag{2.23}$$

Air is a mixture of diatomic molecules and, under many conditions, is well represented by the perfect gas assumption with $\gamma = 1.4$. In this case, c_p is given by

$$c_p = \frac{\gamma}{\gamma - 1} \frac{\tilde{R}}{W} = \frac{1.4}{1.4 - 1} \frac{8314}{28.97} = 1.00 \times 10^3 \text{ J/kg-K}$$

In the following example, we illustrate the usefulness of Eq. (2.14) for reciprocity.

Example 1

We are to determine the most general possible relations for e, c_v, and c_p that are consistent with the equation of state

$$p = \frac{RT}{v - b} - \frac{a}{T(v + c)^2} \tag{2.24}$$

where a, b, and c are empirically determined constants. This relation is called a Clausius-II state equation.

We first determine the partial derivative

$$\left.\frac{\partial p}{\partial T}\right|_v = \frac{R}{v - b} + \frac{a}{T^2(v + c)^2}$$

Equation (2.14) thus becomes

$$\left.\frac{\partial e}{\partial v}\right|_T = \frac{RT}{v - b} + \frac{a}{T(v + c)^2} - \frac{RT}{v - b} + \frac{a}{T(v + c)^2} = \frac{2a}{T(v + c)^2}$$

Upon integration, we have

$$e = f(T) + \frac{2a}{T} \int \frac{dv}{(v + c)^2} = f - \frac{2a}{T(v + c)} \tag{2.25a}$$

where f is a function of integration that only depends on T. In this form, the internal energy only depends on v and T. Hence, by differentiation with respect to T, we have

$$c_v = \left.\frac{\partial e}{\partial T}\right|_v = \frac{df}{dT} + \frac{2a}{T^2(v + c)} \tag{2.25b}$$

In order to determine c_p, we first evaluate h as follows:

$$h = e + pv = f - \frac{2a}{T(v + c)} + pv$$

Here, the enthalpy depends not only on p and T, which is necessary for determining c_p, but also on v. Consequently, differentiation with respect to T yields

$$c_p = \left.\frac{\partial h}{\partial T}\right|_p = \frac{df}{dT} + \frac{2a}{T^2(v+c)} + \frac{2a}{T(v+c)^2}\left.\frac{\partial v}{\partial T}\right|_p + p\left.\frac{\partial v}{\partial T}\right|_p$$

$$= \frac{df}{dT} + \frac{2a}{T^2(v+c)} + \left[\frac{RT}{v-b} + \frac{a}{T(v+c)^2}\right]\left.\frac{\partial v}{\partial T}\right|_p$$

By implicit differentiation of Eq. (2.24), we obtain the partial derivative $(\partial v/\partial T)_p$

$$0 = \frac{R}{v-b} - \frac{RT}{(v-b)^2}\left.\frac{\partial v}{\partial T}\right|_p + \frac{a}{T^2(v+c)^2} + \frac{2a}{T(v+c)^3}\left.\frac{\partial v}{\partial T}\right|_p$$

or

$$\left.\frac{\partial v}{\partial T}\right|_p = \frac{[R/(v-b)] + \{a/[T^2(v+c)^2]\}}{\{RT/[(v-b)^2]\} - \{2a/[T(v+c)^3]\}}$$

We thus have for c_p

$$c_p = \frac{df}{dT} + \frac{\dfrac{R^2T}{(v-b)^2} - \dfrac{3a^2}{T^3(v+c)^4} + \dfrac{2aR(2v+c-b)}{T(v-b)^2(v+c)^2}}{\dfrac{RT}{(v-b)^2} - \dfrac{2a}{T(v+c)^3}} \quad (2.25c)$$

Equations (2.25) constitute the answers originally called for. The function f is generally determined by spectroscopic data in conjunction with statistical mechanics.

Equation (2.24) is not much more complicated than the thermally perfect equation of state, which is a special case of Eq. (2.24). Aside from $f(T)$, reciprocity determines the functional form of the other state variables, such as the specific heats. These other variables depend on both v and T and are often quite complex.

In the following example, we illustrate how the first law is applied to a perfect gas.

Example 2

Air is compressed in a cylinder from a value of 1-atm pressure and 300 K to 10% of its initial volume. The heat transfer is given by

$$dq = -h_f R\,d(T - T_w)$$

for which the dimensionless film coefficient h_f is 2 and the wall temperature T_w is 300 K. We are to determine the final pressure and temperature.

We assume air to be a perfect gas with

$$\gamma = 1.4, \qquad R = 287 \text{ J/kg-K}$$

Subscripts 1 and 2 denote, respectively, the initial and final equilibrium states of the gas. A reversible process is assumed, so that we can utilize Eq. (2.7)

$$de = dq - p\, dv$$

Notice that the following relations apply:

$$e = c_v T = \frac{R}{\gamma - 1} T$$

$$de = \frac{R}{\gamma - 1} dT$$

$$dq = -2R\, dT$$

$$p = RT/v$$

Thus, the first law becomes

$$\frac{R}{\gamma - 1} dT = -2R\, dT - \frac{RT}{v} dv$$

which rearranges to

$$\left(\frac{2\gamma - 1}{\gamma - 1} \right) \frac{dT}{T} = - \frac{dv}{v}$$

We integrate from state 1 to state 2, to obtain

$$\left(\frac{2\gamma - 1}{\gamma - 1} \right) \ln \frac{T_2}{T_1} = \ln \frac{v_1}{v_2}$$

With $(v_1/v_2) = 10$, T_2 is given by

$$T_2 = T_1 \left(\frac{v_1}{v_2} \right)^{[(\gamma - 1)/(2\gamma - 1)]} = 300(10)^{0.4/1.8} = 500.4 \text{ K}$$

The pressure p_2 is determined by

$$\frac{p_2}{p_1} = \frac{T_2}{T_1} \frac{v_1}{v_2}$$

or
$$p_2 = 1 \times \frac{500.4}{300} \times 10 = 16.68 \text{ atm}$$

The heat transfer q can be determined by integrating $dq = -2R\,dT$. Since q is negative, the gas is cooled by heat transfer during the compression process.

2.7 ISENTROPIC PROCESS

Let us consider a reversible process from state 1 to state 2. The change in entropy is given by Eq. (2.10), so that

$$ds = \frac{(dq)_{\text{rev}}}{T}$$

or, on integration,

$$\Delta s = s_2 - s_1 = \int_1^2 \frac{(dq)_{\text{rev}}}{T}$$

By means of the first law, this becomes

$$\Delta s = \int_1^2 \frac{de}{T} + \int_1^2 \frac{p}{T}\,dv$$

With the aid of Eqs. (2.12b), (2.14), and (2.21a), we obtain

$$\Delta s = \int_1^2 \frac{1}{T}\left[\left.\frac{\partial e}{\partial T}\right|_v dT + \left.\frac{\partial e}{\partial v}\right|_T dv\right] + \int_1^2 \frac{p}{T}\,dv$$

$$= \int_1^2 \frac{c_v}{T}\,dT + \int_1^2 \frac{1}{T}\left[\left.\frac{\partial e}{\partial v}\right|_T + p\right]dv = \int_1^2 \frac{c_v}{T}\,dT + \int_1^2 \left.\frac{\partial p}{\partial T}\right|_v dv \quad (2.26)$$

which is a general thermodynamic result, not restricted to a perfect gas or to a reversible process. However, derivatives of both state equations are required in order to evaluate the rightmost integrals.

For a perfect gas, we have

$$c_v = \frac{R}{\gamma - 1} = \text{const}$$

and

$$\left.\frac{\partial p}{\partial T}\right|_v = \frac{R}{v}$$

Hence, Eq. (2.26) becomes

$$\Delta s = \frac{R}{\gamma-1}\int_1^2 \frac{dT}{T} + R\int_1^2 \frac{dv}{v} = \frac{R}{\gamma-1}\ln\left(\frac{T_2}{T_1}\right) + R\ln\left(\frac{v_2}{v_1}\right)$$

$$\frac{\Delta s}{R} = \ln\left(\frac{T_2}{T_1}\right)^{1/(\gamma-1)} - \ln\frac{\rho_2}{\rho_1} = \ln\left[\frac{\rho_1}{\rho_2}\left(\frac{T_2}{T_1}\right)^{1/(\gamma-1)}\right]$$

This result is simplified by setting

$$\frac{s_1}{R} = -\ln\left[\rho_1/T_1^{1/(\gamma-1)}\right] + \text{const}$$

and dropping the state 2 subscript. Thus, we obtain

$$\frac{s}{R} = \ln\left[T^{1/(\gamma-1)}/\rho\right] + \text{const} \tag{2.27}$$

for a perfect gas in which the above constants are the same.

For a real flow, an important property of s is that it often is approximately constant. In this circumstance, $\Delta s = 0$ and, from Eq. (2.27), we have

$$\frac{\rho_2}{\rho_1} = \left(\frac{T_2}{T_1}\right)^{1/(\gamma-1)} \tag{2.28a}$$

From the state equation $p = \rho RT$, we obtain

$$\frac{p_2}{p_1} = \frac{\rho_2}{\rho_1}\frac{T_2}{T_1} = \left(\frac{T_2}{T_1}\right)^{\gamma/(\gamma-1)} \tag{2.28b}$$

Other forms can be derived as needed. Equations (2.28) are called the isentropic relations; a reversible, adiabatic process is isentropic.

2.8 SECOND LAW OF THERMODYNAMICS

There are many versions of the second law, such as the one due to Clausius: "Heat cannot pass spontaneously from a lower to a higher temperature level." The version most appropriate to compressible flow is

$$ds \geq \frac{dq}{T} \tag{2.29}$$

where dq is the actual heat transfer. If the transfer is reversible, the equality sign applies. For a real, adiabatic process, $ds > 0$, i.e., the entropy increases monotonically for such a process. If dq is negative, as in Example 2, then s may decrease during the process.

2.9 SPEED OF SOUND

The speed of sound is denoted by a. By means of acoustic theory, one can show that

$$a^2 = \left.\frac{\partial p}{\partial \rho}\right|_s \tag{2.30}$$

For a perfect gas, this relation will be derived in Sec. 3.5. Since a is a thermodynamic state variable, however, it is convenient to introduce it here.

For a perfect gas, Eq. (2.27) can be written as

$$\frac{s}{R} = \ln\left[\left(\frac{p}{R}\right)^{1/(\gamma-1)} \bigg/ \rho^{\gamma/(\gamma-1)}\right] + \text{const}$$

when T is replaced with the thermal equation of state. This equation is solved for p, with the result

$$p = (p_1/\rho_1^\gamma)\rho^\gamma \exp[(s - s_1)/c_v]$$

Consequently, the partial derivative in Eq. (2.30) is

$$\left.\frac{\partial p}{\partial \rho}\right|_s = (p_1/\rho_1^\gamma)\exp[(s - s_1)/c_v]\gamma\rho^{\gamma-1} = \gamma\frac{p}{\rho}$$

and the speed of sound for a perfect gas is

$$a = \left(\gamma\frac{p}{\rho}\right)^{\frac{1}{2}} = (\gamma RT)^{\frac{1}{2}} \tag{2.31}$$

This relation is often used in later chapters. Air at room temperature has the value

$$a_{\text{air}} = \left(1.4 \times \frac{8314}{28.97} \times 288\right)^{\frac{1}{2}} = 340 \text{ m/s}$$

In the final example, we synthesize much of the foregoing discussion.

Example 3

We are to determine c_p, s, and a for the equations of state

$$p = (RT/v) - \alpha \tag{2.32a}$$

and

$$e = c_v T + \alpha v + \beta \tag{2.32b}$$

where R, c_v, α, and β are constants.

We first show that Eqs. (2.32) are compatible with each other. From the state equations we obtain

$$\left.\frac{\partial e}{\partial v}\right|_T = \alpha, \qquad \left.\frac{\partial p}{\partial T}\right|_v = \frac{R}{v}$$

This is substituted into Eq. (2.14), which yields an identity, and verifies the compatibility of Eqs. (2.32).

We next determine the enthalpy, as follows:

$$h = e + pv = c_v T + \alpha v + \beta + RT - \alpha v = (c_v + R)T + \beta$$

so that

$$c_p = \left.\frac{\partial h}{\partial T}\right|_p = c_v + R = \text{const} \qquad (2.33a)$$

Hence, both c_v and c_p are constant and satisfy Eq. (2.22).

To determine s, we start with Eq. (2.11)

$$de = T\,ds - p\,dv$$

With the use of Eqs. (2.32), this becomes

$$c_v\,dT + \alpha\,dv = T\,ds - RT\frac{dv}{v} + \alpha\,dv$$

which simplifies to

$$ds = c_v \frac{dT}{T} + R\frac{dv}{v}$$

We integrate to obtain

$$s = s_1 + R\,\ell n\left[\left(\frac{v}{v_1}\right)\left(\frac{T}{T_1}\right)^{c_v/R}\right] \qquad (2.33b)$$

for the entropy, where the subscript 1 refers to the initial state.

For the speed of sound, replace v and T in Eq. (2.33b) with

$$v = \frac{1}{\rho}, \qquad T = \frac{p+\alpha}{R\rho}$$

to obtain

$$s = s_1 + R\,\ell n\left[\left(\frac{\rho_1}{\rho}\right)^{c_p/R}\left(\frac{p+\alpha}{p_1+\alpha}\right)\right] = s_1 + c_v\gamma\,\ell n(\rho_1/\rho) + \ell n\frac{p+\alpha}{p_1+\alpha}$$

This relation is solved for p, to obtain

$$p = -\alpha + (p_1 + \alpha)(\rho/\rho_1)^\gamma \exp[(s - s_1)/c_v]$$

From Eq. (2.30), we have

$$a = (\gamma RT)^{\frac{1}{2}} \tag{2.33c}$$

which coincides with the earlier perfect gas result. Equations (2.33) are the answers originally called for.

Problems

2.1 Argon ($W = 40$ kg/kmole) is isentropically expanded from $V_1 = 0.3$ m³ to $V_2 = 0.8$ m³. Determine T_2/T_1, p_2/p_1, and $s_2 - s_1$.

2.2 Air at 290 K and 1-atm pressure is adiabatically compressed from 0.35 m³ to 0.025 m³. Denote the initial state by 1 and the final state by 2. Compute p_2, T_2, ρ_2, $e_2 - e_1$, the work w, $(s_2 - s_1)/R$, and the mass of the gas.

2.3 Air at 290 K and 1 atm is compressed from 0.35 m³ to 0.025 m³ and a pressure p_2 of 7 atm. Except for p_2 and w, compute the quantities called for in Problem 2.2. Why is w not called for?

2.4 Helium ($W = 4$ kg/kmole) is adiabatically compressed from 1 atm and 280 K to 500-atm pressure. Determine ρ_1, ρ_2, T_2, and $s_2 - s_1$.

2.5 The gas in Problem 2.4 has an initial volume of 10^3 m³. Determine the final volume, the mass of gas m, the work w, per unit mass, done on the gas, and the change in internal energy $e_2 - e_1$.

2.6 A reversible isothermal process is used to compress air from 1 atm and 10^3 m³ to 50 m³ at 300 K. Determine the mass of gas, the final pressure p_2, w, the heat transfer per unit mass into the gas q, $e_2 - e_1$, and $s_2 - s_1$.

2.7 With van der Waals' equation of state,

$$p = \frac{RT}{v - \beta} - \frac{\alpha}{v^2}$$

determine

$$\left.\frac{\partial p}{\partial v}\right|_T, \left.\frac{\partial p}{\partial T}\right|_v, \left.\frac{\partial v}{\partial p}\right|_T, \left.\frac{\partial v}{\partial T}\right|_p, \left.\frac{\partial T}{\partial p}\right|_v, \left.\frac{\partial T}{\partial v}\right|_p$$

2.8 Use reciprocity to determine the most general possible state equation for the internal energy e of a van der Waals' gas (see Problem 2.7).

2.9 Extend your Problem 2.8 result by determining the most general possible equations for c_v, c_p, and the enthalpy h. Determine the two specific heats as functions only of T and v.

2.10 Extend your Problem 2.9 results by determining the entropy in the form $s = s(v, T)$, where the temperature part is an indefinite integral.

2.11 Continue with the van der Waals' gas of Problems 2.7–2.10 by determining the speed of sound a as a function only of v and T. Start by differentiating s and the thermal state equation.

2.12 For the virial equation of state

$$\frac{pv}{RT} = 1 + \frac{B(T)}{v} + \frac{C(T)}{v^2} + \frac{D(T)}{v^3} + \cdots$$

determine the constant volume specific heat $c_v = c_v(v, T)$.

2.13 Derive an equation of state in the form $p = f(h, s, \gamma)$ for a thermally and calorically perfect gas, where p is pressure, h enthalpy, s entropy, and γ the ratio of specific heats.

2.14 Use the Redlich-Kwong equation of state,

$$p = \frac{RT}{v-b} - \frac{a}{T^{\frac{1}{2}}(v^2 + vb)}$$

where a, b, and R are constants, to determine

$$\left.\frac{\partial p}{\partial v}\right|_T, \left.\frac{\partial p}{\partial T}\right|_v, \left.\frac{\partial v}{\partial p}\right|_T, \left.\frac{\partial v}{\partial T}\right|_p, \left.\frac{\partial T}{\partial p}\right|_v, \left.\frac{\partial T}{\partial v}\right|_p$$

2.15 Determine the most general form possible for the caloric equation of state, $e = e(v, T)$, that is compatible with the Redlich-Kwong equation of state of Problem 2.14.

2.16 For the Dieterici equation of state

$$p = \frac{RT}{v - \beta} \exp\left[-\frac{\alpha}{RTv}\right]$$

where α, β, and R are constants, determine a simple formula for $(\partial e/\partial v)_T$. Derive $e = e(v, T)$. [Hint: use the substitution $u = 1 - (\beta/v)$.]

2.17 Show that the following state equations

$$p = \rho RT - \hat{a}\rho^2$$

$$s = -R \ln \rho + \tfrac{3}{2} R \ln T + \text{const}$$

are compatible, where $\hat{a}$ is a constant. Then determine the speed of sound a.

2.18 Derive the general thermodynamic formula for the speed of sound

$$a = v \left[\frac{T}{c_v} \left(\frac{\partial p}{\partial T} \bigg|_v \right)^2 - \frac{\partial p}{\partial v} \bigg|_T \right]^{\frac{1}{2}}$$

for a substance with Eqs. (2.12) as state equations.

2.19 Consider an arbitrary process x. Show that the specific heat for a constant x process is given by

$$c_x = \frac{dq}{dT}\bigg|_x = c_v + (c_p - c_v) \frac{(\partial v/\partial T)_x}{(\partial v/\partial T)_p}$$

Does the relation require $e = e(T)$?

2.20 Determine the most general thermal equation of state that is consistent with

$$e = f(T) + g(\rho)$$

and determine c_v and c_p. Finally, determine the entropy $s = s(\rho, T)$ and the speed of sound a.

3. ONE-DIMENSIONAL CONSERVATION EQUATIONS

Our objective in this chapter is to derive a complete system of equations for one-dimensional flow. This system will be used, for example, to determine how the pressure and speed in a flow vary with position and time. The equations will express the conservation laws as differential equations, and algebraic thermodynamic state relations. We first discuss the streamtube concept and the substantial, or Eulerian, derivative, since these topics simplify the subsequent derivations. After the governing equations have been obtained, Sec. 3.5 takes up a simple but important example of how the equations are utilized. The acoustic approximation is used to show that the thermodynamic speed of sound represents the speed with which a weak disturbance travels. A summary section concludes the chapter.

3.1 STREAMTUBE FLOW

The motion of a gas is fully described by the velocity field

$$\mathbf{V} = \mathbf{V}(\mathbf{r}, t) \tag{3.1}$$

where $\mathbf{r}$ is the position vector and t is time. If we know how $\mathbf{V}$ varies with $\mathbf{r}$ and t, the problem is partly solved, since other variables can be found. In writing Eq. (3.1), we are utilizing the Eulerian approach. This is much more convenient in fluid mechanics than the Lagrangian approach, which follows the motion of an individual fluid particle. In the Lagrangian approach, $\mathbf{V}$ is a function of t and the initial position of the fluid particles.

It is difficult to obtain a solution to a problem in fluid mechanics in terms of Eq. (3.1). We thus introduce various approximations or simplifications. One simplification is a steady flow, in which case $\mathbf{V}$ depends only on $\mathbf{r}$. Another approximation—the one we pursue in this and subsequent chapters—is to assume that $\mathbf{V}$ depends only on a single spatial coordinate. In this circumstance, we have a quasi-one-dimensional, or streamtube, flow.

We define streamlines as the loci of points that are tangent to the velocity field at any fixed instant of time. A streamtube is a surface formed by streamlines that intersect a simple closed curve, as shown in Fig. 3.1. The surface of the streamtube may be steady or unsteady; it consists of streamlines that pass through the defining curve. The cross-sectional area of the streamtube may be large, as in a rocket nozzle flow. Since a streamline

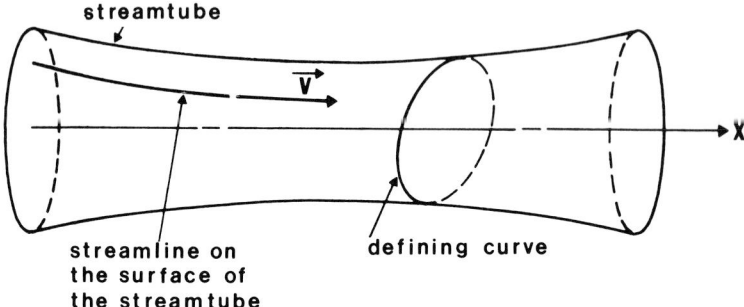

Fig. 3.1 Streamline and streamtube sketch.

is tangent to **V**, the streamtube surface is also tangent to the velocity field. Consequently, there is no mass flux across the streamtube's surface. It is this property that makes a streamtube useful, as we will see when we derive the conservation equations.

The one-dimensional coordinate associated with the streamtube is x. For the streamtube approximation to be valid, conditions cannot vary appreciably across the streamtube in any plane perpendicular to x. Thus, the component of **V** perpendicular to x is neglected. This is equivalent to choosing the location of the x coordinate such that the net momentum in any plane perpendicular to x can be neglected in comparison with the momentum along x. In short, we treat variables, such as p and the flow speed V, as uniform across any transverse cross-sectional plane. If not exactly uniform, they must be nearly uniform, and an average cross-sectional value is then utilized.

3.2 SUBSTANTIAL DERIVATIVE

Another useful concept is the Eulerian or substantial derivative. Consider a thin slice of fluid in the streamtube shown in Fig. 3.1. The particle of fluid travels a distance dx in a time dt. Thus the flow speed

$$V(x, t) = \frac{dx}{dt} \qquad (3.2)$$

is a function of position and time. Any quantity, say pressure, is also a function of position and time

$$p = p(x, t)$$

We take the differential of p, to obtain

$$dp = \frac{\partial p}{\partial x} dx + \frac{\partial p}{\partial t} dt$$

Dividing by dt and rearranging the previous equation yields

$$\frac{dp}{dt} = \frac{\partial p}{\partial t} + \frac{\partial p}{\partial x}\frac{dx}{dt} = \frac{\partial p}{\partial t} + V\frac{\partial p}{\partial x}$$

The derivative on the left is the substantial or Eulerian derivative, and hereafter is written as D( )/Dt, where

$$\frac{D}{Dt} = \frac{\partial}{\partial t} + V\frac{\partial}{\partial x} \qquad (3.3)$$

Thus, the time rate of change of $p(x,t)$ depends on two terms. The first provides the unsteady variation of p at a fixed location. The second term provides the convective change, which consists of the gradient of p times the speed with which a fluid particle moves past the fixed location. The sum of the two terms represents the way the pressure of a fluid particle changes.

Because both x and t are independent variables, the substantial derivative of the position x is given by

$$\frac{Dx}{Dt} = \frac{\partial x}{\partial t} + V\frac{\partial x}{\partial x} = 0 + V \cdot 1 = V \qquad (3.4)$$

in which the partial derivative, $\partial x/\partial t$, is zero. Note that Eqs. (3.2) and (3.4) coincide.

A fluid particle of infinitesimal size at position x has a speed V. As indicated by the discussion leading to Eq. (3.2), Dx/Dt provides the time rate of change of the position of a moving fluid particle. As we have seen, Dp/Dt provides the time rate of change of pressure for a moving fluid particle. We thus "follow a fluid particle" by using the substantial derivative. This notion is often used in fluid dynamics. For instance, we occasionally visualize a fluid particle as it moves through a flowfield. Shortly, extensive use will be made of the derivative when the equations of motion are derived.

As a second illustration, we consider the acceleration of a fluid particle, given by

$$\frac{DV}{Dt} = \frac{\partial V}{\partial t} + V\frac{\partial V}{\partial x}$$

This equation consists of an unsteady term, $\partial V/\partial t$, and a convective term, $V(\partial V/\partial x)$. The convective term depends on time, if $\partial V/\partial t$ is not zero. In a steady flow, $V, p, \rho, \ldots$, depend only on x, so that

$$\frac{\partial V}{\partial t} = 0, \qquad \frac{\partial p}{\partial t} = 0, \qquad \frac{\partial \rho}{\partial t} = 0, \ldots$$

and

$$\frac{DV}{Dt} = V\frac{\partial V}{\partial x} = \frac{d}{dx}\left(\frac{1}{2}V^2\right), \qquad \frac{Dp}{Dt} = V\frac{dp}{dx}, \ldots$$

As a final illustration, consider the substantial derivative of an infinitesimal element of length dx. Let the two endpoints of the element be x_1 and x_2, so that

$$dx = x_2(t) - x_1(t)$$

The infinitesimal length is known at time t, as indicated by the argument of x_1 and of x_2. We now compute the substantial derivative

$$\frac{D(dx)}{Dt} = \frac{Dx_2}{Dt} - \frac{Dx_1}{Dt} = V(x_2, t) - V(x_1, t)$$

where the two speeds are at the same time but at slightly different locations. A Taylor series expansion is used with respect to x_1 to obtain

$$V(x_2, t) = V(x_1, t) + \frac{\partial V}{\partial x}\bigg|_{\substack{x=x_1 \\ t=t}} dx + \cdots$$

With this, our final result is

$$\frac{D(dx)}{Dt} = \frac{\partial V}{\partial x} dx \qquad (3.5)$$

which is used in deriving the equation for mass conservation.

3.3 CONSERVATION EQUATIONS

The physical principles, or laws, that govern the motion of any fluid are:
(1) Conservation of mass.
(2) Conservation of momentum.
(3) Conservation of energy.

These principles result in three differential equations, which are referred to as the equations of motion.

At this point, we itemize the various assumptions that apply to the resulting equations of motion:
 (1) Unsteady, quasi-one-dimensional flow.
 (2) No body forces, such as gravity or electromagnetic forces.
 (3) No wall shear (inviscid flow).
 (4) No heat transfer, i.e., an adiabatic flow.

Despite what appears to be severe restrictions, these assumptions have many important practical applications, which are discussed in subsequent chapters.

We begin by considering a thin slice of a streamtube at a single instant of time. As shown in Fig. 3.2, the differential element can be thought of as a frustum of a cone with a volume given by

$$\text{volume} = A\, dx$$

and a mass given by

$$\text{mass} = dm = \rho A\, dx \qquad (3.6)$$

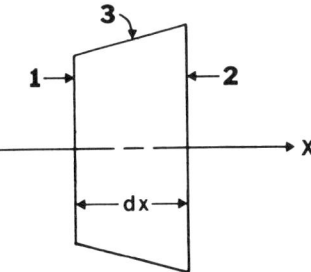

Fig. 3.2 Differential element of a streamtube.

where A is the cross-sectional area of the streamtube. In the conservation law derivations, we will consider a fixed mass dm, whose position and volume change with time. Specifically, no mass crosses surface 3, which remains part of a streamtube, with changing time. Because ρ and A are not constant, the element's volume changes with both position and time. This aspect is accounted for by using Eq. (3.5).

The equations of motion are sometimes derived by considering the flow through a control volume, which is a fixed volume that is fixed in space. (This approach is further discussed later in this section.) Of course, the two approaches result in the same equations. However, for an inviscid, adiabatic flow, the fluid particle derivation is considerably simpler and easier to visualize. This approach follows a fluid particle, and is conceptually close to the formulation used in mechanics. Nevertheless, our final form for the equations of motion is Eulerian, in which x and t are the independent variables.

Mass

Since dm is constant with time, the vanishing of the substantial derivative of dm yields conservation of mass

$$\frac{D(dm)}{Dt} = 0 \tag{3.7}$$

We now utilize Eqs. (3.5) and (3.6) to obtain

$$\frac{D(\rho A\, dx)}{Dt} = A\, dx \frac{D\rho}{Dt} + \rho\, dx \frac{DA}{Dt} + \rho A \frac{D(dx)}{Dt} = 0$$

$$= A\, dx \frac{D\rho}{Dt} + \rho\, dx \frac{DA}{Dt} + \rho A\, dx \frac{\partial V}{\partial x} = 0$$

or, upon division by $\rho A\, dx$,

$$\frac{1}{\rho}\frac{D\rho}{Dt} + \frac{1}{A}\frac{DA}{Dt} + \frac{\partial V}{\partial x} = 0 \tag{3.8a}$$

If Eq. (3.3) is used to replace the substantial derivatives, a second useful form,

$$\frac{\partial(\rho A)}{\partial t} + \frac{\partial(\rho A V)}{\partial x} = 0 \tag{3.8b}$$

is obtained. Either of Eqs. (3.8) represents conservation of mass, or continuity, as it is usually called. (The continuity designation means that the fluid must be continuous and not double-valued.)

Momentum

The equation we will derive is sometimes called conservation of momentum. This, however, is a misnomer, since the momentum of the fluid particle $V \, dm$ is not conserved. Our basic equation is Newton's second law, which states that the time rate of change of momentum equals the net external force on the fluid element. In other words,

$$\frac{D(V \, dm)}{Dt} = dF$$

where $V \, dm$ is the component of momentum in the x direction.

This relation becomes

$$dF = dm \frac{DV}{Dt} + V \frac{D(dm)}{Dt} = dm \frac{DV}{Dt} = \rho \frac{DV}{Dt} A \, dx$$

where Eqs. (3.6) and (3.7) are used. What remains is to evaluate dF where dF comprises the components in the x direction of all external forces.

By assumption, there are no body or viscous forces. Hence, only pressure forces act on the three surfaces of the element. For surfaces 1 and 2, which are perpendicular to x, we have

$$F_1 = pA$$
$$F_2 = -\left[(pA) + \frac{\partial(pA)}{\partial x} dx\right]$$

as shown in Fig. 3.3. The force F_2 is negative because it is oriented in the negative x direction. (Because the streamtube's cross-sectional area A is not a differential, the forces F_1 and F_2 are not differential forces, as is dF_3, which is given below.)

The x component of the normal pressure force on surface 3 is dF_3. This component is given by the average of the pressures acting on surfaces 1 and 2 times the annular area that is perpendicular to the x axis:

$$dF_3 = \left\{\left[p + \left(p + \frac{\partial p}{\partial x} dx\right)\right]/2\right\}\left[\left(A + \frac{\partial A}{\partial x} dx\right) - A\right]$$

$$= \left(p + \frac{1}{2}\frac{\partial p}{\partial x} dx\right)\frac{\partial A}{\partial x} dx = p\frac{\partial A}{\partial x} dx$$

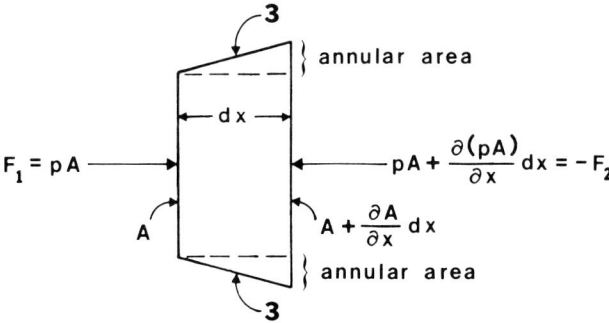

Fig. 3.3 Forces on the differential element of Fig. 3.2.

where $\mathcal{O}((dx)^2)$ terms are ignored. This component is zero only when $(\partial A / \partial x) = 0$.

The surface pressure forces sum to

$$dF = F_1 + F_2 + dF_3 = pA - pA - \frac{\partial (pA)}{\partial x} dx + p \frac{\partial A}{\partial x} dx$$

$$= -A \frac{\partial p}{\partial x} dx - p \frac{\partial A}{\partial x} dx + p \frac{\partial A}{\partial x} dx = -\frac{\partial p}{\partial x} A \, dx$$

Consequently, conservation of momentum becomes

$$-\frac{\partial p}{\partial x} A \, dx = \rho \frac{DV}{Dt} A \, dx$$

or

$$\frac{DV}{Dt} = -\frac{1}{\rho} \frac{\partial p}{\partial x} \qquad (3.9)$$

Observe that the acceleration DV/Dt is positive when the pressure gradient is negative. A decreasing pressure, called a favorable pressure gradient, thus accelerates a particle of fluid.

Energy

The elemental slice of material in Fig. 3.2 is a particle of fluid that constitutes a closed adiabatic system for which the first law of thermodynamics applies. This holds even when the particle is moving and has translational kinetic energy. With $dq = 0$ in Eq. (2.7), we have

$$de = -p \, dv$$

The differentials de and dv are for a particle of fluid that is in motion and whose state is changing with time. These differentials, therefore, represent changes previously denoted by the substantial derivative; hence,

$$\frac{De}{Dt} = -p\frac{Dv}{Dt} = \frac{p}{\rho^2}\frac{D\rho}{Dt} \tag{3.10a}$$

This form for conservation of energy does not explicitly contain a term for the kinetic energy of translation. However, we can obtain an equation for the total energy per unit mass $e + V^2/2$ by adding to Eq. (3.10a) V times Eq. (3.9). We also note that Eqs. (3.9) and (3.10a) do not contain the streamtube's cross-sectional area. This variable appears only in the continuity equation.

A closed thermodynamic system, which is a fluid particle, was consistently used in the derivation of each of the conservation equations. This approach greatly simplifies the derivation, especially the one for the energy equation. In the control volume derivation for the energy equation, an energy balance would involve the following terms:

(1) Enthalpy transport across a transverse surface, such as surfaces 1 and 2 in Fig. 3.2, $= \rho A V[e + (p/\rho) + \frac{1}{2}V^2]$. Here, $\rho A V$ is the mass flow rate across a transverse surface, $e + V^2/2$ is the energy, and p/ρ represents the work needed to move the gas across a transverse surface.

(2) Rate of work done by the element on its surroundings by expanding laterally

$$= p\frac{\partial A}{\partial t}dx$$

This is just the rate of work, $pd[(\rho A\,dx)v]/dt$, with x fixed, as given by Eq. (2.6).

(3) Decrease with time in the total energy of the element

$$= -\frac{\partial}{\partial t}\left[\rho A\,dx\left(e + \frac{1}{2}V^2\right)\right]$$

After considerable manipulation, the foregoing terms, when added, would yield Eq. (3.10a).

The presence of $e + (p/\rho) + \frac{1}{2}V^2$ in item (1) suggests introducing the stagnation enthalpy, defined by

$$h_0 = h + \frac{1}{2}V^2 = e + \frac{p}{\rho} + \frac{1}{2}V^2 \tag{3.11}$$

into the energy equation. (The concept of a stagnation condition is discussed in Chap. 4.) With this change, the energy equation becomes

$$\frac{De}{Dt} = \frac{D}{Dt}\left(h_0 - \frac{p}{\rho} - \frac{1}{2}V^2\right) = \frac{p}{\rho^2}\frac{D\rho}{Dt}$$

or

$$\frac{Dh_0}{Dt} - \frac{1}{\rho}\frac{Dp}{Dt} + \frac{p}{\rho^2}\frac{D\rho}{Dt} - V\frac{DV}{Dt} = \frac{p}{\rho^2}\frac{D\rho}{Dt}$$

ONE-DIMENSIONAL CONSERVATION EQUATIONS

The p/ρ^2 terms cancel and the DV/Dt term is replaced with Eq. (3.9), to yield

$$\frac{Dh_0}{Dt} - \frac{1}{\rho}\frac{Dp}{Dt} - V\left(-\frac{1}{\rho}\frac{\partial p}{\partial x}\right) = 0$$

By expanding Dp/Dt, we have

$$\frac{Dh_0}{Dt} = \frac{1}{\rho}\frac{\partial p}{\partial t} \qquad (3.10b)$$

which is an alternate form for the energy equation. Thus, in a steady adiabatic flow, the stagnation enthalpy of a fluid particle is constant.

3.4 CHANGE IN ENTROPY

Since we are following a fixed particle of fluid, we can write the first law, Eq. (2.11), as

$$T\frac{Ds}{Dt} = \frac{De}{Dt} + p\frac{Dv}{Dt}$$

in which the ordinary derivatives are replaced by substantial derivatives. By conservation of energy, Eq. (3.10a), the right side is zero; hence,

$$\frac{Ds}{Dt} = 0 \qquad (3.12)$$

and a particle of fluid has a constant entropy, i.e., the flow is isentropic. (A flow in which all fluid particles have the same constant entropy is referred to as homentropic.) The flow is isentropic because the derivation of the energy and momentum equations did not include heat transfer or viscous stresses that cause the entropy to change. While s is constant following a particle, it need not necessarily have the same value for different particles. This is evident if we write

$$\frac{Ds}{Dt} = \frac{\partial s}{\partial t} + V\frac{\partial s}{\partial x} = 0$$

which shows that s can still vary with x and t.

3.5 ACOUSTIC APPROXIMATION

By way of illustration, the equations of motion are used to investigate the propagation of a small disturbance. We thereby demonstrate that the disturbance travels with the thermodynamically defined speed of sound. The analysis also shows how the equations of motion plus thermodynamic state equations are combined to effect a solution. A constant-area duct, shown in Fig. 3.4, is envisioned that contains a motionless, perfect gas. The initial pressure and density are p_∞ and ρ_∞, respectively. On the left side of the duct is a diaphragm that starts to vibrate sinusoidally at time $t = 0$. As a result, an oscillatory motion is induced in the gas, and V, p, and ρ change from their initial values. We assume that the magnitude of the displacement of the diaphragm from its neutral position is small. As a consequence, the gas experiences a small perturbation of its state, and we can write

$$V = V'(x, t), \qquad p = p_\infty + p'(x, t), \qquad \rho = \rho_\infty + \rho'(x, t) \quad (3.13)$$

where a prime denotes a perturbation quantity.

Based on the earlier discussion, substitute

$$A = \text{const}, \qquad e = \frac{1}{\gamma - 1} \frac{p}{\rho}$$

into Eqs. (3.8b), (3.9), and (3.10a) to obtain

$$\frac{\partial \rho}{\partial t} + V \frac{\partial \rho}{\partial x} + \rho \frac{\partial V}{\partial x} = 0$$

$$\frac{\partial V}{\partial t} + V \frac{\partial V}{\partial x} + \frac{1}{\rho} \frac{\partial p}{\partial x} = 0$$

$$\frac{1}{p} \frac{\partial p}{\partial t} + \frac{V}{p} \frac{\partial p}{\partial x} - \frac{\gamma}{\rho} \frac{\partial \rho}{\partial t} - \gamma \frac{V}{\rho} \frac{\partial \rho}{\partial x} = 0$$

for the equations of motion. Equations (3.13) are now introduced, to yield

$$\frac{\partial \rho'}{\partial t} + V' \frac{\partial \rho'}{\partial x} + (\rho_\infty + \rho') \frac{\partial V'}{\partial x} = 0$$

$$\frac{\partial V'}{\partial t} + V' \frac{\partial V'}{\partial x} + \frac{1}{\rho_\infty + \rho'} \frac{\partial p'}{\partial x} = 0$$

$$\frac{1}{p_\infty + p'} \frac{\partial p'}{\partial t} + \frac{V'}{p_\infty + p'} \frac{\partial p'}{\partial x} - \frac{\gamma}{\rho_\infty + \rho'} \frac{\partial \rho'}{\partial t} - \frac{\gamma V'}{\rho_\infty + \rho'} \frac{\partial \rho'}{\partial x} = 0$$

These relations can be simplified, since the product of two primed quantities, for instance, is negligibly small. Thus, we have

$$\left| \frac{\partial \rho'}{\partial t} \right| \gg \left| V' \frac{\partial \rho'}{\partial x} \right|$$

$$(\rho_\infty + \rho') \frac{\partial V'}{\partial x} \cong \rho_\infty \frac{\partial V'}{\partial x}$$

$$\frac{1}{p_\infty + p'} \frac{\partial p'}{\partial t} \approx \frac{1}{p_\infty} \frac{\partial p'}{\partial t}, \ldots$$

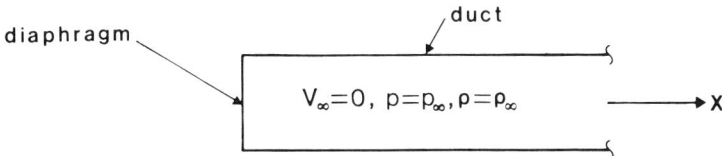

Fig. 3.4 Duct schematic.

Hence, the equations of motion become

$$\frac{\partial \rho'}{\partial t} + \rho_\infty \frac{\partial V'}{\partial x} = 0$$

$$\frac{\partial V'}{\partial t} + \frac{1}{\rho_\infty} \frac{\partial p'}{\partial x} = 0$$

$$\frac{1}{p_\infty} \frac{\partial p'}{\partial t} - \frac{\gamma}{\rho_\infty} \frac{\partial \rho'}{\partial t} = 0$$

The density derivative terms are easily eliminated, after which the pressure terms are eliminated by cross differentiation. We thereby obtain

$$\frac{\partial^2 V}{\partial t^2} - \gamma \frac{p_\infty}{\rho_\infty} \frac{\partial^2 V}{\partial x^2} = 0 \qquad (3.14)$$

where we have set $V' = V$. This relation is called a wave equation for reasons that will soon be apparent. It is a linear, second-order partial differential equation, and superposition of solutions apply. It is linear as a result of the small-perturbation approximation embodied in Eqs. (3.13).

The general solution of Eq. (3.14), by direct substitution, is shown to be

$$V(x, t) = f(x - ct) + g(x + ct) \qquad (3.15)$$

where the constant

$$c = \left(\gamma \frac{p_\infty}{\rho_\infty}\right)^{\frac{1}{2}} \qquad (3.16)$$

has units of speed. The functions f and g are arbitrary. They are determined by initial and boundary conditions. For a constant value of $x - ct$, f is a constant on the right-running lines

$$x = ct + \text{const} \qquad (3.17a)$$

since $c > 0$. Similarly, g is a constant on the left-running lines

$$x = -ct + \text{const} \qquad (3.17b)$$

Thus, f corresponds to right-running waves, while g corresponds to left-running waves.

For the duct problem originally posed, only right-running waves are permissible; hence, $g = 0$. To evaluate f, the sinusoidal motion of the diaphragm is written as

$$x_d = \ell \sin(\omega t) \tag{3.18}$$

where ℓ is a small constant with units of length and ω is a constant frequency. The speed of the diaphragm, transferred to its neutral position, is

$$V(0, t) = \frac{dx_d}{dt} = \ell\omega \cos(\omega t) \tag{3.19}$$

By comparison with Eq. (3.15), we see that the solution of Eq. (3.14) is given by

$$V(x, t) = 0, \qquad x - ct > 0$$

$$= \ell\omega \cos\left[\frac{\omega}{c}(x - ct)\right], \qquad x - ct \leq 0 \tag{3.20}$$

By direct substitution, we can verify that the solution satisfies Eq. (3.14), the boundary condition, Eq. (3.19), and the initial condition, $V(x, 0) = 0$. By utilizing Eq. (3.20), p', ρ', etc., can be determined. These quantities are all zero for $x - ct > 0$, and sinusoidally depend on $x - ct$ when $x - ct \leq 0$.

Figure 3.5 is a sketch in the x-t plane of the waves generated by the diaphragm's motion, Eq. (3.18), which is shown along the t axis. The waves propagate rightward with the slope

$$\frac{dx}{dt} = c$$

which stems directly from Eq. (3.17a). As can be seen from Eq. (3.20), the disturbance does not attenuate as it propagates. This one-dimensional

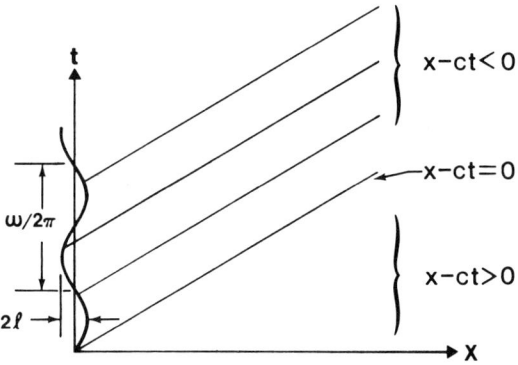

Fig. 3.5 Sketch of the x-t plane for the problem shown in Fig. 3.4.

property is essential for the operation of a stethoscope, which uses a constant-area flexible duct to conduct, undiminished, the sound of the heartbeat to the listener's ear. Fortunately, this property does not hold in two or three dimensions. Thus, the acoustical disturbance associated with rock music attenuates rapidly as it travels outward.

The disturbance sketched in Fig. 3.5 is an acoustic one consisting of weak sound waves. The waves, including the front of the disturbance, propagate with speed c. However, from Eq. (2.31) we see that c, defined by Eq. (3.16), is the same as a_∞, which is the thermodynamic speed of sound in the undisturbed gas. Consequently, we have justified calling a the speed of sound.

3.6 SUMMARY

The three conservation equations are

$$\frac{1}{\rho}\frac{D\rho}{Dt} + \frac{1}{A}\frac{DA}{Dt} + \frac{\partial V}{\partial x} = 0 \qquad (3.8a)$$

$$\frac{DV}{Dt} = -\frac{1}{\rho}\frac{\partial p}{\partial x} \qquad (3.9)$$

$$\frac{Dh_0}{Dt} = \frac{1}{\rho}\frac{\partial p}{\partial t} \qquad (3.10b)$$

where

$$\frac{D}{Dt} = \frac{\partial}{\partial t} + V\frac{\partial}{\partial x}, \qquad h_0 = e + \frac{p}{\rho} + \frac{1}{2}V^2$$

We assume that the area $A = A(x, t)$, which appears only in Eq. (3.8a), is a known quantity. We then have three equations for the four unknown variables, as shown in Table 3.1.

The system is closed by adding one more unknown, T, and two new equations. These are the thermal and caloric equations of state. For a perfect gas, we thus add

$$p = \rho RT, \qquad e = c_v T = [R/(\gamma - 1)]T$$

The final result is a nonlinear system of equations consisting of two algebraic equations and three first-order partial differential equations.

Table 3.1 Unknown Variables

Equation	Unknown Variables
Continuity	ρ, V
Momentum	ρ, V, p
Energy	ρ, V, p, e (or h_0)

The foregoing system of equations is used throughout Chaps. 4–10. The one exception occurs in Chap. 8, where drag and heat-transfer terms are added to the conservation equations. With this one change, we use these equations to study steady flow applications in Chaps. 4–8. Unsteady flow applications are considered in Chaps. 9 and 10.

Problems

3.1 Derive Eq. (3.8b).

3.2 With a equal to the acceleration DV/Dt, determine Da/Dt in terms of partial derivatives with respect to x and t.

3.3 Consider a body in a wind tunnel, as shown in the sketch:

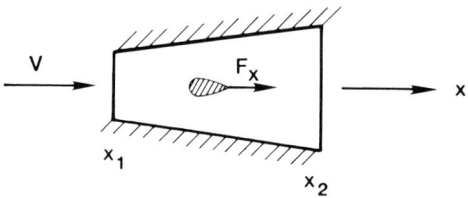

where F_x is the x component of the force on the body by the fluid. Assume steady, one-dimensional flow with no body forces and negligible shear on the wind-tunnel wall, and derive an equation for F_x. Assume that flow conditions are uniform at x_1 and x_2, and start with the second equation in the subsection headed "Momentum," Sec. 3.3, but add the force F_x to the right side. The resulting equation for F_x depends on conditions at x_1 and x_2, $A = A(x)$, and $p = p(x)$. (Since the flow is steady, $\rho V A$ is constant.)

3.4 Start with the unsteady equations of motion. Solve these equations for p, ρ, and e assuming the speed $V = 0$, but do not assume a perfect gas.

3.5 Continue Problem 3.4 with the assumption of a perfect gas. Determine ρ, p, T, and A.

3.6 Consider flow in a duct whose cross-sectional area A varies only with time. If the pressure and density also vary only with time, determine the speed $V(x, t)$. Do not assume adiabatic flow or a perfect gas.

3.7 Derive p', ρ', and T' as functions of x and t for the example considered in Sec. 3.5.

3.8 Derive a one-dimensional wave equation for $V'(x, t)$ for a sound wave that propagates down a tube whose cross-sectional area is

$$A = c_1 + c_2 \sin(x + c_3 t)$$

where the c_i are positive constants, and c_2/c_1 is small compared to unity.

3.9 Derive a one-dimensional wave equation for $V'(x,t)$ for an acoustic disturbance propagating down a tube whose area is a known function of position $A(x)$.

3.10 A cylinder has a closed end at $x = \ell$ and a diaphragm at $x = 0$, as shown in the sketch below. Initial and boundary conditions are

$$V'(x,0) = 0, \qquad 0 \le x \le \ell$$

$$V'(0,t) = \varepsilon a_\infty \sin(\omega t), \qquad t \ge 0$$

where ε is a constant that is small compared to unity.
(a) Determine the boundary condition for the left-running wave $V'_L(\ell, t)$ in the time interval $\ell \le a_\infty t \le 2\ell$.
(b) Determine the solution for $a(x,t)$ in the region where the incident wave and the first reflected wave from the closed end overlap.

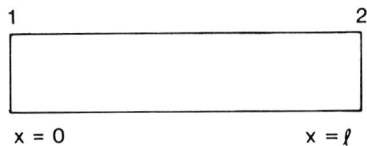

3.11 Use the sketch in Problem 3.10 for a cylinder with diaphragms at both ends that vibrate according to

$$V'(0,t) = \varepsilon a_\infty \sin(\omega_1 t), \qquad V'(\ell, t) = \varepsilon a_\infty \sin(\omega_2 t), \qquad t \ge 0$$

To first order in ε, determine $p(\ell/2, t)$ for the time interval $0 \le t \le (3\ell/2a_\infty)$. What is $p(\ell/2, t)$ if $\omega_1 = \omega_2$?

3.12 A diaphragm at $x = 0$ has the displacement

$$x = 0, \qquad t < 0$$

$$= \varepsilon \sin(\omega_s t)\sin(\omega_f t), \qquad t \ge 0$$

where $\omega_f > \omega_s > 0$ and ε is small compared to unity. Determine the flow speed $V(x,t)$ and the speed of sound $a(x,t)$ in the gas to the right of the diaphragm. Be sure to simplify your results.

3.13 Determine $p(x,t)$ for the flow in Problem 3.12. Note that $(1+z)^n = 1 + nz + \cdots$ for small z.

4. STEADY STREAMTUBE FLOW

The equations from Chap. 3 are next applied to a steady, adiabatic streamtube flow in which

$$\frac{\partial}{\partial t} = 0, \qquad \frac{\partial}{\partial x} = \frac{d}{dx}, \qquad \frac{D}{Dt} = V\frac{d}{dx}$$

For this application, Eq. (3.12) becomes

$$\frac{ds}{dx} = 0 \qquad (4.1)$$

and the entropy is a constant. Recall that with a constant entropy the flow is isentropic. The analysis is simplified by assuming a perfect gas and introducing the Mach number. We thereby obtain algebraic equations for pressure, temperature, etc., with the Mach number as the independent variable. These streamtube results are explored in what follows and are summarized in Appendix C.

4.1 INTEGRATED FORM OF THE GOVERNING EQUATIONS

For steady flow, the continuity equation becomes

$$\frac{d\rho}{\rho} + \frac{dA}{A} + \frac{dV}{V} = 0$$

which integrates to

$$\rho A V = \dot{m} = \text{const} \qquad (4.2)$$

The momentum equation becomes

$$V\frac{dV}{dx} + \frac{1}{\rho}\frac{dp}{dx} = 0$$

or

$$V\,dV + \frac{1}{\rho}\,dp = 0 \qquad (4.3)$$

Observe that V is constant if dp is zero. In other words, a pressure gradient is required for a change in flow speed. If we integrate this equation

$$\frac{1}{2}V^2 + \int \frac{dp}{\rho} = \text{const}$$

the compressible form of the Bernoulli equation is obtained. However, actual integration requires that we know the thermal state equation

$$p = p(\rho, s)$$

where, in view of Eq. (4.1), $s = s_0 = \text{const}$. The Bernoulli equation thus becomes

$$\frac{1}{2}V^2 + \int a^2(\rho, s_0)\frac{d\rho}{\rho} = \text{const}$$

since $a^2 = dp/d\rho$ when s is constant.

Finally, the energy equation becomes

$$V\frac{dh_0}{dx} = 0$$

which integrates to

$$h_0 = h + \tfrac{1}{2}V^2 = \text{const} \qquad (4.4)$$

where h_0 is the stagnation enthalpy. Thus, when a flow is steady, two of the equations of motion can be integrated leaving only momentum as an ordinary differential equation.

To integrate the momentum equation, we assume a perfect gas, so that

$$p = \rho RT, \qquad h = \frac{\gamma}{\gamma - 1}RT = \frac{\gamma}{\gamma - 1}\frac{p}{\rho} \qquad (4.5)$$

Equation (4.4) becomes

$$\frac{\gamma}{\gamma - 1}\frac{p}{\rho} + \frac{1}{2}V^2 = h_0$$

or, after differentiation,

$$\frac{\gamma}{\gamma - 1}\frac{dp}{\rho} - \frac{\gamma}{\gamma - 1}\frac{p}{\rho^2}d\rho + V\,dV = 0$$

We use Eq. (4.3) to eliminate $V\,dV$, thereby obtaining

$$\frac{\gamma}{\gamma - 1}\frac{dp}{\rho} - \frac{\gamma}{\gamma - 1}\frac{p}{\rho^2}d\rho - \frac{dp}{\rho} = 0$$

or
$$\frac{dp}{p} - \gamma \frac{d\rho}{\rho} = 0$$

which integrates to
$$p/\rho^\gamma = \text{const} \tag{4.6}$$

The momentum, or Bernoulli, equation has thus been integrated. Equation (4.6) also can be obtained from $s = \text{const}$ for a perfect gas.

4.2 ISENTROPIC RELATIONS

Since the gas is perfect, the speed of sound is
$$a = \left(\gamma \frac{p}{\rho}\right)^{\frac{1}{2}} = (\gamma RT)^{\frac{1}{2}}$$

We introduce a dimensionless speed, called the Mach number,
$$M = V/a \tag{4.7}$$

which is simply the ratio of the local flow speed to the local speed of sound. This will be the dominant parameter throughout this book. When $M < 1$, V is less than a and the flow is referred to as subsonic. Similarly, when $M = 1$, the flow is sonic and, when $M > 1$, the flow is supersonic. The importance of the sonic demarcation condition will become apparent in this and subsequent chapters. Aside from its physical significance, the introduction of the Mach number into the analysis also provides a considerable simplification of the equations.

The Mach number is introduced by writing
$$V^2 = a^2 M^2 = \gamma RT M^2$$

We eliminate h using Eq. (4.5), and eliminate V^2 from the energy equation, in order to obtain
$$\frac{\gamma}{\gamma - 1} RT + \frac{\gamma}{2} RT M^2 = h_0$$

Replace h_0 with $\gamma RT_0/(\gamma - 1)$, where T_0 is the stagnation temperature, and simplify
$$\frac{T}{T_0} = \frac{1}{1 + [(\gamma - 1)/2] M^2} \tag{4.8}$$

Since h_0 is a constant, T_0 is also constant.

In a similar way, we introduce the reference constants ρ_0 and p_0:
$$\frac{p}{\rho^\gamma} = \frac{p_0}{\rho_0^\gamma}, \qquad p_0 = \rho_0 RT_0$$

We now obtain p and ρ in terms of M as follows:

$$\frac{p}{p_0} = \left(\frac{\rho}{\rho_0}\right)^\gamma = \frac{(p/T)^\gamma}{(p_0/T_0)^\gamma} = \left(\frac{p}{p_0}\right)^\gamma \left(\frac{T}{T_0}\right)^{-\gamma}$$

or

$$\left(\frac{p}{p_0}\right)^{1-\gamma} = \left(\frac{T}{T_0}\right)^{-\gamma}$$

so that

$$\frac{p}{p_0} = \left(\frac{T}{T_0}\right)^{\gamma/(\gamma-1)} = \left(1 + \frac{\gamma-1}{2} M^2\right)^{-[\gamma/(\gamma-1)]} \qquad (4.9)$$

$$\frac{\rho}{\rho_0} = \left(\frac{p}{p_0}\right)^{1/\gamma} = \left(1 + \frac{\gamma-1}{2} M^2\right)^{-[1/(\gamma-1)]} \qquad (4.10)$$

In the above equations, h_0, T_0, p_0, and ρ_0 correspond to h, T, p, and ρ when the Mach number, or the speed, is isentropically decreased to zero. Hence, these parameters are referred to as stagnation quantities. For the present situation, they are constants. Consequently, static conditions, i.e., h, T, p, and ρ, depend on only the Mach number squared, and do so in a rather simple way. Equations that provide this dependence, (4.8)–(4.10), are called isentropic relations, since $s =$ const.

Figure 4.1 shows the variations of p/p_0 and T/T_0 vs M for $\gamma = 1.4$. The curve for p/p_0 falls slowly at first. For instance, at $M = 0.4$ we have $p/p_0 = 0.8956$. The decrease in T/T_0 is substantially less, while the ρ/ρ_0 curve would fall in between the pressure and temperature curves. When the change in p (or ρ) from p_0 (or ρ_0) is not large, the flow is effectively

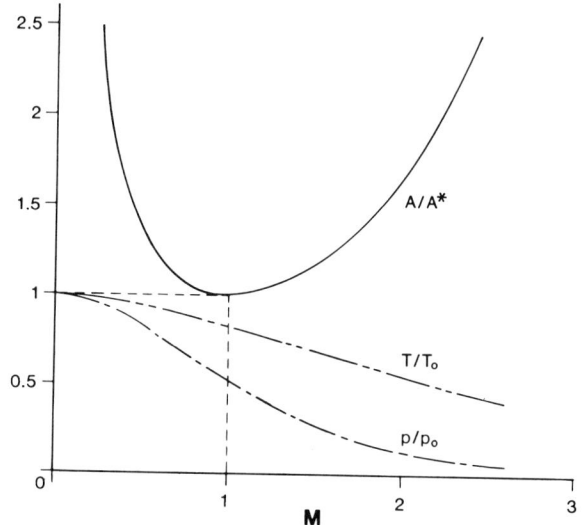

Fig. 4.1 p/p_0, T/T_0, and A/A^* vs Mach number for $\gamma = 1.4$.

incompressible. Thus, the statement in Chap. 1 that a flow is nearly incompressible when $M \leq 0.4$ is verified. As the Mach number increases, however, the changes become more pronounced. At $M = 1$, for example, we have $p/p_0 = 0.5283$. The curve for A/A^* in the figure will be discussed in the next section.

The dynamic pressure $\rho V^2/2$ is often used in incompressible flow. In the next example, we compare $p + \rho V^2/2$ with p_0.

Example 1

Compare $p + \rho V^2/2$ with p_0 when the Mach number is small, and determine the incompressible counterpart to T_0.

We begin with the perfect-gas energy equation

$$T\left(1 + \frac{\gamma - 1}{2} M^2\right) = T_0$$

which holds whether the flow is compressible or incompressible. Thus, the incompressible counterpart of T_0 is T. By way of contrast, for an incompressible flow, Eq. (4.3) becomes

$$d\left(\frac{1}{2} V^2\right) + d\left(\frac{p}{\rho}\right) = 0$$

which integrates to

$$p + \tfrac{1}{2}\rho V^2 = p + q = Q = \text{const}$$

where q is the dynamic pressure. Replacing V with the Mach number then yields

$$Q = p\left(1 + \frac{\gamma}{2} M^2\right)$$

Thus, Q (low Mach number flow) and p_0 (compressible flow) are not the same. They are, however, consistent with each other, as can be shown by expanding

$$p_0 = p\left(1 + \frac{\gamma - 1}{2} M^2\right)^{\gamma/(\gamma - 1)}$$

in the $M^2 \to 0$ limit. We thereby obtain

$$p_0 = p\left[1 + \frac{\gamma}{\gamma - 1} \frac{\gamma - 1}{2} M^2 \right.$$
$$\left. + \frac{[\gamma/(\gamma - 1)]\{[\gamma/(\gamma - 1)] - 1\}}{2!} \left(\frac{\gamma - 1}{2}\right)^2 M^4 + \cdots \right]$$
$$= p\left(1 + \frac{\gamma}{2} M^2 + \frac{\gamma}{8} M^4 + \cdots \right)$$

which agrees with Q when M^2 is small. Since higher-order terms are positive, we observe that $p_0 \geq Q$.

4.3 AREA AND MASS FLOW RATE

Our immediate task is to relate the area $A(x)$ to the Mach number and to determine the mass flow rate $\dot{m}$. Starting with continuity, we first determine $\dot{m}$ and introduce M wherever possible:

$$\dot{m} = \rho V A = \frac{\rho}{\rho_0} \rho_0 \frac{V}{a} a A = \left(1 + \frac{\gamma - 1}{2} M^2\right)^{-[1/(\gamma - 1)]} M\sqrt{\gamma R(T/T_0)T_0}\, \rho_0 A$$

$$= M\left(1 + \frac{\gamma - 1}{2} M^2\right)^{-[1/(\gamma - 1)]} \left(1 + \frac{\gamma - 1}{2} M^2\right)^{-\frac{1}{2}} \sqrt{\gamma R T_0} \left(\frac{p_0}{RT_0}\right) A$$

$$= \frac{M}{\{1 + [(\gamma - 1)/2] M^2\}^{(\gamma + 1)/2(\gamma - 1)}} \left(\frac{\gamma}{RT_0}\right)^{\frac{1}{2}} p_0 A \qquad (4.11)$$

For Eqs. (4.8)–(4.10), the reference condition is a stagnation condition, where $M = 0$. Here, we obtain a more convenient reference condition by setting $M = 1$. This reference condition is denoted by an asterisk, so that $A = A^*$, $p = p^*$, etc., when $M = 1$. Since the mass flow rate is a constant, it is not altered by this choice. Hence, at $M = 1$, we have, from Eq. (4.11),

$$\dot{m} = \left(\frac{2}{\gamma + 1}\right)^{(\gamma + 1)/[2(\gamma - 1)]} \left(\frac{\gamma}{RT_0}\right)^{\frac{1}{2}} p_0 A^* \qquad (4.12)$$

We eliminate $\dot{m}$ from Eqs. (4.11) and (4.12) to obtain

$$\frac{A}{A^*} = \left(\frac{2}{\gamma + 1}\right)^{(\gamma + 1)/[2(\gamma - 1)]} \frac{1}{M} \left(1 + \frac{\gamma - 1}{2} M^2\right)^{(\gamma + 1)/[2(\gamma - 1)]} \qquad (4.13)$$

Equations (4.11)–(4.13) are the desired results.

Equation (4.13) is an isentropic relation, which is shown in Fig. 4.1. Note that A/A^* is double-valued relative to M, with a minimum when $M = 1$. The location of this minimum is called the throat. Equation (4.13) determines how the cross-sectional area of a streamtube varies with Mach number in an isentropic flow. If inlet conditions to the duct are subsonic, then, for supersonic exit flow, the duct must have a throat. Upstream of the throat, the flow accelerates to $M = 1$, while downstream of the throat, it accelerates from $M = 1$ to its final supersonic value. This type of flow is referred to as isentropic nozzle flow.

In the foregoing description, it is presumed that the flow can continuously accelerate through the converging/diverging nozzle. This is possible only if the pressure at the exit of the nozzle is sufficiently low relative to p_0.

STEADY STREAMTUBE FLOW

When this is the case, the geometrical duct area at the throat does equal A^* and $M^* = 1$. Alternative possibilities are discussed in Chap. 7.

The foregoing discussion is illustrated in the next example.

Example 2

At the location where the diameter of an insulated duct is 10 cm, air flows with a speed of 30 m/s and has a pressure of 3 atm. At a downstream section the pressure is 1.5 atm and the temperature is 200 K. We are to determine the mass flow rate, the Mach number at each section, and the diameter of the downstream section.

Since the duct is insulated, the flow is assumed to be isentropic. The upstream and downstream sections are denoted by subscripts 1 and 2, respectively. Since the pressures p_1 and p_2 are known, the isentropic relation, Eq. (2.28b), is used for the temperature T_1:

$$T_1 = T_2 \left(\frac{p_1}{p_2}\right)^{(\gamma-1)/\gamma} = 200(2)^{0.4/1.4} = 243.8 \text{ K}$$

Hence, M_1 is given by

$$M_1 = \frac{V_1}{\sqrt{\gamma R T_1}} = \frac{30}{\sqrt{1.4 \times 287 \times 243.8}} = 0.0959$$

Although not asked for, T_0 and p_0 are useful. They are given by

$$1 + \frac{\gamma-1}{2} M_1^2 = 1.0018$$

$$T_0 = T_1\left(1 + \frac{\gamma-1}{2} M_1^2\right) = 243.8 \times 1.0018 = 244.2 \text{ K}$$

and

$$p_0 = p_1\left(1 + \frac{\gamma-1}{2} M_1^2\right)^{\gamma/(\gamma-1)} = 3 \times 1.0018^{3.5} = 3.017 \text{ atm}$$

To find M_2, we solve Eq. (4.9) as follows:

$$\left(\frac{p_0}{p_2}\right)^{(\gamma-1)/\gamma} = 1 + \frac{\gamma-1}{2} M_2^2$$

or

$$M_2 = \left\{\frac{2}{\gamma-1}\left[\left(\frac{p_0}{p_2}\right)^{(\gamma-1)/\gamma} - 1\right]\right\}^{\frac{1}{2}} = 5\left[\left(\frac{3.017}{1.5}\right)^{0.4/1.4} - 1\right]^{\frac{1}{2}} = 1.051$$

The mass flow rate is evaluated at section 1 where the diameter is known. From Eq. (4.11) we have

$$\dot{m} = \frac{M_1}{\left(1 + \frac{\gamma-1}{2} M_1^2\right)^{(\gamma+1)/2(\gamma-1)}} \left(\frac{\gamma}{RT_0}\right)^{\frac{1}{2}} p_0 A$$

$$= \frac{0.0959}{1.0018^3} \left(\frac{1.4}{287 \times 244.2}\right)^{\frac{1}{2}} 3.017 \times 1.013 \times 10^5 \times \frac{\pi}{4} \times 10^{-2}$$

$$= 1.023 \text{ kg/s}$$

It is worth mentioning that for SI units the diameter must be in meters, and the pressure, which is in atmospheres, must be converted to Pascals, which is accomplished by the 1.013×10^5 conversion factor. Generally, unknown Mach numbers should be determined early in a problem. Once established, other quantities are then easily found. We again use Eq. (4.11) but now solve for the diameter D_2.

$$D_2 = \left[\frac{4}{\pi} \frac{\left(1 + \frac{\gamma-1}{2} M_2^2\right)^{(\gamma+1)/2(\gamma-1)} \dot{m}}{M_2(\gamma/RT_0)^{\frac{1}{2}} p_0}\right]^{\frac{1}{2}}$$

$$= \left[\frac{4}{\pi} \frac{(1 + 0.2 \times 1.051^2)^3 \times 1.023}{1.051(1.4/287 \times 244.2)^{\frac{1}{2}} \times 3.017 \times 1.013 \times 10^5}\right]^{\frac{1}{2}}$$

$$= 4.064 \times 10^{-2} \text{ m}$$

Since $M_1 < 1 < M_2$, there is a throat in the streamtube somewhere between sections 1 and 2. The diameter D^* at this location is slightly smaller than D_2. It is easily found using Eq. (4.12).

Nozzle flow will be examined in detail in Chap. 7. We first discuss shock and expansion waves in Chaps. 5 and 6 because the flow in a nozzle is not always isentropic. Examining these waves first will make the nozzle flow presentation more realistic and complete.

Problems

4.1 Nitrogen ($W = 28$ kg/kmole) has a stagnation pressure of 2 atm and a stagnation speed of sound of 300 m/s. Determine the static pressure and static enthalpy at $M = 3$.

4.2 Air expands from 5 atm and 500 K to 1 atm in a steady flow process without heat transfer. Assume the initial velocity V_1 is negligible, and compute the final Mach number M_2 and the flow rate $\dot{m}$, assuming the final duct size is 10 cm in diameter.

4.3 A stream of helium ($W = 4$ kg/kmole) flows in a duct of 0.15 m diameter at a rate of 1.5 kg/s. The stagnation temperature is 300 K, and at one section the static pressure is 0.3 atm. Calculate the Mach number at this section.

4.4 Evaluate Eqs. (4.8)–(4.10) when $\gamma = 1$. (Hint: Use the binomial expansion.)

4.5 With fixed stagnation conditions and for a steady isentropic flow of a perfect gas, determine the speed when $M \to \infty$.

Computational Problems:

4.6 Prepare two tables, one with $\gamma = 1.4$ and the other with $\gamma = 5/3$, that show:

$$p/p_0, \quad T/T_0, \quad \rho/\rho_0, \quad A/A^*$$

vs M. The spacing on M should be

$$0(0.01)\,5(0.1)\,10$$

which means

$$M = 0, 0.01, 0.02, \ldots, 4.99, 5, 5.1, 5.2, \ldots, 9.9, 10$$

Use these tables during the rest of the course and for the following problems.

4.7 The mass flow rate in Problem 4.1 is to be 0.13 kg/s. Determine the streamtube's cross-sectional area when $M = 0.1$, 1, and 3. What is the flow speed when $M = 3$?

4.8 For conditions in Problem 4.2, determine the throat area, and the speed and temperature at the exit when $M = M_2$.

5. NORMAL AND OBLIQUE SHOCK WAVES

When a subsonic flow slows down, it does so gradually. A supersonic flow, however, can decelerate either gradually or abruptly. An abrupt change can occur only when the flow speed exceeds the sound speed. In this circumstance, the gas is unable to signal the upstream flow that a change is imminent. Typically, the change is caused by a turn in the wall that compresses the flow.

While a gradual deceleration in a supersonic flow is possible, it usually occurs abruptly. The resultant discontinuity in the flow is called a shock wave. Shocks are the most distinctive feature of a supersonic flow and are important building blocks for establishing the overall nature of the flowfield. In Sec. 5.1 we discuss steady, normal shock waves in which the discontinuity is perpendicular to the velocity. This is the simplest type of shock wave. However, in most flowfields, the shocks are at an oblique angle relative to the velocity. Section 5.2 considers this type of shock. Principal results of this chapter are summarized in Appendix D.

5.1 STEADY, NORMAL SHOCK WAVES

We consider flow in a constant cross-sectional area duct. For the flow, Eq. (4.3) can be integrated with the use of the continuity equation

$$\rho V \, dV + dp = 0$$

$$\frac{\dot{m}}{A} dV + dp = 0$$

$$\frac{\dot{m}}{A} V + p = \text{const}$$

We now replace $\dot{m}/A$ with ρV, to obtain

$$p + \rho V^2 = \text{const} \tag{5.1a}$$

The above constant is not the stagnation pressure. Why? Equation (5.1a), along with

$$\rho V = \frac{\dot{m}}{A} = \text{const} \tag{5.1b}$$

and

$$h + \tfrac{1}{2}V^2 = h_0 = \text{const} \qquad (5.1c)$$

are the one-dimensional conservation equations in a constant-area duct.

Let us now inquire into the possibility of a discontinuity in the flow, as shown by the vertical double line in Fig. 5.1. Regardless of whether there is a discontinuity separating regions 1 and 2, Eqs. (5.1) must hold in both regions. Furthermore, the constants on the right sides of the equations must be the same in both regions. We thus have

$$(\rho V)_1 = (\rho V)_2$$
$$(p + \rho V^2)_1 = (p + \rho V^2)_2 \qquad (5.2)$$
$$(h + \tfrac{1}{2}V^2)_1 = (h + \tfrac{1}{2}V^2)_2$$

Equations (5.2) are referred to as the normal shock equations or shock jump conditions.

Equations (5.2) are trivially satisfied if $\rho_2 = \rho_1$, $V_2 = V_1$, etc. Our interest will focus on the discontinuous solution wherein ρ jumps from ρ_1 to ρ_2, V from V_1 to V_2, and so on. In reality, a shock wave is not a discontinuity but a smooth transition from state 1 to state 2 with very steep gradients. These gradients involve both viscous stress and heat transfer. Both processes occur only in the flow direction, and result in large pressure, temperature, and velocity changes. However, the transition takes place over a few molecular mean free paths, which for sea-level air means that the shock is only about 10^{-7} m thick. External to the shock, the viscous stress and heat transfer are zero; hence, the external flow is inviscid and adiabatic. The idealization of a shock as a discontinuity in the flow is thus appropriate.

If we assume that all conditions in region 1 are known, we then have as unknowns ρ_2, V_2, p_2, and h_2. With four unknowns and three equations, it becomes necessary to make a statement about the thermal and caloric states of the gas. The simplest assumption is made, namely, that the gas is perfect:

$$p = \rho RT \qquad (5.3a)$$
$$h = [\gamma/(\gamma - 1)] RT \qquad (5.3b)$$
$$a^2 = \gamma RT \qquad (5.3c)$$

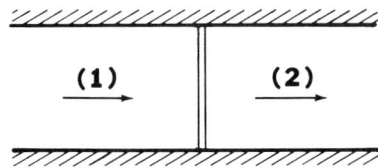

Fig. 5.1 One-dimensional flow in a constant area duct; the double line is a normal shock wave.

The most convenient form for the solution involves introducing the upstream and downstream Mach numbers:

$$M_1 = V_1/a_1, \qquad M_2 = V_2/a_2 \tag{5.4}$$

where M_1 is referred to as the shock Mach number.

To begin with, let us manipulate the energy equation using Eq. (5.3b) and

$$V^2 = a^2 M^2 = \gamma RTM^2 \tag{5.5}$$

to obtain

$$\left(\frac{\gamma}{\gamma - 1} RT + \frac{1}{2} \gamma RTM^2 \right)_2 = \left(\frac{\gamma}{\gamma - 1} RT + \frac{1}{2} \gamma RTM^2 \right)_1$$

By factoring out the enthalpy term, we have

$$T_2 \left(1 + \frac{\gamma - 1}{2} M_2^2 \right) = T_1 \left(1 + \frac{\gamma - 1}{2} M_1^2 \right)$$

Since the stagnation temperature T_0 equals $T[1 + (\gamma - 1)M^2/2]$, we have this important result:

$$T_{02} = T_{01} \tag{5.6}$$

Thus, the stagnation temperature does not change across a steady shock wave, a result we might have anticipated from Eq. (5.1c), which is for an adiabatic flow. The speed of sound and enthalpy are obtainable from

$$\frac{a_2^2}{a_1^2} = \frac{h_2}{h_1} = \frac{T_2}{T_1} = \frac{1 + [(\gamma - 1)/2] M_1^2}{1 + [(\gamma - 1)/2] M_2^2} \tag{5.7}$$

We now substitute Eqs. (5.3a) and (5.5) into momentum in order to obtain

$$\left(\rho RT + \gamma \rho RTM^2 \right)_2 = \left(\rho RT + \gamma \rho RTM^2 \right)_1$$

By factoring out ρRT and using the energy equation result, we get

$$\rho_2 \left\{ \frac{1 + \gamma M^2}{1 + [(\gamma - 1)/2] M^2} \right\}_2 = \rho_1 \left\{ \frac{1 + \gamma M^2}{1 + [(\gamma - 1)/2] M^2} \right\}_1$$

In contrast to our result for temperature, this result differs from the isentropic density relation, Eq. (4.10). The speed and density ratios across the shock are given by

$$\frac{V_1}{V_2} = \frac{\rho_2}{\rho_1} = \frac{1 + \gamma M_1^2}{1 + [(\gamma-1)/2] M_1^2} \frac{1 + [(\gamma-1)/2] M_2^2}{1 + \gamma M_2^2} \quad (5.8)$$

Since the equation of state can be written as

$$\left(\frac{p}{\rho T}\right)_2 = \left(\frac{p}{\rho T}\right)_1$$

we have for the pressure

$$[p(1 + \gamma M^2)]_2 = [p(1 + \gamma M^2)]_1$$

or

$$\frac{p_2}{p_1} = \frac{1 + \gamma M_1^2}{1 + \gamma M_2^2} \quad (5.9)$$

We now square both sides of the continuity equation and substitute Eq. (5.5) to obtain

$$(\rho^2 T M^2)_2 = (\rho^2 T M^2)_1$$

The density and temperature are replaced to yield, after simplification,

$$\frac{M_2^2\{1 + [(\gamma-1)/2] M_2^2\}}{(1 + \gamma M_2^2)^2} = \frac{M_1^2\{1 + [(\gamma-1)/2] M_1^2\}}{(1 + \gamma M_1^2)^2} \quad (5.10)$$

which relates the two Mach numbers defined by Eqs. (5.4).

Equation (5.10) is a quadratic in M_2^2 and therefore has two roots. These are found to be

$$M_2^2 = \frac{1 + [(\gamma-1)/2] M_1^2}{\gamma M_1^2 - [(\gamma-1)/2]} \quad (5.11)$$

and

$$M_2^2 = M_1^2$$

The second solution can be disregarded since there is no shock in this case. In Eq. (5.11), if $M_1 = 1$ then $M_2 = 1$, and vice versa. Thus, if $M_1 > 1$, then $M_2 < 1$, or if $M_1 < 1$, then $M_2 > 1$. These two cases are later analyzed by means of the second law of thermodynamics.

Equations (5.7)–(5.9) can be simplified by eliminating M_2 from them. From Eq. (5.11), we obtain

$$1 + \frac{\gamma-1}{2} M_2^2 = \left(\frac{\gamma+1}{2}\right)^2 \frac{M_1^2}{\gamma M_1^2 - [(\gamma-1)/2]} \quad (5.12)$$

and

$$1 + \gamma M_2^2 = \frac{\gamma + 1}{2} \frac{1 + \gamma M_1^2}{\gamma M_1^2 - [(\gamma - 1)/2]} \quad (5.13)$$

which simplifies the subsequent algebra. We thereby have

$$\frac{p_2}{p_1} = \frac{2}{\gamma + 1}\left(\gamma M_1^2 - \frac{\gamma - 1}{2}\right) \quad (5.14a)$$

$$\frac{T_2}{T_1} = \left(\frac{2}{\gamma + 1}\right)^2 \frac{1}{M_1^2}\left(1 + \frac{\gamma - 1}{2} M_1^2\right)\left(\gamma M_1^2 - \frac{\gamma - 1}{2}\right) \quad (5.14b)$$

$$\frac{\rho_2}{\rho_1} = \frac{V_1}{V_2} = \frac{\gamma + 1}{2} \frac{M_1^2}{1 + [(\gamma - 1)/2] M_1^2} \quad (5.14c)$$

Other dimensionless ratios are of use. For instance,

$$\frac{T_{02}}{T_{01}} = \frac{T_2}{T_1} \frac{1 + [(\gamma - 1)/2] M_2^2}{1 + [(\gamma - 1)/2] M_1^2} = 1 \quad (5.15a)$$

and

$$\frac{p_{02}}{p_{01}} = \frac{p_2}{p_1}\left\{\frac{1 + [(\gamma - 1)/2] M_2^2}{1 + [(\gamma - 1)/2] M_1^2}\right\}^{\gamma/(\gamma - 1)}$$

$$= \frac{[(\gamma + 1)/2]^{(\gamma + 1)/(\gamma - 1)} M_1^{2\gamma/(\gamma - 1)}}{\{1 + [(\gamma - 1)/2] M_1^2\}^{\gamma/(\gamma - 1)} \{\gamma M_1^2 - [(\gamma - 1)/2]\}^{1/(\gamma - 1)}} \quad (5.15b)$$

Equations (5.15a) and (5.6) are the same.

Equations (5.1) and (5.2) are for an adiabatic, inviscid flow. Since these equations are devoid of viscous or heat-transfer terms, they cannot represent the continuous transition of the flow from state 1 to state 2. However the extreme gradients inside the shock represent a dissipative process. The second law of thermodynamics thus requires that the downstream entropy s_2 exceed s_1. To determine the entropy difference, we utilize Eq. (2.27) in the form

$$\frac{s_2 - s_1}{R} = \ell n \left[\left(\frac{T_2}{T_1}\right)^{1/(\gamma - 1)} \left(\frac{\rho_1}{\rho_2}\right)\right]$$

With the aid of Eqs. (5.14), this becomes

$$\frac{s_2 - s_1}{R} = \ln\left[\left(\frac{2}{\gamma+1}\right)^{(\gamma+1)/(\gamma-1)}\right.$$

$$\left. \times \frac{\{1 + [(\gamma-1)/2]M_1^2\}^{\gamma/(\gamma-1)}\{\gamma M_1^2 - [(\gamma-1)/2]\}^{1/(\gamma-1)}}{(M_1^2)^{\gamma/(\gamma-1)}}\right] \quad (5.16a)$$

$$= -\ln\left(\frac{p_{02}}{p_{01}}\right) \quad (5.16b)$$

where the last result stems from Eq. (5.15b). One can readily show that

$$\frac{s_2 - s_1}{R} < 0, \quad M_1 < 1$$

$$= 0, \quad M_1 = 1$$

$$> 0, \quad M_1 > 1$$

$$= \infty, \quad M_1 \to \infty$$

Since the flow is adiabatic, the second law of thermodynamics, Eq. (2.29), reduces to $ds \geq 0$, or $s_2 - s_1 \geq 0$. This holds only when $M_1 \geq 1$, while the equality sign occurs when there is no discontinuity. Therefore, the upstream flow must be supersonic for a normal shock wave to occur.

The denominator of Eq. (5.11) is never zero, since $M_1 \geq 1$. We also have, for M_2,

$$M_2 = 1, \quad M_1 = 1$$

$$= \left(\frac{\gamma - 1}{2\gamma}\right)^{\frac{1}{2}}, \quad M_1 = \infty$$

For $M_1 > 1$, M_2 is subsonic. When $\gamma = 1.4$, M_2 decreases monotonically from 1 to 0.378 as M_1 increases from unity to infinity.

The following examples illustrate the above discussion.

Example 1

Argon (a monatomic gas with a molecular weight of 40) encounters a normal shock. The temperature ahead of the shock is 200 K, and the Mach number downstream of the shock is 0.5. We are to determine the speed V_2 downstream of the shock.

Since Eq. (5.10) is symmetric, its solution for M_1^2 is Eq. (5.11) with the indices interchanged. Hence, we have

$$M_1 = \left\{ \frac{1 + [(\gamma-1)/2] M_2^2}{\gamma M_2^2 - [(\gamma-1)/2]} \right\}^{\frac{1}{2}} = 3.606$$

where $\gamma = 5/3$ and $M_2 = 0.5$. From Eq. (5.7), we obtain

$$T_2 = \left\{ \frac{1 + [(\gamma-1)/2] M_1^2}{1 + [(\gamma-1)/2] M_2^2} \right\} T_1 = 984.6 \text{ K}$$

The speed V_2 is given by

$$V_2 = (\gamma R T_2)^{\frac{1}{2}} M_2 = \left(\frac{5}{3} \times \frac{8314}{40} \times 984.6 \right)^{\frac{1}{2}} \times 0.5 = 292.0 \text{ m/s}$$

Example 2

As a second example, let us develop the Rayleigh pitot pressure formula. We then determine the Mach number and speed for an aircraft flying at an altitude of 20 km with a measured pitot pressure of 5.53×10^4 N/m².

The pitot probe is a blunted cylinder with a small inner tube (Fig. 5.2), which measures p_3 at a point where the flow speed is negligible. When the probe is aligned with the oncoming supersonic flow, a symmetric bow shock wave is located a short distance upstream of the probe. Consider the streamline that passes through the normal part of the shock. This streamline goes through point 2 where the subsonic Mach number M_2 is given by Eq. (5.11). The streamline then stagnates at point 3 where $M_3 = 0$. Between points 2 and 3 the flow is isentropic since no shear or heat transfer occurs with respect to this streamline. Hence, the measured pressure p_3 is

$$p_3 = p_{03} = p_{02}$$

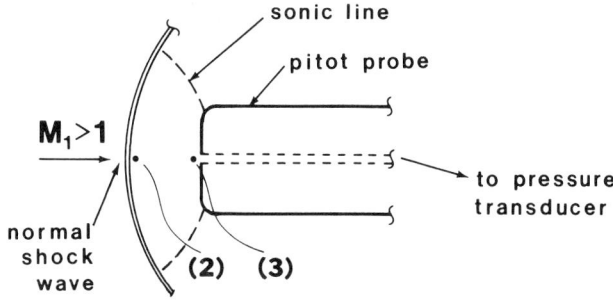

Fig. 5.2 Pitot probe aligned with a supersonic flow.

The pressure ratio p_{03}/p_1 is given by

$$\frac{p_{03}}{p_1} = \frac{p_{02}}{p_2}\frac{p_2}{p_1} = \left(1 + \frac{\gamma-1}{2}M_2^2\right)^{\gamma/(\gamma-1)} \frac{2}{\gamma+1}\left(\gamma M_1^2 - \frac{\gamma-1}{2}\right)$$

With the use of Eq. (5.12), this becomes the Rayleigh pitot formula

$$\frac{p_{03}}{p_1} = \left(\frac{\gamma+1}{2}\right)^{(\gamma+1)/(\gamma-1)} \frac{M_1^{2\gamma/(\gamma-1)}}{\{\gamma M_1^2 - [(\gamma-1)/2]\}^{1/(\gamma-1)}} \quad (5.17)$$

At an altitude of 20 km, a standard atmosphere has a pressure p_1 of 5.53×10^3 N/m² and a temperature T_1 of 217 K. Consequently,

$$\frac{p_{03}}{p_1} = \frac{5.53 \times 10^4}{5.53 \times 10^3} = 10$$

and, from Eq. (5.17) or a table based on this equation, we have $M_1 = 2.72$. The Mach number just downstream of the shock M_2 is 0.492, with $(p_2/p_1) = 8.66$. Thus, there is a small but significant pressure increase along the streamline between points 2 and 3. Since we know the freestream temperature T_1, we can determine the aircraft's speed V_1 as

$$V_1 = a_1 M_1 = \sqrt{\gamma R T_1}\, M_1 = (1.4 \times 287 \times 217)^{\frac{1}{2}} \times 2.72 = 803 \text{ m/s}$$

Although the transition across the shock is not isentropic, nevertheless, isentropic point relations can be used, either upstream or downstream of the shock. The use of an isentropic point relation is equivalent to removing a small particle of fluid and isentropically (without friction or heat transfer) bringing it to rest. In this state, the pressure equals the stagnation pressure of the moving fluid at the point where the particle was removed. Similar results hold for density and temperature. We will often use isentropic point relations, as in the derivation of Eq. (5.17), where p_{02}/p_2 is used.

The above example illustrates another concept. If an aircraft is flying in quiescent air at a steady speed, we are at liberty to analyze the flow about the vehicle by bringing the aircraft to rest and having air flow about it. This velocity transformation is the principle behind most wind tunnels.

5.2 OBLIQUE SHOCK WAVES

We first consider a steady normal shock, as shown in Fig. 5.3(a). The ratio of speeds u_1/u_2 across the shock is given by Eq. (5.14c)

$$\frac{u_1}{u_2} = \frac{\rho_2}{\rho_1} = \frac{\gamma+1}{2} \frac{M_{1n}^2}{1 + [(\gamma-1)/2]M_{1n}^2} \quad (5.18)$$

On both sides of the shock a constant tangential velocity v is added to the flow, as shown in Fig. 5.3(b). This sketch shows that the magnitude of the actual downstream velocity V_2 is less than the upstream velocity V_1. The downstream flow is also turned more toward the shock than is the upstream flow. Figure 5.4 is a more elaborate sketch, which shows the angles β and θ. In the figure, β is the acute angle between V_1 and the shock, and θ is the acute angle between V_1 and V_2.

By trigonometry, we establish from Fig. 5.4

$$u_1 = V_1 \sin \beta = v \tan \beta$$
$$u_2 = V_2 \sin(\beta - \theta) = v \tan(\beta - \theta)$$
(5.19)

Now, $M_1 = V_1/a_1$, so that

$$M_{1n} = \frac{u_1}{a_1} = \frac{V_1}{a_1} \sin \beta = M_1 \sin \beta$$
(5.20)

An n subscript denotes the normal component of the velocity, or Mach number, relative to the shock. Thus, M_1 in the normal shock equations, such as Eqs. (5.14)–(5.16), is replaced by M_{1n} or by $M_1 \sin \beta$. This replacement holds for the velocity components normal to the shock. It does not apply to the tangential component or to V_1/V_2, but does apply to u_1/u_2. Thus, Eq. (5.14c) becomes

$$\frac{u_1}{u_2} = \frac{\rho_2}{\rho_1} = \frac{\gamma + 1}{2} \frac{M_1^2 \sin^2 \beta}{1 + [(\gamma - 1)/2] M_1^2 \sin^2 \beta}$$
(5.21)

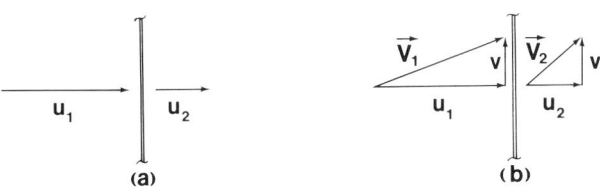

Fig. 5.3 Normal and tangential velocity components for a steady, oblique shock wave.

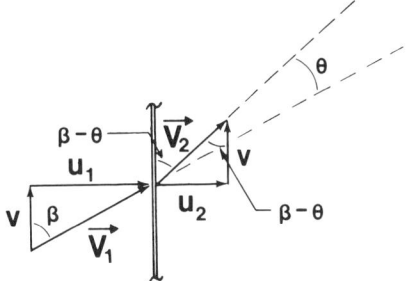

Fig. 5.4 Nomenclature for an oblique shock wave where the upstream velocity is V_1.

We also have

$$M_2 = \frac{V_2}{a_2} = \frac{u_2}{a_2}\frac{1}{\sin(\beta-\theta)}$$

which becomes

$$M_{2n} = \frac{u_2}{a_2} = M_2\sin(\beta-\theta) \tag{5.22}$$

Hence, M_2 in most normal shock relations is replaced by M_{2n} or by $M_2\sin(\beta-\theta)$. The solution of normal shock problems is expedited by the use of a normal shock table whose generation is called for in the homework. Such a table also can be used for oblique shocks with M_1 and M_2 becoming M_{1n} and M_{2n}, respectively.

Any shock wave is a result of boundary conditions that force the flowfield to readjust. Consequently, Fig. 5.1 is unrealistic in that the sketch shows no reason for the shock to be at its particular location. By way of contrast, Fig. 5.2 is realistic; the pitot probe is the boundary condition for the flowfield with a detached bow shock. For an oblique shock, Fig. 5.4 indicates that the velocity has an abrupt turn, of angle θ, at the shock. In making this turn, the flow is adjusting to an abrupt turn of angle θ in the adjacent wall, as shown in Fig. 5.5. Thus, β is the shock angle relative to $\mathbf{V}_1$, while θ is the wall turn angle. As previously noted, the downstream velocity is turned closer to the shock than is the upstream velocity.

If the oncoming flow in Fig. 5.5 were subsonic, the flow would start to adjust well upstream of the turn. The adjustment is a continuous one, with the pressure gradually changing both upstream and downstream of the turn. However, in a supersonic flow, the adjustment is abrupt because the disturbance propagates with a speed less than that of the upstream flow V_1. Hence, the upstream flow first learns of the change in wall slope when it meets the shock wave.

The largest value that β can attain is 90 deg, in which case $v=0$ and the shock is a normal one. The minimum value is obtained from the supersonic condition, $M_{1n} \geq 1$, which leads to

$$M_1\sin\beta \geq 1$$

or

$$\beta \geq \sin^{-1}(1/M_1)$$

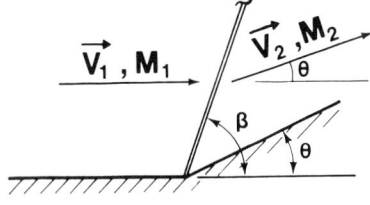

Fig. 5.5 Wall shape for an oblique shock wave.

When the equality sign holds, $M_{1n} = 1$, and Eq. (5.11) shows that $M_{2n} = 1$. From Eqs. (5.20)–(5.22), we then obtain

$$u_1 = u_2 = a_1 = a_2$$

In this circumstance, the shock has zero strength in that the velocity, pressure, etc., do not change abruptly across it. Furthermore, the component of the velocity perpendicular to the wave equals the speed of sound. This type of wave is called a Mach wave. It represents a shock wave of zero strength.

An additional relation between the quantities γ, M_1, β, and θ is obtained by eliminating u_1/u_2 from Eq. (5.21) and from

$$\frac{u_1}{u_2} = \frac{\tan \beta}{\tan(\beta - \theta)}$$

which comes from Eqs. (5.19). We thus obtain

$$\tan \theta = \tan \beta \frac{M_1^2 \sin^2\beta - 1}{\tan^2\beta + \{[(\gamma+1)/(\gamma-1)] + \tan^2\beta\}[(\gamma-1)/2] M_1^2 \sin^2\beta}$$

which simplifies to

$$\tan \theta = \cot \beta \frac{M_1^2 \sin^2\beta - 1}{1 + \{[(\gamma+1)/2] - \sin^2\beta\} M_1^2} \quad (5.23)$$

For a given value of γ, Eq. (5.23) results in a family of curves, one per M_1 value, which we discuss next.

As shown in Fig. 5.6, each $M_1 = $ const curve has an inverted U shape and is thus double-valued for θ. The dashed curve that passes through the peaks is labeled θ_d for reasons to be discussed shortly. The solid line curves at the left of the peak are called the weak solution, whereas the dashed curves at the right are called the strong solution. Also shown is the curve for which $M_2 = 1$. To the left of this curve, M_2 is supersonic; to the right, M_2 is subsonic. Accurate plots of Eq. (5.23) for $\gamma = 1.4$ and $5/3$, along with curves for various M_2 values, are shown in B. M. Argrow's figures in Appendix E.

Suppose we have $M_1 = 3$, $\gamma = 1.4$, and $\theta = 10$ deg. Then, from Appendix E, we have

$$\beta = 27.3 \text{ deg, weak solution}$$

$$= 86.5 \text{ deg, strong solution}$$

As shown in Figs. 5.6 and 5.7, two possibilities occur: the weak and strong solutions. For the strong solution, the shock is nearly normal. Consequently, it causes a more intense disturbance than its weak-solution coun-

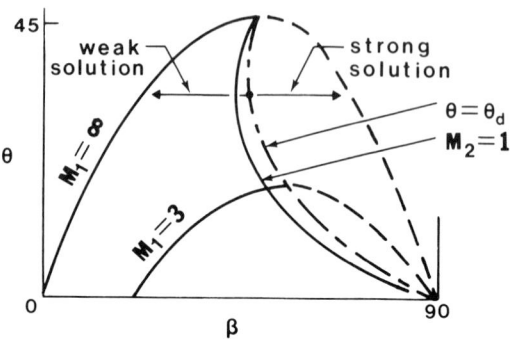

Fig. 5.6 β-θ curves based on Eq. (5.23).

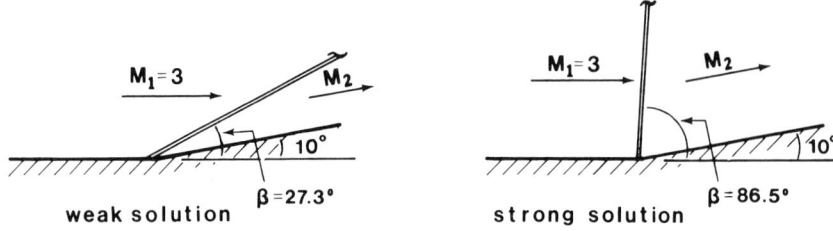

Fig. 5.7 Sketches, to scale, showing the weak and strong oblique shock waves when $\gamma = 1.4$, $M_1 = 3$, and $\theta = 10$ deg.

terpart. For instance, the pressure and entropy jumps satisfy

$$1 < \left(\frac{p_2}{p_1}\right)_{\text{weak}} < \left(\frac{p_2}{p_1}\right)_{\text{strong}}$$

$$1 < \left(\frac{s_2}{s_1}\right)_{\text{weak}} < \left(\frac{s_2}{s_1}\right)_{\text{strong}}$$

Which of the above solutions is correct? In some circumstances, both are correct. For instance, for the bow shock wave in Fig. 5.2, the normal part of the shock is at the $\theta = 0$ deg, $\beta = 90$ deg point in Fig. 5.6. Moving outward along the shock, we follow to the left of Fig. 5.6 a constant, strong-solution M_1 curve. Thus, at some point on the downstream side of the shock (Fig. 5.2), sonic conditions are encountered. As distance increases from the symmetry axis, the shock gradually weakens, tending toward the weak-solution point, where $\theta = 0$ deg and $M_1 \sin \beta = 1$ in Fig. 5.6. As previously mentioned, the zero-strength shock wave is then called a Mach wave.

On the other hand, Fig. 5.7 is an either/or situation. Experiments have confirmed that for an attached shock only the weak solution is expected to occur. (See also Sec. 19.2.) If the shock is detached, i.e., a bow shock wave, it appears as in Fig. 5.8. Except for the narrow crescent-shaped region in

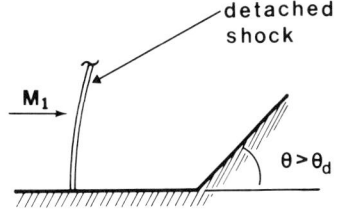

Fig. 5.8 Sketch of a detached shock wave.

Fig. 5.6, the flow downstream of the weak-solution shock is supersonic.
We determine in Example 3 the expected solution for the flow shown in Fig. 5.7.

Example 3

We are to determine M_2 and p_2/p_1 for the weak-solution case in Fig. 5.7. By means of Eqs. (5.11), (5.14a), (5.20), and (5.22), we have

$$M_{1n} = M_1 \sin\beta = 3\sin 27.3° = 1.376$$

$$M_{2n} = \left\{ \frac{1 + [(\gamma-1)/2] M_{1n}^2}{\gamma M_{1n}^2 - [(\gamma-1)/2]} \right\}^{\frac{1}{2}} = 0.7501$$

$$M_2 = \frac{M_{2n}}{\sin(\beta - \theta)} = \frac{0.7501}{\sin 17.3°} = 2.522$$

$$\frac{p_2}{p_1} = \frac{2}{\gamma + 1}\left(\gamma M_{1n}^2 - \frac{\gamma - 1}{2}\right) = 2.042$$

Hence, a 10-deg ramp causes the pressure to double and the Mach number to decrease to 2.5.

Let us now consider a flow where M_1 is fixed, and where we gradually increase the ramp angle θ. Examination of Fig. 5.6 shows a maximum value θ_d for θ. If θ is increased past this value, the shock detaches (hence the d subscript) and becomes a bow shock as sketched in Fig. 5.8. Although the shock is shown as touching the wall, it is considered a detached shock. In any event, the reason for a shock is the ramp, and an attached shock would start there.

Planar, oblique shock waves are useful for describing the flow about a slender wedge where angles θ and θ' need not be the same, as shown in Fig. 5.9. The angles need not be the same because the two flowfields are independent of each other. Conditions behind the shocks are established by considering two wedges in a flow with Mach number M_1. One wedge has a ramp angle θ and the other an angle θ'. Note that the angles are measured relative to the freestream velocity and not to the wedge's centerline. If the larger of the two ramp angles, say θ', exceeds its detachment value, then both shocks detach in favor of an asymmetric bow shock wave.

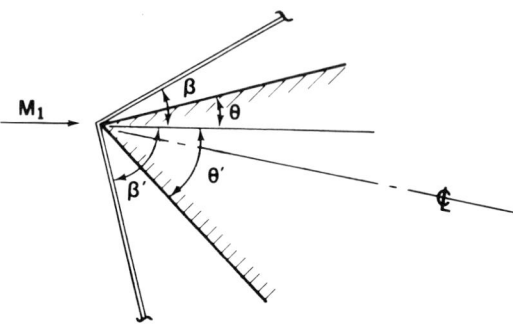

Fig. 5.9   Flow past a wedge with attached oblique shock waves.

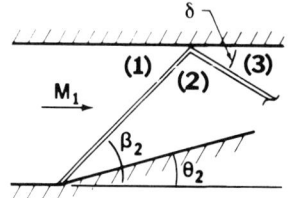

Fig. 5.10   Flow schematic for Example 4.

A more complex type of problem involves shock reflection at a wall, as shown in Fig. 5.10. The method of solution is illustrated in the next example, which uses this figure.

Example 4

A two-dimensional wind tunnel uses carbon monoxide (CO, $\gamma = 1.4$) and has a planar ramp on the lower wall with a 15-deg angle. The incident shock has a 40-deg angle relative to the freestream. We are to determine M_1, M_2, M_3, δ, p_3/p_1, and p_{03}/p_{01}.

We are given

$$\theta_2 = 15 \text{ deg}, \qquad \beta_2 = 40 \text{ deg}, \qquad \gamma = 1.4$$

so that, from Appendix E, $M_1 = 2.28$. As in Example 3, we have

$$M_{1n} = M_1 \sin \beta_2 = 1.466$$

$$M_{2n} = 0.713$$

$$M_2 = \frac{M_{2n}}{\sin(\beta_2 - \theta_2)} = 1.69$$

$$\frac{p_2}{p_1} = 2.34$$

From Eq. (5.15b), we obtain

$$\frac{p_{02}}{p_{01}} = \left(\frac{\gamma+1}{2}\right)^{(\gamma+1)/(\gamma-1)}$$
$$\times \frac{M_{1n}^{2\gamma/(\gamma-1)}}{\{1+[(\gamma-1)/2]M_{1n}^2\}^{\gamma/(\gamma-1)}\{\gamma M_{1n}^2-[(\gamma-1)/2]\}^{1/(\gamma-1)}}$$
$$= 0.940$$

It is worth noting that the Rayleigh-pitot formula for p_{02}/p_1, Eq. (5.17), cannot be used for any nonnormal shock wave. Why?

We now consider the flow across the reflected shock that separates regions 2 and 3. As shown in Fig. 5.11, the dashed line is parallel to V_2 and thus parallel to the ramp. Hence, we have $\theta_3 = 15$ deg and $\beta_3 = \delta + 15$ deg. Since M_2 and θ_3 are known, Appendix E yields $\beta_3 = 55.5$ deg and, consequently,

$$\delta = \beta_3 - 15 \text{ deg} = 40.5 \text{ deg}$$

In order not to confuse the normal component of the Mach number ahead of the β_3 shock with M_{2n}, we call the component $\hat{M}_{2n}$. We thus have the following:

$$\hat{M}_{2n} = M_2 \sin\beta_3 = 1.69 \sin 55.5° = 1.39$$

$$M_{3n} = \left\{\frac{1+[(\gamma-1)/2]\hat{M}_{2n}^2}{\gamma\hat{M}_{2n}^2 - [(\gamma-1)/2]}\right\}^{\frac{1}{2}} = 0.744$$

$$\frac{p_3}{p_2} = \frac{2}{\gamma+1}\{\gamma\hat{M}_{2n}^2 - [(\gamma-1)/2]\} = 2.10$$

$$\frac{p_{03}}{p_{02}} = \left(\frac{\gamma+1}{2}\right)^{(\gamma+1)/(\gamma-1)}$$
$$\times \frac{\hat{M}_{2n}^{2\gamma/(\gamma-1)}}{\{1+[(\gamma-1)/2]\hat{M}_{2n}^2\}^{\gamma/(\gamma-1)}\{\gamma\hat{M}_{2n}^2-[(\gamma-1)/2]\}^{1/(\gamma-1)}}$$
$$= 0.961$$

$$M_3 = \frac{M_{3n}}{\sin(\beta_3 - \theta_3)} = \frac{0.744}{\sin 40.5°} = 1.15$$

$$\frac{p_3}{p_1} = \frac{p_3}{p_2}\frac{p_2}{p_1} = 2.1 \times 2.34 = 4.91$$

$$\frac{p_{03}}{p_{01}} = \frac{p_{03}}{p_{02}}\frac{p_{02}}{p_{01}} = 0.961 \times 0.94 = 0.903$$

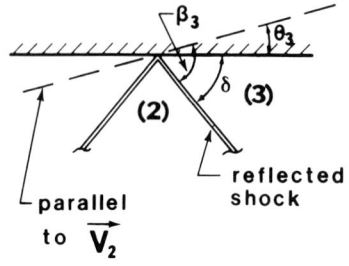

Fig. 5.11 Nomenclature at the wall reflection point for Example 4.

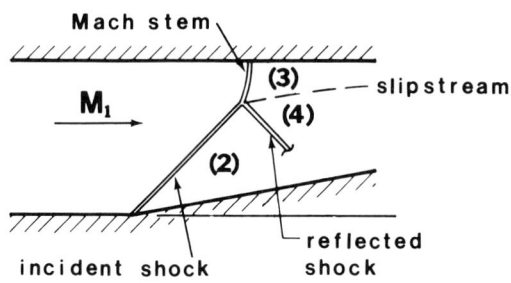

Fig. 5.12 Mach reflection pattern shown in contrast to the regular reflection pattern in Fig. 5.10.

We observe that $M_1 > M_2 > M_3$, as expected. While p_3 exceeds p_1 by almost a factor of 5, the overall drop in stagnation pressure is only 10%. Also, note that $\delta \neq \beta_2$ although the difference between δ and β_2 is small. It is evident from Fig. 5.6 that $\delta > \beta_2$ since $M_1 > M_2$. Moreover, although not explicitly mentioned, the shock waves in Examples 3 and 4 are attached shock waves. Hence, only the weak solution in Appendix E has been used.

The shock reflection pattern shown in Fig. 5.10 is called regular reflection. As the Mach number M_1 decreases, so does M_2. It is evident from Fig. 5.6 that an M_2 value can be reached such that the reflected shock will detach. As discussed in Sec. 19.5, actual detachment occurs slightly before θ_{3d} is reached. Once the incident shock detaches from the upper wall, we have a Mach reflection pattern, which is a three-shock pattern as depicted in Fig. 5.12. The shock wave labeled "Mach stem" is a strong-solution curved shock wave normal to the upper wall. (Although the Mach stem reaches the wall, it is not considered an attached shock.) The three shocks meet at a point called a triple point, where a new type of discontinuity, called a slipstream, begins. The discontinuity starts at the triple point because the entropy jump across the two weak-solution shocks $s_4 - s_1$ is less than $s_3 - s_1$ and, consequently, $s_3 > s_4$. As with a shock wave, there are transition conditions across a slipstream, which is also a streamline. These conditions are that $p_3 = p_4$ and that $\mathbf{V}_3$ be parallel to $\mathbf{V}_4$. Thus, across the slipstream, V_3 differs from V_4, T_3 from T_4, and so forth.

NORMAL AND OBLIQUE SHOCK WAVES

Problems

The tables generated in Problem 5.31 will expedite the solution to many of the following problems.

5.1 Air at 500 K encounters a normal shock. The pressure ratio across the shock is 5. Determine M_1, M_2, T_2, p_{02}/p_{01}, ρ_2/ρ_1, and $\Delta s/R$.

5.2 Air at 500 K encounters a normal shock. The Mach number M_1 is 3. Determine V_1, V_2, and $\Delta s/R$.

5.3 Air at 1-atm pressure encounters a normal shock. The Mach number M_1 is 4. Determine M_2, p_2, p_{02}/p_{01}, T_{02}/T_{01}, and $\Delta s/R$.

5.4 Nitrogen encounters a normal shock where $s_2 - s_1 = 500$ N-m/kg-K. Determine M_1, M_2, p_2/p_1, and p_{02}/p_{01}.

5.5 Nitrogen encounters a normal shock. The downstream pressure and density are 8×10^5 N/m² and 2.7 kg/m³. The upstream temperature is 350 K. Determine M_1 and M_2.

5.6 For conditions in Problem 5.5, determine p_1, ρ_1, V_1, p_{01}, T_{01}, T_2, V_2, p_{02}, T_{02}, and Δs.

5.7 If a pitot pressure tube is aligned with the flow in Problem 5.5, what pressure value would it read?

5.8 Helium at 500 K encounters a normal shock with a pressure ratio of 5. Determine M_1, M_2, V_1, and V_2.

5.9 Determine the temperature ratio T_2/T_1 and the stagnation temperature ratio T_{02}/T_{01} across a steady, normal shock wave with Mach number $M_1 = 3$ in helium gas.

5.10 Helium at 1-atm pressure encounters a normal shock. The Mach number M_1 is 4. Determine M_2, p_2, p_{02}/p_{01}, T_{02}/T_{01}, and $\Delta s/R$.

5.11 The Mach number downstream of a normal shock in argon is $M_2 = 0.5$. We also know that $T_2 = 2000$ K and $p_{01} = 3$ atm. Determine M_1, ρ_1, ρ_2, V_1, and V_2.

5.12 Determine the enthalpy ratio h_2/h_1 across a normal shock in a monatomic gas when $M_1 = 3M_2$.

5.13 A steady, normal shock wave in argon has $V_2 = 500$ m/s and $T_2 = 10^3$ K. Determine M_1, T_1, and V_1.

5.14 A reentry missile has a speed of 5 km/s at an altitude of 17 km, where the atmosphere's pressure and temperature are 0.15 atm and 218 K, respectively. Determine the pressure, temperature, and Mach number at point 2 in Fig. 5.2, which is behind the shock and on the centerline.

5.15 A normal shock has a pressure ratio $p_2/p_1 = 2.5$ and a temperature ratio $T_2/T_1 = 1.4$. Determine γ.

5.16 A pitot tube measures a pressure of 3×10^5 N/m² in helium. The downstream static pressure is 2×10^5 N/m². Determine the freestream pressure and Mach number.

5.17 The pitot pressure behind a normal shock wave is 3.1×10^5 N/m². The gas is helium and the temperature at the entrance to the pitot tube is 350 K, while the gas speed ahead of the shock is 1500 m/s. Determine M_1, p_1, and T_1.

5.18 Air flows over a hinged flap as shown in Fig. 5.5. The freestream Mach number is 3 and p_1 is 10^4 Pa. If the hinge angle θ is 13 deg, determine M_2, β, and p_2.

5.19 For conditions in Problem 5.18, determine the hinge angle such that $M_2 = 1$. At what hinge angle will the shock wave detach? If the exposed surface of the hinge is 3 cm $\times$ 10 cm, determine the normal pressure force on the hinge when $M_2 = 1$. What is the magnitude of the pressure force on the hinge when $\theta = 0$ deg?

5.20 Determine the two entropy changes $\Delta s = s_2 - s_1$ and $\Delta s' = s'_3 - s_1$ for the two ramps shown in the sketches, where one ramp has a single turn of 20 deg, while the other has two turns each of 10 deg. The gas is air and M_1 is 4. Also determine p_{02}/p_1, p'_{03}/p_1, p_2/p_1, and p'_3/p_1.

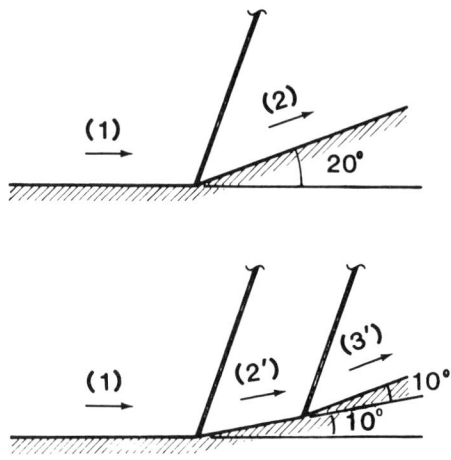

5.21 A sharp wedge aligned with the flow has a half-angle θ of 7 deg. The oncoming air has a Mach number M_1 of 3.2, and $p_1 = 1$ atm, $T_1 = 300$ K.

Determine the shock angle β, and M_2, p_2, and T_2 behind the shock. If the shock wave were normal, with $M_1 = 3.2$, what would be M_2', p_2', and T_2'?

5.22 Air with $M_1 = 3$ encounters a sharp wedge. Across the shock $\Delta s = 200$ J/kg-K. Determine θ, β, and M_2.

5.23 Air flows by a 25-deg wedge as shown in the sketch. The Mach number in region 3 is unity. Determine M_1 and M_2.

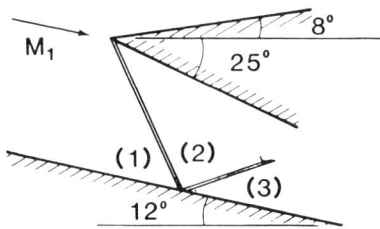

5.24 Consider an oblique shock wave in air with a shock wave angle β of 20 deg. Upstream of the wave, we have $p_1 = 7$ atm, $T_1 = 300$ K, and $V_1 = 2$ km/s. Determine M_2, p_2, T_2, V_2, and the flow deflection angle.

5.25 For the flow shown in the sketch: $M_1 = 2.6$, $T_1 = 400$ K, $p_1 = 1$ atm, and $\gamma = 1.4$. Find M_2, T_2, p_2, and M_3, T_3, p_3.

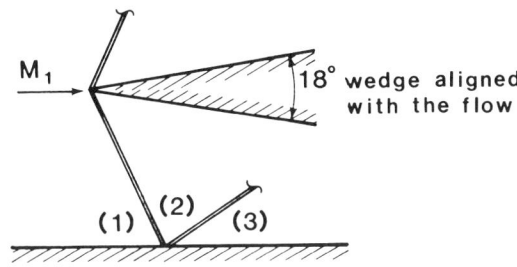

5.26 Ahead of an oblique shock, air has a Mach number of 4. Behind the oblique shock the Mach number is 2. Determine β and θ in degrees.

5.27 Air flows around a 29-deg wedge aligned with the flow, as in Problem 5.25. With $M_1 = 3$, determine M_2 and M_3.

5.28 A monatomic gas flowing at a supersonic speed encounters a ramp with a 5-deg angle. The ramp causes a shock wave that has an angle of 20 deg relative to the ramp. What is the preshock Mach number?

5.29 A duct has a 5-deg ramp as shown in the sketch. Air with a Mach number $M_1 = 2.5$ enters the duct. Determine the shock wave impingement point on the lower wall, M_2, M_3, p_2/p_1, and p_3/p_1.

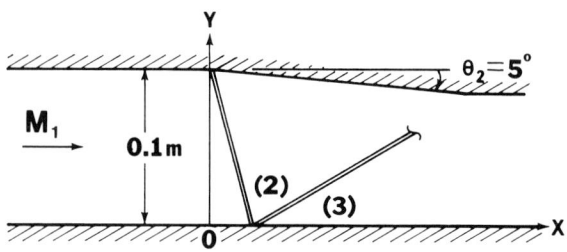

5.30 A wedge with a half-angle α of 7.5 deg is misaligned in a wind tunnel where $M_1 = 3$. The wedge is known to be misaligned because wall pressure measurements show $p_5/p_1 = 2.07$ and $p_3/p_1 = 3.86$. Determine the misalignment angle ϕ.

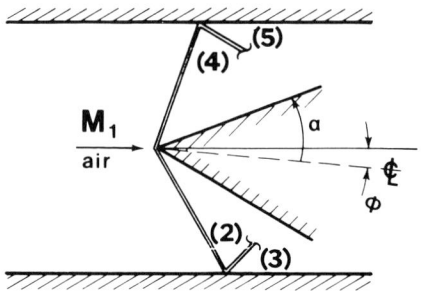

Computational Problems:

5.31 For $\gamma = 1.4$ and $5/3$ and for $M_1 = 1(0.01)\ 5(0.1)\ 10$, generate normal shock tables for M_2, p_2/p_1, T_2/T_1, ρ_2/ρ_1, p_{02}/p_{01}, p_{02}/p_1.

5.32 Generate the Appendix E β-θ figure when $\gamma = 5/3$ or $\gamma = 1.1$. Show lines for $M_2 = 0.5, 1, 2, 4$.

6. PRANDTL-MEYER FLOW AND SHOCK-EXPANSION THEORY

We considered in Chap. 5 the generation of an oblique shock by a sharp turn in the wall that compresses the flow (Fig. 5.5). If θ is negative, we need an expansion to turn the supersonic flow. This is accomplished with a Prandtl-Meyer expansion fan. Such an expansion can occur only in a supersonic or sonic flow and is the subject of Sec. 6.1. Along with shock waves, expansion waves frequently occur in a sonic or supersonic flow, and they too represent essential building blocks for establishing the overall flowfield. Principal results of this section are summarized in Appendix F.

In shock-expansion theory, Sec. 6.2, we combine the Prandtl-Meyer expansion with the oblique shock wave of Chap. 5 to determine the lift and drag of simple, two-dimensional airfoils in a supersonic flow. Appendix G summarizes and extends these results.

6.1 PRANDTL-MEYER FLOW

Mach Lines or Waves

With M_1 fixed and supersonic, let us consider the limit of $\theta \to 0$ using Eq. (5.23). As $\theta \to 0$, we have $\tan \theta \to 0$ and, consequently, the numerator on the right side of Eq. (5.23) must also go to zero. Hence,

$$M_1^2 \sin^2 \beta = 1$$

or

$$\mu = \sin^{-1} \frac{1}{M} \tag{6.1}$$

where the subscript on M is dropped and β is relabeled as μ. As mentioned in Sec. 5.2, the shock wave is called a Mach wave in this limit and μ is called the Mach angle. From Eq. (6.1), we observe that μ is defined only when $M \geq 1$. Figure 6.1 is a sketch showing the Mach waves and Mach angles when the flow is uniform. For reasons explained in Chap. 9, Mach waves are often called characteristics. However, in a two-dimensional flow, they are usually called Mach lines.

Shortly, we shall need other trigonometric functions of μ. Based on Eq. (6.1), we construct the right triangle shown in Fig. 6.2. From this diagram,

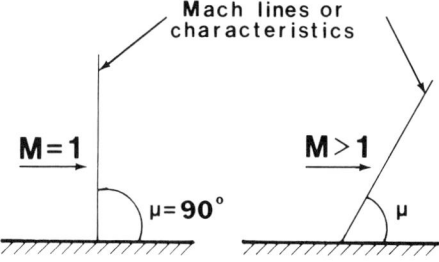

Fig. 6.1 Mach angles and lines.

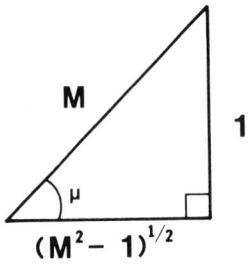

Fig. 6.2 Mach angle right triangle.

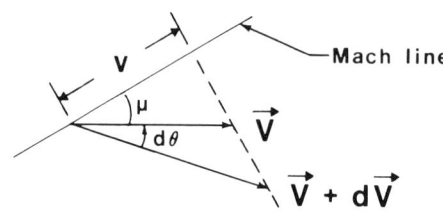

Fig. 6.3 Relation between V and V + dV for the change across a Mach line.

we readily obtain

$$\cos\mu = \frac{(M^2-1)^{\frac{1}{2}}}{M}, \qquad \tan\mu = \frac{1}{(M^2-1)^{\frac{1}{2}}} \qquad (6.2)$$

Prandtl-Meyer Function

Consider a Mach line that causes an infinitesimal change in magnitude and direction of the velocity in a two-dimensional flow, as shown in Fig. 6.3. Our first objective is to relate dV to $d\theta$. We know that the component of the velocity tangential to the Mach wave is unchanged. This statement follows directly from oblique shock theory, even when the shock is weak. The figure schematically shows how **V** and **V** + d**V** have the same tangential component v. From this figure, we obtain

$$\frac{v}{V} = \cos\mu = \frac{(M^2-1)^{\frac{1}{2}}}{M} \qquad (6.3\text{a})$$

$$\frac{v}{V+dV} = \cos(\mu+d\theta) \qquad (6.3\text{b})$$

where Eq. (6.2) is used. With $\cos(d\theta) = 1$, $\sin(d\theta) = d\theta$, and with Eqs. (6.1) and (6.2), we can expand the cosine term in Eq. (6.3b) as follows:

$$\cos(\mu + d\theta) = \cos\mu \cos(d\theta) - \sin\mu \sin(d\theta) = \frac{(M^2-1)^{\frac{1}{2}}}{M} - \frac{d\theta}{M}$$

This result is used for the right side of Eq. (6.3b), and v is eliminated from Eqs. (6.3) to yield

$$\frac{V}{V+dV} \frac{(M^2-1)^{\frac{1}{2}}}{M} = \frac{(M^2-1)^{\frac{1}{2}}}{M} - \frac{d\theta}{M}$$

With

$$\frac{V}{V+dV} = 1 \bigg/ \left(1 + \frac{dV}{V}\right) = 1 - \left(\frac{dV}{V}\right) + \cdots$$

we obtain

$$\frac{dV}{V} = \frac{d\theta}{(M^2-1)^{\frac{1}{2}}} \tag{6.4}$$

This relation does not apply across a shock wave where the flow angle and speed change in a discontinuous manner. It should be clear from the Mach number term in the denominator that the flow must be sonic or supersonic. With these provisos, the relation partly governs the dynamics of the flow over a curved wall with either a positive or a negative slope change. It also holds locally anywhere in a two-dimensional supersonic flow, not just along a wall. However, integration of Eq. (6.4) requires elimination of one of the three variables, which we now proceed to do.

Consider a supersonic flow that is being turned by a wall. The turn may result in an expansion, with an increasing Mach number, or a compression, with a decreasing Mach number. In either case, the flow is isentropic, providing shock waves are not present. (In the compression case, the turn must be gradual. As we shall see, this restriction does not apply to an expansive turn.) Remember that a continuous flow with no heat transfer or viscous forces has a constant entropy. From the relation $V = aM$ and the isentropic equation

$$\frac{a}{a_0} = \left(\frac{T}{T_0}\right)^{\frac{1}{2}} = \left(1 + \frac{\gamma-1}{2}M^2\right)^{-\frac{1}{2}}$$

where a_0 is the constant stagnation speed of sound, we have

$$V = a_0 M \left(1 + \frac{\gamma-1}{2}M^2\right)^{-\frac{1}{2}} \tag{6.5a}$$

By logarithmic differentiation, this becomes

$$\frac{dV}{V} = \frac{1}{1 + [(\gamma - 1)/2] M^2} \frac{dM}{M} \quad (6.5b)$$

Elimination of dV/V from Eqs. (6.4) and (6.5b) yields

$$d\theta = \frac{(M^2 - 1)^{\frac{1}{2}}}{1 + [(\gamma - 1)/2] M^2} \frac{dM}{M} \quad (6.6)$$

We choose as a reference condition $M = 1$ when $\theta = 0$. Equation (6.6) is integrated from this reference condition to an angle ν that corresponds to a Mach number M:

$$\int_0^\nu d\theta = \int_1^M \frac{(M^2 - 1)^{\frac{1}{2}}}{1 + [(\gamma - 1)/2] M^2} \frac{dM}{M} \quad (6.7)$$

The Mach number integral is performed by changing variables to $z = M^2$ and then using the method of partial fractions. We thus obtain

$$\nu(M) = \sqrt{(\gamma + 1)/(\gamma - 1)} \tan^{-1} \sqrt{[(\gamma - 1)/(\gamma + 1)](M^2 - 1)}$$

$$- \tan^{-1} \sqrt{M^2 - 1} \quad (6.8)$$

where $\nu(M)$ is called the Prandtl-Meyer function and is in radians. A streamline must turn by an angle ν if the Mach number is to change along it from unity to M, where $M > 1$. The direction of the turn must be away from the Mach line for the Mach number to increase, as indicated in Fig. 6.3. In this situation, the turn is an expansive one.

Let us consider an abrupt wall turn of angle θ, as shown in Fig. 6.4. Upstream of the leading edge (LE) of the expansion, we have a uniform supersonic flow with Mach number M_1 and a Mach angle μ_1. A typical streamline, which is parallel to the upstream wall, is gradually turned until, at the trailing edge (TE) of the expansion, it is parallel to the downstream wall. Downstream of the trailing edge, the flow is uniform with a Mach number M_2, where $M_2 > M_1$, and a Mach angle μ_2. The radial lines within the expansion are Mach lines of the same family as those upstream and downstream of the expansion. Thus, the leading and trailing edges of the expansion are also Mach lines whose Mach angle is shown in the figure. The expansion itself is referred to as a centered Prandtl-Meyer expansion. Since neither heat transfer nor viscous shear occurs, the expansion is an isentropic process.

Examination of Fig. 6.4 shows that there is no length scale. Consequently, the flowfield inside the expansion can depend only on an angular coordinate, such as η. In turn, this means that velocity, Mach number, pressure, etc., are constant along a given Mach line.

PRANDTL-MEYER FLOW AND SHOCK-EXPANSION THEORY 73

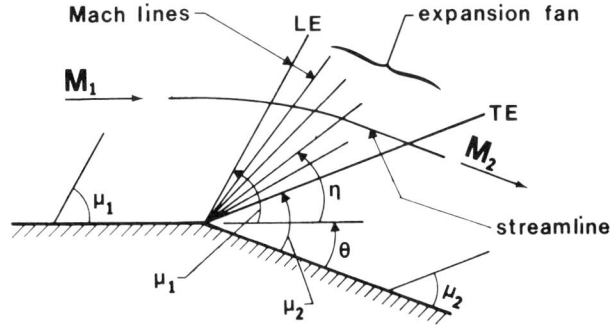

Fig. 6.4 Centered Prandtl-Meyer expansion for an abrupt wall turn of angle θ.

To determine M_2, consider Eq. (6.7) again, as follows:

$$\int_0^\theta d\theta = \int_{M_1}^{M_2} \frac{(M^2-1)^{\frac{1}{2}}}{1+[(\gamma-1)/2]M^2} \frac{dM}{M} = \int_1^{M_2} \cdots - \int_1^{M_1} \cdots$$

which becomes

$$\theta = \nu(M_2) - \nu(M_1)$$

or

$$\nu(M_2) = \nu(M_1) + \theta \qquad (6.9)$$

The solution of Eq. (6.9) for either M_1, M_2, or θ is expedited by the use of a ν vs M table, which is the subject of homework problem 6.32.

The Mach angle and Prandtl-Meyer functions not only apply to a sharp turn but also to a smooth one, as shown in Fig. 6.5. Suppose we know the Mach number M_1 at some point P_1 on the wall. We then compute $\mu(M_1)$ and $\nu(M_1)$, where $\mu(M_1)$ is the angle relative to the wall of the Mach line labeled 1. The angle the wall makes at P_1 relative to an arbitrary reference is denoted by θ_1. Let us move downstream to a point P_2 on the wall where the known wall slope is θ_2. To compute M_2, we first determine

$$\nu(M_2) = \nu_2 = \nu_1 + (\theta_2 - \theta_1)$$

where θ is now positive when measured clockwise, as shown in Fig. 6.5. Once ν_2 is known, Eq. (6.8) provides M_2. The figure also shows a typical streamline. Where the streamline crosses the Mach lines, velocities $\mathbf{V}_1$ and $\mathbf{V}_2$ are parallel to the wall at points P_1 and P_2, respectively. (Recall that $\mathbf{V}$ is constant along a Mach line.)

The foregoing calculation can be done in a stepwise fashion along the wall to determine the location of various Mach lines and the Mach number along each of these lines. Since the flow is isentropic, the temperature and

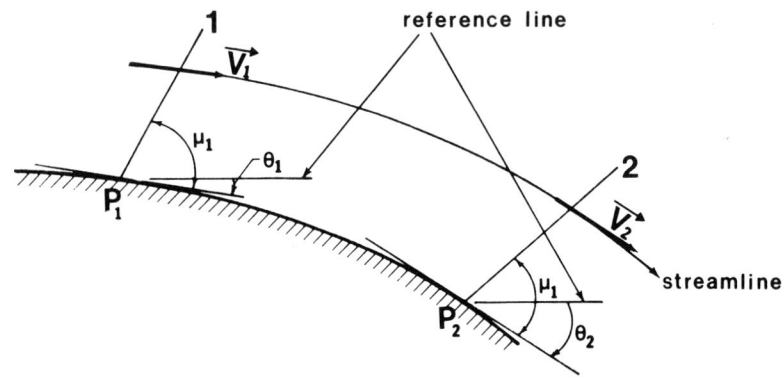

Fig. 6.5 Prandtl-Meyer expansion for a gradual turn; the reference line from which θ is measured is arbitrary.

pressure along line 2 are given by

$$\frac{T_2}{T_1} = \frac{T_2/T_0}{T_1/T_0} = \frac{1 + [(\gamma-1)/2] M_1^2}{1 + [(\gamma-1)/2] M_2^2}$$

and

$$\frac{p_2}{p_1} = \left\{ \frac{1 + [(\gamma-1)/2] M_1^2}{1 + [(\gamma-1)/2] M_2^2} \right\}^{\gamma/(\gamma-1)}$$

where stagnation quantities are constant and where T_1 and p_1 are assumed to be known. This is generally the case, since initial conditions usually correspond to a known uniform flowfield.

The method of solution for an expansive wall turn is given in Example 1, which follows. Example 2 then provides the method for a compressive turn.

Example 1

Air is flowing along a planar wall with $M_1 = 2$. A sharp convex corner, as in Fig. 6.4, is encountered where $\theta = 20$ deg. We are to determine the angles associated with the leading and trailing edges, the angular extent ϕ of the expansion fan, the downstream Mach number M_2, and the pressure and temperature ratios across the expansion. A Mach number table, in degrees, is used for ν and μ, as is one for the isentropic values of p/p_0 and T/T_0.

Since $M_1 = 2$, we have

$$\nu_1 = \nu(M_1) = 26.4 \text{ deg}, \qquad \mu_1 = \mu(M_1) = 30 \text{ deg}$$

Downstream of the trailing edge, Eq. (6.9) yields

$$\nu_2 = \nu_1 + \theta = 26.4 + 20 = 46.4 \text{ deg}$$

and, consequently,

$$M_2 = 2.83, \quad \mu_2 = 20.7 \text{ deg}$$

A sketch of the flowfield is shown in Fig. 6.6. The expansion fan angle is seen to be given by

$$\phi = \theta + \mu_1 - \mu_2 = 20 + 30 - 20.7 = 29.3 \text{ deg}$$

Finally, the pressure and temperature ratios are

$$\frac{p_2}{p_1} = \frac{p_2/p_0}{p_1/p_0} = \frac{0.03467}{0.1278} = 0.271$$

and

$$\frac{T_2}{T_1} = \frac{T_2/T_0}{T_1/T_0} = \frac{0.3827}{0.5556} = 0.689$$

Let us suppose that M_1 is kept fixed but that the turn angle θ is gradually increased. In this situation, conditions on the leading Mach line, including μ_1, are unaltered. However, the expansion fan angle, as well as M_2, increase. A limiting value for $\nu(M_2)$ as $M_2 \to \infty$ is obtained from the Prandtl-Meyer function. Hence,

$$\nu_2 = \nu(\infty) = \left[\sqrt{(\gamma+1)/(\gamma-1)} - 1\right]90° = 180 \text{ deg}, \quad \gamma = 1.25$$

$$= 130.5 \text{ deg}, \quad \gamma = 1.4$$

$$= 90 \text{ deg}, \quad \gamma = 5/3$$

and θ in Example 1 becomes

$$\theta_\infty = \nu(\infty) - \nu_1 = 130.5 - 26.4 = 104 \text{ deg}$$

In this case, the trailing edge is along the downstream wall, since $\mu(\infty) = 0$ deg. On this Mach line, we also have $p_2 = \rho_2 = T_2 = 0$.

If the turn increases further, a vacuum occurs between the trailing edge and the wall, as shown in Fig. 6.7. The expansion is no longer affected by the downstream wall when this occurs.

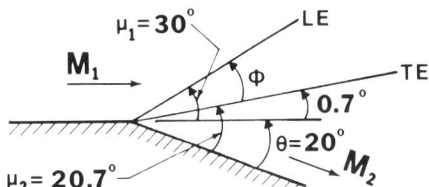

Fig. 6.6 Flowfield of Example 1.

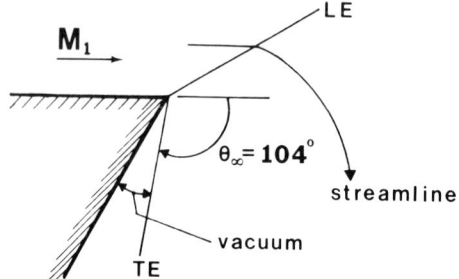

Fig. 6.7 Prandtl-Meyer flow when θ exceeds θ_∞.

Small vernier rockets are used to change the orientation of an orbiting satellite. When a vernier rocket is firing, the jet leaving the rocket expands into a vacuum. The foregoing discussion describes how the initial angle of the edge of the jet, relative to the nozzle's centerline, is found. This information is necessary to decide where the vernier rockets are to be located since the expanding jet should not impinge on sensitive hardware, such as optical lenses or solar cells.

To some extent, real gas effects should be considered when the temperature becomes small. These effects, however, are not as severe as might be thought since the gas is generally unable to condense in a very low-density, high-speed flow.

Example 2

We reconsider Example 1 but with a gradual 20-deg turn by the wall into the flow. The turn must be gradual if there is to be a shock-free region adjacent to the wall.

As in Example 1, we have

$$\nu_1 = 26.4 \text{ deg}, \qquad \mu_1 = 30 \text{ deg}$$

but now θ must be taken as negative in Eq. (6.9); hence,

$$\nu_2 = 26.4 - 20 = 6.4 \text{ deg}$$

Consequently, we have $M_2 = 1.31$ and $\mu_2 = 49.8$ deg. A sketch of the flowfield is shown in Fig. 6.8. Finally, the pressure and temperature ratios

Fig. 6.8 Compressive turn of 20 deg (see Example 2).

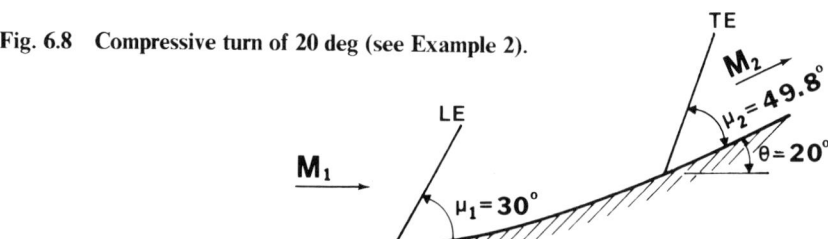

are

$$\frac{p_2}{p_1} = \frac{p_2/p_0}{p_1/p_0} = \frac{0.3560}{0.1278} = 2.79$$

and

$$\frac{T_2}{T_1} = \frac{0.7445}{0.5556} = 1.34$$

In contrast to Example 1, where pressure in an expansive turn decreases, pressure in a compressive turn increases. Away from the compressive corner, the Mach lines converge to form an oblique shock wave. Far from the corner, the oblique shock corresponds approximately to the weak-solution oblique shock wave discussed in Sec. 5.2 and shown in Fig. 5.5. (This flowfield is reconsidered, in considerable detail, in Sec. 19.3.)

If the wall turn angle in Example 2 exceeds 26.4 deg, then ν_2 would be negative, which is impossible. Once again, the overall flowfield must change sharply from the one shown in Fig. 6.8 to one that resembles Fig. 5.8.

6.2 SHOCK-EXPANSION THEORY

An excellent illustration of the use of oblique shock and Prandtl-Meyer theory is their application to supersonic airfoils, known as shock-expansion theory. All pertinent quantities except viscous drag are obtainable from the theory. These include lift, drag, center of pressure, and turning moments. However, even for viscous drag, the theory is useful because it establishes the pressure along the airfoil, which is needed for determining the viscous drag. We will discuss only lift and drag, given as dimensionless lift and drag coefficients per unit span for a steady supersonic flow about a two-dimensional airfoil (Fig. 6.9).

In the figure, the freestream Mach number is denoted as M_∞. The chord length is ℓ, and the airfoil is shown at a positive angle of attack α. The angle of attack denotes the orientation of the airfoil's chord line relative to the freestream velocity $\mathbf{V}_\infty$. By definition, the drag D is the retarding force in the direction of motion. Hence, a positive drag is oriented parallel to $\mathbf{V}_\infty$, as shown in the side sketch. The lift L is defined as perpendicular to $\mathbf{V}_\infty$. For a thin airfoil, the net force can be taken as perpendicular to the chord, as indicated in the side sketch.

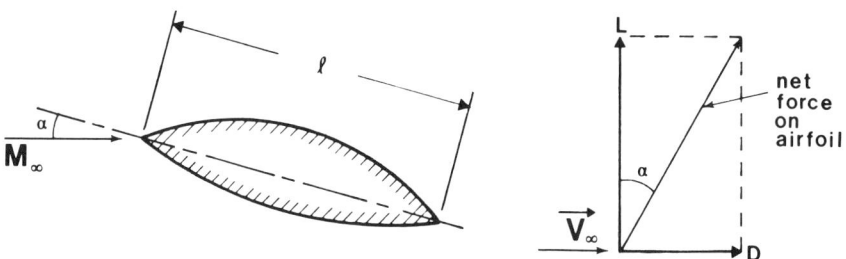

Fig. 6.9 Two-dimensional airfoil at angle of attack α.

The lift and drag coefficients are defined as the lift and drag divided by the freestream dynamic pressure times the planform area. Here, the planform area per unit span is the chord length ℓ. Consequently, we have

$$C_L = \frac{L}{\frac{1}{2}(\rho V^2)_\infty \ell} \tag{6.10a}$$

$$C_D = \frac{D}{\frac{1}{2}(\rho V^2)_\infty \ell} \tag{6.10b}$$

which are dimensionless. By our earlier definition of the Mach number for a perfect gas, we obtain

$$(\rho V^2)_\infty = (\rho a^2)_\infty M_\infty^2 = \gamma p_\infty M_\infty^2 \tag{6.11}$$

which is used in the denominator of the coefficients.

The drag caused by the pressure force is called the wave drag. We would like to have a rough idea of how this drag compares with the viscous drag. The boundary layer on an airfoil moving at a supersonic speed is typically turbulent. Hence, we obtain an estimate for a turbulent drag coefficient, to be denoted as $C_{D\text{visc}}$. Figure 8.9 provides a turbulent skin-friction coefficient C_f for incompressible pipe flow. Nevertheless, the magnitude of C_f, say 0.003, is typical for a flat plate in a supersonic flow at zero angle of attack. (In this flow, the turbulent skin-friction coefficient actually decreases slowly with increasing Mach number.) The skin-friction coefficient is defined by

$$C_f = \frac{\bar{\tau}_w}{\frac{1}{2}(\rho V^2)_\infty}$$

where $\bar{\tau}_w$ is an average wall shear stress given by

$$\bar{\tau}_w = \frac{1}{\ell} \int_0^\ell \tau_w \, dx$$

To convert C_f into a drag coefficient, multiply the numerator and denominator of the C_f definition by the surface area ℓ per unit span. Hence, $C_D = C_f$, and Eq. (6.10b) holds for the viscous stress with $D = \ell \bar{\tau}_w$. Thus, a typical value for $C_{D\text{visc}}$ is 0.003.

Let us suppose that we have a flat plate of zero thickness and at zero angle of attack. Clearly, there is no lift and the only drag is frictional. Consequently, this airfoil has $C_D = C_{D\text{visc}}$ with an approximate magnitude of 0.006 for the two sides.

We now consider the above flat plate at a positive angle of attack α, as shown in Fig. 6.10. The flow below the plate adjusts to the compressive turn of angle α by means of a weak-solution oblique shock wave. Similarly, the flow above the plate adjusts to an expansive turn of angle α by means of a Prandtl-Meyer expansion. On the top (bottom) surface, the pressure p_1 (p_2)

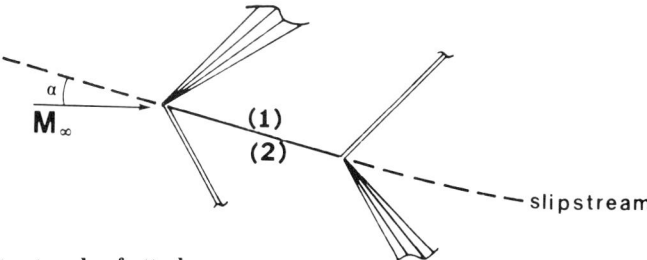

Fig. 6.10 Flat plate at angle of attack α.

is uniform and results in a force that is normal to the plate. Since $p_1 < p_\infty < p_2$, there is positive lift and wave drag on the plate.

At the downstream end of the plate (see Fig. 6.10), a shock and an expansion reoccur. These later waves enable the flow to become horizontal somewhere downstream of the plate. A particle of fluid that passes above the plate experiences a slightly greater jump in entropy than occurs below the plate. This happens because the downstream shock is slightly stronger than the upstream one since $M_2 < M_\infty < M_1$. Consequently, there is a slipstream—defined at the end of Chap. 5—from the plate's trailing edge. In terms of the lift and drag computation, the downstream disturbances, including the slipstream, are immaterial. Henceforth, they are disregarded.

We have, from Figs. 6.9 and 6.10,

$$L = (p_2 - p_1)\ell \cos \alpha, \qquad D = (p_2 - p_1)\ell \sin \alpha$$

Consequently, Eqs. (6.10), with the aid of Eq. (6.11), become

$$C_L = \frac{2}{\gamma M_\infty^2}\left(\frac{p_2}{p_\infty} - \frac{p_1}{p_\infty}\right)\cos \alpha, \qquad C_D = C_L \tan \alpha \qquad (6.12)$$

What remains is to determine p_2/p_∞ and p_1/p_∞ where these pressure ratios depend only on γ, M_∞, and α. When $\alpha = 0$, $p_2 = p_1$, and the lift is zero. When $\alpha \neq 0$, the drag is called wave drag due to lift, and the corresponding drag coefficient is denoted as $C_{D\text{lift}}$. In contrast to viscous drag, which occurs whenever there is motion, this pressure drag occurs only when the airfoil has lift. The next example illustrates the method of computation.

Example 3

Consider a flat plate airfoil at 8-deg angle of attack (Fig. 6.10) in an airstream at $M_\infty = 4$. We are to determine C_L, C_D, M_1, and M_2.

For side 2, we have a compressive turn with $\theta = 8$ deg and $M_\infty = 4$. From Appendix E, we see that the shock angle relative to $\mathbf{V}_\infty$ is $\beta = 20.5$ deg. Hence, we have

$$M_{\infty n} = M_\infty \sin \beta = 1.40$$

and, from a normal shock table,

$$\frac{p_2}{p_\infty} = 2.12, \qquad M_{2n} = 0.740$$

so that

$$M_2 = \frac{M_{2n}}{\sin(\beta - \theta)} = \frac{0.740}{\sin 12.5°} = 3.42$$

For side 1, we readily establish that

$$\nu(M_\infty) = 65.78 \text{ deg}$$

$$\nu(M_1) = \nu(M_\infty) + \theta = 73.78 \text{ deg}$$

$$M_1 = 4.68$$

$$\frac{p_\infty}{p_0} = 6.59 \times 10^{-3}$$

$$\frac{p_1}{p_0} = 2.76 \times 10^{-3}$$

$$\frac{p_1}{p_\infty} = \frac{p_1}{p_0} \frac{p_0}{p_\infty} = \frac{2.76 \times 10^{-3}}{6.59 \times 10^{-3}} = 0.419$$

From Eqs. (6.12), we obtain

$$C_L = \frac{2}{1.4 \times 16}(2.12 - 0.419)\cos 8° = 0.150$$

and

$$C_D = 0.150 \times \tan 8° = 0.0211$$

Observe that the pressure drag is appreciably greater than a typical viscous drag.

We now consider an airfoil with finite thickness. To avoid undue complexity, we assume that the airfoil is a symmetrical diamond, or a double-wedge, airfoil at zero angle of attack, as shown in Fig. 6.11. Since the flow is symmetric about the centerline, $C_L = 0$. Hence, $C_{D\text{lift}}$ is also zero. However, C_D itself is not zero, since

$$p_1 = p_3 > p_2 = p_4$$

The drag consists of four terms,

$$D = \sum_{i=1}^{4} D_i$$

where

$$D_1 = D_3 = p_1 \ell_s \sin\phi, \qquad D_2 = D_4 = p_2 \ell_s \sin(-\phi) = -p_2 \ell_s \sin\phi$$

The D_2 and D_4 forces are in the upstream direction; hence, they are negative. Overall, we have

$$D = 2\ell_s(p_1 - p_2)\sin\phi \qquad (6.13a)$$

By geometry, ℓ_s is given by

$$\ell_s = \frac{\ell}{2\cos\phi} \qquad (6.13b)$$

Equations (6.10), (6.11), and (6.13) combine to yield

$$C_D = \frac{2}{\gamma M_\infty^2}\left(\frac{p_1}{p_\infty} - \frac{p_2}{p_\infty}\right)\tan\phi$$

or

$$C_D = \frac{2}{\gamma M_\infty^2}\frac{p_1}{p_\infty}\left(1 - \frac{p_2}{p_1}\right)\tan\phi \qquad (6.14)$$

The ratio p_1/p_∞ is just the static pressure ratio across an oblique shock, while p_2/p_1 is the pressure ratio across the Prandtl-Meyer expansion. As shown in Fig. 6.11, the expansion is for a wall turn angle of 2ϕ. It is important to note that this C_D is special to the assumed airfoil shape and to $\alpha = 0$ deg. The drag here is due solely to the finite thickness of the airfoil and is denoted as $C_{D\,\text{thick}}$. It is called the wave drag due to thickness.

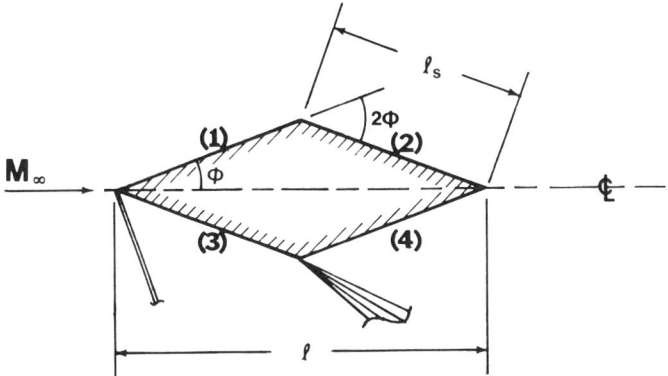

Fig. 6.11 Diamond-shaped airfoil at zero angle of attack; the flow is symmetric about the centerline.

In summary, we see that the drag coefficient in a supersonic flow can be decomposed into three components:

$$C_D = C_{D\text{ visc}} + C_{D\text{ lift}} + C_{D\text{ thick}}$$

(There also is a fourth component associated with camber.) Shock-expansion theory provides accurate values for the induced lift and thickness drags. (A small error occurs if the Mach waves that reflect from the shock wave intersect the airfoil. For the simple shapes under consideration, it is permissible to neglect the effect of wave reflection.) The theory breaks down when a detached shock wave, a subsonic Mach number downstream of a shock wave, or a detached boundary layer occurs. Note that, at a given angle of attack, $C_{D\text{ lift}} + C_{D\text{ thick}}$ are determined as a single entity.

In Appendix G, formulas are provided for lift and drag coefficients for an arbitrary quadrilateral airfoil at an arbitrary angle of attack. These formulas also hold for a triangular or a flat plate airfoil.

Problems

Use the table generated in Problem 6.32 to solve many of the following problems.

6.1 Air approaches a sharp 10-deg convex corner with a Mach number of 3, a temperature of 300 K, and a pressure of 1 atm. Determine the Mach number, static and stagnation temperatures, and static and stagnation pressures downstream of the corner.

6.2 As shown in the sketch below, air approaches a sharp 15-deg convex corner with $M_1 = 2.5$ and $T_1 = 200$ K. Determine M_2, ϕ_1, and ϕ_2.

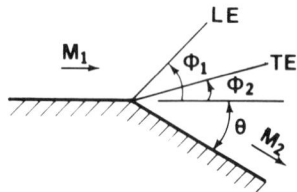

6.3 Use the data in Problem 6.2 to determine T_{02}, the speed V_2, and the Mach number on a Mach line through the origin, which is at an angle $\phi = 15$ deg relative to $\mathbf{V}_1$.

6.4 As shown in the sketch below, a monatomic gas flows around the convex corner. Determine M_1 and M_2.

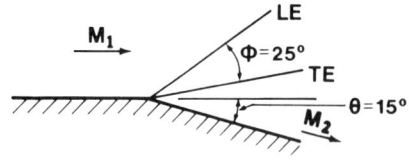

6.5 A schlieren photo for an airflow shows the angles in the sketch below. Determine M_1, M_2, and ϕ.

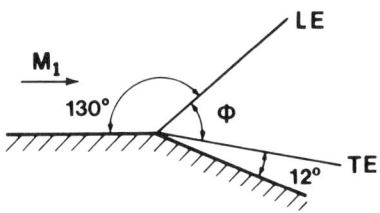

6.6 Nitrogen approaches a sharp convex corner with a Mach number M_1 of 2 and a temperature of 250 K. Determine the Mach number M_2 and speed V_2 on the downstream edge of the fan if the turn angle is (a) 30 deg, (b) 60 deg. Also, determine the fan expansion angle ϕ for both cases.

6.7 Repeat Problem 6.6 with a wall turn angle of 112 deg.

6.8 A $M_1 = 4$ missile has a two-dimensional air intake with a 15-deg smooth ramp. Determine M_2 and p_2/p_1. Determine M_2 and p_2/p_1 if the turn is sharp and causes an oblique shock.

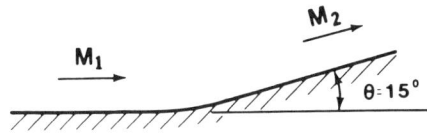

6.9 Helium with $M_1 = 3.5$ encounters a gradual compressive turn as shown in the Problem 6.8 sketch. The pressure ratio p_2/p_1 is 2.1. Determine M_2 and θ.

6.10 A wall has a gradual bend as shown in the sketch below with an angle $\phi = 20$ deg. The air flowing over the bend has a pressure ratio of $(p_2/p_1) = 0.2$. Determine M_1 and M_2.

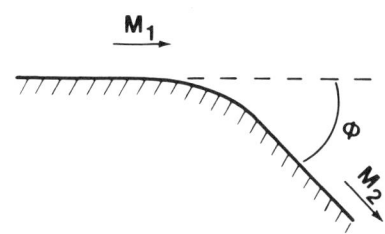

6.11 Helium flows around a circular-arc bend, as shown in the sketch below. The following parameters are known: $M_1 = 3$, $r = 0.1$ m, and $\alpha = 15$ deg. Determine M_2 and the angle ϕ of the expansion fan.

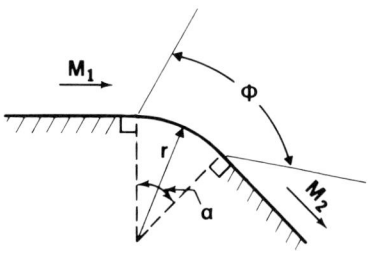

6.12 Helium encounters a jog in the wall as shown in the sketch below. We are given: $M_1 = 3$ and $\beta = 28$ deg. Determine M_3.

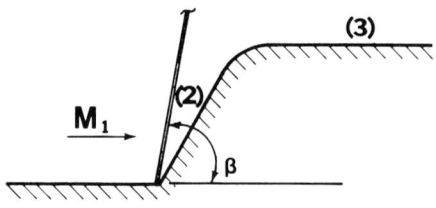

6.13 Air encounters a wall with two sharp turns as shown in the sketch below. Line A is a Mach line. Determine M_3 and p_3/p_1.

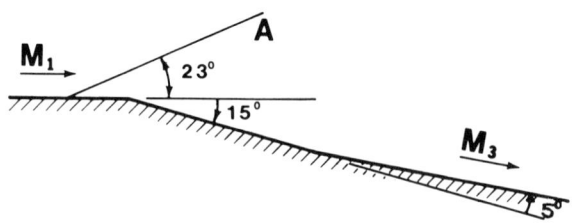

6.14 Air flows over a contoured wall as shown in the sketch below, where walls A and B are parallel. Determine p_2/p_1 and p_3/p_1.

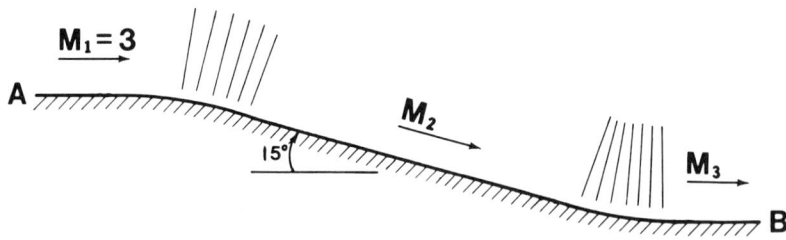

6.15 Air flows over a ramp as shown in the sketch below. We are given $M_1 = 3$ and $\delta = 10$ deg. Determine M_2, M_3, θ_1, θ_2, and ϕ_{23}.

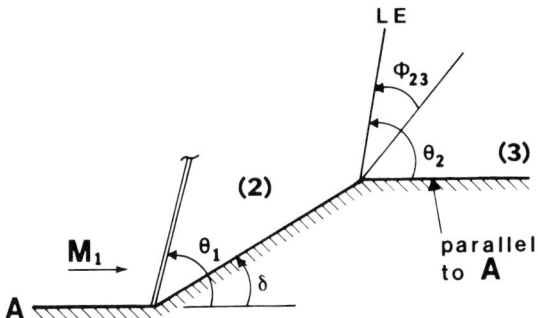

6.16 Air flows over the double ramp, as shown in the sketch below, where $M_1 = 1.5$ and $\delta = 20$ deg. Determine M_2, M_3, ϕ_{12}, and ϕ_{23}.

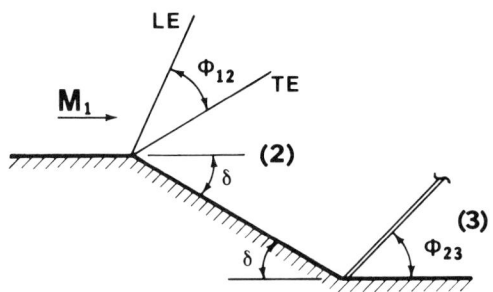

6.17 As shown in the sketch below, air flows over a wall with a compound bend. Known information is: $M_1 = 3$, $\theta_1 = 10$ deg, and $\theta_3 = 15$ deg. Determine α_1, α_2, α_3, α_4, M_2, and M_4.

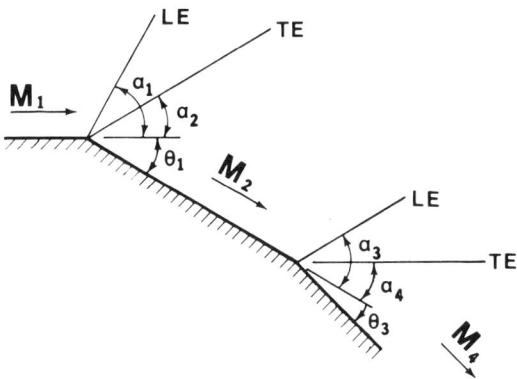

6.18 A gas with $\gamma = 1$ and $M_1 = 1$ expands around a 90-deg corner. Determine M_2, T_2/T_1, and p_2/p_1.

6.19 The fan angle ϕ for a Prandtl-Meyer expansion in air is 78.53 deg, and the pressure ratio p_2/p_1 across the fan is 9.714×10^{-4}. Determine M_1 and M_2.

6.20 Determine the equation of a streamline in a centered Prandtl-Meyer expansion of a perfect gas. (Hints: Use the corner as the origin and the length of a ray as a parameter. Note that the mass flow rate between the wall and a streamline is fixed.)

6.21 A planar flow has a curved wall downstream of the origin, as shown in the sketch below. Determine the equations for the streamline that starts at height y_1. Assume a perfect gas, and simplify your results. (Hint: Consider x_w to be the independent variable.)

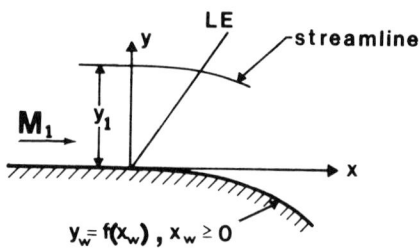

6.22 A flat plate in air has $M_\infty = 3$ and for the lower side $M_2 = 1$. Determine α and C_L.

6.23 Consider a flat plate in a supersonic airflow at $M_\infty = 3$. Determine the angle of attack α at which shock-expansion theory no longer holds. Determine C_L and C_D just before the theory breaks down.

6.24 Use shock-expansion theory to determine C_L for a symmetrical diamond airfoil in air when $M_\infty = 3$, $\phi = 10$ deg, and $\alpha = 0, 15$ deg (two cases).

6.25 Use the data in Problem 6.24 to evaluate C_D when $\alpha = 0, 15$ deg.

6.26 The quadrilateral airfoil shown in the sketch below has $p_3 = 2.064 p_1$. Determine the angle of attack α.

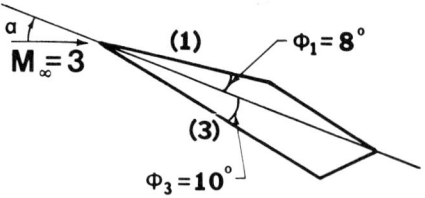

6.27 With $\alpha = 0$ deg, $M_\infty = 4$, and $\phi = 15$ deg, determine C_L and C_D for the airfoil in air shown below.

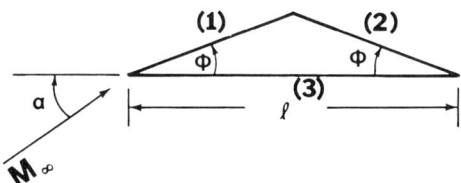

6.28 For the airfoil in Problem 6.27, determine C_L and C_D when $\alpha = 10$ deg.

6.29 Determine C_L and C_D for the airfoil in the accompanying sketch when $\alpha = \phi = 7$ deg.

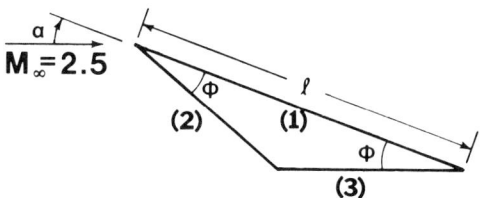

6.30 A double flat plate of equal-length segments is tested in air at $M_\infty = 2$. The angle α is 10 deg. Use the numbering shown in the sketch below, and determine C_L and C_D with ℓ as the reference length.

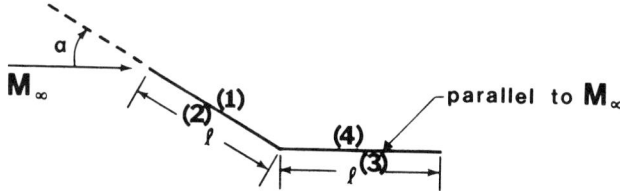

6.31 Develop equations for C_L and C_D for a flat plate when the angle of attack α is small. Your answer for C_L and C_D should only involve α and M_∞.

Computational Problems:

6.32 For $\gamma = 1.4$ and $5/3$ and for $M = 1(0.01)\ 5(0.1)\ 10$, generate a table showing μ and ν. This result should be added to the supersonic part of Problem 4.6.

6.33 Develop a computer code for C_L and C_D for the quadrilateral airfoil of Appendix G. Use this code to help solve the triangular and quadrilateral airfoil problems.

7. NOZZLE AND DIFFUSER FLOW

Section 4.3 showed that a duct with a throat is required if a supersonic flow is to be produced. This type of duct is called a converging/diverging nozzle or, for short, nozzle. Typical applications involve supersonic wind tunnels, rocket nozzles, and certain gas lasers. Generally, a uniform flow is desired at the exit plane of the nozzle. A uniform flow is one in which no variation exists in any flow property transverse to the nozzle's axis and the velocity is parallel to the axis.

In this chapter, we will examine conditions under which a nozzle produces isentropic subsonic or supersonic flow. A key element in this investigation is the ambient, or back pressure, p_∞. Depending on the value of p_∞, the nozzle may have nonisentropic flow; i.e., shock waves may be present. Within the quasi-one-dimensional approximation, we systematically examine all flow possibilities in Sec. 7.1. For this examination, it will be necessary to discuss not only the flow inside the nozzle but also the jet that issues from the nozzle. Principal results are summarized in Appendix H, which is to be used in conjunction with Appendices C through F.

In actual fact, nozzle flows are two-dimensional or three-dimensional. However, in a majority of applications, the one-dimensional approximation yields excellent agreement with experiment, even in situations where agreement is not expected. For example, with a nozzle of minimum length, as discussed in Chap. 17, one might expect poor agreement because the flow in the neighborhood of the throat experiences a large transverse pressure gradient. Nevertheless, the one-dimensional equations yield accurate results for this case.

Why are the one-dimensional equations accurate? In most applications, the nozzle is designed to provide a uniform flow at its exit. Often, only exit conditions are of interest and, when the exit flow is uniform, the one-dimensional approximation is exact. At an intermediate nozzle location, the flow is not uniform and the one-dimensional solution then represents average conditions.

In a converging/diverging duct, the flow can go from subsonic to supersonic or the reverse. A duct in which the flow goes from supersonic to subsonic is called a diffuser, which is the subject of Sec. 7.2. Typical applications include wind tunnels, gas lasers, and supersonic engine inlets.

7.1 NOZZLE FLOW

We assume steady, isentropic, quasi-one-dimensional flow of a perfect gas. In this situation, stagnation conditions, such as $T_0, p_0, s_0, \ldots$, are constants. The equations for T, p, ρ, $\dot{m}$, and A are given by Eqs. (4.8)–(4.13) or in Appendix C.

Figure 7.1 shows a sketch of a typical converging/diverging nozzle. The nature of the flow in the nozzle is governed by three dimensionless parameters: γ, p_∞/p_0, and A_3/A_2. The pressure ratio is the ambient pressure divided by the stagnation pressure p_0 in the reservoir. The parameter A_3/A_2 is referred to as the nozzle's area ratio. The inlet area ratio A_1/A_2 is usually not a fundamental parameter for reasons to be explained shortly.

If $(p_\infty/p_0) = 1$, no flow occurs and the pressure equals p_0 inside the nozzle. For a flow to occur, the speed V must be nonzero, which requires that $\mathrm{d}p/\mathrm{d}x$ be nonzero. Thus, p_∞ must be less than p_0. Let us suppose that p_∞ is slightly less than p_0. In this case, the flow will be subsonic everywhere, as illustrated in the next example.

Example 1

An air nozzle has the following parameters:

$$\frac{A_1}{A_2} = 6, \qquad \frac{A_3}{A_2} = 2.5, \qquad \frac{p_\infty}{p_0} = 0.97, \qquad T_1 = 300 \text{ K}, \qquad p_1 = 2 \text{ atm}$$

We are to determine M_1, M_2, M_3, p_1/p_0, p_2/p_0, the speed V_3 in the exit plane, and the mass flow rate per unit area $\dot{m}/A_1$. The numbering here refers to Fig. 7.1.

The flow is assumed to be subsonic, which we verify later, and uniform across the exit plane at station 3. Consequently, the jet leaving the nozzle is subsonic, and the exit plane pressure must equal the ambient pressure,

$$p_3 = p_\infty \tag{7.1}$$

When $M_3 < 1$, the flow in the nozzle adjusts until Eq. (7.1) is satisfied. If, for instance, the pressure in the exit plane at the centerline is different from

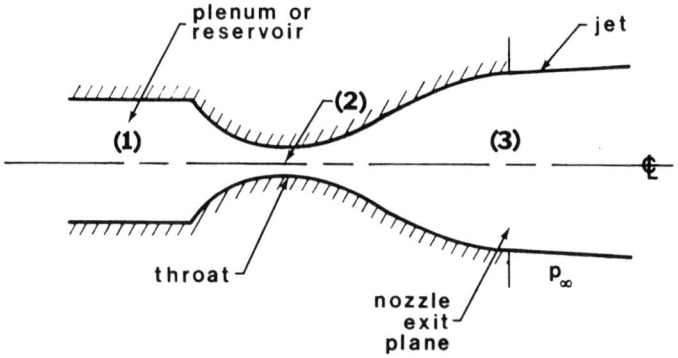

Fig. 7.1 Sketch and nomenclature for a converging/diverging nozzle.

p_∞, then the flow in this plane is not uniform. While the uniform flow assumption is essential for Eq. (7.1), it is generally an understood assumption. To be consistent with the quasi-one-dimensional assumption, we henceforth assume uniform exit flow. In view of Eq. (7.1), we have, from the Chap. 4 tables with $\gamma = 1.4$,

$$\frac{p_3}{p_0} = 0.97, \quad M_3 = 0.21, \quad \frac{A_3}{A^*} = 2.83, \quad \frac{T_3}{T_0} = 0.991$$

and, consequently,

$$\frac{A_2}{A^*} = \frac{A_2}{A_3}\frac{A_3}{A^*} = \frac{2.83}{2.5} = 1.132$$

This area ratio then provides the following conditions at the throat:

$$M_2 = 0.65, \quad \frac{p_2}{p_0} = 0.753$$

To determine plenum conditions, we first evaluate

$$\frac{A_1}{A^*} = \frac{A_1}{A_2}\frac{A_2}{A^*} = 6 \times 1.132 = 6.792$$

so that

$$M_1 = 0.085, \quad \frac{p_1}{p_0} = 0.995$$

Of the three Mach numbers, the one at the throat, M_2, is the largest, a general result for subsonic nozzle flow. This can be verified by examining the A/A^* curve in Fig. 4.1. Since $M_2 < 1$, we have established as correct the earlier subsonic flow assumption.

In the Chap. 4 analysis, the flow was assumed to accelerate continuously through the nozzle. As Example 1 shows, this is not the case for subsonic nozzle flow. In this situation, the flow accelerates in the converging section but decelerates in the diverging section. Furthermore, the Mach number M_2 at the throat is less than unity, and the geometrical throat area A_2 exceeds A^*. Only when sonic throat conditions prevail is A_2 equal to A^*.

Another aspect of the solution involves the fact that M_1 is quite small. When A_1/A_2 is sufficiently large, the smallness of M_1 is a general feature of nozzle flow, independent of M_2. Since M_1 is small, we can take

$$p_0 = p_1, \quad T_0 = T_1, \quad \rho_0 = \rho_1, \ldots \tag{7.2}$$

with only a negligible error. We utilize Eqs. (7.2) in all subsequent analyses. The area ratio A_1/A_2 is used only to establish M_1. When A_1/A_2 is sufficiently large, then $M_1 \ll 1$ and this area ratio is no longer essential.

The exit speed can now be found

$$V_3 = a_3 M_3 = [\gamma R(T_3/T_0)T_0]^{\frac{1}{2}} M_3$$

$$= (1.4 \times 287 \times 0.991 \times 300)^{\frac{1}{2}} \times 0.21 = 72.6 \text{ m/s}$$

The mass flow rate is given by Eq. (4.11) and can be most accurately evaluated at the throat. (It can also be evaluated at station 1; however, using tables, M_1 is not as accurate as M_2.) Hence, Example 1 concludes with

$$\frac{\dot{m}}{A_1} = \frac{M_2}{\{1 + [(\gamma-1)/2]M_2^2\}^{(\gamma+1)/2(\gamma-1)}} \left(\frac{\gamma}{RT_0}\right)^{\frac{1}{2}} p_0 \frac{A_2}{A_1}$$

$$= \frac{0.65}{(1 + 0.2 \times 0.65^2)^3} \left(\frac{1.4}{287 \times 300}\right)^{\frac{1}{2}} 2 \times 1.013 \times 10^5 \times \frac{1}{6} = 69.4 \text{ kg/m}^2\text{-s}$$

A series of flows is now envisioned in which all parameters except p_∞ are kept fixed. (We could equally as well vary p_0 with p_∞ fixed.) The left side of Fig. 7.2 shows a sketch of the nozzle as well as pressure and Mach number traces for different cases, which are labeled (a) through (g). Case (a), in which the flow is subsonic, is the uppermost sketch on the right and was the subject of Example 1. A Venturi meter, which measures the flow rate in a pipe, operates in this regime. The measurement requires not only a knowledge of the gas but also T_1, p_1, A_1/A_2, and p_2.

As p_∞ is decreased to p_b, a limiting case (b) is encountered. We still have $M_3 < 1$ and $p_3 = p_\infty$, but now $M_2 = 1$ at the throat. That is, the flow is subsonic everywhere except at the throat. For this flow, we have

$$p_3 = p_b, \qquad A_2 = A^*, \qquad M_2 = 1$$

Equations (4.9) and (4.13) determine this limiting condition, as follows:

$$\frac{p_b}{p_0} = \left(1 + \frac{\gamma-1}{2} M_3^2\right)^{-\gamma/(\gamma-1)}$$

$$\frac{A_3}{A_2} = \frac{A_3}{A^*} = \left(\frac{2}{\gamma+1}\right)^{(\gamma+1)/2(\gamma-1)} \frac{1}{M_3} \left(1 + \frac{\gamma-1}{2} M_3^2\right)^{(\gamma+1)/2(\gamma-1)} \quad (7.3)$$

We eliminate M_3 from these equations to obtain

$$\left(\frac{p_b}{p_0}\right)^{2/\gamma} - \left(\frac{p_b}{p_0}\right)^{(\gamma+1)/\gamma} = \frac{\gamma-1}{2}\left(\frac{2}{\gamma+1}\right)^{(\gamma+1)/(\gamma-1)} \left(\frac{A_2}{A_3}\right)^2 \quad (7.4)$$

This is an implicit equation for p_b/p_0 that depends only on γ and A_3/A_2.

NOZZLE AND DIFFUSER FLOW

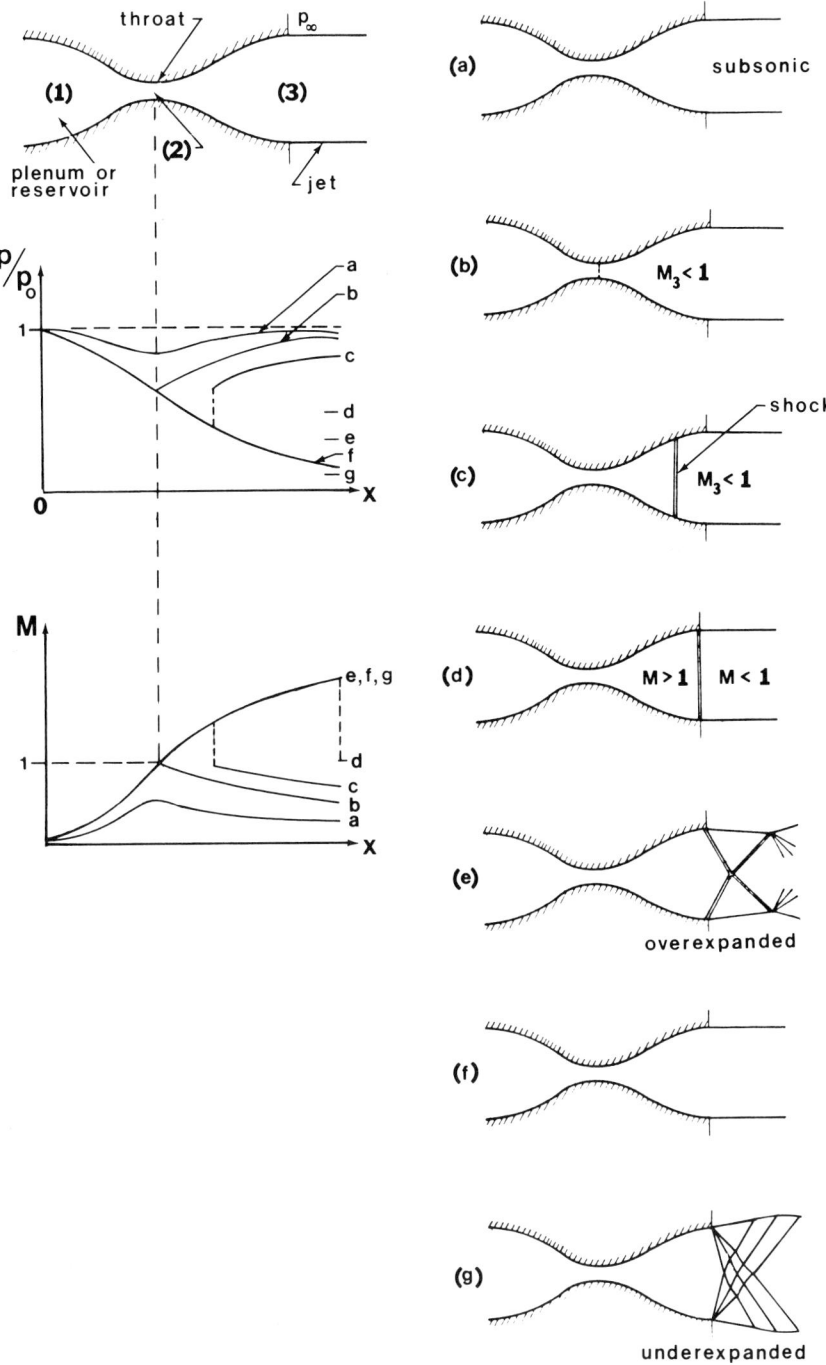

Fig. 7.2 Nozzle flow regimes.

Isentropic tables can also be used to establish p_b/p_0 since $A_2 = A^*$. For conditions in Example 1, where $A_3/A_2 = 2.5$, we have

$$M_3 = 0.24, \qquad p_b/p_0 = 0.9607$$

Observe that p_∞ need not be much below p_0 for a sonic condition to occur at the throat. If the nozzle only converged, then A_3 would equal A_2, and we would find that $p_b/p_0 = 0.5283$. Thus, p_b/p_0 varies rapidly with A_3/A_2 when this area ratio is just above unity.

The mass flow rate per unit area is most conveniently given by Eq. (4.12) as

$$\frac{\dot{m}}{A_1} = \left(\frac{2}{\gamma+1}\right)^{(\gamma+1)/2(\gamma-1)} \left(\frac{\gamma}{RT_0}\right)^{\frac{1}{2}} p_0 \frac{A_2}{A_1}$$

$$= \frac{1}{1.2^3} \left(\frac{1.4}{287 \times 300}\right)^{\frac{1}{2}} 2 \times 1.013 \times 10^5 \times \frac{1}{6} = 78.8 \text{ kg/m}^2\text{-s}$$

This $\dot{m}/A_1$ value and M_3 are larger than their counterparts in Example 1.

When the Mach number at the throat is sonic, the flow is said to be choked. Any further decrease in p_∞ below p_b has no effect on M_2, which is unity, on $\dot{m}$, or on the flow in the converging part of the nozzle. Thus, $\dot{m}/A_1$ remains at 78.8 kg/m²-s, and the pressure at the throat is fixed at

$$\frac{p_2}{p_0} = 0.5283$$

Henceforth, all adjustments in the flow occur downstream of the throat.

For the condition $p_\infty = p_c$, a supersonic region occurs downstream of the throat. However, in order that $p_3 = p_\infty$, a normal shock occurs inside the nozzle as indicated in Fig. 7.2(c). On the left side of the figure, the normal shock is indicated by vertical dashed lines.

In an actual nozzle, an internal normal shock is an approximation for a more complex flow. Normally, an asymmetric system of oblique shocks occurs. These may oscillate back and forth in the axial direction, i.e., buzz, or spin about the axis if the nozzle is axisymmetric. In addition, the viscous boundary layer on the wall in the divergent part of the nozzle may separate. Such separation leads to a very nonuniform flow in the nozzle's exit plane. (See Ref. 1, p. 166.)

A normal shock, however, is a reasonable way of modeling average conditions, such as the loss in stagnation pressure, providing the exit plane flow is approximately uniform. The strength of the shock depends on its location. When it is near the throat, the Mach number ahead of it is only slightly greater than unity and the shock is weak. This situation occurs when p_∞ is slightly less than p_b. As p_∞ decreases, the shock moves downstream and becomes steadily stronger.

Thus, the location and strength of the shock depend on p_∞/p_0. By location, we mean the area ratio A_3/A_2 (or A_4/A_2) as shown in Fig. 7.3.

NOZZLE AND DIFFUSER FLOW

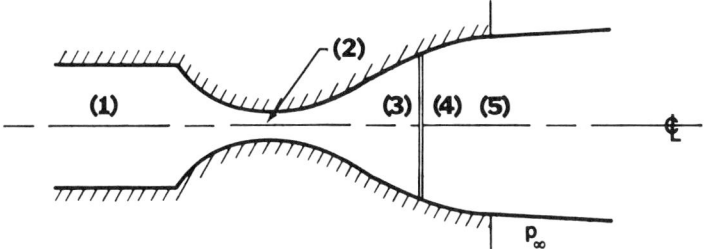

Fig. 7.3 Nozzle flow with an internal normal shock wave.

Determination of A_3/A_2 is greatly simplified by an isentropic area-pressure relation of the form

$$\frac{pA}{p_0 A^*} = f(\gamma, M)$$

Since T_0 and $\dot{m}$ are constant in the nozzle and across a shock wave, so must $p_0 A^*$ be constant to be in accord with Eq. (4.12). Although p_0 and A^* individually change across a shock, their product does not. With the numbering in Fig. 7.3, we have

$$p_{03} A_3^* = p_{04} A_4^* \tag{7.5a}$$

However, the flows from 1 to 3 and 4 to 5 are isentropic. As a consequence,

$$A_3^* = A_1^*, \quad A_4^* = A_5^*, \quad p_{03} = p_{01}, \quad p_{04} = p_{05}$$

so that

$$p_{05} A_5^* = p_{01} A_1^* \tag{7.5b}$$

Divide $p_5 A_5$ by Eq. (7.5b) to obtain

$$\frac{p_5}{p_{05}} \frac{A_5}{A_5^*} = \frac{p_5}{p_{01}} \frac{A_5}{A_1^*} = \frac{p_5}{p_{01}} \frac{A_5}{A_2} \tag{7.6a}$$

The two ratios on the left are isentropic point functions that depend only on M_5,

$$\frac{p_5}{p_{05}} \frac{A_5}{A_5^*} = \left(1 + \frac{\gamma-1}{2} M_5^2\right)^{-\gamma/(\gamma-1)} \frac{1}{M_5} \left(\frac{2}{\gamma+1}\right)^{(\gamma+1)/2(\gamma-1)}$$

$$\times \left(1 + \frac{\gamma-1}{2} M_5^2\right)^{(\gamma+1)/2(\gamma-1)}$$

$$= \left(\frac{2}{\gamma+1}\right)^{(\gamma+1)/2(\gamma-1)} \frac{1}{M_5 \{1 + [(\gamma-1)/2] M_5^2\}^{\frac{1}{2}}} \tag{7.6b}$$

Once the right side of Eq. (7.6b) is tabulated, the Mach number M_5 is found by means of Eqs. (7.6). In Eq. (7.6a), the ratios p_5/p_{01} and A_5/A_2 are generally known. By combining the two equations and solving Eq. (7.6b), which is quadratic in M_5^2, we obtain

$$M_5^2 = \frac{1}{\gamma - 1}\left\{\left[1 + 2(\gamma - 1)\left(\frac{2}{\gamma + 1}\right)^{(\gamma+1)/(\gamma-1)}\left(\frac{p_{01}}{p_5}\frac{A_2}{A_5}\right)^2\right]^{\frac{1}{2}} - 1\right\} \quad (7.7)$$

After M_5 is established, M_3, M_4, and A_3/A_2 are easily found, as shown in the next example.

Example 2

We reconsider the nozzle flow in Example 1 after decreasing the pressure ratio to $(p_\infty/p_{01}) = 0.6$. Determine M_3, M_4, M_5, A_3/A_2, and p_∞ when $p_1 = 2$ atm.

The nozzle area ratio from Example 1 is $A_5/A_2 = 2.5$. Assume $p_5 = p_\infty$, so that $(p_5/p_{01}) = 0.6$ and $(p_5/p_{01})(A_5/A_2) = 1.5$. From tables or Eq. (7.7), we obtain

$$M_5 = 0.38, \qquad \frac{p_5}{p_{05}} = 0.905$$

Next, the stagnation pressure ratio across the shock is found:

$$\frac{p_{04}}{p_{03}} = \frac{p_{05}}{p_{01}} = \frac{p_{05}}{p_5}\frac{p_\infty}{p_{01}}\frac{p_{01}}{p_1} = \frac{0.6 \times 1}{0.905} = 0.663$$

By means of a normal shock table, we have

$$M_3 = 2.12, \qquad M_4 = 0.558$$

With an isentropic table, and using M_3 as the entry, the area ratio at the shock is

$$\frac{A_3}{A_2} = 1.87$$

Finally, p_∞ is given by

$$p_\infty = \frac{p_\infty}{p_{01}}p_1 = 0.6 \times 2 = 1.20 \text{ atm}$$

Note that the solution is self-consistent in that the following inequalities hold:

$$M_5 < M_4 < 1 < M_3, \qquad \frac{p_{04}}{p_{03}} < 1, \qquad 1 < \frac{A_3}{A_2} < \frac{A_5}{A_2}$$

A second limiting condition occurs when the shock wave is at the exit plane, shown in Fig. 7.2(d). Again, we have $p_\infty = p_d$. This case is analyzed using the numbering shown in Fig. 7.4. Equation (7.3) applies and determines M_3. Next, we use M_3 and Eq. (4.9) to determine p_3/p_{01}. Equation (5.14a) yields

$$\frac{p_d}{p_3} = \frac{2}{\gamma+1}\left(\gamma M_3^2 - \frac{\gamma-1}{2}\right)$$

Thus, the limiting pressure ratio is given by

$$\frac{p_d}{p_0} = \frac{p_d}{p_3}\frac{p_3}{p_{01}}$$

The same result can be obtained by using isentropic and normal shock tables.

For $p_\infty < p_d$, the flow inside the nozzle is isentropic and unaltered. Furthermore, this solution is accurate, as discussed at the beginning of the chapter. An additional reason for this accuracy is the decreasing nozzle pressure. This represents a favorable pressure gradient and, consequently, the viscous boundary layer is thin and attached to the wall.

With $p_\infty < p_d$, the supersonic flow in the nozzle's exit plane is uniform, and all adjustments now occur in the external jet. With p_∞ moderately less than p_d, the normal shock becomes oblique shocks, shown in Fig. 7.2(e). Because the nozzle exit pressure is below p_∞, the nozzle flow is referred to as overexpanded. This case is considered in the next example.

Example 3

We reconsider Examples 1 and 2 but with $(p_\infty/p_0) = 0.1$. We are to determine p_d/p_0, M_4, and the angles β and θ as shown in Fig. 7.5. Both angles are measured relative to a line that is parallel to the centerline, and the nozzle exit flow is assumed to be uniform and two-dimensional.

From isentropic tables with $(A_3/A^*) = 2.5$ as the entry, we obtain

$$M_3 = 2.44, \quad \frac{p_3}{p_{01}} = 0.0643 \tag{7.8}$$

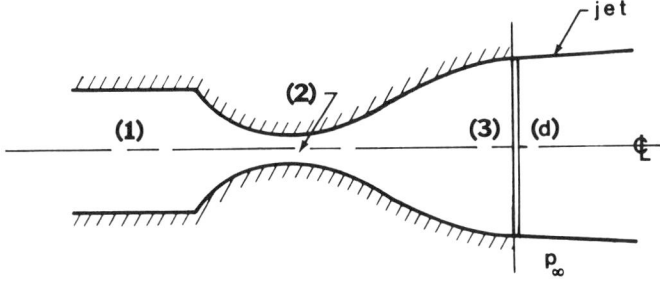

Fig. 7.4 Nozzle flow with a normal shock in the exit plane.

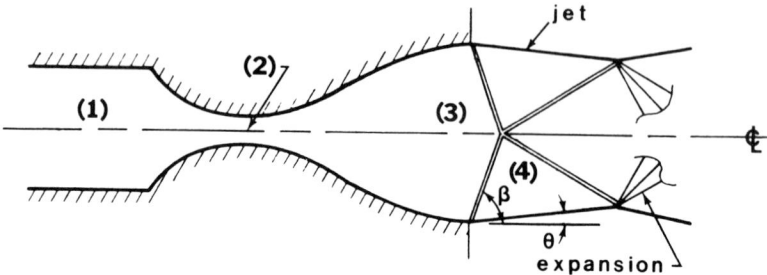

Fig. 7.5 Overexpanded nozzle flow with a simple oblique shock system.

Since $p_\infty > p_3$, the nozzle flow is overexpanded. With M_3 and a normal shock table, we have $(p_d/p_3) = 6.78$, so that

$$\frac{p_d}{p_0} = \frac{p_d}{p_3}\frac{p_3}{p_{01}} = 6.78 \times 0.0643 = 0.436$$

With $p_d > p_\infty$, the actual shock system for the overexpanded flow must be external to the nozzle.

Region 4 of Fig. 7.5 is a uniform flow region in which the pressure is given by

$$p_4 = p_\infty \tag{7.9}$$

Hence,

$$\frac{p_4}{p_3} = \frac{p_\infty}{p_0}\frac{p_0}{p_3} = \frac{0.1}{0.0643} = 1.555$$

and with the oblique shock relation

$$\frac{p_4}{p_3} = \frac{2}{\gamma+1}\left(\gamma M_3^2 \sin^2\beta - \frac{\gamma-1}{2}\right)$$

we obtain

$$\beta = \sin^{-1}\left[\frac{1}{M_3}\left(\frac{\gamma+1}{2\gamma}\frac{p_4}{p_3} + \frac{\gamma-1}{2\gamma}\right)^{\frac{1}{2}}\right] = 29.9 \text{ deg}$$

Thus, the shock wave, which starts at the lip of the nozzle, is a weak-solution shock. If β is large enough, a strong-solution shock wave would occur. With M_3 and β, we use Appendix E to determine

$$\theta = 7 \text{ deg}$$

To find M_4, we compute the following:

$$M_{3n} = M_3 \sin \beta = 2.44 \sin 29.9° = 1.215$$

$$M_{4n} = 0.833$$

$$M_4 = \frac{M_{4n}}{\sin(\beta - \theta)} = \frac{0.833}{\sin 22.9°} = 2.14$$

where M_{4n} comes from a normal shock table.

Figure 7.2(f) illustrates another limiting flow in which the pressure in the nozzle's exit plane equals p_∞. Conditions in the exit plane are given by Eqs. (7.8) and $p_\infty = p_3 = p_f$. The jet is a uniform supersonic flow that remains unaltered except for mixing with the ambient gas along its periphery. For the applications mentioned at the start of the chapter, this case represents the desired flow condition.

In Fig. 7.2(g), the exit pressure is greater than p_∞. Hence, this nozzle flow is termed underexpanded. The flow adjusts to the ambient pressure by means of expansions that start at the lips of the nozzle (Fig. 7.6). When the flow is two-dimensional, these are centered Prandtl-Meyer expansions.

Our discussion has indicated the importance of p_∞ in determining the nature of the flow in the nozzle and in the jet. For a converging/diverging nozzle, the exit pressure equals the ambient pressure whenever the exit flow is subsonic. When the exit flow is supersonic, however, the exit pressure may exceed, equal, or be less than the ambient pressure. For a converging-only nozzle, the exit flow may be subsonic or sonic, and the exit pressure may only equal or exceed the ambient pressure. When the exit pressure exceeds p_∞, the jet adjusts by expansion waves that start from a sonic condition. This latter result will prove useful in Chap. 8 when we examine Rayleigh and Fanno flows. The next example further illustrates this discussion.

Example 4

We reconsider Examples 1–3, but with $(p_\infty/p_0) = 0.04$. We are to determine θ, as shown in Fig. 7.6, M_4, and V_3.

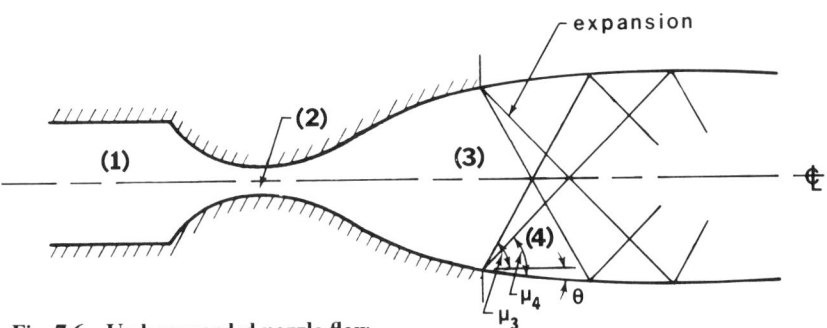

Fig. 7.6 Underexpanded nozzle flow.

Conditions in the exit plane are given by Eqs. (7.8). Thus, V_3 is determined as follows:

$$\frac{T_3}{T_0} = 0.456$$

$$V_3 = [\gamma R(T_3/T_0)T_0]^{\frac{1}{2}} M_3$$

$$= (1.4 \times 287 \times 0.456 \times 300)^{\frac{1}{2}} \times 2.44 = 572 \text{ m/s}$$

The exit speed V_3 is a factor of 8 larger than its subsonic counterpart in Example 1.

Region 4 is a uniform flow region in which Eq. (7.9) holds. Since the flow is isentropic across the expansion, the pressure ratio $(p_\infty/p_0) = (p_4/p_0)$ yields $M_4 = 2.75$. For θ, we have

$$\theta = \nu(M_4) - \nu(M_3) = 44.69° - 37.71° = 6.98 \text{ deg}$$

For convenience, Fig. 7.6 also shows the Mach angles μ_3 and μ_4.

Our treatment of Example 4 tacitly assumes that the nozzle and exit flow are two-dimensional. If the nozzle were axisymmetric, the above solution would hold in the immediate vicinity of the lip. Away from the lip, the expansions are no longer Prandtl-Meyer expansions, which are defined only for a two-dimensional flow.

Besides the ones sketched in Figs. 7.2 and 7.5, several different shock patterns are possible when the flow is overexpanded. The pattern shown in these figures is accurate when p_∞ is above but close to p_f. For a higher value of p_∞, the pattern shown in Fig. 7.7 occurs. A nearly normal shock, called a Mach disk, occurs in the central part of the jet. Downstream of the Mach disk, the flow is subsonic, and this flow is bordered by slipstreams. The oblique incident and reflected shocks join the Mach disk at the triple

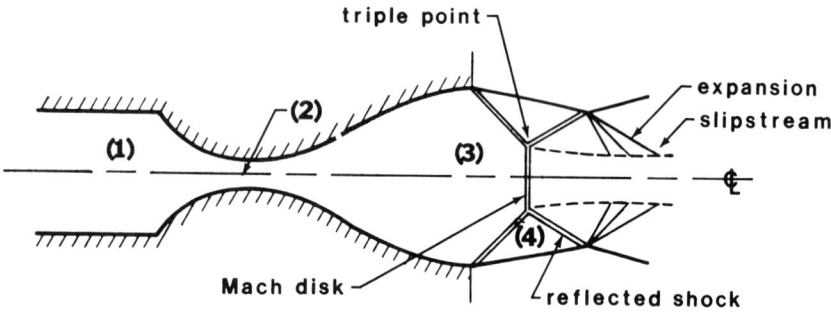

Fig. 7.7 Overexpanded nozzle flow with a Mach disk.

point. As shown in Figs. 7.2, 7.5, and 7.7, the reflected shock reflects from the edge of the jet as an expansion. The reason for this is that the pressure downstream of the reflected shock exceeds p_∞, since $p_4 = p_\infty$. Consequently, an expansion is needed to adjust this pressure to the ambient. In both Figs. 7.5 and 7.7, region 4 is uniform and supersonic. The supersonic qualification is essential if a reflected shock is to occur on the downstream side of the region.

At a still higher value for p_∞, a pattern approaching that in Fig. 7.4 must occur.[2] Figure 7.8 shows a slightly curved strong-solution shock wave. The jet downstream of the shock is subsonic and nonuniform. The demarcation condition between the flows in Figs. 7.7 and 7.8 is probably $M_4 = 1$, since only when $M_4 > 1$ is the Mach disk pattern possible. Experimental evidence is lacking, however, so that the condition is only a probable one. For the foregoing examples, one can show that $(p_4/p_0) = 0.344$ when $M_4 = 1$. Thus, the pattern in Fig. 7.8 is limited to

$$0.344 < \frac{p_\infty}{p_0} \leq 0.436$$

where the upper bound is p_d/p_0.

7.2 DIFFUSER FLOW

The function of a diffuser is to convert translational kinetic energy into thermal energy as efficiently as possible. For instance, the inlet to a jet engine is a diffuser regardless of the flight speed. Good engine performance requires a high pressure inside the combustor and adequate time for the air and fuel to mix and fully react. For supersonic flight, the inlet diffuser significantly slows the flow and raises the static pressure. In this case, the diffuser is particularly important, since the loss in stagnation pressure across a normal shock can be quite large. Such a loss is equivalent to a reduced engine efficiency. Thus, supersonic inlets are designed to decelerate the flow with a minimum of shock losses.

When the inlet flow is supersonic, the simplest concept for doing this is to have a duct with a flow opposite to that in Fig. 7.2(f). With p_∞/p_0 properly adjusted, the flow in Fig. 7.2(f) is isentropic and free of shock losses. Frictional losses in the nozzle are confined to a viscous boundary layer that

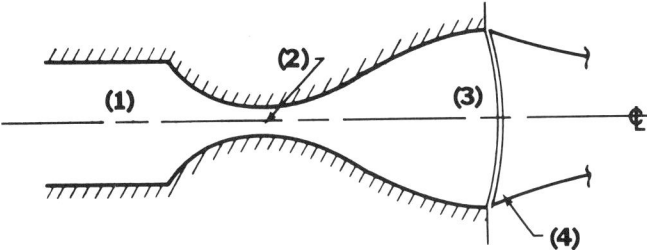

Fig. 7.8 Overexpanded nozzle flow with a strong-solution shock.

is very thin because of the favorable pressure gradient. (A favorable pressure gradient is one in which dp/dx is negative.) Unfortunately, a diffuser based on Fig. 7.2(f) does not work. A discussion of the reasons for this is one of the main concerns of this section.

The fact that the viscous boundary layer experiences an adverse pressure gradient in any diffuser represents a major difficulty. Furthermore, the adverse gradient becomes larger as the inlet Mach number increases. (Imagine that a particle of fluid is isentropically moving from right to left along the p/p_0 curve of Fig. 4.1 as it moves through the diffuser.) For a sufficiently large adverse pressure gradient, the boundary layer will separate from the wall. The separated boundary layer then acts as an effective wall. The flow external to the separated boundary layer thus encounters a compressive ramp. If the external flow is supersonic, then the ramp causes an oblique shock wave that further alters the overall flowfield. In short, supersonic flowfields with and without separated boundary layers are quite different from each other. While the role of the boundary layer is usually benign in a nozzle flow, it is of dominating importance in a diffuser flow.

Large supersonic wind tunnels require a large flow rate and generally operate in a closed-cycle configuration, as shown schematically in Fig. 7.9. For steady flow operation, a cooler is required to remove the increase in stagnation temperature caused by the compressor. For a high test section Mach number, a dryer is required to avoid water vapor condensation in the downstream part of the nozzle. A compressor is required to make up stagnation pressure losses associated with:

(1) Wall frictional losses.
(2) Flow turning losses.
(3) Cooler, dryer, and screen pressure drop losses.
(4) Diffuser losses.
(5) Losses due to the model and its supporting sting.

Suppose the diffuser is a straight duct, in which case stations 3 through 6 in Fig. 7.9 have a constant cross-sectional area. We would then assume that the supersonic flow at the end of the test section, station 3, becomes

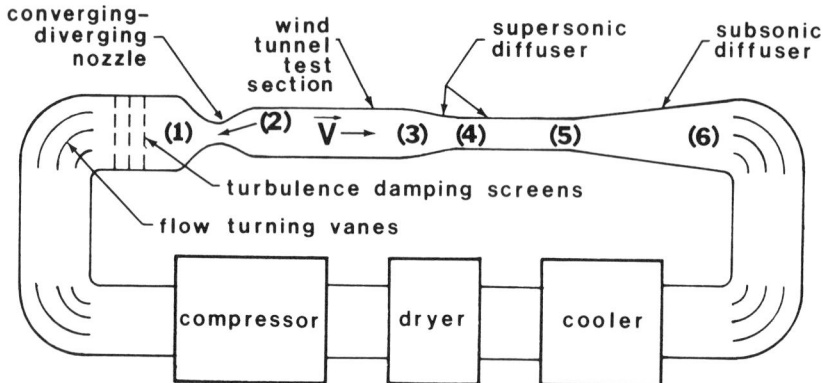

Fig. 7.9 Schematic of a closed-cycle, supersonic wind-tunnel facility.

subsonic by means of a normal shock wave. From a normal shock table, we see that the loss in stagnation pressure can be considerable, especially at a high Mach number. For example, if $M_3 = 5$ the loss is 94% of p_{01}. More important, the compressor must provide a compression ratio of 16.2 to cover this loss alone.

A second diffuser application involves ground testing a rocket nozzle designed for exoatmospheric operation. At an altitude where $p_\infty \cong 0$, the flow fills the nozzle. When tested at sea level, however, flow conditions resemble Fig. 7.2(c), except that the flow will separate from the nozzle wall downstream of the shock. An effective way to ground-test the engine without flow separation is to attach a diffuser to the nozzle's exit plane,[3] as shown in Fig. 7.10. The numbering in Figs. 7.9 and 7.10 coincides; thus, station 3 is the inlet to the diffuser. Although not as common as a wind-tunnel diffuser, this application is somewhat simpler. It serves as a focus for the subsequent discussion.

We assume isentropic flow in the rocket nozzle and that $p_{01} \cong p_1$. At the diffuser exit, station 6, the subsonic Mach number is small and $p_{06} \cong p_6$. Furthermore, the ambient pressure, p_∞, which is 1 atm, also equals p_6.

Downstream of the diffuser's inlet is a converging section whose function is to compress the flow to a smaller supersonic Mach number. The diffuser therefore has a throat between stations 4 and 5, where the cross-sectional area is constant. A system of weak-solution shock waves starts near station 4. These shocks crisscross each other as they repeatedly reflect from the wall. This section of the diffuser has a length-to-diameter ratio of about 10 to allow room for the reflection process. The final shock wave is a normal one; however, it is relatively weak because the Mach number upstream of it has been reduced to a low supersonic value. Consequently, the Mach number at station 5 has a high subsonic value. Between stations 5 and 6 is a subsonic diffuser that further reduces the Mach number to a small subsonic value. This section is a lengthy, diverging duct with a shallow half-angle to avoid boundary-layer separation. The angle is usually 3–6 deg.

On occasion we will also consider the diffuser to consist of a constant cross-sectional area duct over its entire length. In this case, the diffuser does not have a throat.

A frequently used[4] performance measure of a diffuser is p_{06}/p_{03}, i.e., diffuser outlet stagnation pressure divided by the inlet stagnation pressure. For the rocket engine application, this ratio is approximately the same as p_∞/p_{01}.

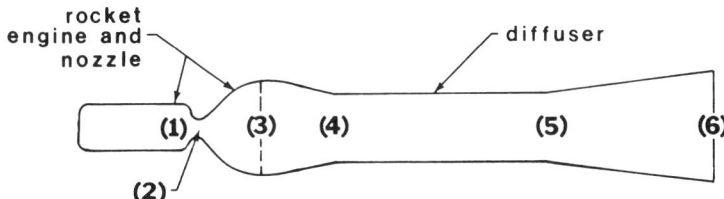

Fig. 7.10 Schematic of a diffuser for an exoatmospheric rocket engine.

A normal shock model is the simplest one for estimating the pressure recovery provided by a diffuser. The model utilizes a normal shock at the inlet, station 3, with isentropic flow in the rest of the diffuser. Thus, determination of p_{04}/p_{01}, which approximately equals p_∞/p_{01}, is identical to that of Fig. 7.2(d). (The stagnation pressure p_{04} is just downstream of the shock, which is located at station 3.) We need only know γ and the nozzle area ratio A_3/A_2 for this computation. Furthermore, the result for p_{04}/p_{01} does not depend on whether the diffuser has a throat.

The normal shock model would appear to be most appropriate for a constant cross-sectional area diffuser. However, experiments[4] show that p_∞/p_{01} is overestimated by the model by as much as a factor of 2. Part of the overestimation is due to the neglect of frictional losses, which are small except at low Reynolds numbers. A usually more important reason would be a blockage effect[5] caused by the rapidly thickening boundary layer. The blockage causes a nonuniform velocity profile in the turbulent flow, which results in a sizable loss in stagnation pressure in the bulk of the flow.

On the other hand, the normal shock model is reasonably accurate[3,4] for diffusers with a throat, even though the actual oblique shock system is between stations 4 and 5 and not at the inlet. Hence, this type of diffuser significantly outperforms a straight duct diffuser and therefore is generally used.

We now discuss the starting process for the Fig. 7.10 diffuser when it is a straight duct without a throat. Actually, we need to discuss the starting process for the nozzle-diffuser system under the proviso that p_∞ is constant. We thus imagine gradually increasing the plenum pressure p_{01}, starting with a value of p_∞. For p_{01} slightly greater than p_∞, the flow is subsonic with frictional losses and diffuser blockage as the primary loss mechanisms. At a higher value for p_{01}, the flow chokes at the nozzle's throat, and a shock wave occurs between stations 2 and 3. By further increasing p_{01}, the shock is driven downstream of station 3. It is now an oblique shock system inside the constant-area diffuser duct. The flow in the nozzle is isentropic and M_3 is supersonic. At the exit of the diffuser, the Mach number is subsonic, $p_6 = p_\infty$, and $p_6 > p_3$. Under these conditions, the diffuser is started and provides some pressure recovery. If p_{01} is further increased, then p_3 increases, the shock system is driven downstream, and M_6 may become supersonic. In this situation, p_6 exceeds p_∞, and the flow further expands after it leaves the diffuser.

The foregoing discussion is illustrated in Fig. 7.11, which shows how p_3 varies in response to changes in p_{01}, where both are normalized by the ambient pressure. Ideal flow conditions are assumed, i.e., inviscid, unseparated flow with a normal shock. Initially all pressures equal p_∞, which is point B. As p_{01} increases, the flow at station 3 is subsonic with $p_3 = p_\infty$. At point C the shock is just upstream of station 3.

Once the shock is downstream of station 3, p_3 becomes proportional to p_{01}, since M_3 has a constant supersonic value. Thus, p_3/p_∞ is on the straight line DE that passes through the origin. Consequently, p_3/p_∞ changes abruptly from point C to point D as the shock passes station 3. A further increase in p_{01} increases p_3/p_∞ along the line DE. The diffuser is

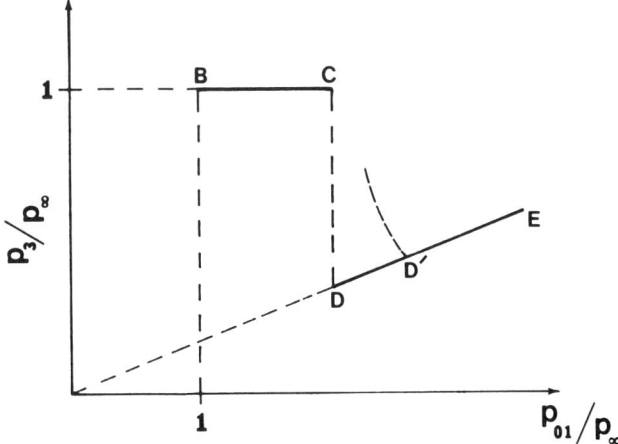

Fig. 7.11 Performance curves for a constant cross-sectional area diffuser.

started when point D is achieved, which is the optimum operating point, since the recovered pressure ratio p_∞/p_{01} is a maximum. Recall that this constant duct area model overpredicts the pressure recovery. Actual data will follow the dashed line to a point D' on line DE. This line is unaltered by diffuser losses.

We next discuss starting the diffuser shown in Fig. 7.10 when it has a throat. For simplicity, consider station 5 as having the smallest cross-sectional area. (For the duct between stations 4 and 5, station 5 has the thickest boundary layer and effectively the smallest cross-sectional area.) Suppose A_5 equals the nozzle's throat area A_2. Figure 7.2(b) is achieved as p_{01} increases. In this condition, the flow is sonic at both stations 2 and 5 and subsonic elsewhere. A further increase in p_{01} causes the flow downstream of station 5 to become supersonic followed by a shock. However, the flow upstream of station 5 remains subsonic, with station 2 sonic. Thus, with $A_5 = A_2$, the diffuser will not start.

Consider increasing A_5 so that it slightly exceeds A_2. Supersonic flow followed by a shock will occur in the rocket nozzle. However, the nozzle shock will stop moving downstream once the flow chokes at station 5. A further increase in p_{01} simply increases the overall pressure level from stations 1 to 5 without altering the Mach number variation in this part of the diffuser. Downstream of station 5, the flow is supersonic, followed by a shock wave such that the pressure at station 6 equals the ambient pressure.

For isentropic supersonic flow throughout the nozzle, the minimum diffuser area must be large enough for a normal shock in the diverging part of the nozzle to move to station 3. This can happen if Eq. (7.5a) is satisfied on both sides of the shock and the flow at stations 2 and 5 is sonic. Thus, for A_5 we have

$$A_5^* = \frac{p_{01}}{p_{06}} A_2^*$$

or
$$A_5 = \frac{p_{01}}{p_\infty} A_2 \tag{7.10}$$

This relation is for a normal shock at station 3. To provide optimum pressure recovery, the shock must move to a location downstream of station 4. However, a normal shock is not stable in the converging duct between stations 3 and 4. To drive the shock into the duct between stations 4 and 5, where it becomes an oblique shock system, the pressure p_{01} is momentarily increased above its nominal value. After the shock system is stably located between stations 4 and 5, a further increase in p_{01} tends to drive it into the subsonic diffuser, while a decrease may cause it to jump into the nozzle upstream of station 3.

Equation (7.10) provides the minimum diffuser throat area for starting. Once started, this area can be decreased to improve the diffuser's efficiency. Such a diffuser is referred to as a variable-geometry diffuser. On an ideal basis, A_5 can be reduced to A_5'', in which case a normal shock is situated in the diffuser's throat. [Recall that A_5 in Eq. (7.10) is based on a normal shock at station 3. A normal shock between stations 4 and 5 is a weaker shock since the flow is isentropically compressed between stations 3 and 4.] The amount of compression is determined by A_5''/A_3, which is not uniquely defined. It depends on several factors, such as the design of the converging section between stations 3 and 4. Too large a reduction in A_5 will cause the shock to jump upstream into the divergent part of the nozzle.

The foregoing description is clarified by Fig. 7.12, which is restricted to an ideal flow, i.e., inviscid, unseparated flow with a normal shock. As in Fig. 7.11, operation starts at point B and moves to point C as p_{01} increases. (The figure ignores the very brief regime to the right of point B, when only subsonic flow exists.) Again, the shock moves from the nozzle into the diffuser when conditions jump from point C to point D. The line from the origin to point E is the same as in Fig. 7.11. At point D, the oblique shock system in the duct between stations 4 and 5 provides an overall change in stagnation pressure equivalent to that of a normal shock at station 3. By

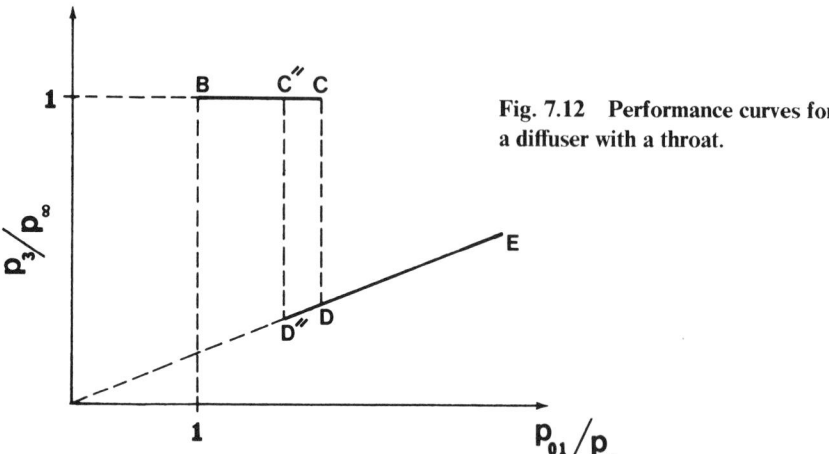

Fig. 7.12 Performance curves for a diffuser with a throat.

reducing A_5 to A_5'', the losses are reduced to that for a normal shock at station 5. Thus, point D'' is to the left of D since the normal shock is weaker when situated in the throat. If A_5 is further reduced, operation jumps to point C'', thereby providing a hysteresis effect. Since p_{01}/p_∞ is a minimum at point D'', this condition represents the most efficient operating point for the diffuser.

References

[1] Crocco, L., "One-Dimensional Treatment of Steady Gas Dynamics," *Fundamentals of Gas Dynamics, High Speed Aerodynamics and Jet Propulsion*, Vol. 3, edited by H. W. Emmons, Princeton University Press, Princeton, NJ, 1958.

[2] Illingworth, C. R., "Shock Waves," Chap. IV, Sec. 8 in *Modern Developments in Fluid Dynamics*, Vol. I, edited by L. Howarth, Clarendon Press, Oxford, England, 1953.

[3] Roschke, E. J., Massier, P. F., and Gier, H. L., "Experimental Investigation of Exhaust Diffusers for Rocket Engines," TR-32-210, Jet Propulsion Laboratory, California Institute of Technology, Pasadena, March 1962.

[4] Lukasiewicz, J., "Diffusers for Supersonic Wind Tunnels," *Journal of the Aeronautical Sciences*, Vol. 20, Sept. 1953, pp. 617–626.

[5] Reneau, L. R., Johnston, J. P., and Kline, S. J., "Performance and Design of Straight, Two-Dimensional Diffusers," *Journal of Basic Engineering*, Vol. 89, March 1967, pp. 141–150; see also Greywall, M. S., "Performance Prediction of Straight Two-Dimensional Diffusers," ME-MG80-1, Wichita State University, Wichita, KS, Sept. 1980.

Problems

7.1 An air nozzle has a throat area of 10 cm², a reservoir temperature of 400 K, an area ratio of 5, and a back pressure of 76 torr (760 torr = 1 atm). The emerging supersonic jet is both shock- and expansion-free. Determine the mass flow rate $\dot{m}$.

7.2 Hydrogen expands from 7.3 atm and 515 K to 1 atm in a steady flow process without heat transfer. Compute the final velocity assuming that the initial velocity is negligible, and compute the flow rate if the final duct size is 10 cm in diameter. Assume that nozzle operation is isentropic.

7.3 Air flows steadily in a converging/diverging nozzle. With the known area ratios

$$\frac{A_1}{A_2} = \frac{A_3}{A_2} = 10$$

determine M_i, p_i/p_0, T_i/T_0, and ρ_i/ρ_0 ($i = 1, 2, 3$) for the nozzle shown in Fig. 7.1. Assume supersonic flow at station 3.

7.4 For the nozzle in Problem 7.3, the throat radius is 1 cm, $T_2 = 600$ K, and $p_3 = 0.5 \times 10^5$ N/m². Determine $\dot{m}$.

7.5 Argon ($\gamma = 5/3$, $W = 40$ kg/kmole) flows isentropically through a nozzle with an area ratio of 10. Determine the exit supersonic Mach number M_3 and p_3/p_0.

7.6 A converging/diverging nozzle discharges air into a receiver with a static pressure of 1 atm. A duct of radius 0.1 m feeds the nozzle with air at 6 atm, 600 K, and a Mach number $M_1 = 0.3$. Flow is steady and isentropic, and the nozzle exit pressure p_3 matches the receiver pressure. Determine $\dot{m}$, A_2 (throat area), and A_3. (The 6-atm and 600 K values are static conditions.)

7.7 A stream of argon flows in a duct 0.1 m in diameter at a rate of 1 kg/s. The stagnation temperature is 350 K, and at one section the static pressure is 0.5 atm. Calculate the Mach number, flow speed, stagnation pressure, and static temperature at this section.

7.8 For isentropic flow, show that the energy equation can be written as

$$V^2 + \frac{2}{\gamma - 1}a^2 = \frac{\gamma + 1}{\gamma - 1}(a^*)^2$$

where a^* is the speed of sound when $M = 1$.

7.9 A blowdown wind tunnel consists of a tank of volume V that feeds the tunnel through a duct containing a rapidly opening valve. The tank is pressurized, after which the valve opens. During operation the flow of the perfect gas is choked, and heat transfer from the tank's wall to the gas is negligible. Derive an algebraic equation for the time dependence of the density $\rho(t)$ of the gas in the tank. Assume quasi-steady flow through the section where the flow is choked.

7.10 Assume for the blowdown wind tunnel of Problem 7.9 that the valve of area A^* opens at time $t = 0$. Determine an equation for the length of time t^* for which the flow through A^* remains choked. Let A_3/A_2 be the area ratio for the tunnel, and assume a constant ambient pressure p_∞.

7.11 Air at $p_0 = 1$ atm and $T_0 = 500$ K flows through a nozzle (Fig. 7.1), where

$$\frac{A_1}{A_2} = 5, \qquad \frac{A_3}{A_2} = 8$$

and $p_\infty \cong 0$. Determine p_1, M_1, p_3, M_3, V_2, and $\dot{m}/A_2$.

7.12 A converging/diverging nozzle with an area ratio of 1.5 has air flowing through it. The ambient pressure is 3×10^4 N/m².

(a) Determine the inlet pressure p_1 when the flow is isentropic and the exit Mach number is 0.4.

(b) Determine p_1 when the flow is not isentropic and the exit Mach number is 0.6.

7.13 A converging/diverging nozzle discharges into a receiver in which the pressure is 1 atm. A 0.5-m² duct feeds the nozzle with air at 6 atm, 900 K, and a Mach number of 0.5. The exit area is such that the nozzle exit pressure matches the receiver pressure. Calculate the flow rate, throat area A_2, exit area A_3, M_3, and T_3.

7.14 Helium ($\gamma = 5/3$, $W = 4$ kg/kmole) isentropically flows through a converging/diverging nozzle. Section 3 is downstream of the throat, and section 4 is downstream of section 3. The pressure ratio for the two sections is (p_3/p_4) = 3. Consider three cases for the area ratio $A_4/A_3 = 1.8$, 1.9, and 2. Determine M_3 for each case. (Hint: Determine M_3 analytically.)

7.15 A two-dimensional nozzle has a steady flow of air with a plenum pressure of 7 atm. Use the numbering system of Fig. 7.3 with $A_5/A_2 = 9.666$ and $p_\infty = 0.5, 0.03$ atm (2 cases). Determine p_5 and M_5. Assume the flow is uniform at the nozzle's exit plane. Determine the angle θ, relative to the jet's centerline, that the initial jet has. Finally, determine the Mach number M_6 in the external part of the jet near the nozzle's lip.

7.16 A converging/diverging nozzle has an internal normal shock as shown in Fig. 7.3. Determine the stagnation pressure ratio p_{01}/p_{05} as a function of γ and p_4/p_3. Simplify your results.

7.17 An air nozzle has an internal shock as shown in Fig. 7.3. Flow conditions are $p_0 = 2$ atm, $T_0 = 400$ K, $A_3/A_2 = 6$, and $A_5/A_2 = 8$. Determine p_∞, M_5, and $\dot{m}/A_2$.

7.18 An air nozzle has an internal shock as shown in Fig. 7.3. Flow conditions are $p_0 = 4$ atm, $p_\infty = 1$ atm, $A_2 = 10$ cm², $T_5 = 300$ K, and $A_5/A_2 = 10$. Determine M_3, M_4, M_5, A_3/A_2, and $\dot{m}$.

7.19 Air flows in a converging/diverging nozzle with an area ratio of 7. The stagnation pressure p_0 is 1 atm, and the measured pressure ratio across the shock in Fig. 7.3 is (p_4/p_3) = 8. Determine M_2, M_3, M_4, M_5, and p_5.

7.20 You are to design a convergent/divergent nozzle for compressed air with a plenum pressure of 15 atm. A normal shock is to sit at the exit of the nozzle when the ambient pressure is 1 atm. Determine the area ratio A_5/A_2, where A_2 is the throat area and A_5 is the nozzle exit area. After the nozzle is built, it is found that only 5-atm plenum pressure is available. With a 1-atm ambient pressure, determine the shock's location A_3/A_2.

7.21 Air flows through a nozzle (see Fig. 7.3), where

$$\frac{A_3}{A_2} = 3, \qquad \frac{A_5}{A_2} = 5, \qquad p_\infty = 1 \text{ atm}$$

Determine p_{01}, p_{04}, p_2, p_3, p_4, p_5, M_2, M_3, M_4, and M_5.

7.22 Nitrogen flows through a nozzle (see Fig. 7.3) where

$$p_\infty = 1 \text{ atm}, \qquad p_0 = 1.005 \text{ atm}, \qquad T_0 = 400 \text{ K}, \qquad \frac{A_5}{A_2} = 5$$

Is the flow choked? Is there an internal shock, and, if there is, where is it located? Determine $\dot{m}/A_2$; $M_2, \ldots, M_5$; $p_2, \ldots, p_5$.

7.23 Redo Problem 7.22 with one change; set $p_0 = 1.1$ atm.

7.24 An air nozzle has an area ratio of 6.79. The back pressure p_∞ is 1 atm, and the plenum pressure is (a) 90 atm, (b) 50 atm, (c) 4 atm, (d) 1.002 atm. For each case, determine the pressure and Mach number at the exit plane of the nozzle.

7.25 Determine simplified equations for the pressure ratio p_d/p_0.

7.26 A gas with $\gamma = 1.5$ flows through a converging/diverging nozzle that has an area ratio of 3. Determine the pressure ratios p_b/p_0, p_d/p_0, and p_f/p_0.

7.27 The exit of a nozzle has uniform flow of He at $M_1 = 3$ and $p_1 = 100$ torr. The receiver pressure p_∞ is 200 torr. Determine β and θ (Fig. 7.5) and M_2 (1 atm = 760 torr).

7.28 A two-dimensional nozzle has an area ratio of 3. The nozzle operates with air at an inlet pressure of 2 atm and a receiver pressure of 50 torr. Determine the nozzle's exit Mach number and pressure, in torr, and the angle, in degrees, of the exiting flow at the lip of the nozzle relative to the nozzle's centerline.

7.29 You are to design an air nozzle with a plenum pressure of 18 atm. The nozzle is to operate at the case (f) condition (Fig. 7.2) when the ambient pressure is 1 atm. After the nozzle is built, it is found that only 2.3-atm plenum pressure is available. Determine the nozzle's area ratio, exit Mach number, and, if there is a shock wave inside the nozzle, the area ratio where it might be located.

7.30 A two-dimensional nozzle has a uniform exit flow as shown in Fig. 7.6. The gas is N_2, $p_1 = 2$ atm, $T_1 = 600$ K, $A_3/A_2 = 5.9$, and $\theta = 35$ deg. Determine M_4, p_∞, and the flow speed in region 4, V_4.

7.31 At the exit plane of a nozzle, there is a uniform flow of He with $M_3 = 3$, $p_3 = 100$ torr (see Fig. 7.5). The back pressure is 200 torr. Determine β, θ, and M_4.

7.32 All conditions are the same as in Problem 7.31 except that $p_\infty = 50$ torr. Determine β and θ, in degrees, and M_4, where region 4 is downstream of the expansion and β is the angle the leading edge has with respect to the centerline.

7.33 A two-dimensional air nozzle has a 5-atm stagnation pressure. The nozzle has an area ratio of 6, and the flow in the exit plane is uniform. When the back pressure is 1 atm, determine:
(a) Is the jet over- or underexpanded?
(b) Are the lip shocks given by the weak or strong solutions?
(c) The angle β that the shock has relative to the nozzle's centerline.

Computational Problems:

7.34 Add Eq. (7.6b) to the tables of Problem 4.6.

7.35 Develop a code to model the weak-solution, oblique shock wave, reflection process shown in the sketch below. The final shock should be a normal one. For given values of γ, θ, and M_∞, determine the largest value n_{max} that n can take. For given values of γ, θ, M_∞, and $1 \leq n \leq n_{max}$, determine y_f, x_f, M_{n+1}, and the overall stagnation pressure ratio. Compare your results with normal shock pressure recovery at the freestream Mach number M_∞.

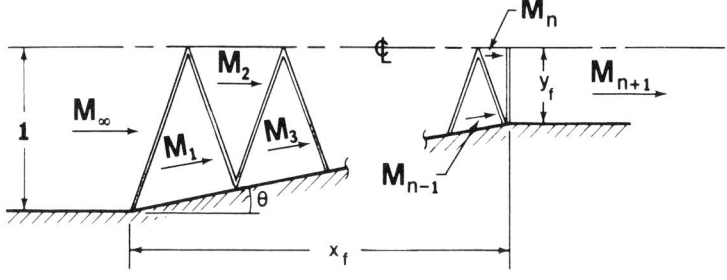

8. DUCTS WITH AREA CHANGE, HEAT TRANSFER, AND FRICTION

Initially, this chapter considers an unsteady, quasi-one-dimensional flow confined to a streamtube. The cross-sectional area of the streamtube need not be small and may change with both time and axial position. Its rate of change with time or distance cannot be too rapid, however, if the one-dimensional approximation is to be valid. Heat transfer into or out of the streamtube and viscous shear on its surface are both allowed. For consistency with the one-dimensional approximation, these effects must be averaged across the streamtube.

Under the above assumptions, the equations for conservation of mass (continuity), Newton's second law, and conservation of energy are derived in Sec. 8.1. They are then specialized for steady flow and a perfect gas. As usual, the equations are simplified by introducing the Mach number and using it as the independent variable. In subsequent sections, we examine Rayleigh flows (no friction or area change) and Fanno flows (no heat transfer or area change).

Appendices I, J, and K contain summaries of the general case and the two special cases. The standard reference for this one-dimensional approach is by Shapiro,[1] who pioneered it in the late 1940s. For more details than can be given here, consult Chaps. 6–8 of Ref. 1.

8.1 INFLUENCE COEFFICIENT METHOD

The procedure in Chap. 3 is used to derive the equations of motion by considering a thin slice of a streamtube (Fig. 8.1). Only now, heat dq per unit mass is transferred to the streamtube, and a shear stress τ operates on surface 3. This contributes a shear force

$$dF_s = -\tau \cos\phi\, dS_3$$

in the x direction, where ϕ is the angle between surface 3 and the x axis. Inclusion of shear and heat transfer does not in any way alter continuity. Thus, Eq. (3.8b) is unchanged.

The surface pressure forces F_1, F_2, and dF_3 discussed in Sec. 3.3 also remain unaltered. To these forces we need to add dF_s. The area dS_3 can be derived by treating surface 3 as a frustum of a right circular cone. The frustum surface area is given by

$$dS_3 = \pi\left[r + \left(r + \frac{\partial r}{\partial x}dx\right)\right]\left\{(dx)^2 + \left[r - \left(r + \frac{\partial r}{\partial x}dx\right)\right]^2\right\}^{\frac{1}{2}}$$

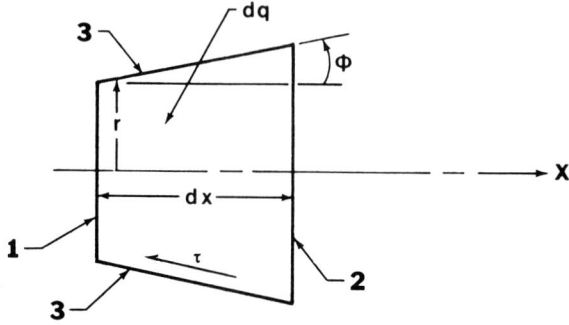

Fig. 8.1 Differential element of a streamtube.

which simplifies to

$$dS_3 = \frac{c\,dx}{\cos\phi}$$

where the circumference c is

$$c = 2\pi r \tag{8.1}$$

and

$$\cos\phi = \left[1 + \left(\frac{\partial r}{\partial x}\right)^2\right]^{-\frac{1}{2}}$$

Hence, dF_s becomes

$$dF_s = -c\tau\,dx$$

We add dF_s to $F_1 + F_2 + dF_3$. Therefore, Newton's second law is given by

$$\rho\frac{DV}{Dt}A\,dx = -\frac{\partial p}{\partial x}A\,dx - c\tau\,dx$$

or

$$\frac{DV}{Dt} = -\frac{1}{\rho}\frac{\partial p}{\partial x} - \frac{c\tau}{\rho A}$$

For convenience, the skin-friction coefficient C_f and hydraulic diameter D are introduced by means of

$$C_f = \frac{\tau}{\rho V^2/2}, \quad D = \frac{4A}{c} \tag{8.2}$$

to obtain the final form

$$\frac{DV}{Dt} = -\frac{1}{\rho}\frac{\partial p}{\partial x} - 2\frac{C_f}{D}V^2 \qquad (8.3)$$

The substantial derivative is still given by Eq. (3.3).
In line with Eq. (3.10a), the energy equation can be written as

$$\frac{De}{Dt} = -p\frac{Dv}{Dt} + \dot{q}$$

where $\dot{q}(x,t)$ represents heat transferred by conduction, radiation, or combustion, per unit mass and per unit time. $\dot{q}$ is positive when the surroundings supply the heat. By the same derivation in Chap. 3, we obtain

$$\frac{Dh_0}{Dt} = \frac{1}{\rho}\frac{\partial p}{\partial t} + \dot{q} \qquad (8.4)$$

for the energy equation.
For a steady flow, the equations of motion simplify to

$$\rho VA = \dot{m} \qquad (8.5a)$$

$$VdV + \frac{1}{\rho}dp = -\frac{1}{2}V^2\left(4C_f\frac{dx}{D}\right) \qquad (8.5b)$$

$$dh_0 = dq \qquad (8.5c)$$

where $dq = \dot{q}\,dx/V$ and the mass flow rate $\dot{m}$ is a constant. Equation (8.5c) integrates to

$$q = h_{02} - h_{01} \qquad (8.6)$$

where subscripts 1 and 2 denote upstream and downstream sections of the streamtube, respectively.
We now assume a perfect gas, introduce the Mach number, and differentiate Eq. (8.5a). After some further algebra, the equations of motion become

$$\frac{d\rho}{\rho} + \frac{1}{2}\frac{dT}{T} + \frac{1}{2}\frac{dM^2}{M^2} = -\frac{dA}{A}$$

$$\frac{d\rho}{\rho} + \left(1 + \frac{\gamma}{2}M^2\right)\frac{dT}{T} + \frac{\gamma}{2}M^2\frac{dM^2}{M^2} = -\frac{\gamma}{2}M^2\left(4C_f\frac{dx}{D}\right) \qquad (8.7)$$

$$\frac{d\rho}{\rho} - \frac{1}{\gamma-1}\frac{dT}{T} = -\frac{dq}{RT} - \frac{\gamma}{2}M^2\left(4C_f\frac{dx}{D}\right)$$

For convenience, we use Eq. (8.5c) to replace dq/RT as follows:

$$\frac{dq}{RT} = \frac{[\gamma R/(\gamma-1)]\,dT_0}{RT_0\{1+[(\gamma-1)/2]M^2\}^{-1}} = \frac{\gamma}{\gamma-1}\left(1+\frac{\gamma-1}{2}M^2\right)\frac{dT_0}{T_0} \quad (8.8)$$

Thus, the heat-transfer process, which can heat or cool the flow in the duct, is associated with a change in the stagnation temperature of the gas.

The terms on the right side of Eqs. (8.7) constitute forcing functions such that the wall shear, heat transfer, or an area change requires changes in flow conditions. Of course, the forcing functions can act individually or in combination to alter the flow.

Equations (8.7) are linear in $d\rho/\rho$, dT/T, and dM^2/M^2. They are solved by standard matrix methods for these variables. With Eq. (8.8) used to replace dq/RT, the results are given in Appendix I, along with similar equations for $dp/p, dp_0/p_0, \ldots$. The equations are easily derived for variables other than $d\rho/\rho$, dT/T, and dM^2/M^2. For instance, the equation for dp/p comes from the perfect gas thermal state equation, while the equation for dp_0/p_0 stems from the logarithmic differentiation of the isentropic point relation

$$p_0 = p\left(1+\frac{\gamma-1}{2}M^2\right)^{1/(\gamma-1)}$$

To obtain the equation for the change in entropy ds, we can start with the thermodynamic relation

$$ds = \frac{1}{T}\,dh - \frac{1}{\rho T}\,dp$$

which is taken from Appendix B. For a perfect gas, this becomes

$$\frac{ds}{R} = \frac{\gamma}{\gamma-1}\frac{dT}{T} - \frac{dp}{p}$$

By using the dT/T and dp/p equations in Appendix I, we obtain the given ds/R equation.

The coefficients of dA/A, dT_0/T_0, and $4C_f\,dx/D$ depend only on γ and M^2. Because of its elegant structure, the Appendix I formulation is referred to as an influence coefficient approach. The solution of these equations requires a knowledge of how dA/A, dT_0/T_0, and C_f/D vary with x or with the Mach number. Since Eq. (8.5b) for momentum has no general solution, these equations have no general closed-form solution. (Given sufficient data, however, the equations can be numerically integrated.) There are three important special cases that can be solved in closed form. These are shown in Table 8.1. Chapter 4 addresses isentropic flow. Sections 8.2 and 8.3 cover Rayleigh and Fanno flows.

Table 8.1 Cases with Closed-Form Solution

Flow	dA/A	dT_0/T_0	$4C_f\,dx/D$
Isentropic	Given	0	0
Rayleigh	0	Given	0
Fanno	0	0	Given

Of particular interest in Appendix I is the entropy equation. An area change does not effect ds, i.e., this type of change is isentropic. If $C_f = 0$, then ds has the same sign as dT_0. Heat transfer into the gas increases both T_0 and s. Since the shear term is always positive, it can only increase the entropy, where the rate of increase varies as M^2.

The equation for the Mach number can be rewritten as

$$dM^2 = \frac{M^2\{1 + [(\gamma-1)/2]M^2\}}{M^2 - 1}$$

$$\times \left[2\frac{dA}{A} - (1+\gamma M^2)\frac{dT_0}{T_0} - \gamma M^2 \left(4C_f \frac{dx}{D}\right)\right] \qquad (8.9)$$

We observe that when $M = 1$, the right side is indeterminate. In isentropic flow, where $dT_0 = C_f = 0$, the indeterminacy is resolved by requiring the duct to have a throat, i.e., $dA = 0$ when $M = 1$. In the general case, the bracketed term must be set equal to zero

$$\frac{1}{A}\frac{dA}{dx} = \frac{\gamma + 1}{2}\frac{1}{T_0}\frac{dT_0}{dx} + 2\gamma\frac{C_f}{D}$$

when $M = 1$. This is necessary if the flow is to make a smooth transition across $M = 1$ in either direction.

Normal shock waves can occur. Their method of incorporation is the same as that for an internal shock in a nozzle. The jump conditions of Sec. 5.1 apply along with $p_0 A^* = $ const across the shock.

The net thrust $\mathcal{T}$ on the wall in the x direction is determined by the momentum equation. By convention, $\mathcal{T}$ is positive when the direction of the force on the wall is opposite to the fluid's velocity. This force is just $dF_3 + dF_s$. For a duct between sections 1 and 2, we have

$$\mathcal{T} = \int_{x_1}^{x_2} (dF_3 + dF_s)$$

From the derivation in Sec. 3.3, we obtain

$$\mathcal{T} = \int_{x_1}^{x_2} \left(\rho \frac{DV}{Dt} A\, dx - F_1 - F_2 \right) = \int_{x_1}^{x_2} \left(\rho V \frac{dV}{dx} A\, dx - pA + pA + \frac{\partial(pA)}{\partial x} dx \right)$$

$$= \int_{x_1}^{x_2} \left(\dot{m}\, dV + \frac{\partial(pA)}{\partial x} dx \right) = \dot{m}(V_2 - V_1) + (pA)_2 - (pA)_1$$

$$= (pA + \rho V^2 A)_2 - (pA + \rho V^2 A)_1 = F_2 - F_1 \qquad (8.10)$$

where the impulse function F is

$$F = pA + \rho V^2 A = pA(1 + \gamma M^2) \qquad (8.11)$$

(The forces F_1 and F_2 of Chap. 3 should not be confused with the impulse function.) Equations (8.10) and (8.11) are useful for determining the thrust of a rocket or a jet engine. They provide the thrust whether or not an area change, heat transfer, or a shock wave occurs between sections 1 and 2. In the following example, the impulse function is put in the influence coefficient format.

Example 1

We are to derive an equation for dF/F in the influence coefficient format.

Differentiate Eq. (8.11) logarithmically to obtain

$$\frac{dF}{F} = \frac{dp}{p} + \frac{dA}{A} + \frac{\gamma M^2}{1 + \gamma M^2} \frac{dM^2}{M^2}$$

Use Appendix I to replace dp/p and dM^2/M^2, to obtain the following result:

$$\frac{dF}{F} = \frac{1}{1 + \gamma M^2} \frac{dA}{A} - \frac{\gamma}{2} \frac{M^2}{1 + \gamma M^2} \left(4C_f \frac{dx}{D} \right) \qquad (8.12)$$

Note that dF/F is independent of the heat transfer. However, in the next example, a relation in which F does depend on T_0 is obtained.

Example 2

An integrated form is to be obtained for F, in terms of M^2 and T_0, for a duct with a constant cross-sectional area. An asterisk is used to denote the reference $M = 1$ condition.

DUCTS WITH AREA CHANGE, HEAT TRANSFER, AND FRICTION

With $dA = 0$, we obtain from Eqs. (8.9) and (8.12)

$$\frac{dM^2}{M^2} = \frac{(1+\gamma M^2)\{1+[(\gamma-1)/2]M^2\}}{M^2-1}\frac{dT_0}{T_0}$$

$$-\frac{\gamma M^2\{1+[(\gamma-1)/2]M^2\}}{M^2-1}\left(4C_f\frac{dx}{D}\right)$$

and

$$\frac{dF}{F} = -\frac{\gamma}{2}\frac{M^2}{1+\gamma M^2}\left(4C_f\frac{dx}{D}\right)$$

Eliminate the C_f terms from these equations to obtain

$$\frac{dF}{F} = -\frac{1}{2}\frac{M^2-1}{M^2(1+\gamma M^2)\{1+[(\gamma-1)/2]M^2\}}dM^2 + \frac{1}{2}\frac{dT_0}{T_0}$$

The Mach number term can be integrated with the use of partial fractions. We thereby obtain the answer

$$\frac{F}{F^*} = \frac{1+\gamma M^2}{(2(\gamma+1)M^2\{1+[(\gamma-1)/2]M^2\})^{\frac{1}{2}}}\left(\frac{T_0}{T_0^*}\right)^{\frac{1}{2}} \quad (8.13)$$

The thrust is given by

$$\mathcal{T} = F^*\left(\frac{F_2}{F^*} - \frac{F_1}{F^*}\right)$$

where F^* and T_0^* are constant reference conditions. Although Eq. (8.13) does not contain a C_f factor, the thrust is due solely to friction, since $dA = 0$.

In general, friction causes the stagnation pressure p_0 to decrease along a duct. However, this need not occur if the gas is cooled sufficiently. This idea is the subject of the next example.

Example 3

Consider a constant cross-sectional area duct with a constant stagnation pressure. Further, we assume C_f is constant, which holds for subsonic, fully developed, turbulent pipe flow. We are to determine how q and M vary with x.

With $dp_0 = dA = dC_f = 0$, we have, from Appendix I,

$$\frac{dp_0}{p_0} = -\frac{\gamma M^2}{2}\left(\frac{dT_0}{T_0} + 4C_f\frac{dx}{D}\right) = 0$$

or

$$\frac{dT_0}{T_0} = -4C_f \frac{dx}{D}$$

Thus, the flow must be cooled if p_0 is to be constant. This relation is integrated from $x = x^*$, where $M = 1$ and $T_0 = T_0^*$, to obtain

$$\frac{T_0}{T_0^*} = \exp(-4C_f L_m/D)$$

and where $L_m = x - x^*$. We now differentiate this equation

$$dT_0 = -4\frac{C_f}{D} T_0^* \exp(-4C_f L_m/D)\, dx$$

and combine it with

$$dq = c_p\, dT_0 = \frac{\gamma}{\gamma - 1} R\, dT_0$$

to obtain

$$dq = -\frac{4\gamma R}{\gamma - 1} T_0^* \frac{C_f}{D} \exp(4C_f x^*/D) \exp(-4C_f x/D)\, dx$$

The first answer is obtained by integrating this relation to yield

$$\frac{q}{RT_0^*} = \frac{\gamma}{\gamma - 1} \left[\exp(-4C_f L_{m2}/D) - \exp(-4C_f L_{m1}/D) \right]$$

where x_1 and x_2 are the duct's inlet and outlet locations, respectively. We use Eq. (8.9) with $dA = 0$ to obtain

$$\frac{dM^2}{M^2} = -\frac{1 + [(\gamma - 1)/2]\, M^2}{M^2 - 1} \left[(1 + \gamma M^2) \frac{dT_0}{T_0} + \gamma M^2 \left(4C_f \frac{dx}{D} \right) \right]$$

Elimination of dT_0/T_0 then yields

$$\frac{M^2 - 1}{1 + [(\gamma - 1)/2]\, M^2} \frac{dM^2}{M^2} = 4C_f \frac{dx}{D}$$

If M_1 is subsonic, then the Mach number decreases along the duct. If M_1 is supersonic, the Mach number increases along the duct. This relation integrates to

$$\frac{1}{M^2} \left(\frac{2}{\gamma + 1} \{ 1 + [(\gamma - 1)/2]\, M^2 \} \right)^{(\gamma+1)/(\gamma-1)} = \exp(4C_f L_m/D)$$

DUCTS WITH AREA CHANGE, HEAT TRANSFER, AND FRICTION   121

for the final answer. Clearly q/RT_0^* can be written in terms of γ, M_1, and M_2.

8.2 RAYLEIGH FLOW

A constant-area duct without friction is assumed. Heat can be removed or added to the gas by heat exchange through the wall of the duct, by radiative heat transfer, by combustion in the gas, by evaporation, or by condensation. With the use of continuity, Eq. (8.5b) is integrated to yield the impulse function

$$p + \rho V^2 = \frac{F}{A}$$

We use continuity, Eq. (8.5a), to eliminate V, thereby obtaining

$$p + \frac{G^2}{\rho} = \frac{F}{A} \tag{8.14}$$

where the mass flux G is

$$G = \rho V = \frac{\dot{m}}{A} = \text{const} \tag{8.15}$$

Equation (8.14) involves only constants and thermodynamic variables. It is referred to as the Rayleigh line equation. Since only thermodynamic variables occur, Eq. (8.14) represents a curve on a Mollier diagram, as shown in Fig. 8.2 for a perfect gas. This curve is for specific values of $\dot{m}/A$ and F/A.

At the asterisk state, the entropy has a maximum, i.e., $ds = 0$. We differentiate Eq. (8.14) to obtain

$$\left.\frac{\partial p}{\partial \rho}\right|_R = \left(\frac{G}{\rho}\right)^2 = V^2$$

where the R subscript means the derivative is restricted to being along the Rayleigh line. At the asterisk point, $ds = 0$; hence, the derivative on the left side is a^2. At this state, $V = a$, and the Mach number is unity. It is easy to see that the upper branch of the Rayleigh curve is subsonic while the lower one is supersonic.

From Appendix I, we have

$$\frac{dT}{T} = \frac{(\gamma M^2 - 1)\{1 + [(\gamma - 1)/2] M^2\}}{M^2 - 1} \frac{dT_0}{T_0}$$

and

$$\frac{ds}{R} = \frac{\gamma}{\gamma - 1} \{1 + [(\gamma - 1)/2] M^2\} \frac{dT_0}{T_0}$$

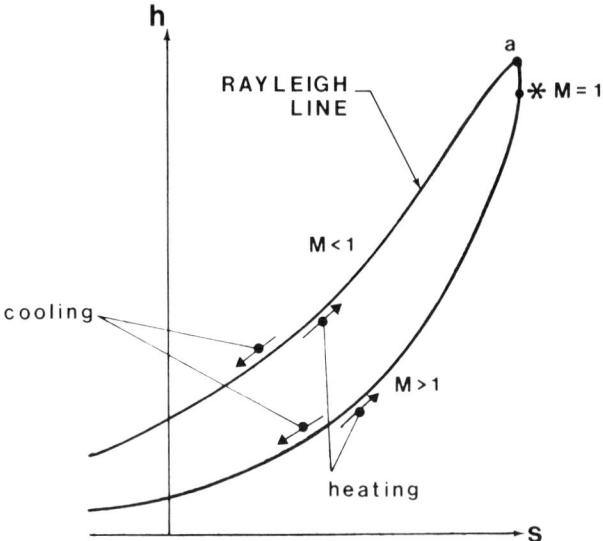

Fig. 8.2 Mollier diagram for a perfect-gas Rayleigh line that shows heating and cooling process directions.

As noted in the preceding section, the entropy increases with heating and decreases with cooling. The static temperature also increases with heating, except when $\gamma^{-1} < M^2 < 1$, and the gas actually cools. The lower limit of this Mach number range stems from the $\gamma M^2 - 1$ factor in the numerator of the dT/T equation. Likewise, the temperature decreases with cooling, except for the same Mach number range. This region falls between the sonic point and point a, as shown in Fig. 8.2. Point a can be shown to occur when $dh = 0$. When the flow is subsonic, the heat transferred into the gas normally increases both the kinetic and internal energy. However, when the Mach number is in the above range, some thermal energy, as well as the heat transferred into the gas, goes into the $V^2/2$ kinetic energy term.

The second law of thermodynamics, Eq. (2.29), must be satisfied. In this inequality, either Eq. (8.5c) or Eq. (8.8) provides dq. With either equation, the heating and the cooling directions are in accord with the arrows shown in Fig. 8.2. From this figure, we observe that adding heat causes the Mach number to increase toward unity if the flow is subsonic initially and to decrease toward unity if the flow is supersonic initially. If sufficient heat is added, the flow chokes thermally, once $M = 1$ occurs at the exit of the duct. When heat is removed, the arrow directions reverse in Fig. 8.2.

Let us consider the choking condition under the proviso that inlet conditions at the start of the duct are fixed. As more heat is added to the flow, which may be subsonic or supersonic, outlet conditions move along the Rayleigh line toward the sonic point. Once outlet conditions reach the sonic point, adding more heat moves the outlet state to the left on the Rayleigh line. But such movement constitutes a second law violation. In order to avoid a violation, inlet conditions must change. As a consequence,

F/A, G, or both, change. When one or both change, the flow shifts to a different Rayleigh curve on the Mollier diagram. In a subsonic flow, the change involves a decrease in G.

As shown in Fig. 8.3, supersonic flow at the inlet of the duct can be generated by a converging/diverging nozzle. If the duct flow is supersonic at its inlet, then the flow in the nozzle is realistically treated as isentropic, see Fig. 7.2(f). (We assume uniform flow at the exit of the nozzle. If the flow is not uniform, then oblique shock waves will occur in the duct.) In this case, choking requires a normal shock wave at, or upstream of, the nozzle's exit plane. Until the flow in the nozzle becomes fully subsonic, the nozzle flow is choked, $\dot{m}/A$ is fixed, and only F/A can change as a result of thermal choking. Once the nozzle flow is subsonic, then thermal choking changes both parameters.

The equations in Appendix I hold here by setting $C_f = \mathrm{d}A = 0$. The resulting differential equations are integrated to yield the results shown in Appendix J, where we retain the notation introduced at the bottom of Appendix I. The asterisk refers to the $M = 1$ reference condition. The parameters $p^*, T^*, \ldots$ are thus constants for a given Rayleigh line. In addition, we use Eq. (8.6) for the heat transfer. A more useful Mach number form is obtained by writing

$$q = (h_0^* - h_{01}) - (h_0^* - h_{02}) = q_{m1} - q_{m2}$$

which is made nondimensional by dividing by RT_0^*. The parameter q_m/RT_0^*, which is defined in Appendix J, depends only on γ and M^2. It represents the heat addition that would choke a flow having an inlet Mach number M. The formula given in Appendix J for q_m/RT_0^* is derived by using the T_0^*/T_0 equation.

The following example is of fundamental importance for supersonic Rayleigh flow.

Example 4

Consider a frictionless, constant-area duct with heat addition or removal q. The inlet Mach number M_1 is supersonic. We are to prove that the location of a normal shock in the duct does not alter subsonic exit conditions. We also determine how these exit conditions are related to exit conditions with no shock present.

Consider two flows that are otherwise identical, as shown in Fig. 8.4. We have $M_1 > 1$ and $q = q_a + q_b$. All inlet conditions and q are presumed

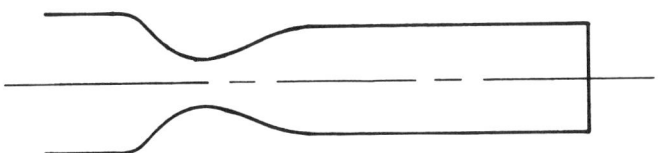

Fig. 8.3 Duct fed by a converging/diverging nozzle.

known. In case (A), the supersonic exit Mach number M_4' is given by

$$\frac{q}{RT_0^*} = \frac{q_{m1}}{RT_0^*} - \frac{q_{m4}'}{RT_0^*}$$

In case (B), we have

$$\frac{q_a}{RT_0^*} = \frac{q_{m1}}{RT_0^*} - \frac{q_{m2}}{RT_0^*}, \qquad \frac{q_b}{RT_0^*} = \frac{q_{m3}}{RT_0^*} - \frac{q_{m4}}{RT_0^*}$$

and, since $q = q_a + q_b$,

$$\frac{q}{RT_0^*} = \frac{q_a}{RT_0^*} + \frac{q_b}{RT_0^*} = \frac{q_{m1}}{RT_0^*} - \frac{q_{m2}}{RT_0^*} + \frac{q_{m3}}{RT_0^*} - \frac{q_{m4}}{RT_0^*} = \frac{q_{m1}}{RT_0^*} - \frac{q_{m4}'}{RT_0^*}$$

where case (A) provides the rightmost equality. Thus, q_{m4}' is given by

$$\frac{q_{m4}'}{RT_0^*} = \frac{q_{m4}}{RT_0^*} + \frac{q_{m2}}{RT_0^*} - \frac{q_{m3}}{RT_0^*}$$

For q_{m3}/RT_0^* (Appendix J), we can write

$$\frac{q_{m3}}{RT_0^*} = \frac{\gamma}{\gamma-1}\left(\frac{M_3^2-1}{1+\gamma M_3^2}\right)^2$$

where M_3^2 is given by the normal shock relation

$$M_3^2 = \frac{1+[(\gamma-1)/2]\,M_2^2}{\gamma M_2^2 - [(\gamma-1)/2]} \tag{8.16}$$

Upon substitution, q_{m3}/RT_0^* becomes

$$\frac{q_{m3}}{RT_0^*} = \frac{\gamma}{\gamma-1}\left(\frac{M_2^2-1}{1+\gamma M_2^2}\right)^2 = \frac{q_{m2}}{RT_0^*}$$

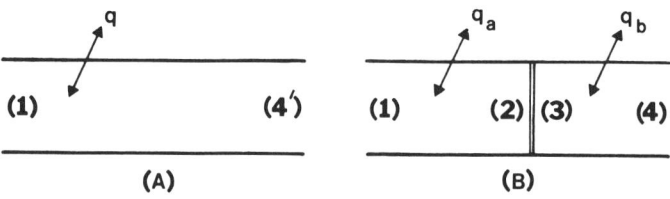

Fig. 8.4 Two ducts with identical conditions, except that duct (A) is supersonic, while duct (B) has a normal shock.

Thus, q_m/RT_0^* is constant across a normal shock. At the two exits, we have

$$\frac{q'_{m4}}{RT_0^*} = \frac{q_{m4}}{RT_0^*}$$

which means that the two Mach numbers are related by the normal shock equation, where $M_4 \leq 1 \leq M'_4$. Furthermore, M_4 depends only on γ, M_1, and q/RT_0^*. Since M_4 does not depend on q_a/q or q_b/q, M_2 cannot be determined. Consequently, the location of the shock wave cannot be determined.

Consider a duct with heat addition and an inlet Mach number M_1 that is subsonic. As we have seen, the outlet Mach number M_2 is restricted to the range $M_1 < M_2 \leq 1$. Of course, if $M_2 = 1$, the flow is thermally choked. Our earlier discussion of choking presumed that inlet conditions were fixed. This presumption, however, is not realistic. The duct exit pressure p_2 must satisfy the conditions discussed just before Example 4 in Chap. 7. Thus, if $M_2 < 1$, $p_2 = p_\infty$ and inlet conditions must adjust to this downstream boundary condition. (As noted earlier, inlet conditions mean G and F/A.) As more or less heat is added to the flow, the $p_2 = p_\infty$ condition remains, which means that inlet conditions change. Each time G and/or F/A changes, we shift to a different Rayleigh line.

Once choking has occurred, with $M_1 < 1$, the exit condition becomes $p_2 \geq p_\infty$. Increasing q in this case can result in a new Rayleigh line where the flow is no longer choked. Thus, we have the unexpected result that increasing q can unchoke the flow. On the other hand, decreasing q also results in a new Rayleigh line but one that remains choked. Both results are demonstrated in later examples.

Significant differences occur if the inlet Mach number M_1 is supersonic. With heat addition and a shock-free flow, the exit Mach number falls in the range $1 \leq M_2 < M_1$. As long as $M_2 > 1$, p_2 can be less or greater than p_∞. The flow outside the duct adjusts to the ambient pressure in the same way a nozzle flow does. Fixed inlet conditions when $M_1 > 1$ are now realistic. As q is increased, or decreased, the flow remains on the same Rayleigh line. If sufficient heat is added to choke the flow, then inlet conditions must change. This change can occur only by means of a shock wave at, or upstream of, the inlet. As demonstrated by Example 4, a shock wave anywhere inside the duct will not alleviate the choked condition. The reason for this is that T_0 is unaltered by the shock.

We do not mean to suggest that a shock wave cannot occur inside the duct, because it can. Consider case (B) of Fig. 8.4, where heat addition occurs both upstream and downstream of the shock. The Rayleigh line for the flow is shown in Fig. 8.5. Point 1 is on the lower branch, since $M_1 > 1$. A normal shock occurs between points 2 and 3 and, consequently, $s_3 > s_2$. The subsonic Mach number increases downstream of the shock and at the exit $p_4 = p_\infty$. In this situation, the Mach numbers are bounded as follows:

$$M_3 \leq M_4 \leq 1 \leq M_2 \leq M_1$$

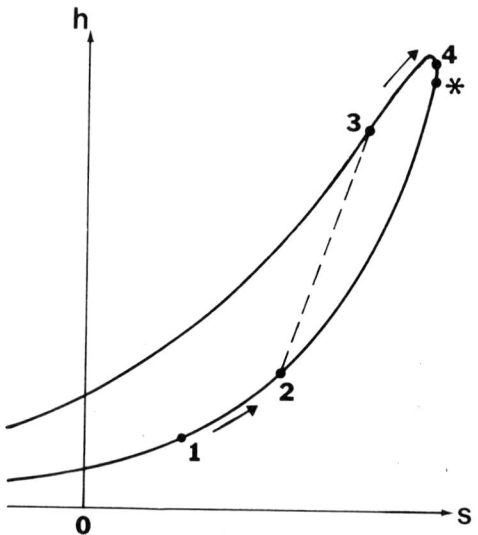

Fig. 8.5 Rayleigh line when M_1 is supersonic and the duct has a normal shock wave. The numbering corresponds to case (B) in Fig. 8.4.

It is worth mentioning again that parameters, such as $T^*, T_0^*, p^*, \ldots$, are constant on the Rayleigh line and, therefore, have the same value at all four states shown in Fig. 8.5. The following three examples illustrate the foregoing discussion.

Example 5

Nitrogen flows through an isentropic converging nozzle that feeds a straight, frictionless duct, as shown in Fig. 8.6. Plenum conditions are $p_1 = 6 \times 10^5$ N/m² and $T_1 = 500$ K. At the start of the duct, we have $M_2 = 0.5$ and at the exit, $M_3 = 1$. We are to determine p_3, q, and $\dot{m}/A$.

From a Rayleigh flow ($\gamma = 1.4$) table at $M_2 = 0.5$, we have

$$\frac{p_2}{p^*} = 1.778, \qquad \frac{p_{02}}{p_0^*} = 1.114, \qquad \frac{T_{02}}{T_0^*} = 0.6914$$

and from an isentropic table, at $M_2 = 0.5$, we have

$$\frac{p_2}{p_{01}} = 0.8430, \qquad \frac{T_2}{T_{01}} = 0.9524$$

Since $M_3 = 1$, $p_3 = p^*$, and we have

$$p_3 = \frac{p_3}{p^*} \frac{p^*}{p_2} \frac{p_2}{p_{01}} \frac{p_{01}}{p_1} p_1 = \frac{0.843 \times 6 \times 10^5}{1.778} = 2.845 \times 10^5 \text{ N/m}^2$$

DUCTS WITH AREA CHANGE, HEAT TRANSFER, AND FRICTION

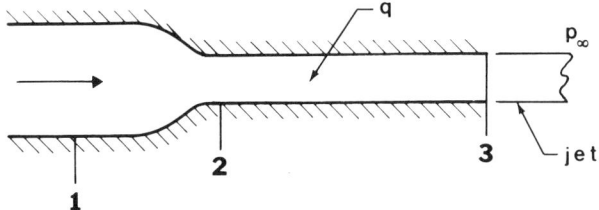

Fig. 8.6 Frictionless duct fed by an isentropic converging nozzle.

As discussed in Chap. 7, we have used the approximation $p_{01} \cong p_1$. With M_2 and M_3 known, we have for q

$$\frac{q_{m2}}{RT_0^*} = \frac{\gamma}{\gamma-1}\left(\frac{M_2^2-1}{1+\gamma M_2^2}\right)^2 = 1.080, \qquad \frac{q_{m3}}{RT_0^*} = 0$$

and

$$q = 1.08\, RT_0^*$$

For N_2, the gas constant is 296 J/kg-K. Since $T_{02} = T_{01} \cong T_1$, we have

$$T_0^* = \frac{T_0^*}{T_{02}}T_{02} = \frac{500}{0.6914} = 723\ \text{K}$$

Hence, q is given by

$$q = 1.08 \times 296 \times 723 = 2.31 \times 10^5\ \text{J/kg}$$

Next, we determine p_2 and T_2 as follows:

$$p_2 = \frac{p_2}{p_{01}}p_{01} = 0.843 \times 6 \times 10^5 = 5.059 \times 10^5\ \text{N/m}^2$$

$$T_2 = \frac{T_2}{T_{01}}T_{01} = 0.9524 \times 500 = 476.2\ \text{K}$$

Then, continuity can be written as

$$\frac{\dot{m}}{A} = \rho V = \frac{p}{RT}aM = \frac{p}{RT}(\gamma RT)^{\frac{1}{2}}M = p\left(\frac{\gamma}{RT}\right)^{\frac{1}{2}}M \qquad (8.17)$$

This relation holds at any station in the flow and is not limited to Rayleigh flow. We apply it at station 2 to obtain

$$\frac{\dot{m}}{A} = 5.059 \times 10^5 \left(\frac{1.4}{296 \times 476.2}\right)^{\frac{1}{2}} \times 0.5 = 7.97 \times 10^2\ \text{kg/m}^2\text{-s}$$

Example 5 provides a fairly comprehensive design-point analysis. In practice, two off-design situations are common for a given duct configuration and a given gas. In the first, inlet thermodynamic conditions, i.e., p_1 and/or T_1, change, with q and p_∞ fixed. This change alters $\dot{m}/A$ and F/A, thereby shifting the flow to a new Rayleigh line, and the analysis proceeds along the lines of Example 5. In the second situation, only q changes. We analyze this case in the next two examples.

Example 6

Conditions that change from Example 5 are denoted with a prime. The heat added to the duct is increased by 20% so that $q' = 1.2q$. Assume that p_∞ is fixed and is equal to p_3 in Example 5 and that p_1 and T_1 are also fixed. We are to determine M_2' and M_3'.

With p_1 and T_1 fixed, h_1 and s_1 are also fixed. Station 1, however, is not the inlet for the duct. Conditions at the inlet, station 2, are not fixed.

We assume $M_3' < 1$, and justify this assumption later. In this subsonic situation, $p_3' = p_\infty = 2.845 \times 10^5$ N/m^2; however, in contrast to Example 5, we have $p^{*\prime} \neq p_3'$. We can compute the known pressure ratio p_3'/p_{01} as

$$\frac{p_3'}{p_{01}} = \frac{p_3'}{p^{*\prime}} \frac{p^{*\prime}}{p_2'} \frac{p_2'}{p_{01}} = \frac{\gamma+1}{1+\gamma M_3'^2}\left(\frac{1+\gamma M_2'^2}{\gamma+1}\right)\left(1+\frac{\gamma-1}{2}M_2'^2\right)^{-[\gamma/(\gamma-1)]}$$

$$= \frac{2.845 \times 10^5}{6 \times 10^5} = 0.4742$$

This is solved for $M_3'^2$ to obtain

$$M_3'^2 = \frac{1}{\gamma}\left[\frac{p_{01}}{p_3'} \frac{1+\gamma M_2'^2}{\{1+[(\gamma-1)/2]M_2'^2\}^{\gamma/(\gamma-1)}} - 1\right] \qquad (8.18)$$

Equation (8.18) constitutes one equation for the two unknown Mach numbers. It represents a pressure balance for the duct shown in Fig. 8.6. A second equation is needed and is provided by the energy equation

$$T_{03}' - T_{02}' = \frac{q'}{c_p}$$

This equation can be manipulated into the form

$$\frac{T_{03}'}{T_{02}'} = 1 + \frac{\gamma-1}{\gamma} \frac{1.2q}{RT_1}$$

DUCTS WITH AREA CHANGE, HEAT TRANSFER, AND FRICTION 129

where $T_{02} = T_1 = 500$ K, and the right side of the equation is known. With the aid of Appendix J for $(T'_{03}/T_0^{*\prime})/(T_{02}/T_0^{*\prime})$, we obtain

$$\frac{M_3'^2}{M_2'^2}\left(\frac{1+\gamma M_2'^2}{1+\gamma M_3'^2}\right)^2 \left\{\frac{1+[(\gamma-1)/2]M_3'^2}{1+[(\gamma-1)/2]M_2'^2}\right\} = 1 + \frac{\gamma-1}{\gamma}\frac{1.2q}{RT_1} = 1.536 \qquad (8.19)$$

With $\gamma = 1.4$ and $(p_3'/p_{01}) = 0.4742$, Eqs. (8.18) and (8.19) can be solved for M_2' and M_3'.

The solution process is simplified by first examining a Mollier diagram, Fig. 8.7. The Rayleigh line for Example 5 is the inner one. State 1 provides plenum conditions for both examples. The isentropic process from 1 to 2 or 2' is a vertical line because s is a constant. The heat transfer q is sufficient to choke the flow, and state 3 is a sonic condition with $p_3 = p_3' = p_\infty$. Since p_∞ is fixed and by assumption that the outlet flow is subsonic, state 3' must fall on the $p = p_\infty$ curve and must intersect a Rayleigh line on the upper (subsonic) branch. One can show that the new Rayleigh line must be outside the old one. As a consequence, M_2' is bounded according to $0 < M_2' < M_2 = 0.5$.

To solve Eqs. (8.18) and (8.19), we first guess a value for M_2' that is less than 0.5. Through Eq. (8.18), this value yields an estimate for M_3'. Substitute both Mach numbers into the left side of Eq. (8.19), which will not equal 1.536. We choose a new value for M_2' and repeat the process until the left side of Eq. (8.19) equals 1.536. This occurs approximately when

$$M_2' = 0.465, \qquad M_3' = 0.989$$

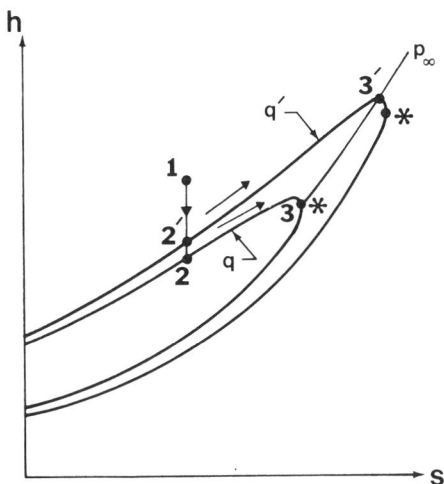

Fig. 8.7 Rayleigh lines for Examples 5 and 6. The numbering corresponds to Fig. 8.6.

Thus, we obtain a solution consistent with the $M_3' < 1$ assumption and with Fig. 8.7. Observe that a 20% increase in q requires a much smaller percent decrease in the Mach numbers. It is also worth mentioning that Mach number tables are of minimal value when the Mach numbers are unknown. The analysis is appreciably more difficult in this situation.

Example 7

Conditions that change from Example 5 are denoted with a double prime. Twenty percent less heat is added, i.e., $q'' = q/1.2$. Further, p_∞, p_1, and T_1 are fixed, as in Example 6. We are to determine M_2'', M_3'', and p_3''.

We first consider retaining the Example 5 Rayleigh line along with $p_\infty = 2.845 \times 10^5$ N/m². With $q'' < q$, the flow would no longer be choked, so that $M_3'' < 1$ and $p_3'' > p_\infty$. The exit pressure result, however, is inconsistent with a subsonic value for M_3''. The flow thus requires a new Rayleigh line. In accord with Examples 5 and 6, we assume that the new line is inside that for Example 5. The $p = p_\infty$ curve on the Mollier diagram will intersect the new Rayleigh line on the lower (supersonic) branch, as is evident from Fig. 8.7. To avoid a second law violation, the flow remains choked, as in Example 5, and

$$M_3'' = 1, \qquad p_3'' > p_\infty$$

A supersonic expansion occurs in the jet outside the duct.

Equation (8.19) now has the form

$$\frac{M_3''^2}{M_2''^2}\left(\frac{1+\gamma M_2''^2}{1+\gamma M_3''^2}\right)^2 \left\{\frac{1+[(\gamma-1)/2]M_3''^2}{1+[(\gamma-1)/2]M_2''^2}\right\} = 1 + \frac{\gamma-1}{\gamma}\frac{q}{1.2 RT_1} = 1.372$$

With $\gamma = 1.4$ and $M_3'' = 1$, this equation is a quadratic in $M_2''^2$, which is easily solved. Of the two roots, we must choose the subsonic one. Hence,

$$M_2'' = 0.5267$$

which is slightly larger than M_2, as expected.

A pressure balance equation is written that is similar to Eq. (8.18),

$$\frac{p_3''}{p_{01}} = \frac{p_3''}{p^{*''}}\frac{p^{*''}}{p_2''}\frac{p_2''}{p_{01}} = \frac{\gamma+1}{1+\gamma M_3''^2}\left(\frac{1+\gamma M_2''^2}{\gamma+1}\right)\left(1+\frac{\gamma-1}{2}M_2''^2\right)^{-[\gamma/(\gamma-1)]}$$

With M_2'' and M_3'' given above, the pressure ratio is 0.4789, and

$$p_3'' = 0.4789 \times 6 \times 10^5 = 2.873 \times 10^5 \text{ N/m}^2$$

which slightly exceeds p_∞, again as expected.

DUCTS WITH AREA CHANGE, HEAT TRANSFER, AND FRICTION   131

Since p_1 and T_1, and therefore ρ_1, are fixed in Examples 5–7, V_1 and $\dot{m}$ must differ from example to example. One can readily show that

$$\dot{m}_6 < \dot{m}_5 < \dot{m}_7$$

for the three examples. With p_∞ fixed and q increasing, we can reach a situation in which the Rayleigh line in Fig. 8.7 passes through state 1, and the duct from 1 to 2 then has a constant area. Any further increase in q will require a change in state 1 conditions.

Heat transfer in ducts can be examined from a quite different viewpoint. In heat- and mass-transfer courses, forced convection in pipes is treated by using correlation formulas involving nondimensional variables, such as the Nusselt and Reynolds numbers. We examine briefly how this approach is related to Rayleigh theory.

For pipe flow, a convective heat-transfer coefficient (or film coefficient) h_f is used. The coefficient has units of J/m²-s-K and can be defined by

$$\frac{q}{A_s} = h_f(T_w - T_0)/\dot{m} \tag{8.20}$$

where T_w is the inside wall temperature, T_0 is the average cross-sectional stagnation temperature, and A_s is the inside wetted surface area. This area is given by cL, where L is the length of the duct, and c is the inside perimeter, which is related to the hydraulic diameter D by Eq. (8.2).

For a differential length element dx of the duct, we replace q with dq and A_s with $\pi D\, dx$. Equation (8.20) becomes

$$\frac{dq}{dx} = \frac{\pi D}{\dot{m}} h_f(T_w - T_0) \tag{8.21}$$

where $D/\dot{m}$ is a known constant, h_f will be given by a correlation formula, and T_w is a known function of x. To obtain a relation for T_0, eliminate T from Eq. (8.8) and the T/T^* equation in Appendix J to obtain

$$\frac{dT_0}{dx} = \frac{\gamma - 1}{2\gamma(\gamma + 1)} \frac{T_0}{RT_0^*} \frac{(1 + \gamma M^2)^2}{M^2\{1 + [(\gamma - 1)/2]M^2\}} \frac{dq}{dx} \tag{8.22}$$

This relation can be combined with Eq. (8.9), with $C_f = dA = 0$, to yield

$$\frac{dM}{dx} = \frac{\gamma - 1}{4\gamma(\gamma + 1)} \frac{1}{RT_0^*} \frac{(1 + \gamma M^2)^3}{M(1 - M^2)} \frac{dq}{dx} \tag{8.23}$$

for the Mach number.

We use the correlation formula in Ref. 2, which holds for turbulent airflow in a smooth circular duct. The formula is based on different inlet geometries, different inlet gas temperatures, different wall temperatures (for both heating and cooling), and different length-to-diameter ratios L/D.

Only subsonic flow was investigated, but sonic exit conditions often occurred. Hence, h_f is given by

$$h_f = 0.023 \frac{k(T)}{D} \left(\frac{4\dot{m}}{\pi D \mu(T)}\right)^{0.8} \left(\frac{T_0}{T_w}\right)^{0.8} [Pr(T)]^{0.4} \qquad (8.24)$$

where

$$k = \text{thermal conductivity, J/s-m-K}$$

$$Re = \text{Reynolds number} = \frac{\rho D V}{\mu} = \frac{4\dot{m}}{\pi D \mu}$$

$$Pr = \text{Prandtl number} = \frac{c_p \mu}{k}$$

The quantities k, μ, γ, and Pr depend on the film temperature. The appropriate film temperature for this correlation[2] is $(T_w + T_0)/2$. Furthermore, the thermal conductivity is presumed to vary as $T^{\frac{1}{2}}$. With tables or curve fit air property data for the above parameters, Eqs. (8.21)–(8.24) are simultaneously integrated numerically. Once $T_0(x)$, $M(x)$, and $q(x)$ are known, all other quantities of interest are readily obtainable.

In the experiments of Ref. 2, an average heat-transfer coefficient was determined for a duct of length L. It was found to vary as $(L/D)^{-0.1}$. This factor is not included on the right side of Eq. (8.24), since the equations must be formulated for a differential length. The skin friction was also measured and fit with a nondimensional correlation formula. Our approach is thus approximate in that Eq. (8.24) is not for a frictionless flow. Here we discuss convective heat transfer for purposes of simplicity. A more systematic approach would be to use both the heat-transfer and skin-friction correlation formulas in conjunction with the more general influence coefficient approach of Sec. 8.1.

We conclude this section by briefly discussing the validity, as well as some applications, of Rayleigh flow analysis. The theory requires uniform inlet and outlet conditions, or conditions that can be averaged. In a subsonic duct flow, there often is adequate flow time to establish a suitable outlet condition. Rayleigh theory does not require a known distribution for the heat addition $q(x)$. The computation is then of a "black-box" type. In other words, the details of the flowfield between the inlet and the outlet stations are not required. Average conditions at the outlet are based on the exact, one-dimensional conservation equations.

The theory is of less value for a shock-free supersonic flow, in part because such duct flows are difficult to maintain shock-free over any appreciable distance. Lateral heat transfer is also slow in this type of flow. An oblique shock system would enhance the lateral energy transfer. Even though the shocks are oblique, instead of a single normal shock, Rayleigh, as well as Fanno, theory still provides correct average exit conditions since flowfield details are not required. Nevertheless, the analysis is applicable to the type of supersonic flow that occurs in chemical and gasdynamic lasers where the heat addition process is internal to the gas flow. In fact, thermal choking has been observed in the chemical laser, where its onset causes a severe drop in laser power. The Rayleigh flow analysis for this application can be found in Ref. 3. Other areas of application are laser-induced chemistry and laser isotope separation. In both technologies, a gas at low temperature and low pressure is generally required for the process. These conditions are achieved by expanding the gas through a converging/diverging nozzle. Consequently, laser energy is absorbed by the gas in a supersonic flow.[4]

Another application is the analysis of detonation waves. For instance, a gaseous mixture of methane and air will explode when ignited by a spark. The explosive process consists of a normal shock wave immediately followed by a brief but intense zone in which energy is released by the oxidation of the methane. A Mollier diagram for the process would resemble Fig. 8.5, except that points 1 and 2 are coincident on the supersonic branch, and point 4 is choked, i.e., $M_4 = 1$. The heat addition that occurs between points 3 and 4 provides the energy that drives the shock wave. The shock Mach number M_1 is not known a priori but is determined by the choking condition. In other words, the speed and strength of the shock adjusts so that state 4 is sonic.

8.3 FANNO FLOW

A constant-area duct without heat transfer is assumed. Equation (8.5c) yields

$$h + \tfrac{1}{2}V^2 = h_0 = \text{const} \tag{8.25}$$

As in Rayleigh flow, we eliminate V from this equation and from continuity to obtain the Fanno line equation

$$h + \frac{G^2}{2}\frac{1}{\rho^2} = h_0 \tag{8.26}$$

in which only thermodynamic variables appear. Figure 8.8 shows a Fanno line on a Mollier diagram for a perfect gas. The curve is for specific values of G and h_0, where $h = h_0$ is the dashed horizontal line. As was done for Rayleigh flow, we can show that $M = 1$ at the asterisk point, where the entropy is a maximum. The upper branch is subsonic and is asymptotic to h_0 as s decreases. It is worth noting that the Fanno curve shifts horizontally to the left as G increases.

As anticipated, we see from Appendix I that entropy can only increase as a result of friction. Whether on the upper or lower branch, the state point can move only to the right as the flow progresses down the duct. As a subsonic flow moves down the duct, friction increases the Mach number and, surprisingly, the flow speed V. The opposite trends occur on the supersonic branch.

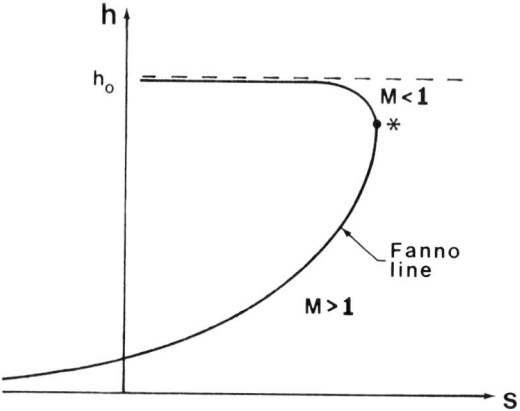

Fig. 8.8 Mollier diagram for a perfect gas Fanno line.

Under the proviso that inlet conditions are fixed, we consider subsonic or supersonic duct flow. As the duct length L increases, the outlet point moves along the Fanno line toward the sonic point. Once outlet conditions reach this point, the flow is frictionally choked. As in Rayleigh flow, the second law of thermodynamics requires that $ds \geq 0$; hence, the outlet point cannot move leftward. If a further increase in duct length occurs, inlet conditions must change or, when the inlet flow is supersonic, a normal shock wave must occur inside the duct. If inlet conditions change, the flow shifts to a new Fanno line. If the adjustment is the result of a shock, the flow stays on the same line, as we will discuss shortly.

The equations in Appendix I hold here with $dT_0 = dA = 0$. Except for the dM^2 equation, the resulting differential equations are integrated to yield the results shown in Appendix K, where the convenient notation of Appendix I is retained.

Equation (8.9) for Fanno flow becomes

$$\frac{1}{\gamma}\frac{(1-M^2)\,dM^2}{M^4\{1+[(\gamma-1)/2]M^2\}} = 4C_f\frac{dx}{D}$$

The method of partial fractions is used to perform the Mach number integral. The result is

$$4\bar{C}_f\frac{x^*-x}{D} = \frac{\gamma+1}{2\gamma}\ell n\left\{\frac{[(\gamma+1)/2]M^2}{1+[(\gamma-1)/2]M^2}\right\} + \frac{1}{\gamma}\left(\frac{1}{M^2}-1\right) \quad (8.27)$$

where x^* is the value for x when $M = 1$, and $\bar{C}_f$ is an average skin-friction coefficient

$$\bar{C}_f = \frac{1}{x-x^*}\int_{x^*}^{x} C_f\,dx$$

The value for $\overline{C}_f$ depends on whether the flow is laminar, transitional, or turbulent. It thus depends on the Reynolds number, $\rho DV/\mu$, and, when turbulent, the relative wall roughness factor e/D. It also depends on the amount of heat transfer. For subsonic, adiabatic flow, a Moody chart provides $\overline{C}_f$, as shown in Fig. 8.9. No equivalent chart exists for supersonic flow.[5]

It is useful to define a duct length sufficient for choking as

$$L_m = x^* - x \tag{8.28}$$

Using this definition in combination with Eq. (8.27), we have the first $4\overline{C}_f L_m/D$ equation in Appendix K. For a duct starting at x_1 of length L, we have

$$L = x_2 - x_1 = (x^* - x_1) - (x^* - x_2) = L_{m1} - L_{m2}$$

so that

$$4\overline{C}_f \frac{L}{D} = \left(4\overline{C}_f \frac{L_m}{D}\right)_1 - \left(4\overline{C}_f \frac{L_m}{D}\right)_2 \tag{8.29}$$

The two terms on the right depend, respectively, only on M_1 and M_2 via Eq. (8.27), which is readily tabulated. Equation (8.29) is the frictional counterpart to the q/RT_0^* equation in Appendix J.

Much of the Rayleigh flow discussion applies directly to a Fanno flow. The role of the ambient pressure p_∞ is identical in the two flows, as summarized in Chap. 7. Discussions of the black box and an oblique shock system vs a single normal shock also apply here. One different aspect is that

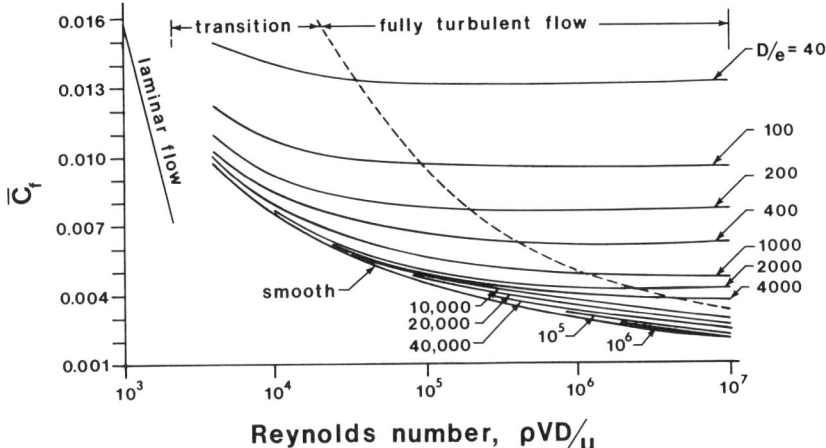

Fig. 8.9 Average skin friction coefficient as a function of Reynolds number and relative roughness.

the location of a normal shock inside a duct can be determined. The shock can be located because, unlike T_0, $4\bar{C}_f L_m/D$ is not conserved across a shock. In fact, if there is a shock inside the duct, we can move it upstream by increasing p_∞ or downstream by decreasing p_∞. In the next two examples, supersonic duct flow is analyzed with and without a shock wave.

Example 8

Air exits a supersonic axisymmetric nozzle with $M_1 = 3$ and enters an insulated, circular duct, as shown in Fig. 8.10. The plenum pressure and temperature are 3 atm and 300 K, respectively. For the duct, the skin-friction coefficient is 0.0025. In this example, the duct flow is taken as shock-free, and a subscript 2 denotes exit conditions. In Example 9, we will consider an internal shock. We are to determine L/D, $\Delta s/R$, $\mathcal{T}/A$, and $\dot{m}/A$ when the exit Mach number $M_2 = 2$.

By means of a Fanno flow table at $M_1 = 3$, we have

$$\frac{p_1}{p^*} = 0.2182, \quad \frac{p_{01}}{p_0^*} = 4.235, \quad \left(4\bar{C}_f \frac{L_m}{D}\right)_1 = 0.5222, \quad \frac{F_1}{F^*} = 1.237$$

and at $M_2 = 2$, we have

$$\frac{p_2}{p^*} = 0.4083, \quad \frac{p_{02}}{p_0^*} = 1.687, \quad \left(4\bar{C}_f \frac{L_m}{D}\right)_2 = 0.3050, \quad \frac{F_2}{F^*} = 1.123$$

Hence, L/D is given by

$$4\bar{C}_f \frac{L}{D} = \left(4\bar{C}_f \frac{L_m}{D}\right)_1 - \left(4\bar{C}_f \frac{L_m}{D}\right)_2 = 0.5222 - 0.3050 = 0.2172$$

or

$$\frac{L}{D} = \frac{0.2172}{4\bar{C}_f} = 21.7$$

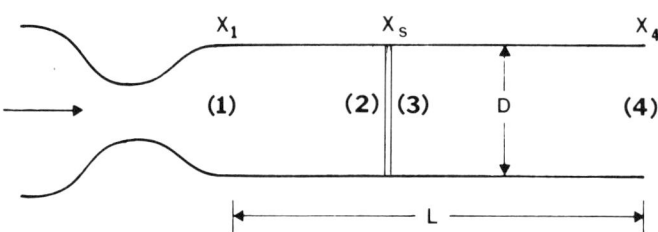

Fig. 8.10 Supersonic nozzle that feeds an insulated frictional duct containing an internal normal shock wave.

DUCTS WITH AREA CHANGE, HEAT TRANSFER, AND FRICTION 137

For $\Delta s/R$, we have

$$\frac{\Delta s}{R} = \frac{s_2 - s^*}{R} - \frac{s_1 - s^*}{R} = -\ln\frac{(p_{02}/p_0^*)}{(p_{01}/p_0^*)} = 0.920$$

Shortly, we will need the isentropic results at station 1:

$$\frac{p_1}{p_{01}} = 0.02722, \qquad \frac{T_1}{T_{01}} = 0.3571$$

The impulse function F^* [see Eq. (8.11)] is given by

$$F^* = (\gamma + 1)Ap^*$$

where $A^* = A$, since the duct has a constant cross-sectional area, and

$$p^* = \frac{p^*}{p_1}\frac{p_1}{p_{01}}p_{01} = \frac{0.02722 \times 3 \times 1.013 \times 10^5}{0.2182} = 3.792 \times 10^4 \text{ N/m}^2$$

Thus, we have

$$\frac{\mathcal{T}}{A} = \frac{F_2 - F_1}{A} = \frac{F^*}{A}\left(\frac{F_2}{F^*} - \frac{F_1}{F^*}\right)$$

$$= 2.4 \times 3.792 \times 10^4 \, (1.123 - 1.237) = -1.034 \times 10^4 \text{ N/m}^2$$

The thrust is negative, since the frictional force is in the direction of the flow. For the mass flow rate, we utilize Eq. (8.17) as follows:

$$\frac{\dot{m}}{A} = p_1\left(\frac{\gamma}{RT_1}\right)^{\frac{1}{2}}M_1 = \frac{p_1}{p_{01}}p_{01}\left(\frac{\gamma}{RT_{01}}\frac{T_{01}}{T_1}\right)^{\frac{1}{2}}M_1$$

$$= 0.02722 \times 3 \times 1.013 \times 10^5\left(\frac{1.4}{287 \times 300 \times 0.3571}\right)^{\frac{1}{2}} \times 3 = 167.5 \text{ kg/m}^2\text{-s}$$

Example 9

Consider a duct with $L/D = 60$ and an ambient pressure of 0.5 atm. The values for $\bar{C}_f$, M_1, and p_{01} are the same as in Example 8. We are to determine the exit Mach number and the location x_s/D of a normal shock.

Figure 8.10 provides a schematic of the nozzle and duct, while Fig. 8.11 is the corresponding Mollier diagram. Observe that s_3 exceeds s_2, as required by the second law. We assume the flow is not frictionally choked; hence, M_4 is subsonic. In this case,

$$p_4 = p_\infty = 0.5 \text{ atm}$$

and

$$\frac{p_4}{p^*} = \frac{p_4}{p_{01}} \frac{p_{01}}{p_1} \frac{p_1}{p^*} = \frac{0.5 \times 0.2182}{3 \times 0.02722} = 1.336$$

where p_1/p_{01} and p_1/p^* come from Example 8. With the aid of a Fanno flow table, p_4/p^* yields

$$M_4 = 0.775, \quad \left(4\bar{C}_f \frac{L_m}{D}\right)_4 = 0.100$$

The M_4 value verifies our earlier subsonic assumption.

To determine x_s/D, we need to determine M_2 or M_3. To this end, we write

$$4\bar{C}_f \frac{x_s - x_1}{D} = \left(4\bar{C}_f \frac{L_m}{D}\right)_1 - \left(4\bar{C}_f \frac{L_m}{D}\right)_2$$

$$4\bar{C}_f \frac{x_4 - x_s}{D} = \left(4\bar{C}_f \frac{L_m}{D}\right)_3 - \left(4\bar{C}_f \frac{L_m}{D}\right)_4 \tag{8.30}$$

There are three unknowns in these two equations; namely, x_s/D, M_2, and M_3. We eliminate x_s/D from the left side by adding the equations to obtain

$$4\bar{C}_f \frac{L}{D} = \left(4\bar{C}_f \frac{L_m}{D}\right)_1 - \left(4\bar{C}_f \frac{L_m}{D}\right)_2 + \left(4\bar{C}_f \frac{L_m}{D}\right)_3 - \left(4\bar{C}_f \frac{L_m}{D}\right)_4$$

With known values for $\bar{C}_f$, L/D, $(4\bar{C}_f L_m/D)_1$, and $(4\bar{C}_f L_m/D)_4$, we obtain

$$\left(4\bar{C}_f \frac{L_m}{D}\right)_3 = \left(4\bar{C}_f \frac{L_m}{D}\right)_2 + 0.1778$$

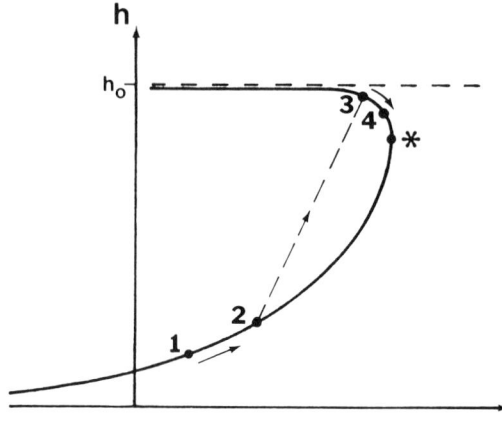

Fig. 8.11 Fanno line for Example 8. The numbering corresponds to Fig. 8.10.

DUCTS WITH AREA CHANGE, HEAT TRANSFER, AND FRICTION

A second relation between M_2 and M_3 is the normal shock equation [Eq. (8.16)], for which tables can be used. An iterative solution of the two equations, where M_3 and M_2 are bounded as follows:

$$M_3 < M_4 = 0.775, \quad 1 < M_2 < M_1 = 3$$

yields

$$M_2 = 1.74, \quad M_3 = 0.630$$

We now utilize the first of Eqs. (8.30) to obtain x_s/D as

$$\frac{x_s}{D} = \frac{1}{4\overline{C}_f}\left[\left(4\overline{C}_f \frac{L_m}{D}\right)_1 - \left(4\overline{C}_f \frac{L_m}{D}\right)_2\right] = 10^2 (0.5225 - 0.2216) = 30.1$$

The shock is thus halfway down the duct.

Recall that a normal shock satisfies

$$G_1 = G_2, \quad (F/A)_1 = (F/A)_2, \quad h_{01} = h_{02}$$

where these conditions are equivalent to $q = \tau = 0$. Consequently, for given values of G, F/A, and h_0, the Fanno and Rayleigh lines simultaneously satisfy the normal shock relations. Figure 8.12, which is for a perfect gas, indicates how this is possible. The lower branch intersection point corresponds to the upstream (supersonic) state for a normal shock, while the

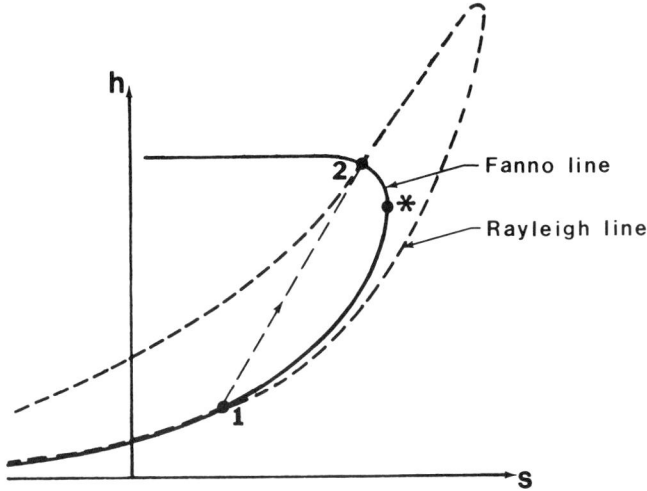

Fig. 8.12 Mollier diagram showing the Rayleigh and Fanno lines for a perfect gas. Intersection points 1 and 2 represent, respectively, the upstream and downstream states for a normal shock.

upper branch intersection point corresponds to the downstream (subsonic) state. Note that $M_2 < 1 < M_1$ and that $s_2 > s_1$ in accord with the second law.

The validity of Fanno flow is limited by the friction factor. The Moody chart is for incompressible flow. In other words, the relations

$$C_{f,\text{laminar}} = \frac{16}{Re}$$

and, when $Re > 2300$,

$$C_{f,\text{turbulent}} = C_f(Re, e/D)$$

hold for steady, fully developed, incompressible pipe flow. The data base for subsonic and supersonic compressible pipe flow is extremely limited. The von Kármán-Nikuradse turbulent relation for a smooth-walled pipe,[2,5]

$$\frac{1}{\sqrt{C_f}} = 4\log_{10}\left(\sqrt{C_f}\,Re\right) - 0.4$$

holds for a subsonic, adiabatic flow in which the Reynolds number is based on a cross-sectional average temperature of the gas. Thus, the smooth-wall, turbulent friction factor apparently does not depend on M for subsonic Fanno flow.

An additional difficulty in Fanno flow is the entrance region, where a Reynolds number based on the length along the duct is important.

As noted earlier, shock-free supersonic flow is difficult to maintain for any appreciable distance. Viscous effects are relegated to a thin boundary layer, and the outlet flow may be far from uniform. In this circumstance, a fully developed laminar or turbulent flow does not occur. The shock waves that frequently occur are a system of mostly oblique shocks distributed over several duct diameters and not a single, normal shock as the theory assumes. Regions of separated flow adjacent to the wall are also present. We know of no theory for C_f in the supersonic case, other than the assumption that it is constant at an empirically determined average value. It is worth noting that the main difficulty in a supersonic flow is the a priori inability to estimate C_f, not the presence of a shock system in lieu of a normal shock. Under the assumption of an average C_f, the Fanno theory equations are an exact solution of the conservation equations between any two uniform flow stations. In this case, a normal shock is a correct one-dimensional representation of the dissipative mechanisms associated with separated flow regions and oblique shock waves.

References

[1] Shapiro, A. H., *The Dynamics and Thermodynamics of Compressible Fluid Flow*, Vol. I, Ronald Press, New York, 1953, Chaps. 6–8.

[2] Humble, L. V., Lowdermilk, W. H., and Desmon, L. G., "Measurements of Average Heat-Transfer and Friction Coefficients for Subsonic Flow of Air in Smooth Tubes at High Surface and Fluid Temperatures," NACA Lewis Flight Propulsion Laboratory, Cleveland, OH, Rept. 1020, 1951.

[3] Emanuel, G., "Choking Analysis for a cw HF or DF Chemical Laser," *AIAA Journal*, Vol. 20, Oct. 1982, pp. 1401–1409.

[4] Emanuel, G., Cline, M. C., and Witte, K. H., "Laser Induced Disturbance with Application to a Low Reynolds Number Flow," *AIAA Journal*, Vol. 19, Feb. 1981, pp. 226–231.

[5] Keenan, J. H. and Neumann, E. P., "Measurements of Friction in a Pipe for Subsonic Flow of Air," *ASME Transactions*, Vol. 68, June 1946, pp. A-91–A-100.

Problems

8.1 Determine explicit equations for the Rayleigh and Fanno lines when the gas is perfect. Denote a fixed reference state with an r subscript.

8.2 Determine an equation for the cross-sectional area $A(x)$ of a duct that keeps M constant when heat transfer $q(x)$ occurs. Ignore friction.

8.3 Determine an equation of the form

$$4\bar{C}_f \frac{L}{D} = f(\gamma, M, q/RT_{01})$$

for a constant-area duct in which the Mach number is constant. In this case, is q positive or negative?

8.4 An air duct has a varying cross-sectional area such that in steady, adiabatic flow the velocity is constant. At the inlet section, the pressure is 5 atm and the temperature is 400 K. Two meters downstream, the pressure is 3 atm. The shear stress is proportional to the duct's radius

$$\tau = br(x)$$

Determine the constant b and the downstream temperature.

8.5 Air flows through a frictionless constant-area duct. At the entrance, the temperature is 300 K and the speed is 50 m/s. At the exit, the flow is thermally choked. Determine q, p_2/p_1, p_{02}/p_{01}, T_2, and T_{02}.

8.6 Air flows through a frictionless constant-area duct. At the entrance, the temperature is 300 K. The pressure ratio for the duct is $(p_2/p_1) = 7$. Determine q, p_{02}/p_{01}, T_{01}, T_{02}, and M_2, when (a) $M_1 = 4$ and (b) $M_1 = 0.1$.

8.7 Helium flows through a frictionless constant-area duct, where the heat transfer is $(q/RT_0^*) = 0.659$ and the pressure ratio p_2/p_1 across the duct is 13.39. Determine the inlet and exit Mach numbers.

8.8 Air flows through a frictionless constant-area duct. The exit Mach number M_2 is 0.7, and the exit stagnation temperature is 580 K. The pressure ratio p_2/p_1 across the duct is 0.6. Determine M_1, q, and T_{02}/T_{01}.

8.9 Air flows through a frictionless constant-area duct. At the entrance, the temperature is 400 K and the Mach number is 0.2. The heat transfer q to the duct is 3×10^5 J/kg. Determine the exit Mach number M_2, T_2/T_1, p_2/p_1, T_{02}/T_{01}, and p_{02}/p_{01}.

8.10 Air flows through a frictionless constant-area duct that contains a normal shock wave with a pressure ratio $p_3/p_2 = 5.48$ [see Fig. 8.4(B)]. The heat addition is

$$q_a = -3 \times 10^4 \text{ J/kg}, \qquad q_b = -8 \times 10^4 \text{ J/kg}$$

and the duct inlet temperature is 200 K. Draw a Mollier diagram for the process, and determine p_4/p_1 and T_4/T_1.

8.11 Inlet conditions T_1 and M_1 to a frictionless constant-area duct are known. Determine an equation of the form

$$\frac{\tilde{q}}{RT_1} = f(\gamma, M_1)$$

where $\tilde{q}$ is the maximum amount of cooling possible. M_1 may be subsonic or supersonic, heating is not allowed, and shocks do not occur.

8.12 Air flows through a frictionless constant-area duct. At the entrance, the temperature is 370 K and the Mach number M_1 is 0.7. At the exit, the Mach number M_2 is 0.24. Determine q, p_{02}/p_{01}, and T_{02}/T_{01}.

8.13 Air flows through a frictionless constant-area duct. At the entrance, the temperature is 370 K and the Mach number M_1 is 2.5. The pressure ratio for the duct is $(p_2/p_1) = 6$. Determine q, p_{02}/p_{01}, T_{02}/T_{01}, and M_2. Sketch the Mollier diagram for the process.

8.14 Consider an idealized ramjet engine as shown in the sketch. The fluid is air, the temperature at 1 is 300 K, and $M_1 = 3$. A normal shock sits at the inlet. Determine q such that $M_3 = 1$. Also determine the speeds V_1 and V_3, T_{03}, and p_3/p_1.

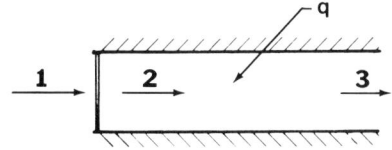

8.15 Air flows through a nozzle with an area ratio of 10.25. There is a normal shock at the exit of the nozzle, which is followed by a smooth constant-area duct with heat addition or removal. The ambient pressure is 1 atm, while the plenum pressure and temperature are 7 atm and 300 K, respectively. Determine q and the exit Mach number.

8.16 Consider a constant cross-sectional area duct with heat addition q to a perfect gas. At the inlet, $M_1 \ll 1$ and p_{01} and T_{01} are fixed. At the outlet, p_∞ is also fixed. With these constraints, derive a formula with the form

$$q = q\left(\gamma, \frac{p_{01}}{p_\infty}, \frac{\dot{m}}{p_\infty A}\right)$$

8.17 The flow of air at the end of a frictionless duct is choked due to a heat addition of 3×10^5 J/kg. At the end of the duct, the temperature is 800 K. Determine the inlet temperature T_1 and Mach number M_1.

8.18 Air enters the entrance to a 20-cm-diam duct with 2-atm pressure, 400 K temperature, and a speed of 100 m/s. How much heat, in units of kJ/s, can be added so that the flow achieves its maximum possible static temperature? Also, determine the final Mach number, temperature, pressure, and mass flow rate.

8.19 Supersonic combustion occurs in a gas with $\gamma = 1.5$. The initial Mach number is $M_1 = 4$. Determine q/RT_1, T_2/T_1, p_2/p_1, and p_{02}/p_{01} such that the exit flow is thermally choked.

8.20 A mixture of monatomic gases enters a frictionless constant-area duct with $p_{01} = 2$ atm, $T_{01} = 500$ K, and $(\dot{m}/A) = 63.31$ kg/s-m². It is found that a heat addition of 5×10^5 J/kg just chokes the flow at the exit of the duct. Determine the inlet Mach number M_1 and the molecular weight of the mixture.

8.21 Nitrogen flows in a constant-area duct. Conditions at section 1 are $p_1 = 12$ atm, $T_1 = 300$ K, and $V_1 = 90$ m/s. At section 2, we find $p_2 = 9$ atm. Determine V_2 and the heat transfer, q, between sections.

8.22 Air flows in a frictionless duct that contains a normal shock, as shown in Fig. 8.4(B). The heat transfer between sections 1 and 2 is 3.551×10^4 J/kg, the Mach number just downstream of the shock is 0.441, the pressure ratio p_4/p_1 is 16.87, and T_4 is 700 K. Determine M_4 and q_b.

8.23 For Rayleigh flow with the assumptions: (1) turbulent flow, (2) T_w = const, (3) Pr = const, and (4) μ and k = const, determine an analytical formula for $M = M(x)$. Use $x = x^*$, $M = 1$ as a reference condition.

8.24 Consider combustion of H_2 with air when the mixture has sufficient O_2 to form water vapor (an oxygen-rich mixture). In this case, we have

$$w = \frac{\dot{m}_{H_2}}{\dot{m}_{air}} \leq 0.0072$$

Within this constraint, the heat addition is $q = 1.21 \times 10^8 w$, J/kg, of gas. Since w is a small number, we can ignore any changes in W or γ. Determine equations for the maximum and minimum possible inlet Mach number M_m as a function of w. Assume the static inlet temperature T_1 is a constant. [A maximum for M_m occurs when the combustion is subsonic (ramjet case), while a minimum occurs when combustion is supersonic (scramjet case).]

8.25 For Rayleigh flow you are given

$$\frac{dq}{dx} = aT^{\frac{3}{2}}$$

where a is a constant, and x is the distance along the duct. Determine $M = M(x)$.

8.26 Air flows through an insulated duct of length 4 m and diameter 10 cm. The entrance and exit Mach numbers are 3 and 1.20. Determine $\bar{C}_f$, p_{02}/p_{01}, T_{02}/T_{01}, and p_2/p_1.

8.27 Use the same inlet, duct size, and $\bar{C}_f$ conditions as in Problem 8.26. If there is a shock wave halfway down the duct, determine the exit Mach number M_4 and the ratios p_{04}/p_{01} and p_4/p_1.

8.28 Determine the mass flow rate $\dot{m}$ and the thrust $\mathscr{T}$ on the duct for the flows in Problems 8.26 and 8.27. In both cases, at the inlet we have $p_{01} = 0.1$ atm and $T_{01} = 600$ K.

8.29 Conditions at the entrance of an insulated duct are $M_1 = 5$ and $p_1 = 5 \times 10^4$ N/m². After a certain length, the flow has reached $M_2 = 2$. Determine p_2, $4\bar{C}_f L/D$, T_2/T_1, and T_{02}/T_{01} if $\gamma = 1.4$.

8.30 An insulated 5-m duct, 10 cm in diameter, contains oxygen flowing at a rate of 12 kg/s. Measurements at the inlet give $p_1 = 2.5$ atm, $T_1 = 400$ K and, at the outlet, $p_2 = 3.2$ atm. Determine M_1, M_2, $\bar{C}_f$, T_{02}, p_{02}, and the thrust $\mathscr{T}$ on the duct.

8.31 Do not assume that $\bar{C}_f$ is a constant for steady, adiabatic, laminar flow in a pipe. Assume that viscosity is proportional to the temperature. Derive the counterpart to the $4\bar{C}_f L_m/D$ equation for this situation, using the $M^* = 1$ reference, as before.

8.32 Air flows isentropically through a nozzle with an area ratio of 5 and a stagnation pressure of 7 atm. The insulated downstream duct has $4\bar{C}_f L/D$

= 0.5. With the numbering in Fig. 8.10, determine p_∞ when (a) there is a normal shock at the duct entrance, (b) there is a normal shock at the duct exit. For both cases, sketch the nozzle-duct process on a Mollier diagram.

8.33 Air isentropically flows through a converging/diverging nozzle of area ratio 7.8. An adiabatic duct with a frictional length $(4\bar{C}_f L/D) = 0.6$ is attached to the nozzle. A normal shock wave sits halfway down the duct. Determine the duct exit Mach number and p_∞/p_0, where p_0 is the nozzle's plenum pressure.

8.34 Air with a stagnation temperature of 500 K flows through a nozzle, which is followed by an insulated duct with an $(L/D) = 50$ and a skin-friction coefficient of 1.4×10^{-3}. A normal shock sits halfway down the duct, the static temperature of the gas at the exit of the duct is 450 K, and the ambient pressure is 1 atm. Determine the nozzle's area ratio and stagnation pressure p_0.

8.35 A large chamber contains air at 6.5×10^6 N/m², which passes through an isentropic converging-only nozzle and then into an insulated duct of constant cross-sectional area. The friction length of the duct is $(4\bar{C}_f L/D) = 1.067$, and the exit Mach number M_2 is 0.96.
 (a) Draw and label a Mollier diagram for the system.
 (b) Determine the Mach number M_1 at the duct entrance.
 (c) Determine the ambient pressure p_∞.

8.36 Your boss wants you to design an isentropic converging nozzle to feed an existing long, insulated pipe. The flow in the pipe is turbulent, so that $\bar{C}_f$ = const, the gas is perfect, and stagnation conditions to the nozzle are fixed. You are to design the nozzle so that the mass flow rate through the pipe is a maximum. Derive a relation for the nozzle's exit Mach number that does this. (The ambient pressure cannot be fixed.)

8.37 An inviscid nozzle with an area ratio of 3.5 feeds air into an insulated duct with an L/D of 30 and a skin-friction coefficient of 0.0015. The plenum pressure is 7 atm, and the ambient pressure is 1 atm. Determine the inlet and exit Mach numbers of the duct.

8.38 A constant-area duct with $\bar{C}_f = 0.004$ and $L/D = 50$ is choked at the exit and has a normal shock wave at the entrance. For a monatomic gas, determine the Mach number M_∞ ahead of the shock.

8.39 A 3-m duct that is 10 cm in diameter contains air flowing adiabatically at a rate of 3 kg/s. Measurements at the inlet give $p_1 = 0.2$ atm and $T_1 = 250$ K and, at the outlet, $p_4 = 2$ atm and the skin friction coefficient is $\bar{C}_f = 0.007$. Determine M_1, M_2, M_3, and M_4 as shown in Fig. 8.4(B).

8.40 A natural-gas well (methane; $\gamma = 1.32$, $W = 16$ kg/kmole) is 5 km deep with a diameter of 15 cm. Assume a steady, fully turbulent flow and a wall roughness ε of 3.75×10^{-4} m. At the pipe's inlet, the gas is at 600 K and 600 atm and, at the top of the well, the flow is choked. Determine $\dot{m}$ and the inlet Mach number M_1.

8.41 A natural-gas (methane) pipeline has a pressure of 10^6 N/m² and a 300 K temperature at its inlet, where the average speed is 150 m/s. The pipeline has a 10-cm diameter and is 70 m long with a $\bar{C}_f$ of 1.3×10^{-3}. Determine the entrance Mach number M_1, the exit Mach number M_2, and $\dot{m}$.

8.42 Air enters a duct with $M_1 = 4$. The duct has an $L/D = 50$ and $\bar{C}_f = 0.00316$. Let p_{01} and p_{04} be the inlet and the outlet stagnation pressures, and p_∞ be the ambient pressure. Determine p_∞/p_{01}, p_{04}/p_{01}, and $\mathcal{T}/(p_{01}A)$ when a normal shock sits at (a) the inlet, (b) midway down the duct, and (c) the exit. Tabulate your results.

8.43 Air flows through a constant cross-sectional area duct, where section a has heat addition and section b is insulated and consists of a series of fine screens that produces a constant skin-friction coefficient C_f. Between sections 2 and 3 there is a normal shock. Measurements show that $M_1 = 3$, $p_{01} = 12$ atm, $M_4 = 0.8$, and $p_{04} = 3$ atm. Determine q/RT_0^* and $4\bar{C}_f L/D$.

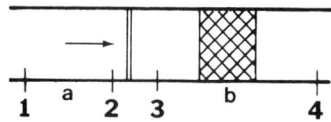

Computational Problems:

8.44 For $\gamma = 1.4$ and $5/3$ and for $M = 0.01$ (0.01) 5(0.1) 10, generate a table for Rayleigh flow, utilizing the equations in Appendix J.

8.45 Repeat Problem 8.44 for Fanno flow, utilizing the equations in Appendix K.

9. UNSTEADY, ONE-DIMENSIONAL FLOW

This chapter is concerned with unsteady motion under the assumptions of inviscid, adiabatic, one-dimensional flow of a perfect gas, with $dA = 0$. We will cover four topics: unsteady normal shock waves in Sec. 9.1, reflected normal shock waves in Sec. 9.2, the method of characteristics in Sec. 9.3, and unsteady expansion waves in Sec. 9.4. The first two topics are important for understanding the role of a moving shock wave in a flow. The third topic is important because the method of characteristics is a mathematical technique whose application extends well beyond unsteady flow. This method is used to obtain the expansion wave solutions of the final section. (Chapter 16 should be consulted for a more complete discussion of the theory of characteristics.) The various flows analyzed are the building blocks for the applications discussed in the next chapter. Appendices L and M contain a brief summary of the unsteady shock wave equations.

9.1 NORMAL SHOCK WAVES

In many situations, the shock wave is moving. For example, a discharge of lightning or an explosion generates a moving shock. If it travels at a constant speed, a simple transformation reduces the moving case to a steady one. Even when the speed of the shock wave is changing, it can be analyzed at each instant as a steady shock because its thickness is negligible. (This aspect is discussed in Chap. 12.)

The simplest way to generate a moving shock is with a piston/cylinder arrangement, as shown in Fig. 9.1. At time $t = 0$, a piston moves impulsively into a quiescent gas with a constant speed V_p'. (A prime is used to distinguish an unsteady flow from a steady one containing a standing shock wave.) The motion of the piston results in a shock wave with a constant speed V_s', which exceeds that of the piston. As a result, a slug of gas develops between the piston's face and the shock wave. This gas moves with the same speed as the piston. An x-t diagram, Fig. 9.2, illustrates this discussion. Region 1 is quiescent, while a uniform flow with speed V_p' occurs in region 2. The figure shows a typical particle path that parallels the piston's path in region 2.

For a finite piston speed, the shock is a finite strength disturbance. Since a disturbance of infinitesimal strength moves with the speed of sound, a finite strength disturbance moves faster. Hence, V_s' exceeds the speed of sound a_1'. We introduce a shock wave Mach number

$$M_s = V_s'/a_1' \tag{9.1}$$

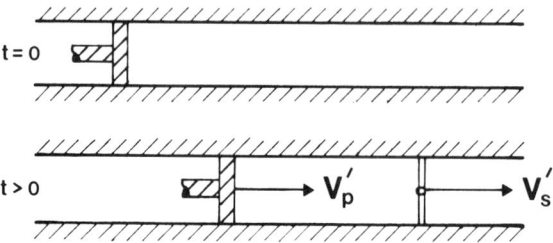

Fig. 9.1 Normal shock wave generated by the motion of a piston.

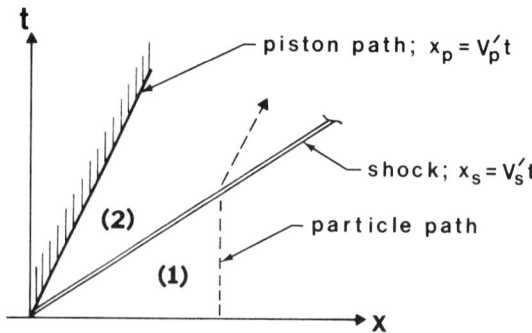

Fig. 9.2 Corresponding x-t diagram to the impulsive piston motion of Fig. 9.1.

which equals or exceeds unity and is used extensively in the subsequent analysis. A prime is not shown on M_s for notational simplicity in the later equations.

As long as the piston has a constant speed, so will the shock wave. The moving shock is converted into a standing shock by using a frame of reference fixed with the shock wave. Consider the left-hand sketch for the moving shock in Fig. 9.3. The speeds V_1' ($=0$) and V_2' ($=V_p'$) are oriented rightward. In the unsteady reference frame, the flow behind the shock moves in the same direction as the shock. On the right side of the figure, we have a conventional, standing normal shock in which the flow is from right to left. The flow behind the shock, with speed V_2, moves away from the shock.

We use the convention that any speed in the rightward, or positive x, direction is positive. To convert the moving shock into a steady one, we subtract the shock speed from all speeds in the moving shock system. Thus, we have

$$V_1 = -V_s', \qquad V_2 = V_2' - V_s'$$

where V_1 and V_2 are negative. In order to avoid negative Mach numbers, and to have the flow oriented as in Fig. 5.1, we interchange regions 1 and 2.

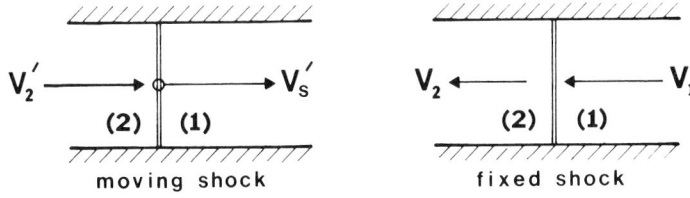

Fig. 9.3 Moving and fixed reference frames for a shock wave.

This is equivalent to multiplying the left side of the above equations by a minus one, to obtain

$$V_1 = V_s', \qquad V_2 = V_s' - V_2' \tag{9.2}$$

Imposition of the speed $-V_s'$ everywhere on the flow is similar to what was done in Sec. 5.2, where, in order to analyze an oblique shock wave, a constant tangential speed was imposed.

The use of a coordinate system fixed with the shock wave transforms an unsteady problem into a steady one. We solve the steady normal shock problem in the usual manner and then revert to the original reference frame. As with oblique shock waves, all static thermodynamic conditions on both sides of the shock are unaltered. Consequently,

$$p_1' = p_1, \qquad p_2' = p_2, \qquad a_1' = a_1, \qquad s_2' = s_2, \ldots \tag{9.3}$$

Usually, we will not use primes on static variables. However, stagnation variables are different in the two reference frames and the use of primes is essential.

As usual, we begin by concentrating on Mach numbers. In shock fixed coordinates, we have only two Mach numbers. These are given by

$$M_1 = \frac{V_1}{a_1} = \frac{V_s'}{a_1'} = M_s$$

and

$$M_2^2 = \left(\frac{V_2}{a_2}\right)^2 = \frac{1 + [(\gamma - 1)/2] M_1^2}{\gamma M_1^2 - [(\gamma - 1)/2]} = \frac{1 + [(\gamma - 1)/2] M_s^2}{\gamma M_s^2 - [(\gamma - 1)/2]} \tag{9.4}$$

Hence, M_1 and M_s are the same. Our procedure here and in Sec. 9.2 will be to use M_s as the independent parameter. We therefore proceed to relate V_p' to M_s as follows:

$$V_p' = V_2' = V_s' - V_2$$

so that

$$\frac{V_p'}{a_1} = \frac{V_s'}{a_1'} - \frac{a_2}{a_1} \frac{V_2}{a_2} = M_s - \left(\frac{T_2}{T_1}\right)^{\frac{1}{2}} M_2$$

where Eq. (9.2) is utilized. We next use Eq. (9.4) and the normal shock equation for T_2/T_1 (see Appendix D)

$$\frac{T_2}{T_1} = \left(\frac{2}{\gamma+1}\right)^2 \frac{1}{M_s^2}\left(1 + \frac{\gamma-1}{2}M_s^2\right)\left(\gamma M_s^2 - \frac{\gamma-1}{2}\right) \quad (9.5)$$

to obtain, after simplification,

$$\frac{V_p'}{a_1} = \frac{2}{\gamma+1}\frac{M_s^2 - 1}{M_s} \quad (9.6)$$

This relation is easily solved for M_s when V_p'/a_1 is known.

In the unsteady frame, the preshock and postshock Mach numbers are

$$M_1' = \frac{V_1'}{a_1} = 0$$

$$M_2' = \frac{V_2'}{a_2} = \frac{a_1}{a_2}\frac{V_p'}{a_1} = \left(\frac{T_1}{T_2}\right)^{\frac{1}{2}}\frac{2}{\gamma+1}\frac{M_s^2-1}{M_s}$$

$$= \frac{M_s^2 - 1}{\{1 + [(\gamma-1)/2]M_s^2\}^{\frac{1}{2}}\{\gamma M_s^2 - [(\gamma-1)/2]\}^{\frac{1}{2}}} \quad (9.7)$$

Next, using p_1 and T_1 as reference conditions, relations are developed for the various pressures and temperatures. In the steady shock frame, we obtain

$$\frac{p_2}{p_1} = \frac{2}{\gamma+1}\left(\gamma M_s^2 - \frac{\gamma-1}{2}\right) \quad (9.8a)$$

$$\frac{p_{01}}{p_1} = \left(1 + \frac{\gamma-1}{2}M_s^2\right)^{\gamma/(\gamma-1)} \quad (9.8b)$$

$$\frac{p_{02}}{p_1} = \frac{p_{02}}{p_2}\frac{p_2}{p_1} = \left(\frac{\gamma+1}{2}\right)^{(\gamma+1)/(\gamma-1)}\frac{M_s^{2\gamma/(\gamma-1)}}{\{\gamma M_s^2 - [(\gamma-1)/2]\}^{1/(\gamma-1)}} \quad (9.8c)$$

where the isentropic pressure relation and Eq. (9.4) are used for p_{02}/p_2. The corresponding results in the unsteady frame are

$$\frac{p_2'}{p_1'} = \frac{p_2}{p_1} \quad (9.9a)$$

$$\frac{p_{01}'}{p_1} = 1 \quad (9.9b)$$

UNSTEADY, ONE-DIMENSIONAL FLOW

$$\frac{p'_{02}}{p_1} = \frac{p_2}{p_1}\frac{p'_{02}}{p'_2} = \frac{p_2}{p_1}\left(1 + \frac{\gamma-1}{2}M'^2_2\right)^{\gamma/(\gamma-1)}$$

$$= \left\{\frac{(\gamma+1)/2}{\gamma M_s^2 - [(\gamma-1)/2]}\right\}^{1/(\gamma-1)} \times \left\{M_s^2\frac{(\gamma-1)M_s^2 + [(3-\gamma)/2]}{1 + [(\gamma-1)/2]M_s^2}\right\}^{\gamma/(\gamma-1)}$$

(9.9c)

where Eqs. (9.7) and (9.8a) are needed for Eq. (9.9c).
Although we can write Eq. (9.8a) as

$$\frac{p_2}{p_1} = \frac{2}{\gamma+1}\left(\gamma M_1^2 - \frac{\gamma-1}{2}\right)$$

the analogous relation

$$\frac{p'_2}{p'_1} = \frac{2}{\gamma+1}\left(\gamma M'^2_1 - \frac{\gamma-1}{2}\right)$$

is incorrect. Why? From Eqs. (9.3), we have

$$p'_1 = p_1, \qquad p'_2 = p_2$$

but the equations

$$p'_{01} = p_{01}, \qquad p'_{02} = p_{02}$$

are incorrect. On the other hand, any isentropic relation, such as

$$\frac{p'_{02}}{p'_2} = \left(1 + \frac{\gamma-1}{2}M'^2_2\right)^{\gamma/(\gamma-1)}, \qquad \frac{p_{02}}{p_2} = \left(1 + \frac{\gamma-1}{2}M_2^2\right)^{\gamma/(\gamma-1)}$$

are point relations and hold in a steady or unsteady flow.

We next consider temperatures, where $(T'_2/T'_1) = (T_2/T_1)$ and where the last ratio is given by Eq. (9.5). For the stagnation temperatures, we have

$$\frac{T'_{01}}{T_1} = 1$$

$$\frac{T'_{02}}{T_1} = \frac{T_2}{T_1}\frac{T'_{02}}{T'_2} = \frac{2}{\gamma+1}\left[(\gamma-1)M_s^2 - \frac{3-\gamma}{2}\right] \qquad (9.10)$$

The stagnation temperature ratio T'_{02}/T'_{01} across an unsteady shock is not unity, except when $M_s = 1$ or $\gamma = 1$. It is essential to keep this difference with a steady flow shock wave in mind. The next example illustrates how many parameters are found for a moving shock wave.

Example 1

A shock travels into quiescent 300 K argon at a speed of 3.225×10^3 m/s. We are to determine:

$$M_s, \quad M_2, \quad M_2'$$

$$p_2/p_1, \quad p_{02}/p_{01}, \quad p_{02}'/p_{01}'$$

$$T_2/T_1, \quad T_{02}/T_{01}, \quad T_{02}'/T_{01}'$$

We first find M_s, as follows:

$$\gamma = \frac{5}{3}, \quad R = \frac{8314}{40} = 208 \text{ J/kg-K}$$

$$a_1 = (\gamma R T_1)^{\frac{1}{2}} = \left(\frac{5}{3} \times 208 \times 300\right)^{\frac{1}{2}} = 322.5 \text{ m/s}$$

$$M_s = \frac{V_s'}{a_1} = 10$$

Appendix L is used and, to simplify the equations, we introduce the approximation $M_s^2 \gg 1$. (This approximation is useful whenever M_s is roughly 7 or larger.) For M_2, Eq. (9.4) is used, with the result

$$M_2 = \left\{ \frac{1 + [(\gamma - 1)/2] M_s^2}{\gamma M_s^2 - [(\gamma - 1)/2]} \right\}^{\frac{1}{2}}$$

$$= \left\{ \frac{[(\gamma - 1)/2] + (1/M_s^2)}{\gamma - [(\gamma - 1)/2](1/M_s^2)} \right\}^{\frac{1}{2}} \cong \left(\frac{\gamma - 1}{2\gamma}\right)^{\frac{1}{2}} = 0.497$$

while Eq. (9.7) yields

$$M_2' \cong \left[\frac{2}{\gamma(\gamma - 1)}\right]^{\frac{1}{2}} = 1.342$$

Observe that M_2' can be below or above unity. For example, when we have $M_s = 1$, then $M_2' = 0$, while the maximum value of M_2' is 1.342 when $\gamma = 5/3$.

The pressure ratios called for are given by

$$\frac{p_2}{p_1} \cong \frac{2\gamma}{\gamma + 1} M_s^2 = 125$$

$$\frac{p_{02}}{p_{01}} \cong \left[\frac{\gamma+1}{2\gamma}\left(\frac{\gamma+1}{\gamma-1}\right)^\gamma \frac{1}{M_s^2}\right]^{1/(\gamma-1)} = 0.0229$$

$$\frac{p'_{02}}{p'_{01}} = \frac{p'_{02}}{p'_1} \cong 2\left(\frac{\gamma+1}{\gamma}\right)^{1/(\gamma-1)} M_s^2 = 404.8$$

It is worth noting that p'_{02}/p'_{01} greatly exceeds unity, whereas p_{02}/p_{01} is well below unity.

The temperature ratios called for are given by

$$\frac{T_2}{T_1} \cong \frac{2\gamma(\gamma-1)}{(\gamma+1)^2} M_s^2 = 31.25$$

$$\frac{T_{02}}{T_{01}} = 1$$

$$\frac{T'_{02}}{T'_{01}} = \frac{T'_{02}}{T_1} \cong \frac{2(\gamma-1)}{\gamma+1} M_s^2 = 50$$

In contrast to p_{02}/p_{01} and T_{02}/T_{01}, both p'_{02}/p_1 and T'_{02}/T_1 greatly exceed unity. (Remember that $p_1 = p'_1 = p'_{01}$ so that $p'_{02}/p_1 = p'_{02}/p'_{01}$, with a similar result for T'_{02}/T_1.) In fact, we have $p'_{02} > p'_2$ and $T'_{02} > T'_2$ because of V'^2_2. More precisely, for the stagnation temperature, we have

$$c_p T'_{02} = c_p T'_2 + \tfrac{1}{2} V'^2_2$$

which becomes

$$\frac{T'_{02}}{T_1} = \frac{T_2}{T_1} + \frac{\gamma-1}{2}\left(\frac{V'_2}{a_1}\right)^2$$

Consequently, T'_{02}/T_1 is the sum of the static temperature increase plus an increase due to the shock-induced velocity. This last equation can be transformed into Eq. (9.10).

9.2 REFLECTED NORMAL SHOCK WAVES

When a shock meets a wall head on, it reflects as a normal shock wave. (This is the simplest type of shock wave reflection process. Other types are discussed in Chap. 19.) Figure 9.4 is an x-t diagram for the flow. We see that the reflected shock moves into the gas previously processed by the incident shock and that the flow speed is zero in both regions 1 and 3. An r or 3 subscript is used to denote reflected conditions. Figure 9.5 shows the flow speeds in a duct that corresponds to Fig. 9.4. Note that the reflected shock speed V'_r is negative.

Normal reflected shock waves occur in a shock tube, which is the subject of Sec. 10.1. They also occur in blast problems, such as grain silo explosions. Although the unsteady pressure levels behind the incident shock, as in Example 1, are quite large, it is the pressure level behind the reflected shock in region 3 that causes severe structural damage. As we shall see, the pressure and temperature in region 3 exceed that in region 2.

As was done for the incident shock, the flow is transformed into a steady one by adding $-V'_r$, which is positive, to the flow speeds on both sides of the shock. Our result for the standing shock in Fig. 9.6 is

$$\hat{V}_2 = V'_2 - V'_r, \qquad V_3 = -V'_r \qquad (9.11)$$

The upstream speed is denoted as $\hat{V}_2$ to distinguish it from the different V_2 speed shown in Fig. 9.3, which is for a fixed incident shock wave. Here, the reflected shock wave is fixed. The V_3 equation means that the region 3 flow speed V'_3 is zero and that

$$p'_{03} = p'_3 = p_3, \qquad T'_{03} = T'_3 = T_3, \ldots \qquad (9.12)$$

for all thermodynamic variables in this region.

Next, we introduce the reflected steady shock Mach number M_r by defining

$$M_r = \hat{V}_2/a_2 \qquad (9.13)$$

The unsteady Mach number for region 2 is M'_2, which is given by Eq. (9.7).

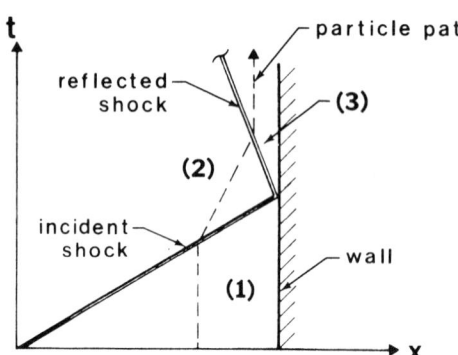

Fig. 9.4 Reflected shock wave x-t diagram.

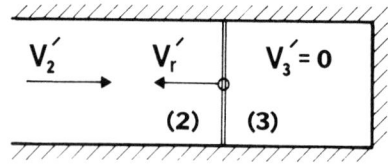

Fig. 9.5 Reflected shock wave in a duct with a closed end.

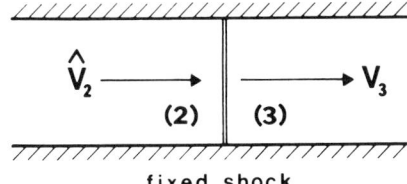

Fig. 9.6 Fixed reference frame for a reflected shock wave.

Our first task is to relate M_r to M_s. This is done by deriving two relations for the speed ratio $\hat{V}_2/V_3$. The first one is

$$\frac{\hat{V}_2}{V_3} = \frac{\hat{V}_2}{\hat{V}_2 - V_2'} = \frac{\hat{V}_2/a_2}{(\hat{V}_2/a_2) - (V_2'/a_2)} = \frac{M_r}{M_r - M_2'}$$

in which Eqs. (9.11) are used. The second equation for the speed ratio is given by the normal shock equation in Appendix D,

$$\frac{\hat{V}_2}{V_3} = \frac{\gamma + 1}{2} \frac{M_r^2}{1 + [(\gamma - 1)/2] M_r^2}$$

Elimination of $\hat{V}_2/V_3$ from these equations results in

$$M_r^2 - \left(\frac{\gamma + 1}{2} M_2'\right) M_r - 1 = 0$$

Since $M_r \geq 1$, the appropriate root is

$$M_r = \frac{\gamma + 1}{4} M_2' + \left[1 + \left(\frac{\gamma + 1}{4} M_2'\right)^2\right]^{\frac{1}{2}}$$

We eliminate M_2' by means of Eq. (9.7) to obtain

$$M_r = \left\{\frac{\gamma M_s^2 - [(\gamma - 1)/2]}{1 + [(\gamma - 1)/2] M_s^2}\right\}^{\frac{1}{2}} \tag{9.14}$$

the desired Mach number relation. Note that $M_r M_2 = 1$, where M_2 is given by Eq. (9.4). The M_r Mach number is rather narrowly bounded, as can be seen from

$$M_r = 1, \qquad M_s = 1$$

$$= \left(\frac{2\gamma}{\gamma - 1}\right)^{\frac{1}{2}}, \qquad M_s \to \infty$$

156   GASDYNAMICS: THEORY AND APPLICATIONS

For subsequent derivations, it is useful to note that

$$\gamma M_r^2 - \frac{\gamma-1}{2} = \frac{\gamma+1}{2} \frac{[(3\gamma-1)/2] M_s^2 - (\gamma-1)}{1 + [(\gamma-1)/2] M_s^2} \qquad (9.15a)$$

$$1 + \frac{\gamma-1}{2} M_r^2 = \frac{\gamma+1}{2} \frac{(\gamma-1) M_s^2 + [(3-\gamma)/2]}{1 + [(\gamma-1)/2] M_s^2} \qquad (9.15b)$$

All pressures and temperatures are again referenced to p_1 and T_1, respectively. In view of Eqs. (9.12) and $V_3' = 0$, we have

$$\frac{p_3}{p_1} = \frac{p_{03}'}{p_1} = \frac{p_3}{p_2} \frac{p_2}{p_1} = \left[\frac{2}{\gamma+1}\left(\gamma M_r^2 - \frac{\gamma-1}{2}\right)\right]\left[\frac{2}{\gamma+1}\left(\gamma M_s^2 - \frac{\gamma-1}{2}\right)\right]$$

$$= \frac{2}{\gamma+1}\left\{\frac{\gamma M_s^2 - [(\gamma-1)/2]}{1 + [(\gamma-1)/2] M_s^2}\right\}\left[\frac{3\gamma-1}{2} M_s^2 - (\gamma-1)\right] \qquad (9.16)$$

where Eq. (9.15a) is used. For the temperature, we have

$$\frac{T_3}{T_1} = \frac{T_{03}'}{T_1} = \frac{T_3}{T_2} \frac{T_2}{T_1}$$

$$= \left[\left(\frac{2}{\gamma+1}\right)^2 \frac{1}{M_r^2}\left(1 + \frac{\gamma-1}{2} M_r^2\right)\left(\gamma M_r^2 - \frac{\gamma-1}{2}\right)\right]$$

$$\times \left[\left(\frac{2}{\gamma+1}\right)^2 \frac{1}{M_s^2}\left(1 + \frac{\gamma-1}{2} M_s^2\right)\left(\gamma M_s^2 - \frac{\gamma-1}{2}\right)\right]$$

$$= \left(\frac{2}{\gamma+1}\right)^2 \frac{1}{M_s^2}\left[(\gamma-1) M_s^2 + \frac{3-\gamma}{2}\right]\left[\frac{3\gamma-1}{2} M_s^2 - (\gamma-1)\right] \qquad (9.17)$$

where Eqs. (9.15) are used. In the following example, we illustrate the above approach by continuing with Example 1.

Example 2

The shock wave of Example 1 reflects normally from a wall. We are to determine M_r, p_3/p_1, and T_3/T_1.

Again, the large M_s^2 approximation is used. Accordingly, we have

$$M_r = \frac{1}{M_2} \cong \left(\frac{2\gamma}{\gamma-1}\right)^{\frac{1}{2}} = 2.24$$

UNSTEADY, ONE-DIMENSIONAL FLOW

Note that M_r is considerably smaller than M_s; thus, a large M_r approximation would not be valid. For the pressure and temperature ratios, we have

$$\frac{p_3}{p_1} \cong \frac{2\gamma(3\gamma - 1)}{\gamma^2 - 1} M_s^2 = 750$$

$$\frac{T_3}{T_1} \cong \frac{2(\gamma - 1)(3\gamma - 1)}{(\gamma + 1)^2} M_s^2 = 75$$

Observe that p_3/p_1 is significantly larger than either the unsteady static or stagnation pressure ratios of Example 1. A similar result holds for T_3/T_1. These ratios are somewhat unrealistic because at these magnitudes real-gas effects may become important. The error is not large for a monatomic gas, such as argon, but would be for any gas composed of polyatomic molecules, such as air.

9.3 METHOD OF CHARACTERISTICS

Inviscid fluid mechanics often deals with a second-order, quasi-linear differential equation in two independent variables. (A quasi-linear partial differential equation is one that is linear in the highest order derivatives.) The general form for such an equation is

$$A\phi_{xx} + 2B\phi_{xy} + C\phi_{yy} = D \qquad (9.18)$$

where ϕ is the dependent variable, x and y are the independent variables,

$$A, B, C, D = f(x, y, \phi, \phi_x, \phi_y)$$

and

$$\phi_x = \frac{\partial \phi}{\partial x}, \qquad \phi_{xx} = \frac{\partial^2 \phi}{\partial x^2}, \ldots$$

Partial differential equations of this type fall into one of three categories:

$$\text{Elliptic if } B^2 - AC < 0$$

$$\text{Parabolic if } B^2 - AC = 0 \qquad (9.19)$$

$$\text{Hyperbolic if } B^2 - AC > 0$$

The nature of the solution changes drastically with type. If A, B, and C depend only on x and y, and D is of the form

$$D = D_0(x, y) + D_1(x, y)\phi + D_2(x, y)\phi_x + D_3(x, y)\phi_y$$

then Eq. (9.18) is linear, and superposition of solutions applies. However, if the equation is quasi-linear, but not linear, then superposition is no longer applicable.

As an illustration, Laplace's equation

$$\frac{\partial^2 \phi}{\partial x^2} + \frac{\partial^2 \phi}{\partial y^2} = 0$$

has

$$A = 1, \quad B = 0, \quad C = 1, \quad D = 0$$

and is elliptic, because

$$B^2 - AC = -1$$

The method of characteristics (MOC) is applicable to a hyperbolic (or parabolic) equation. Let us investigate whether the one-dimensional unsteady equations are hyperbolic and thus susceptible to a solution by the MOC. We start with Eqs. (3.8b), (3.9), and (3.10b), where

$$\frac{D}{Dt} = \frac{\partial}{\partial t} + V\frac{\partial}{\partial x}, \quad \frac{DA}{Dt} = 0$$

We obtain for the conservation equations:

$$\frac{\partial \rho}{\partial t} + \frac{\partial(\rho V)}{\partial x} = 0$$

$$\frac{\partial V}{\partial t} + V\frac{\partial V}{\partial x} + \frac{1}{\rho}\frac{\partial p}{\partial x} = 0 \qquad (9.20)$$

$$\frac{\partial h_0}{\partial t} + V\frac{\partial h_0}{\partial x} - \frac{1}{\rho}\frac{\partial p}{\partial t} = 0$$

where, for a perfect gas,

$$p = \rho RT$$

$$h_0 = \frac{\gamma}{\gamma - 1}RT + \frac{1}{2}V^2$$

$$a = \left(\frac{\partial p}{\partial \rho}\right)_s^{\frac{1}{2}} = (\gamma RT)^{\frac{1}{2}} = (\gamma p/\rho)^{\frac{1}{2}}$$

It is important to note that h_0 and therefore T_0 and a_0 are not constants.

The foregoing relations were the basis of the analysis in Sec. 3.5, where a solution was obtained by using the acoustic approximation, which linearizes

UNSTEADY, ONE-DIMENSIONAL FLOW

the equations. Here, linearization will not be used. Nevertheless, both approaches yield wave solutions.

We can always use the thermodynamic relation $p = p(\rho, s)$ for the pressure. However, the flow is inviscid and adiabatic and, therefore, isentropic except across a shock wave. Hence, with $s = $ const, the pressure only depends on the density

$$p = p(\rho)$$

which can be used in place of the energy equation. For a perfect gas, this equation is the isentropic relation

$$\frac{p}{p_4} = \left(\frac{\rho}{\rho_4}\right)^\gamma \qquad (9.21)$$

where the subscript 4 refers to a fixed, quiescent state for the gas. (The subscript numbering is designed to coincide with that used in Sec. 10.1.) The energy equation is replaced with Eq. (9.21), and then the pressure is eliminated, as follows:

$$\frac{\partial p}{\partial x} = \frac{dp}{d\rho}\frac{\partial \rho}{\partial x} = \left(\frac{\partial p}{\partial \rho}\right)_s \frac{\partial \rho}{\partial x} = a^2 \frac{\partial \rho}{\partial x}$$

Two equations are thereby obtained for the unknowns ρ and V,

$$\frac{\partial \rho}{\partial t} + \frac{\partial (\rho V)}{\partial x} = 0$$

$$\frac{\partial V}{\partial t} + V\frac{\partial V}{\partial x} + \frac{a^2}{\rho}\frac{\partial \rho}{\partial x} = 0 \qquad (9.22)$$

where

$$a^2 = \frac{dp}{d\rho} = \gamma \frac{p_4}{\rho_4^\gamma}\rho^{\gamma-1} = a_4^2\left(\frac{\rho}{\rho_4}\right)^{\gamma-1}$$

A more useful form of the above equations is obtained by replacing ρ with a by means of

$$\frac{\rho}{\rho_4} = \left(\frac{a}{a_4}\right)^{2/(\gamma-1)}$$

$$\frac{\partial \rho}{\partial t} = \frac{2}{\gamma-1}\rho_4\left(\frac{a}{a_4}\right)^{2/(\gamma-1)}\frac{1}{a}\frac{\partial a}{\partial t}$$

$$\frac{\partial \rho}{\partial x} = \frac{2}{\gamma-1}\rho_4\left(\frac{a}{a_4}\right)^{2/(\gamma-1)}\frac{1}{a}\frac{\partial a}{\partial x}$$

After simplification, we have the basic equations for unsteady one-dimensional flow:

$$\frac{\partial a}{\partial t} + V\frac{\partial a}{\partial x} + \frac{\gamma-1}{2}a\frac{\partial V}{\partial x} = 0 \qquad (9.23a)$$

$$\frac{\partial V}{\partial t} + V\frac{\partial V}{\partial x} + \frac{2}{\gamma-1}a\frac{\partial a}{\partial x} = 0 \qquad (9.23b)$$

Although relatively simple in appearance, these are two coupled, quasi-linear, first-order partial differential equations.

Our next step is to show that Eqs. (9.23) can be put in the form of Eq. (9.18) and that the resulting equation is hyperbolic. Moreover, our approach for doing this will make the subsequent application of the MOC relatively easy.

Because of the symmetry of Eqs. (9.23), multiply Eq. (9.23a) by a constant λ, and add the result to Eq. (9.23b) to obtain

$$\frac{\partial}{\partial t}(V+\lambda a) + V\frac{\partial}{\partial x}(V+\lambda a) + \frac{\gamma-1}{2}\lambda a\frac{\partial}{\partial x}\left[V + \left(\frac{2}{\gamma-1}\right)^2\frac{1}{\lambda}a\right] = 0$$

For the rightmost term to conform to the other two, set

$$\left(\frac{2}{\gamma-1}\right)^2\frac{1}{\lambda} = \lambda$$

which results in

$$\lambda = \pm\frac{2}{\gamma-1}$$

With the plus sign for λ, we produce

$$\frac{\partial}{\partial t}\left(V + \frac{2}{\gamma-1}a\right) + (V+a)\frac{\partial}{\partial x}\left(V + \frac{2}{\gamma-1}a\right) = 0$$

while the minus sign produces

$$\frac{\partial}{\partial t}\left(V - \frac{2}{\gamma-1}a\right) + (V-a)\frac{\partial}{\partial x}\left(V - \frac{2}{\gamma-1}a\right) = 0$$

This result suggests that we replace V and a with

$$J_{\pm} = V \pm \frac{2a}{\gamma-1}$$

UNSTEADY, ONE-DIMENSIONAL FLOW

The inverse of these equations results in

$$V = \frac{1}{2}(J_+ + J_-), \qquad a = \frac{\gamma - 1}{4}(J_+ - J_-)$$

With this variable change, the equations of motion become

$$\frac{\partial J_+}{\partial t} + (V+a)\frac{\partial J_+}{\partial x} = \frac{\partial J_+}{\partial t} + \left(\frac{\gamma+1}{4}J_+ + \frac{3-\gamma}{4}J_-\right)\frac{\partial J_+}{\partial x} = 0 \quad (9.24a)$$

$$\frac{\partial J_-}{\partial t} + (V-a)\frac{\partial J_-}{\partial x} = \frac{\partial J_-}{\partial t} + \left(\frac{3-\gamma}{4}J_+ + \frac{\gamma+1}{4}J_-\right)\frac{\partial J_-}{\partial x} = 0 \quad (9.24b)$$

If we eliminate one J, say J_-, after much algebra, we obtain

$$AJ_{tt} + 2BJ_{xt} + CJ_{xx} = D$$

For notational convenience, we have set

$$J = J_+, \qquad J_t = \frac{\partial J_+}{\partial t}, \ldots$$

and

$$A = J_x^2$$

$$B = -\frac{2}{3-\gamma}J_x\left(J_t + \frac{\gamma-1}{2}JJ_x\right)$$

$$C = \frac{\gamma+1}{3-\gamma}J_t\left(J_t + 2\frac{\gamma-1}{\gamma+1}JJ_x\right)$$

$$D = \frac{\gamma^2-1}{2(3-\gamma)}J_x^3(J_t + JJ_x)$$

This second-order, quasi-linear partial differential equation is equivalent to Eqs. (9.23). One can show that

$$B^2 - AC = \left[\frac{\gamma-1}{3-\gamma}J_x(J_t + JJ_x)\right]^2 > 0$$

By comparison with Eqs. (9.19), we conclude that compressible, one-dimensional unsteady flow is hyperbolic. This result also holds for two-dimensional and three-dimensional, compressible, inviscid unsteady flow.

Hyperbolic equations exhibit wave behavior. By this we mean that lines exist along which the solution is a constant. An example would be Prandtl-Meyer flow. Finding these lines in an unsteady flow is our next objective.

We are to find a set of lines in the x-t plane on which J_+ is constant and a different set of lines on which J_- is constant. To do this, write for J_+

$$J_+ = J_+(x, t)$$

and by differentiation

$$dJ_+ = \frac{\partial J_+}{\partial x} dx + \frac{\partial J_+}{\partial t} dt$$

But from Eq. (9.24a) we have

$$\frac{\partial J_+}{\partial t} = -(V+a) \frac{\partial J_+}{\partial x}$$

so that

$$dJ_+ = \frac{\partial J_+}{\partial x} dx - (V+a) \frac{\partial J_+}{\partial x} dt = \frac{\partial J_+}{\partial x} dt \left[\frac{dx}{dt} - (V+a) \right]$$

When $dJ_+ = 0$, J_+ is a constant on the lines

$$\frac{dx_+}{dt} = V + a$$

Note that $(dx_+/dt) \neq V$ and that x_+ does not refer to a particle of fluid. Rather, it denotes a special path in the x-t plane which we call the C_+ characteristic line.

By way of summary, we have

$$J_+ = V + \frac{2}{\gamma - 1} a = \text{const on } C_+ \qquad (9.25a)$$

where the C_+ characteristics are the lines

$$\frac{dx_+}{dt} = V + a \qquad (9.26a)$$

In a similar fashion, we obtain

$$J_- = V - \frac{2}{\gamma - 1} a = \text{const on } C_- \qquad (9.25b)$$

where the C_- characteristics are the lines

$$\frac{dx_-}{dt} = V - a \qquad (9.26b)$$

Along with initial data, Eqs. (9.25) and (9.26) represent an exact solution to Eqs. (9.23). Note that $M = 1$ whenever $dx_-/dt = 0$, but supersonic flow is not required.

Equations (9.25) are an integrated form of the so-called compatibility equations, and the quantities $J_\pm$ are called Riemann invariants. The characteristic coordinates are determined by Eqs. (9.26). These equations provide a nonorthogonal net in the physical, or x-t, plane along which the $J_\pm$ are constants. Furthermore, this net never overlaps itself; i.e., in any x-t region, one and only one set of C_- characteristics exists, and one and only one set of C_+ characteristics exists. (A more extensive discussion of characteristics is provided in Chap. 16.)

In the next section, we analyze several unsteady flows by the MOC.

9.4 UNSTEADY EXPANSION WAVES

Section 9.1 discusses how an impulsively started, constant-speed piston moving into a quiescent gas causes a shock wave. Suppose the piston is impulsively pulled from the gas with a constant positive speed V_p. (Henceforth, we drop the prime notation for designating the unsteady case, which was used to avoid confusing the unsteady case with the standing shock solution. As previously noted, the subscript notation used here is designed to coincide with that in Sec. 10.1.) Figure 9.7 shows an x-t diagram and a piston/cylinder schematic for the flowfield under consideration. Region 4 has a quiescent gas, while region 5 has a uniform flow with a speed of $V_5 = V_p$. Between the leading and trailing edges, there is an isentropic, unsteady expansion.

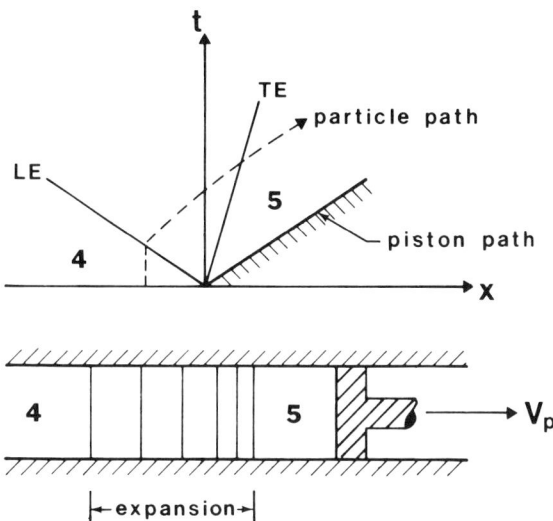

Fig. 9.7 Piston/cylinder schematic and x-t diagram for an unsteady expansion caused by the impulsive motion of a piston.

We observe that the flow speed is nonnegative everywhere when $V_p > 0$. As a consequence, the slope of the C_+ characteristics

$$\frac{dt}{dx_+} = \frac{1}{V+a}, \quad (C_+)$$

is always positive. The slope of the C_- characteristics

$$\frac{dt}{dx_-} = \frac{1}{V-a} = \frac{1}{a}\frac{1}{M-1}, \quad (C_-)$$

is negative when $M < 1$ and positive when $M > 1$. When $M = 1$, the C_- characteristic is a vertical line.

A graphic approach is used for providing an exact solution that is well adapted for computers. We first apply the MOC to the two uniform flow regions, starting with region 4. In this region, $V_4 = 0$, and a_4 is a known constant, so that

$$J_{+4} = \frac{2}{\gamma - 1}a_4 \text{ on } \frac{dx_+}{dt} = a_4$$

$$J_{-4} = -\frac{2}{\gamma - 1}a_4 \text{ on } \frac{dx_-}{dt} = -a_4$$

Consequently, $J_{\pm 4}$ are constants everywhere in region 4, and the characteristics in Fig. 9.8 are a nonorthogonal mesh consisting only of straight lines. Observe that the leading edge of the expansion is a C_- characteristic that passes through the origin. As a consequence, its equation is

$$x_{LE} = -a_4 t \tag{9.27}$$

and all C_{-4} characteristics are parallel to it.

In region 5, the flow is uniform, $V_5 = V_p$, and

$$J_{+5} = V_5 + \frac{2}{\gamma - 1}a_5 = \text{const on } \frac{dx_+}{dt} = V_5 + a_5$$

$$J_{-5} = V_5 - \frac{2}{\gamma - 1}a_5 = \text{const on } \frac{dx_-}{dt} = V_5 - a_5$$

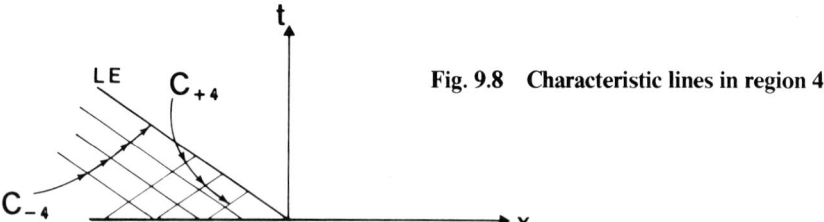

Fig. 9.8 Characteristic lines in region 4.

UNSTEADY, ONE-DIMENSIONAL FLOW   165

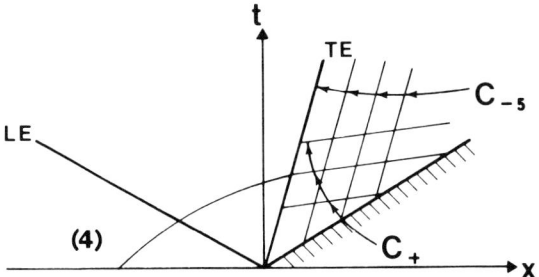

Fig. 9.9 Characteristic lines in region 5. All C_{+5} characteristics originate in region 4.

The characteristic lines are straight, as shown in Fig. 9.9. The trailing edge of the expansion is also a C_- characteristic that passes through the origin. It has the equation

$$x_{TE} = (V_5 - a_5)t = a_5(M_5 - 1)t \qquad (9.28)$$

to which all C_{-5} characteristics are parallel. In Fig. 9.9, we observe that the trailing edge is in the first quadrant when $M_5 > 1$. For a slower piston speed, such that $M_5 < 1$, the trailing edge falls in the second quadrant.

With the speeds V_4, V_5, and a_4 known, we find a_5 by using a C_+ characteristic. As shown in Fig. 9.9, any C_{+4} characteristic will pass through the expansion and end up in region 5. We thus have

$$J_{+4} = J_{+5}$$

$$\frac{2}{\gamma - 1} a_4 = V_5 + \frac{2}{\gamma - 1} a_5$$

or

$$\frac{a_5}{a_4} = 1 - \frac{\gamma - 1}{2} \frac{V_5}{a_4} = 1 - \frac{\gamma - 1}{2} \frac{V_p}{a_4}$$

which determines a_5. Next, eliminate a_5 from Eq. (9.28) to obtain

$$x_{TE} = \left(V_p - a_4 + \frac{\gamma - 1}{2} V_p\right)t = \left(\frac{\gamma + 1}{2} V_p - a_4\right)t$$

Hence, any C_{-5} characteristic has a constant slope

$$\frac{dx_-}{dt} = \frac{\gamma + 1}{2} V_p - a_4 \qquad (9.29)$$

We again use the fact that the C_+ characteristics cross the expansion to determine a solution for the expansion itself. We have

$$J_+ = J_{+4} = \text{const}$$

or

$$V + \frac{2}{\gamma - 1} a = \frac{2}{\gamma - 1} a_4$$

Consequently, the relation

$$\frac{a}{a_4} = 1 - \frac{\gamma - 1}{2} \frac{V}{a_4} \tag{9.30}$$

holds throughout the entire flowfield. Equation (9.25b) provides the value of any J_- invariant. Use Eq. (9.30) to eliminate a in J_-, to obtain

$$J_- = V - \frac{2}{\gamma - 1}\left(a_4 - \frac{\gamma - 1}{2} V\right) = 2V - \frac{2}{\gamma - 1} a_4 = \text{const}$$

Thus, V is a constant on a given C_- characteristic. From Eq. (9.30), observe that a is also a constant on C_- characteristics. Finally, from Eq. (9.26b) we conclude that all C_- characteristics are straight lines. The C_- characteristics inside the expansion have only one special attribute: they pass through the origin and thus constitute a centered expansion fan. As with a Prandtl-Meyer expansion, all flow properties, such as $V, p, \rho, \ldots$, are constant in the fan along a C_- characteristic.

The C_+ characteristics are straight lines in regions 4 and 5 but curve inside the expansion, as shown in Fig. 9.9. The flow is uniform where both C_+ characteristics are straight lines, and the value of any dependent variable, such as flow speed or density, is a constant.

We have a simple wave region where one set of characteristic lines is straight. Here, the dependent variables are constant along the straight characteristics and thus depend only on a single independent variable. For instance, in Prandtl-Meyer flow the independent variable would be an angular coordinate. In the following analysis of an unsteady simple wave region, the nondimensional independent variable is $x/(a_4 t)$. Later, we will discuss a nonsimple wave region in which both sets of characteristic lines curve. In this situation, the solution depends separately on both independent variables.

The flowfield properties inside the expansion are now obtained in terms of x and t. Along a C_- characteristic, V and a are constant; hence, Eq. (9.26b) integrates to

$$x_- = (V - a)t$$

We eliminate a by means of Eq. (9.30) and drop the now superfluous minus subscript to obtain

$$x = \left[V - \left(a_4 - \frac{\gamma-1}{2}V\right)\right]t = \left(\frac{\gamma+1}{2}V - a_4\right)t$$

We solve for V/a_4 and obtain a result that holds only inside the expansion

$$\frac{V}{a_4} = \frac{2}{\gamma+1}\left(1 + \frac{x}{a_4 t}\right) \qquad (9.31\text{a})$$

Hence, V is linear with x, with a discontinuity in slope at the leading and trailing edges, since V is constant in regions 4 and 5. Whereas V itself is discontinuous across a shock wave, here its first derivative is discontinuous on both the leading and trailing edges of the expansion. As will be evident shortly, the first derivatives of a, p, M, etc., are similarly discontinuous. Either type of discontinuity is associated with wave phenomena, which are governed by hyperbolic equations. In subsonic flow or in heat conduction, the equations are elliptic and discontinuities cannot occur.

Other variables can be determined in terms of x and t, as follows:

$$\frac{a}{a_4} = 1 - \frac{\gamma-1}{2}\frac{V}{a_4} = \frac{2}{\gamma+1}\left(1 - \frac{\gamma-1}{2}\frac{x}{a_4 t}\right) \qquad (9.31\text{b})$$

$$\frac{T}{T_4} = \left(\frac{a}{a_4}\right)^2 \qquad (9.32\text{a})$$

$$\frac{p}{p_4} = \left(\frac{a}{a_4}\right)^{2\gamma/(\gamma-1)} \qquad (9.32\text{b})$$

$$\frac{\rho}{\rho_4} = \left(\frac{a}{a_4}\right)^{2/(\gamma-1)} \qquad (9.32\text{c})$$

$$M = \frac{V}{a} = \frac{V/a_4}{a/a_4} = \frac{[2/(\gamma+1)][1 + (x/a_4 t)]}{[2/(\gamma+1)]\{1 - [(\gamma-1)/2](x/a_4 t)\}}$$

$$= \frac{1 + (x/a_4 t)}{1 - [(\gamma-1)/2](x/a_4 t)} \qquad (9.32\text{d})$$

As previously noted, stagnation conditions are not constant inside the expansion. Region 4 conditions are therefore used for nondimensionalizing purposes.

Although stagnation conditions are not constant inside the expansion, the isentropic equations are valid as point relations. For instance,

$$\frac{a}{a_0} = \left(\frac{T}{T_0}\right)^{\frac{1}{2}} = \left(1 + \frac{\gamma-1}{2}M^2\right)^{-\frac{1}{2}}$$

applies and enables us to determine a_0 as follows:

$$\frac{a_0}{a_4} = \frac{a}{a_4}\left(1 + \frac{\gamma-1}{2}M^2\right)^{\frac{1}{2}}$$

$$= \left[\frac{2}{\gamma+1}\left(1 - \frac{\gamma-1}{2}\frac{x}{a_4 t}\right)\right]\left(1 + \frac{\gamma-1}{2}\frac{[1+(x/a_4 t)]^2}{\{1-[(\gamma-1)/2](x/a_4 t)\}^2}\right)^{\frac{1}{2}}$$

$$= \left(\frac{2}{\gamma+1}\right)^{\frac{1}{2}}\left[1 + \frac{\gamma-1}{2}\left(\frac{x}{a_4 t}\right)^2\right]^{\frac{1}{2}} \tag{9.33}$$

From the Mach number relation, Eq. (9.32d), we see that $M = 1$ on $x = 0$, providing

$$V_p \geq \frac{2}{\gamma+1} a_4$$

If this last condition is satisfied, there is sonic or supersonic flow at the trailing edge. If the wave is not centered, then $M = 1$ occurs on a vertically oriented C_- characteristic. From Eq. (9.32d), we also observe that $M = \infty$ when

$$\frac{x}{a_4 t} = \frac{2}{\gamma-1}$$

The corresponding piston speed, Eq. (9.31a), is

$$\frac{V_p}{a_4} = \frac{2}{\gamma-1}$$

If the piston speed exceeds this value, there is a void between the trailing edge and the piston, as shown in Fig. 9.10. In this case, region 5 does not exist, $p = T = \rho = 0$ in the void, and the piston's speed no longer affects the flow. The similarity with Prandtl-Meyer flow is clear.

Regions 4 and 5 are uniform flow regions, while the centered expansion fan is a simple wave region. In this circumstance, Eqs. (9.25) and (9.26) provide an analytic solution to the flowfield. Equations (9.31) are the key equations for the expansion fan since they provide V and a in terms of x and t. Equations (9.32) and (9.33) provide other variables, while Eqs. (9.27)

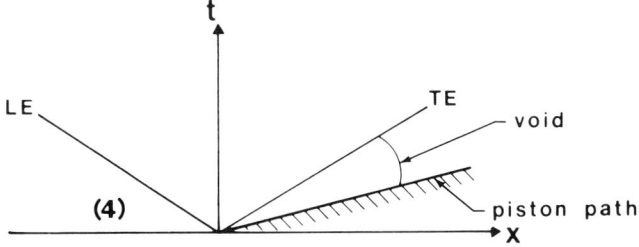

Fig. 9.10 Diagram showing the void region when the piston speed exceeds $2a_4/(\gamma - 1)$.

and (9.28) provide the leading and trailing edges. The following example is designed to clarify the foregoing discussion.

Example 3

Initially, a piston is withdrawn from a quiescent gas impulsively with a constant speed V_{p1}. At time t_c, the piston's speed impulsively increases to V_{p2}. We are to sketch the flowfield in an x-t diagram, determine equations for the various leading and trailing edges, and determine equations for V and a of the various flow regions.

A preliminary x-t diagram is shown in Fig. 9.11. If V_{p2} were smaller than V_{p1}, the V_{p2} piston path would have a steeper slope than that for piston 1, and a shock wave would start at point c. The coordinates of this point satisfy $x_c = V_{p1} t_c$.

To determine the rest of the flow pattern, we need to keep several factors in mind. First, let point c move to the origin. We would then have a single centered expansion caused by a constant piston speed V_{p2}. The leading edge for this expansion would coincide with LE_4, but the trailing edge would be rotated clockwise from TE_4 because $V_{p2} > V_{p1}$.

Second, we note the similarity between steady two-dimensional flow and unsteady one-dimensional flow. Figure 9.12 shows the steady flow that is equivalent to the unsteady double-acceleration flow. The two Mach angles shown in the figure are the same. Consequently, TE_1 and LE_2 are parallel.

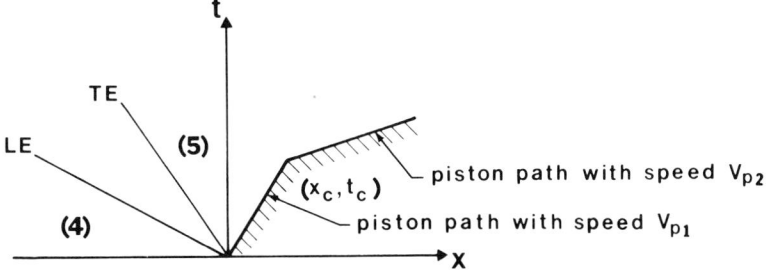

Fig. 9.11 Preliminary sketch for the double-acceleration example.

These observations suggest that the flow pattern will appear as in Fig. 9.13. Regions 4, 5, and 6 are uniform flow regions that border two centered unsteady expansions denoted as E_4 and E_5. The characteristics TE_4 and LE_5 are parallel. Note that a C_+ characteristic starting sufficiently to the left of the origin cuts across both expansions. It is curved inside the expansions, is straight in the uniform flow regions, and has a positive slope everywhere. Where the slope of the C_- characteristics is negative, the flow is subsonic; where positive, the flow is supersonic.

Analysis confirms the sketch in Fig. 9.13. We summarize the solution in the five regions as follows:

Region 4:

$$V = 0, \qquad \frac{a}{a_4} = 1$$

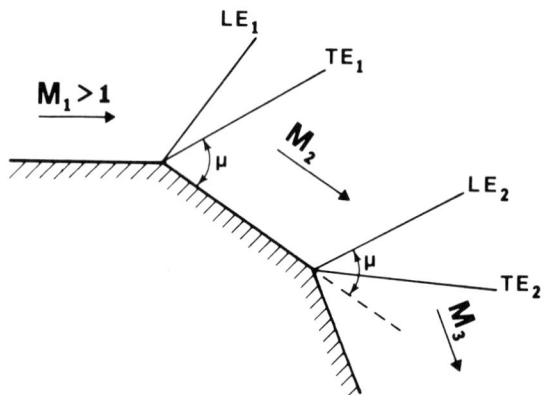

Fig. 9.12 Double acceleration for a steady two-dimensional flow consisting of two centered Prandtl-Meyer expansions.

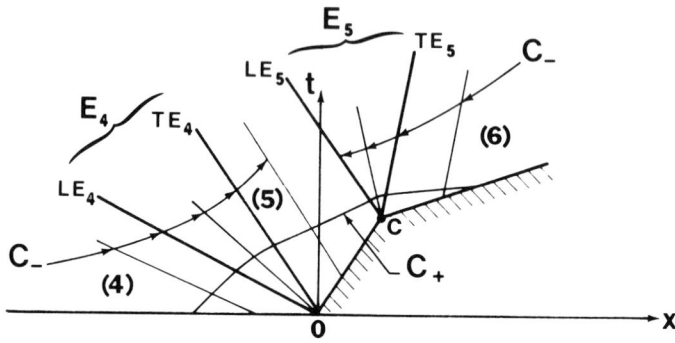

Fig. 9.13 Double-acceleration x-t diagram of Example 3.

Region E_4:

$$x_{LE_4} = -a_4 t, \qquad x_{TE_4} = -a_4 t\left(1 - \frac{\gamma+1}{2}\frac{V_{p1}}{a_4}\right)$$

$$\frac{V}{a_4} = \frac{2}{\gamma+1}\left(1 + \frac{x}{a_4 t}\right), \qquad \frac{a}{a_4} = \frac{2}{\gamma+1}\left(1 - \frac{\gamma-1}{2}\frac{x}{a_4 t}\right)$$

Region 5:

$$\frac{V}{a_4} = \frac{V_{p1}}{a_4}, \qquad \frac{a}{a_4} = 1 - \frac{\gamma-1}{2}\frac{V_{p1}}{a_4}$$

Region E_5:

$$x_{LE_5} = V_{p1} t_c - a_4(t - t_c)\left(1 - \frac{\gamma+1}{2}\frac{V_{p1}}{a_4}\right)$$

$$x_{TE_5} = V_{p1} t_c - a_4(t - t_c)\left(1 - \frac{\gamma+1}{2}\frac{V_{p2}}{a_4}\right)$$

$$\frac{V}{a_4} = \frac{2}{\gamma+1}\left(1 + \frac{1}{a_4}\frac{x - V_{p1} t_c}{t - t_c}\right), \qquad \frac{a}{a_4} = \frac{2}{\gamma+1}\left(1 - \frac{\gamma-1}{2}\frac{1}{a_4}\frac{x - V_{p1} t_c}{t - t_c}\right)$$

Region 6:

$$\frac{V}{a_4} = \frac{V_{p2}}{a_4}, \qquad \frac{a}{a_4} = 1 - \frac{\gamma-1}{2}\frac{V_{p2}}{a_4}$$

Let us verify the solution for region E_5. We start with

$$J_{+4} = J_+$$

which becomes

$$\frac{a}{a_4} = 1 - \frac{\gamma-1}{2}\frac{V}{a_4}$$

For the C_- characteristics, we have

$$\frac{dx_-}{dt} = V - a = \frac{\gamma+1}{2}V - a_4 = \text{const}$$

which integrates to

$$x - x_c = \left(\frac{\gamma+1}{2}V - a_4\right)(t - t_c)$$

where $x_c = V_{p1}t_c$. On the leading and trailing edges, we have

$$V_{LE_5} = V_{p1}, \qquad V_{TE_5} = V_{p2}$$

which yield the edge formulas for E_5. The $x - x_c$ equation in conjunction with the one for a/a_4 then results in the a and V equations for region E_5. We thus conclude Example 3.

So far, the discussion has been limited to uniform and simple wave regions. The simplest case of a nonsimple wave region is provided by the reflection of a centered expansion from a wall, as shown in Fig. 9.14. The uniform flow regions are labeled I, II, and III in order to avoid confusion with the mesh points, which are labeled 1–10. The incident wave consists of C_- characteristics that reflect as C_+ characteristics. In the overlap region we have a nonsimple wave; the solution for this region must be done numerically. The numerical scheme to be described is simply called the MOC.

The region E_4 equations given in Example 3 apply to the centered wave. For purposes of simplicity, let us assume that the TE has a negative slope, as shown in Fig. 9.14. For computational purposes, we subdivide the wave into three parts, each of which subtends an angle $\Delta\theta$, given by

$$\Delta\theta = \frac{(x_{TE}/a_1 t) - (x_{LE}/a_1 t)}{3} = \frac{\gamma + 1}{6} \frac{V_p}{a_1}$$

The centered rays shown in Fig. 9.14 then satisfy the equation

$$\frac{x_i}{a_1 t} = -1 + (i - 1)\Delta\theta, \qquad i = 1,\ldots,4$$

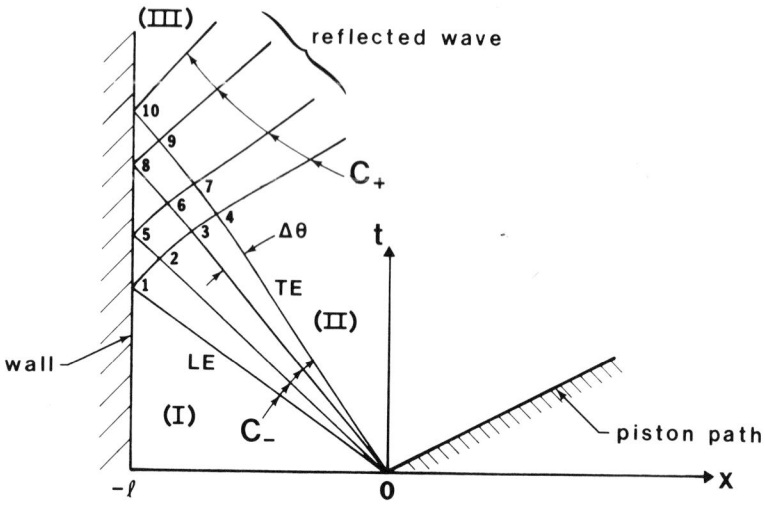

Fig. 9.14 x-t diagram for the reflection from a wall of a centered unsteady expansion.

UNSTEADY, ONE-DIMENSIONAL FLOW

The J_- Riemann invariant is given by the following:

$$J_- = V - \frac{2}{\gamma-1}a = \frac{2}{\gamma+1}a_1\left(1+\frac{x}{a_1 t}\right) - \left(\frac{2}{\gamma-1}\right)\frac{2}{\gamma+1}a_1\left(1-\frac{\gamma-1}{2}\frac{x}{a_1 t}\right)$$

$$= \frac{2}{\gamma+1}a_1\left(-\frac{3-\gamma}{\gamma-1}+2\frac{x}{a_1 t}\right)$$

which becomes, on the four rays,

$$J_{-i} = \frac{2}{\gamma+1}a_1\left[-\frac{\gamma+1}{\gamma-1}+2(i-1)\Delta\theta\right] = -\frac{2}{\gamma-1}a_1 + \frac{2}{3}(i-1)V_p$$

The J_- invariants retain their values in the nonsimple wave region, i.e.,

$$J_{-5} = J_{-2}, \qquad J_{-8} = J_{-6} = J_{-3}, \ldots$$

At the wall, we have $x = -\ell$ and $V = 0$. Consequently, at point 1, we have

$$V_1 = 0, \qquad a_1 = a_I, \qquad J_{-1} = -\frac{2}{\gamma-1}a_I$$

At point 5, we have

$$V_5 = 0$$

$$J_{-5} = -\frac{2}{\gamma-1}a_5 = J_{-2} = -\frac{2}{\gamma-1}a_I + \frac{2}{3}V_p$$

$$a_5 = a_I - \tfrac{1}{3}(\gamma-1)V_p$$

and at point 8,

$$V_8 = 0$$

$$J_{-8} = -\frac{2}{\gamma-1}a_8 = J_{-3} = -\frac{2}{\gamma-1}a_I + \frac{4}{3}V_p$$

$$a_8 = a_I - \tfrac{2}{3}(\gamma-1)V_p$$

For the J_+ invariants, we have

$$J_{+4} = J_{+3} = J_{+2} = J_{+1} = \frac{2}{\gamma-1}a_I, \qquad J_{+7} = J_{+6} = J_{+5} = \frac{2}{\gamma-1}a_5, \ldots$$

Consequently, at point 2, we have

$$V_2 = \frac{1}{2}(J_{+2} + J_{-2}) = \frac{1}{2}\left(\frac{2}{\gamma-1}a_\mathrm{I} - \frac{2}{\gamma-1}a_5\right) = \frac{1}{3}V_p$$

$$a_2 = \frac{\gamma-1}{4}(J_{+2} - J_{-2}) = a_\mathrm{I} - \frac{\gamma-1}{6}V_p$$

while at point 6, we have

$$V_6 = \frac{1}{2}(J_{+6} + J_{-6}) = \frac{1}{2}(J_{+5} + J_{-8}) = \frac{1}{2}\left(\frac{2}{\gamma-1}a_5 - \frac{2}{\gamma-1}a_8\right) = \frac{1}{3}V_p$$

$$a_6 = \frac{\gamma-1}{4}(J_{+6} + J_{-6}) = \frac{\gamma-1}{4}\left(\frac{2}{\gamma-1}a_5 - \frac{2}{\gamma-1}a_8\right) = a_\mathrm{I} - \frac{\gamma-1}{2}V_p$$

In this manner, step by step, the solution in the nonsimple wave region is established.

What remains is to determine the location x_i, t_i of the mesh points. Clearly,

$$x_1 = x_5 = x_8 = x_{10} = -\ell$$

and

$$t_1 = \frac{\ell}{a_\mathrm{I}}$$

For point 2, we approximate the C_+ characteristic that passes through points 1 and 2 by

$$\frac{x_2 - x_1}{t_2 - t_1} \cong \frac{1}{2}\left[\left(\frac{dx_+}{dt}\right)_2 + \left(\frac{dx_+}{dt}\right)_1\right] = \frac{1}{2}(V_2 + a_2 + V_1 + a_1)$$

which becomes

$$\frac{x_2 + \ell}{t_2 - (\ell/a_\mathrm{I})} = a_\mathrm{I} + \frac{3-\gamma}{12}V_p$$

The equation is solved simultaneously with the intersecting ray equation

$$\frac{x_2}{t_2} = -a_\mathrm{I} + \frac{\gamma+1}{6}V_p$$

for x_2, t_2. Thus, the mesh point locations are established after the V, a values are found. While the V, a values are exact, their locations are not. However, the mesh point locations become more exact as the number of rays used for the expansion increases.

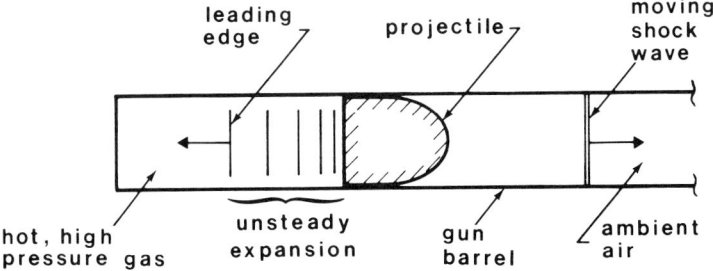

Fig. 9.15 Schematic of a moving projectile in a gun barrel shortly after firing.

For instance, this method of analysis is used in internal ballistics. After ignition, gun powder becomes a hot, high-pressure gas before the projectile has had any time to accelerate. As sketched in Fig. 9.15, once the projectile starts to accelerate, an unsteady expansion propagates into the hot gas, thereby lowering its pressure. The projectile's forward acceleration gradually decreases inside the barrel because of a steadily falling chamber pressure, barrel friction, and an increasing pressure ahead of the projectile. This latter pressure is caused by a shock wave that forms ahead of the projectile. This shock becomes stronger as the projectile's speed increases.

Problems

9.1 A constant-speed piston causes a normal shock in oxygen with a shock Mach number of 3. The temperature of the quiescent gas (region 1) is 300 K. Determine the shock speed, piston speed, V_2', p_2'/p_1', and T_2'/T_1', where region 2 is behind the shock.

9.2 For the conditions in Problem 9.1, determine M_2', p_{02}'/p_1, T_{02}'/T_1, and $(s_2' - s_1')/R$.

9.3 Derive a formula for the work w done by the piston per unit mass of shocked gas in terms of γ, a_1, and M_s. Use this result to determine w for Problem 9.1. Explain why w does not depend on the distance traveled by the piston.

9.4 The shock wave in Problem 9.1 reflects off the cylinder's end wall. Determine V_r', T_3', p_3'/p_1, $(s_3' - s_1')/R$, and M_r.

9.5 For the data in Problem 9.1, the piston travels 1 m. The location x_w of the cylinder's end wall is such that the incident shock reaches it at the same time t_w the piston impulsively stops. Determine this time and the end wall location. When the piston stops, it generates a disturbance that starts at the piston. The leading edge of the disturbance travels with the local sound speed relative to the flow speed. Determine the time t_c and location x_c when this disturbance and the reflected shock wave first meet.

9.6 A normal shock is traveling into still air at 1 atm and 300 K at a velocity of 1.5 km/s. Determine p'_2, T'_2, V'_2, p'_{02}, T'_{02}, and M'_2.

9.7 The shock wave speed from an atomic blast is measured as 60 km/s relative to the ground. The still air is at 1 atm and 300 K. Determine p'_2, T'_2, p'_{02}, and T'_{02}.

9.8 With $\gamma = 5/3$ and $V'_p/a_1 = 1, 2$ (two cases), compute M_s, M'_2, p_2/p_1, T_2/T_1, ρ_2/ρ_1, and p'_{02}/p'_{01}. Tabulate your results.

9.9 A moving normal shock wave is propagating into a quiescent gas. Determine M_s as a function of γ and the pressure ratio p_2/p_1.

9.10 The normal shock wave of Problem 9.6 reflects against a wall. Determine p'_3, T'_3, V_3, p'_{03}, T'_{03}, and M_r.

9.11 Compute p'_{03}, T'_{03}, and M_r for Problem 9.7.

9.12 Derive an equation for $(s_3 - s_1)/R$ that is a function of only γ and M_s.

9.13 The Mach number behind a normal shock that is propagating into a quiescent gas is small compared to unity. Derive perturbation formulas for M_s and for the piston speed that is producing the weak shock.

9.14 A normal shock is traveling into still air ($p_1 = 1$ atm, $T_1 = 300$ K) with a speed of 800 m/s. Determine M_s, M'_2, V'_2, p_2, T_2, p'_{02}, and T'_{02}.

9.15 Determine $(s_2 - s_1)/R$ for an unsteady normal shock as a function of M_s when $\gamma \to 1$.

9.16 Determine a formula for the stagnation pressure ratio across the unsteady centered expansion in Fig. 9.7 in terms of γ and V_p/a_4.

9.17 In terms of x and t, determine the time rate of change of pressure for a particle of fluid inside the unsteady centered expansion in Fig. 9.7.

9.18 A normal shock is moving through quiescent air such that

$$\frac{p_2}{p_1} = 6, \quad T_2 = 900 \text{ K}$$

Determine V'_s and V'_p.

9.19 A piston has a speed of 250 m/s. The quiescent gas is helium ($W = 4$ kg/kmole) at a temperature of 300 K. With p_1 as the pressure of the quiescent gas, determine the pressure ratio p_2/p_1, when the piston is (a) moving into the gas and (b) being withdrawn from the gas.

9.20 Determine the reflected shock wave Mach number M_r as a function of γ and the pressure ratio p_2/p_1, where the flow speed is zero for state 1. Use your result to determine M_r to $\mathcal{O}(\varepsilon)$ when

$$\frac{p_2}{p_1} = 1 + \varepsilon$$

where ε is a small positive number.

9.21 A normal shock is moving into a quiescent gas that has $\gamma = 1.3$, $W = 44$ kg/kmole, and $T_1 = 300$ K. The speed of the gas behind the incident shock is 600 m/s. Determine the temperature T_3 behind the reflected shock wave.

9.22 Argon gas ($W = 40$ kg/kmole) is initially at 300 K. On the downstream side of the reflected shock, the temperature is 1500 K. Determine the incident V'_s and reflected V'_r shock speeds.

9.23 The Mach number behind a normal shock wave that is propagating into 300 K He is 1. Determine the speed of the piston that is producing the shock wave and the shock Mach number M_s.

9.24 Use the MOC to determine a_3, T_3, V_3, and p_3/p_1 for the flow in Problem 9.5. Region 3 is between the piston and expansion as shown in the sketch below.

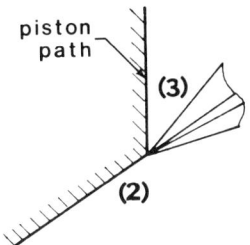

9.25 The density ratio (ρ'_3/ρ'_2) across a reflected shock wave in argon is 2. Determine M'_2.

9.26 The position of a piston changes according to

$$x_p = \alpha e^{\beta t_p} - \alpha$$

where α and β are positive constants. See sketch below. Derive a formula for the pressure ratio $p(x,t)/p_4$, where region 4 is a quiescent gas and (x,t) is any point inside the expansion.

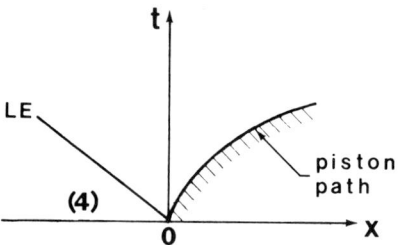

9.27 Derive a formula for Dp_0/Dt inside the unsteady expansion shown in Fig. 9.7. Your final result should be in terms of γ, p_4, a_4, x, and t.

9.28 Determine the piston speed $(V_p/a_4)_m$ that minimizes p_{05}/p_{04}. At this condition, what is M_5? If region 4 consists of air at 300 K, what is $(V_p)_m$? What is the piston speed if the trailing edge coincides with the piston path?

9.29 A piston has the trajectory (see sketch for Problem 9.26)

$$x_p = \alpha t^2$$

where α is a positive constant. Determine an equation for $p(0,t)/p_4$ as a function of γ and $(\alpha t)/a_4$.

9.30 Determine formulas for ρ_5/ρ_4, p_5/p_4, and T_5/T_4 as a function of γ and V_p/a_4 for an unsteady centered expansion.

9.31 Determine formulas for the stagnation ratios ρ_{05}/ρ_{04}, p_{05}/p_{04}, and T_{05}/T_{04} as a function of γ and V_p/a_4 for an unsteady centered expansion.

9.32 With $\gamma = 5/3$ and $V_p/a_4 = 1, 2$ (two cases), compute for an unsteady expansion ρ_5/ρ_4, p_5/p_4, T_5/T_4, ρ_{05}/ρ_{04}, p_{05}/p_{04}, T_{05}/T_{04}, and M_3.

9.33 Consider inviscid, unsteady flow of a perfect gas in a constant cross-sectional area duct. There is heat transfer to the flow in the duct with a known $q(x)$ per unit mass. First, derive the energy equation for this flow. Then, derive the three conservation equations with ρ, V, and a as the dependent variables.

9.34 Start with Eqs. (9.24), and derive the equations for A, B, C, D, and $B^2 - AC$.

UNSTEADY, ONE-DIMENSIONAL FLOW

9.35 A piston with speed

$$V_p = \alpha t_p$$

where α is a positive constant, generates an unsteady compression wave. Derive an explicit formula for $M(x, t)$, where (x, t) is any point inside the isentropic part of the wave.

9.36 A piston is impulsively withdrawn from 300 K argon gas at a speed of 150 m/s. An unsteady expansion wave is generated as shown in the sketch. After traveling a distance ℓ, the piston is impulsively stopped, thereby generating an unsteady normal shock wave. Determine p_3/p_1.

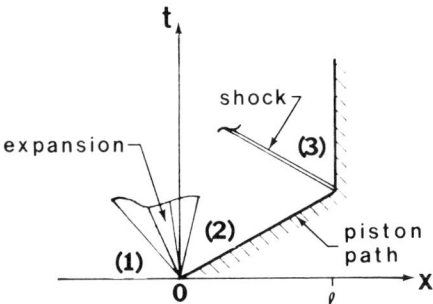

Computational Problems:

9.37 Generate Mach number tables for the equations in Appendices L and M.

9.38 Develop a MOC code to solve the reflected expansion problem of Sec. 9.4. Divide the initial expansion into n segments, where the integer n can be varied. Your final result should be plots of the wall pressure $p(-\ell, t)$ divided by p_1 vs $a_1 t/\ell$, as a function of dimensionless parameters, such as $n, \gamma, V_p/a_1, \ldots$.

10. APPLICATIONS OF UNSTEADY, ONE-DIMENSIONAL FLOW

Our first application is for shock-tube flow. Although discovered around the turn of the century, the shock tube did not become an important research tool until after 1950. It is the device of choice for chemical kinetic, plasma physics, and shock wave investigations. Our second application, which is called a piston expansion tube, would be used in the study of condensation/evaporation physics.

10.1 SHOCK-TUBE FLOW

Initially, two quiescent gases are separated by a diaphragm, as shown in Fig. 10.1(a). Before the diaphragm ruptures, the two gases can, and often do, have different values for T, p, ρ, γ, and R. Different values for T are uncommon, but different values for the other parameters are usual.

When a very strong shock wave is desired, an explosive mixture, such as $H_2 + O_2$, is used for the room-temperature gas in the high-pressure region. The mixture is ignited by a spark, which initiates a detonation wave that greatly increases both pressure and temperature in the high-pressure region. The diaphragm ruptures when the detonation wave reflects from it, and the flow in the tube commences. Because of the extreme pressure levels, a naval gun barrel or its equivalent must be used for the cylindrical wall in the high-pressure region.

After the diaphragm ruptures, the emergent shock wave moves into region 1 at a constant speed V_s. A centered expansion fan (often called a rarefaction wave) occurs whose leading edge (LE) moves into region 4 at a constant speed a_4. A contact surface [Fig. 10.1(b)] provides the demarcation between the two gases initially separated by the diaphragm. This demarcation occurs even if $\gamma_4 = \gamma_1$, $R_4 = R_1$, and $T_4 = T_1$. In this case, we still have $\rho_4 \neq \rho_1$ because $p_4 > p_1$. Basically, a contact surface separates regions of different entropy. The contact surface acts as a piston, pushing on the low-pressure gas and pulling on the high-pressure gas. Diffusion and turbulent mixing smooth out the contact surface with time. But if we neglect mixing, as we do, the contact surface is a discontinuity. The conditions across the contact surface are

$$p_2 = p_5, \quad V_2 = V_5 \qquad (10.1)$$

These conditions will be the key to the subsequent analysis. Other thermo-

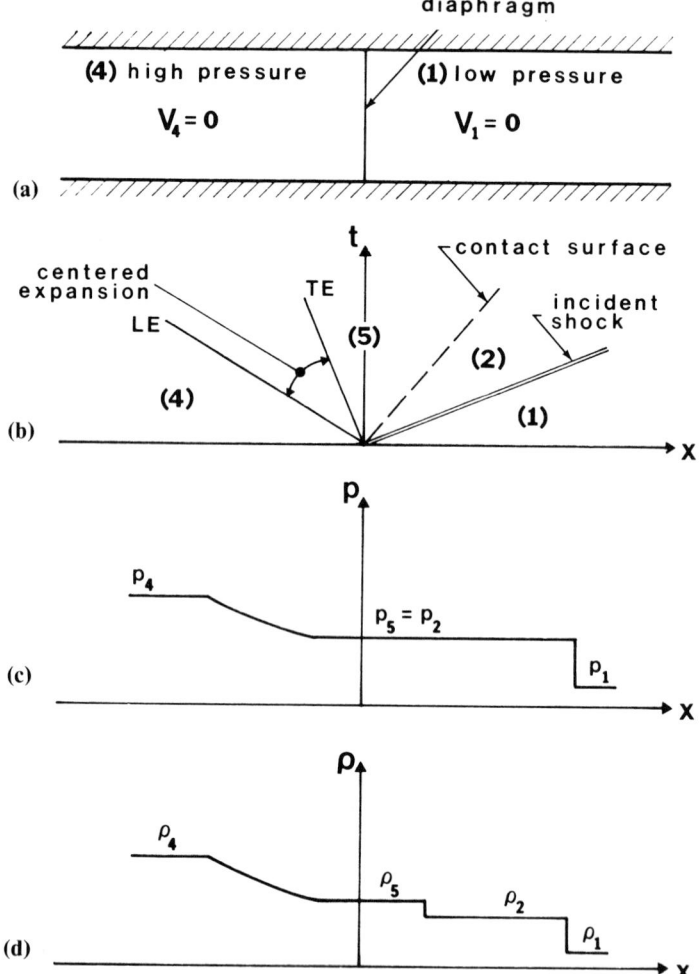

Fig. 10.1 Ideal shock-tube operation: (a) schematic of the shock tube before diaphragm rupture; (b) x-t diagram showing the principal flow features; (c) and (d) pressure and density traces, respectively, at a given time after diaphragm rupture.

dynamic quantities, such as ρ, T, s, and h, go through a jump discontinuity across the contact surface. The variation with x of p and ρ, at a given instant of time, is shown in Figs. 10.1(c) and 10.1(d). Note that the first derivative is discontinuous of both p and ρ on the leading and trailing edges of the expansion.

We consider γ_4, γ_1, a_4/a_1, and p_4/p_1 as known nondimensional initial conditions and determine the incident shock strength p_2/p_1 in terms of these parameters. Thus, p_2/p_1 becomes our primary dependent variable.

We begin with (see Appendix L)

$$\frac{p_2}{p_1} = \frac{2}{\gamma_1 + 1}\left(\gamma_1 M_s^2 - \frac{\gamma_1 - 1}{2}\right)$$

where $M_s = (V_s/a_1)$. This equation yields, on inversion,

$$M_s^2 = \frac{\gamma_1 + 1}{2\gamma_1}\frac{p_2}{p_1} + \frac{\gamma_1 - 1}{2\gamma_1}$$

which is substituted into [see Eq. (9.6)]

$$\frac{V_2}{a_1} = \frac{2}{\gamma_1 + 1}\frac{M_s^2 - 1}{M_s}$$

to obtain

$$V_2 = a_1\left(\frac{p_2}{p_1} - 1\right)\left[\frac{2/\gamma_1}{(\gamma_1 + 1)(p_2/p_1) + (\gamma_1 - 1)}\right]^{\frac{1}{2}} \qquad (10.2)$$

Equations (9.31) and (9.32) are used for the isentropic expansion to obtain

$$\frac{p_5}{p_4} = \left(\frac{a_5}{a_4}\right)^{2\gamma_4/(\gamma_4 - 1)} = \left(1 - \frac{\gamma_4 - 1}{2}\frac{V_5}{a_4}\right)^{2\gamma_4/(\gamma_4 - 1)}$$

We solve for V_5 and write p_5/p_4 as $(p_5/p_1)/(p_1/p_4)$ to yield

$$V_5 = \frac{2}{\gamma_4 - 1}a_4\left[1 - \left(\frac{p_5}{p_1}\frac{p_1}{p_4}\right)^{(\gamma_4 - 1)/2\gamma_4}\right] \qquad (10.3)$$

Equations (10.1)–(10.3) are combined, thereby eliminating p_5 and V_5,

$$\frac{2}{\gamma_4 - 1}a_4\left[1 - \left(\frac{p_2}{p_1}\frac{p_1}{p_4}\right)^{(\gamma_4 - 1)/2\gamma_4}\right]$$

$$= a_1\left(\frac{p_2}{p_1} - 1\right)\left[\frac{2/\gamma_1}{(\gamma_1 + 1)(p_2/p_1) + (\gamma_1 - 1)}\right]^{\frac{1}{2}}$$

Since p_4/p_1 occurs only once, we solve for it

$$\frac{p_4}{p_1} = \frac{p_2}{p_1}\left(1 - \frac{(\gamma_4 - 1)(a_1/a_4)[(p_2/p_1) - 1]}{\{4\gamma_1^2 + 2\gamma_1(\gamma_1 + 1)[(p_2/p_1) - 1]\}^{\frac{1}{2}}}\right)^{-2\gamma_4/(\gamma_4 - 1)} \qquad (10.4)$$

to obtain the desired equation, which is an implicit equation for p_2/p_1.

Once p_2/p_1 is known, other quantities across the moving shock are readily determined. For example, we have

$$M_s = \frac{V_s}{a_1} = \left(\frac{\gamma_1+1}{2\gamma_1}\frac{p_2}{p_1} + \frac{\gamma_1-1}{2\gamma_1}\right)^{\frac{1}{2}} \tag{10.5a}$$

$$M_2 = \frac{V_2}{a_2} = \frac{M_s^2 - 1}{\{1 + [(\gamma_1-1)/2]M_s^2\}^{\frac{1}{2}}\{\gamma_1 M_s^2 - [(\gamma_1-1)/2]\}^{\frac{1}{2}}} \tag{10.5b}$$

and so forth. In a similar fashion, properties involving the expansion are given by

$$\frac{p_5}{p_4} = \frac{p_2/p_1}{p_4/p_1} \tag{10.6a}$$

$$\frac{T_5}{T_4} = \left(\frac{p_5}{p_4}\right)^{(\gamma_4-1)/\gamma_4} \tag{10.6b}$$

$$M_5 = \frac{V_5}{a_5} = \frac{V_2/a_4}{1-[(\gamma_4-1)/2](V_2/a_4)} = \frac{(a_1/a_4)(V_2/a_1)}{1-[(\gamma_4-1)/2](a_1/a_4)(V_2/a_1)}$$

$$= \frac{[2/(\gamma_1+1)](a_1/a_4)[(M_s^2-1)/M_s]}{1-[(\gamma_4-1)/(\gamma_1+1)](a_1/a_4)[(M_s^2-1)/M_s]} \tag{10.6c}$$

and so on.

Both the expansion and incident shock reflect off the end walls of the tube, as shown in an x-t diagram, Fig. 10.2. Except for the contact surface between regions 2 and 5, the diagram is a composite of Figs. 9.4 and 9.14. Regions 3 and 6 are uniform flow regions in which the flow speed is zero. The theory in Chap. 9 is used for these regions. Accordingly, conditions in region 3 are obtainable with the aid of Appendix M.

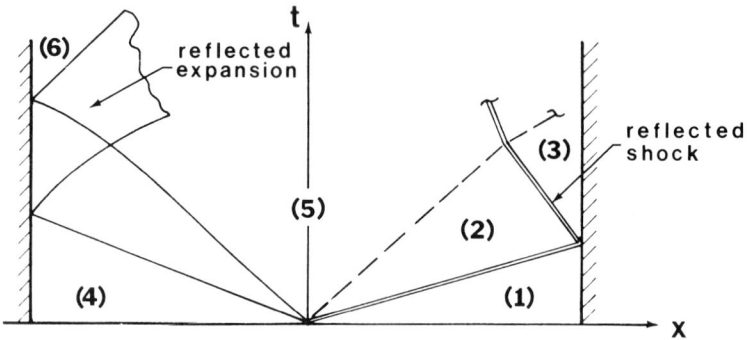

Fig. 10.2 x-t diagram showing the reflected expansion and shock wave.

Shock-tube experiments are performed near the end wall in the low-pressure region. These experiments may be measurements behind the incident shock in region 2 gas. Typically, shock speed measurements are made with heat-transfer gages that are flush with the wall or with pressure transducers, and optical (interferometric or spectroscopic) measurements are made of the hot gas behind the shock. By time-resolving these latter measurements, one obtains relaxation times for various processes behind the incident shock. Often, similar measurements are made behind the reflected shock near the same end wall. There, the gas has a higher temperature and pressure than in region 2.

There are several reasons for carrying out the diagnostic measurements near the low-pressure end wall. The diaphragm rupture process is not instantaneous, and it takes a number of shock-tube diameters before a "clean," normal shock wave is formed. More important, time scales for the measurements, either in region 2 or 3, are a maximum near the end wall. Typically, these scales are about 1 ms in duration, which is adequate for the measurements.

The driving force in a shock tube is p_4/p_1. As this pressure ratio increases, so does p_2/p_1 and M_s. Thus, T_2 and T_3 increase with p_4/p_1. However, a limiting value exists for both p_2/p_1 and M_s as $p_4/p_1 \to \infty$. We obtain the pressure ratio limit from Eq. (10.4) as

$$1 - \frac{(\gamma_4 - 1)(a_1/a_4)[(p_2/p_1)_{max} - 1]}{\{4\gamma_1^2 + 2\gamma_1(\gamma_1 + 1)[(p_2/p_1)_{max} - 1]\}^{\frac{1}{2}}} = 0$$

where $(p_2/p_1)_{max}$ is the maximum possible value for p_2/p_1. The equation is a quadratic in $(p_2/p_1)_{max} - 1$, which solves to

$$\left(\frac{p_2}{p_1}\right)_{max} = 1 + \frac{4\gamma_1}{\gamma_1 + 1}\alpha^2 + \frac{4\gamma_1}{\gamma_1 + 1}\alpha(1 + \alpha^2)^{\frac{1}{2}} \qquad (10.7)$$

where

$$\alpha = \frac{1}{2}\frac{\gamma_1 + 1}{\gamma_4 - 1}\frac{a_4}{a_1} \qquad (10.8)$$

By eliminating $(p_2/p_1)_{max}$ from Eqs. (10.5a) and (10.7), we obtain the maximum value for M_s as

$$M_{s\,max} = \alpha + (1 + \alpha^2)^{\frac{1}{2}} \qquad (10.9)$$

With the use of Eqs. (10.7)–(10.9), other parameter values can be obtained in this limit. For instance, one can show that $M_{5max} = \infty$.

In order to establish experimental time constants, it is useful to establish the paths in the x-t plane for the leading and trailing edges, contact surface, and incident shock wave. For the leading edge, we readily establish

$$\left(\frac{x}{a_4 t}\right)_{LE} = -1 \tag{10.10a}$$

For the shock, we have

$$V_s = \left(\frac{x}{t}\right)_s = a_1 M_s$$

which becomes, with the aid of Eq. (10.5a),

$$\left(\frac{x}{a_4 t}\right)_s = \frac{a_1}{a_4}\left(\frac{\gamma_1 + 1}{2\gamma_1}\frac{p_2}{p_1} + \frac{\gamma_1 - 1}{2\gamma_1}\right)^{\frac{1}{2}} \tag{10.10b}$$

For the contact surface, denoted by a cs subscript, we use Eq. (10.2), to obtain

$$\left(\frac{x}{a_4 t}\right)_{cs} = \frac{a_1}{a_4}\left(\frac{p_2}{p_1} - 1\right)\left[\frac{2/\gamma_1}{(\gamma_1 + 1)(p_2/p_1) + (\gamma_1 - 1)}\right]^{\frac{1}{2}} \tag{10.10c}$$

For the trailing edge, Eq. (9.31a) is rewritten as

$$\frac{x}{a_4 t} = \frac{\gamma_4 + 1}{2}\frac{V}{a_4} - 1$$

At the trailing edge, this equation becomes

$$\left(\frac{x}{a_4 t}\right)_{TE} = \frac{\gamma_4 + 1}{2}\frac{V_{cs}}{a_4} - 1 = \frac{\gamma_4 + 1}{2}\left(\frac{x}{a_4 t}\right)_{cs} - 1$$

$$= \frac{\gamma_4 + 1}{2}\frac{a_1}{a_4}\left(\frac{p_2}{p_1} - 1\right)\left[\frac{2/\gamma_1}{(\gamma_1 + 1)(p_2/p_1) + (\gamma_1 - 1)}\right]^{\frac{1}{2}} - 1 \tag{10.10d}$$

Thus, with Eqs. (10.10), one can show that the contact surface and trailing edge coincide and are located in the first quadrant in the limit, $p_4/p_1 \to \infty$. The next example illustrates how the foregoing results can be extended.

Example 1

Develop equations that determine ρ_5/ρ_2 and T_5/T_2 in terms of p_2/p_1. These equations provide jump conditions across the contact surface. See Fig. 10.1(d) for the density jump.

Across the normal shock, we have [see Appendix L and Eq. (10.5a)]

$$\frac{\rho_2}{\rho_1} = \frac{\gamma_1+1}{2}\frac{M_s^2}{1+[(\gamma_1-1)/2]M_s^2} = \frac{(\gamma_1+1)(p_2/p_1)+(\gamma_1-1)}{(\gamma_1-1)(p_2/p_1)+(\gamma_1+1)} \quad (10.11a)$$

With the aid of Eq. (10.10d), the density ratio across the expansion is given by

$$\frac{\rho_5}{\rho_4} = \left(\frac{a_5}{a_4}\right)^{2/(\gamma_4-1)} = \left\{\frac{2}{\gamma_4+1}\left[1-\frac{\gamma_4-1}{2}\left(\frac{x}{a_4 t}\right)_{TE}\right]\right\}^{2/(\gamma_4-1)}$$

$$= \left(1 - \frac{(\gamma_4-1)(a_1/a_4)[(p_2/p_1)-1]}{\{4\gamma_1^2+2\gamma_1(\gamma_1+1)[(p_2/p_1)-1]\}^{\frac{1}{2}}}\right)^{2/(\gamma_4-1)} \quad (10.11b)$$

The desired density ratio can be written as

$$\frac{\rho_5}{\rho_2} = \frac{\rho_5}{\rho_4}\frac{\rho_4}{\rho_1}\frac{\rho_1}{\rho_2} = \frac{(\rho_5/\rho_4)}{(\rho_2/\rho_1)}\frac{\gamma_4}{\gamma_1}\frac{p_4}{p_1}\left(\frac{a_1}{a_4}\right)^2$$

With the aid of Eqs. (10.11), this becomes

$$\frac{\rho_5}{\rho_2} = 2\gamma_4\left(\frac{a_1}{a_4}\right)^2$$

$$\times \frac{(p_2/p_1)[(\gamma_1-1)(p_2/p_1)+(\gamma_1+1)]}{\left(\{4\gamma_1^2+2\gamma_1(\gamma_1+1)[(p_2/p_1)-1]\}^{\frac{1}{2}}-(\gamma_4-1)(a_1/a_4)[(p_2/p_1)-1]\right)^2}$$

(10.12a)

From the thermal equation of state, we obtain

$$\frac{T_5}{T_2} = \frac{p_5}{p_2}\frac{R_2}{R_5}\frac{\rho_2}{\rho_5} = \frac{W_4}{W_1}\frac{p_2}{\rho_5} \quad (10.12b)$$

for the temperature ratio, where the molecular weight W_5 equals W_4. Equations (10.12) are the desired result. Note that $\rho_5/\rho_2 > 1$ does not imply that T_5/T_2 is greater or less than unity.

The next example illustrates how shock-tube theory is applied.

Example 2

A shock tube has the same monatomic gas and temperature on both sides of the diaphragm, which has a pressure ratio of 100. We are to determine

$$\frac{p_2}{p_1}, \quad \frac{p_5}{p_4}, \quad \frac{p_{02}}{p_1}, \quad \frac{p_{05}}{p_4}, \quad M_s, \quad M_2, \quad M_5$$

From the problem statement, we establish

$$\gamma = \gamma_1 = \gamma_4 = \frac{5}{3}, \quad \frac{W_4}{W_1} = 1, \quad \frac{a_4}{a_1} = 1, \quad \frac{p_4}{p_1} = 10^2$$

Equation (10.4) becomes

$$10^2 = (1+z)\left[1 - \frac{0.2z}{(1+0.8z)^{\frac{1}{2}}}\right]^{-5}$$

where $z = (p_2/p_1) - 1$. This equation is solved iteratively, thereby obtaining

$$\frac{p_2}{p_1} = 1 + z = 1 + 4.8 = 5.8$$

The remaining unknowns are established as follows:

$$\frac{p_5}{p_4} = \frac{p_2/p_1}{p_4/p_1} = 0.058$$

$$M_s = \left[\frac{\gamma+1}{2\gamma}\frac{p_2}{p_1} + \frac{\gamma-1}{2\gamma}\right]^{\frac{1}{2}} = 2.20$$

$$M_2 = \frac{M_s^2 - 1}{\{1 + [(\gamma-1)/2]M_s^2\}^{\frac{1}{2}}\{\gamma M_s^2 - [(\gamma-1)/2]\}^{\frac{1}{2}}} = 0.854$$

$$M_5 = \frac{[2/(\gamma+1)](a_1/a_4)[(M_s^2-1)/M_s]}{1 - [(\gamma-1)/(\gamma+1)](a_1/a_4)[(M_s^2-1)/M_s]} = 2.32$$

$$\frac{p_{02}}{p_1} = \frac{p_2}{p_1}\left(1 + \frac{\gamma-1}{2}M_2^2\right)^{\gamma/(\gamma-1)} = 9.99$$

$$\frac{p_{05}}{p_4} = \frac{p_5}{p_4}\left(1 + \frac{\gamma-1}{2}M_5^2\right)^{\gamma/(\gamma-1)} = 0.757$$

The theory developed here is for an idealized shock tube. The only major nonidealities we have not considered are the diaphragm rupture process and the boundary layer along the wall in regions of gas motion. A well-designed diaphragm minimizes disturbances caused by the rupture process, e.g., scribe lines on the diaphragm that assist in its petaling. Furthermore, the rupture process should result in the diaphragm forming petals that lie flush against the shock-tube wall. In most experiments, the boundary layers are quite thin relative to the tube's diameter and thus have only a minor effect on the flow. As a consequence, the theory of this section is generally accurate.

10.2 PISTON EXPANSION TUBE

The first part of Sec. 9.4 provided the solution for a centered, unsteady expansion whose x-t diagram is shown in Fig. 9.7. The flow is caused by impulsively withdrawing a piston from a quiescent gas at a constant speed V_p. Equations (9.31)–(9.33) largely summarize the solution. Examination reveals that these equations are nondimensional and that the independent variables x and t occur only in the dimensionless combination $\eta = x/(a_4 t)$. This might have been foreseen from Fig. 9.7; the figure shows no intrinsic length scale for the flow. Thus, the only way x can become dimensionless is with $a_4 t$ or $V_p t$.

Equations (9.22) or (9.23) could have been solved by a similarity method instead of the MOC. In the similarity method, η and $\tau = t$ are introduced as new independent variables. Ultimately, the resulting equations depend on η only. The partial differential equations are thereby reduced to ordinary differential equations, which are much easier to solve. With the boundary condition that region 4 is quiescent, we obtain Eqs. (9.31) and (9.32) as the solution to these ordinary differential equations.[1] It will prove convenient in this section to utilize the similarity variable η.

We also drop the shock-tube notation and relabel the regions as shown in the piston expansion tube (PET) x-t diagram, Fig. 10.3. Region 1 is quiescent. A piston with speed V_p moves to the right for a distance ℓ. Region 2 is a uniform flow region between the piston and the trailing edge of the expansion, with $V_2 = V_p$. The gas in region 2 has been cooled by the isentropic expansion to a temperature lower than T_1. The amount of cooling, which can be large, depends on the magnitude of V_p. At $x = \ell$, we

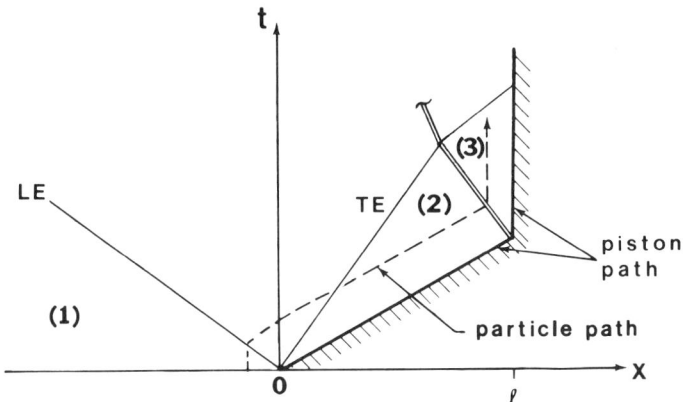

Fig. 10.3 Piston expansion tube x-t diagram.

presume the piston comes to an instantaneous stop. The gas adjacent to the piston, which had been moving to the right with speed V_p, must now stagnate. It does so by means of a leftward-moving shock wave, as shown in the figure. Region 3 is a quiescent gas region, whose conditions differ from region 1 because the gas has passed through an irreversible normal shock.

The utility of the PET would be its ability to perform condensation and evaporation measurements. Both phenomena occur naturally in fog and in clouds, as well as in numerous devices, such as in the last stages of a steam turbine or in a supersonic wind tunnel. The function of the PET is to study the initial phase of condensation when a supersaturated vapor first forms small liquid or solid clusters consisting of a few molecules.

The test gas in region 1 consists of a noncondensable carrier gas plus a room-temperature vapor, with a mole fraction of less than 1%. Hence, the effect of condensation or evaporation is negligible on the gasdynamics of the mixture. As a first approximation, the mixture can be treated as a perfect gas whose molecular weight and ratio of specific heats are those of the carrier gas.

As indicated in Fig. 10.3, region 2 has a uniform flow in which the vapor can condense. Condensation can also occur within the later part of the expansion. The flow in region 3 is a uniform flow but at a temperature above that in region 1. Consequently, the small clusters of condensed vapor that form in the expansion and in region 2 will proceed to evaporate in region 3. Within the same experiment, both the rate of cluster formation and cluster evaporation can be measured. These rates would be measured by Rayleigh scattering of a laser beam by the clusters.

There are a number of other devices for condensation measurements. The oldest of these, invented in 1897, is the Wilson, or expansion, cloud chamber, in which a piston is slowly moved a short distance. A bulk adiabatic cooling of about 25 K can be obtained. A second device is the diffusion cloud chamber, which relies on a vertical thermal gradient. In addition, both wind tunnels[2] and shock tubes[3] have been used in condensation experiments. When the wind tunnel is used, a condensation shock about 1-cm thick occurs in the divergent part of the nozzle. This shock is much thicker than an ordinary shock since the condensation process is relatively slow. Typically, static pressure vs distance and Rayleigh scattering of a laser beam are the diagnostics.

If the high-pressure side of a shock tube contains a carrier gas plus a vapor, then condensation can occur. The condensation is a result of the expansion wave that propagates into the high-pressure gas after the diaphragm is ruptured. We can adjust the strength of the wave by changing the diaphragm pressure ratio.[3] The principal region for optical measurements is between the trailing edge of the expansion and the contact surface. This region is close to the initial diaphragm location, which limits the test time to about 1 ms. The diaphragm rupture process in a shock tube results in a severely perturbed flowfield near the contact surface. This represents a limitation to the approach because this region tends to overlap where condensation occurs.

APPLICATIONS OF UNSTEADY, ONE-DIMENSIONAL FLOW   191

At first glance, it may appear that the piston expansion tube is similar to the Wilson cloud chamber. However, the two are not similar because the PET is not a steady-state device. The PET is closest to the shock-tube approach but differs from it in several ways:

(1) The disturbed flow caused by the diaphragm rupture process is not present.

(2) Rate measurements of cluster formation and cluster evaporation can be made in the same experiment.

(3) Longer test times should be possible.[4]

The most important single difference between the PET and all the foregoing techniques is item (2), which is unique to the PET.

By means of Eqs. (9.31) and (9.32), the similarity solution for the flowfield within the expansion fan can be written as follows:

$$\eta = \frac{x}{a_1 t} \tag{10.13a}$$

$$\frac{V}{a_1} = \frac{2}{\gamma + 1}(1 + \eta) \tag{10.13b}$$

$$\frac{\rho}{\rho_1} = \left(\frac{2}{\gamma + 1} - \frac{\gamma - 1}{\gamma + 1}\eta\right)^{2/(\gamma - 1)} \tag{10.13c}$$

$$\frac{p}{p_1} = \left(\frac{\rho}{\rho_1}\right)^\gamma \tag{10.13d}$$

$$\frac{T}{T_1} = \left(\frac{\rho}{\rho_1}\right)^{\gamma - 1} \tag{10.13e}$$

$$M = \frac{V}{a} = \frac{1 + \eta}{1 - [(\gamma - 1)/2]\eta} \tag{10.13f}$$

The solution for the pressure, temperature, and density is shown in Fig. 10.4 when $\gamma = 1.4$.

By setting V in Eq. (10.13b) equal to V_p, we obtain

$$\eta_{TE} = \frac{\gamma + 1}{2}\frac{V_p}{a_1} - 1 \tag{10.14}$$

Region 2 conditions are found by substituting Eq. (10.14) into Eqs. (10.13c)–(10.13f). For instance, we have for the Mach number

$$M_2 = \frac{V_2}{a_2} = \frac{V_p/a_1}{1 - [(\gamma - 1)/2](V_p/a_1)} \tag{10.15}$$

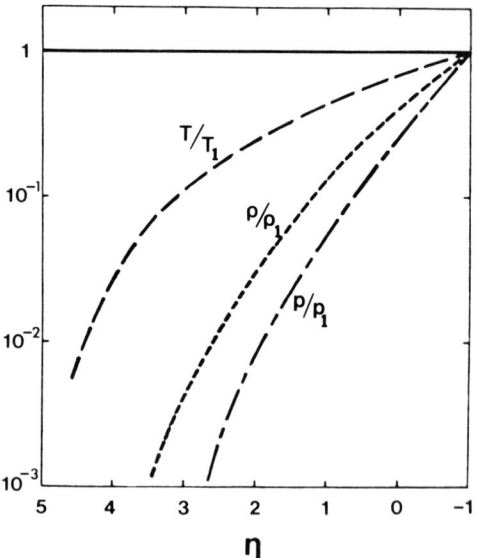

Fig. 10.4 Static pressure, temperature, and density in the expansion region of a PET when $\gamma = 1.4$.

Typical carrier gases would be nitrogen or argon, while H_2O, NH_3, or SF_6 might be the vapor, whose mole fraction is quite small. As mentioned earlier, we can treat the gas as perfect with γ as 1.4 or 5/3. Of course, more realistic calculations can be performed once the vapor and its mole fraction are known.

With any one of the above species as the vapor, typical values for T_1 and T_2 would be 300 and 192 K, respectively. At these temperatures the species are gases at T_1 but condense to a liquid or a solid when in equilibrium at T_2. Thus, with

$$\frac{T_2}{T_1} = \frac{192}{300} = 0.64$$

Eqs. (10.13)–(10.15) yield the results shown in Table 10.1.

As shown in Fig. 10.3, the trailing edge is in the first quadrant and region 2 is supersonic when $\gamma = 1.4$. When $\gamma = 5/3$, however, the trailing edge is in

Table 10.1 Preliminary PET Results

γ	η_{TE}	M_2	V_p/a_1
1.4	0.2	1.25	1
5/3	−0.2	0.75	0.6

the second quadrant and region 2 is subsonic. (The flow relative to the normal shock remains supersonic.)

A particle of gas located near the origin (see Fig. 10.3) experiences a very rapid cooling rate inside the expansion. This cooling rate is on the order of 10^5 K/s. As a consequence, the vapor in the downstream part of the expansion and in region 2 can be highly supersaturated. Nevertheless, cluster formation occurs, and the average cluster size and its rate of growth may be measured by Rayleigh scattering. The clusters will start to disintegrate upon entering region 3, where the temperature exceeds T_1. Rayleigh scattering can also measure the rate of this process.

So far, region 2 conditions have been established. We indicate in the next example how region 3 conditions are found.

Example 3

We are to determine equations for the shock speed V_s, shock Mach number M_r, and temperature ratio T_3/T_1.

A tilde ($\sim$) is used to denote quantities fixed with the shock. From Eqs. (9.11) we obtain

$$\tilde{V}_2 = V_2 - V_s, \qquad \tilde{V}_3 = -V_s \qquad (10.16)$$

where V_s is negative for the leftward-moving shock. We introduce

$$M_2 = \frac{V_2}{a_2}, \qquad M_r = \frac{\tilde{V}_2}{a_2}$$

where M_2 is denoted as M_2' in Sec. 9.2. As in that section, we obtain

$$M_r = \frac{\gamma+1}{4} M_2 + \left[1 + \left(\frac{\gamma+1}{4} M_2\right)^2\right]^{\frac{1}{2}} \qquad (10.17)$$

where M_2 is given by Eq. (10.15). From Appendix D we have

$$\frac{T_3}{T_1} = \left(\frac{2}{\gamma+1}\right)^2 \frac{\{1 + [(\gamma-1)/2] M_r^2\}\{\gamma M_r^2 - [(\gamma-1)/2]\}}{M_r^2} \frac{T_2}{T_1} \qquad (10.18)$$

where this appendix can be used for other region 3-to-region 1 ratios. Finally, the shock speed V_s is given by

$$\frac{V_s}{a_2} = \frac{3-\gamma}{4} M_2 - \left[1 + \left(\frac{\gamma+1}{4} M_2\right)^2\right]^{\frac{1}{2}} \qquad (10.19)$$

where Eqs. (10.16) are used.

If the PET is to become a practical reality, the piston speed must be minimized. However, this minimum is subject to the following constraints:

(1) Nominal temperature values are $T_1 = 300$ K and $T_2 \cong 192$ K. The value for T_2 depends, of course, on the vapor and the purpose of the experiment.
(2) The carrier gas should not be supersaturated at T_2.
(3) The carrier gas should not be radioactive, as is radon.

From Eqs. (10.13c), (10.13e), and (10.14), we obtain, for the piston speed,

$$V_p = \frac{2}{\gamma - 1} \left(\frac{\gamma \tilde{R} T_1}{W} \right)^{\frac{1}{2}} \left[1 - \left(\frac{T_2}{T_1} \right) \right]^{\frac{1}{2}} \qquad (10.20)$$

where $\tilde{R}$ is the universal gas constant, while T_1 and T_2 are fixed. From this equation we observe that a large γ and a large molecular weight W both reduce the magnitude of V_p. With the earlier choice for T_1 and T_2, the results in Table 10.2 are obtained for a variety of possible carrier gases.

Clearly, xenon is the optimum nonradioactive choice. Designing a test facility for a piston speed of 107 m/s is nevertheless a formidable undertaking.

As the molecular weight of the noble gases increase, so does their critical temperature. Hence, item 2 of the foregoing list is of concern. The vapor pressure p_v of xenon is shown in Table 10.3.

Thus, if p_2 does not exceed several atmospheres, the xenon carrier gas will not be supersaturated at a temperature of 192 K.

With xenon and the previous choice for T_1 and T_2, we can determine T_3 and V_s. With $V_p = 107$ m/s and

$$a_1 = \left(\frac{5}{3} \frac{8314}{131.3} \times 300 \right)^{\frac{1}{2}} = 178 \text{ m/s}$$

Eqs. (10.17)–(10.19) yield $M_r = 1.62$, $T_3 = 311$ K, and $V_s = -112$ m/s.

A principal difficulty with the PET is the large piston speed. Of course, the initial acceleration and final deceleration of the piston must be gradual. A demonstration of a practical PET device is still to be done.

Table 10.2 Piston Speed

Carrier Gas	W, kg/kmole	V_p, m/s
N_2	28	353
He	4	612
Ne	20.2	272
Ar	40	193
Kr	83.8	134
Xe	131.3	107

Table 10.3 Xenon Vapor Pressure

T, K	p_v, torr
132	40
140.4	100
156.1	400
165.2	760

APPLICATIONS OF UNSTEADY, ONE-DIMENSIONAL FLOW   195

References

[1] Emanuel, G., "Potential Applications of Piston Generated Unsteady Expansion Waves," *AIAA Journal*, Vol. 19, Aug. 1981, pp. 1015–1018.

[2] For typical example, see Wegener, P. P., Clumper, J. A., and Wu, B. J. C., "Homogeneous Nucleation and Growth of Ethanol Drops in Supersonic Flow," *The Physics of Fluids*, Vol. 15, Nov. 1972, pp. 1869–1876.

[3] Glass, I. I. et al., "Condensation of Water Vapor in Rarefaction Waves: I. Homogeneous Nucleation," *AIAA Journal*, Vol. 14, Dec. 1976, pp. 1731–1737; "Condensation of Water Vapor in Rarefaction Waves: II. Heterogeneous Nucleation," *AIAA Journal*, Vol. 15, Feb. 1977, pp. 215–221; "Condensation of Water Vapor in Rarefaction Waves: III. Experimental Results," *AIAA Journal*, Vol. 15, May 1977, pp. 686–693.

[4] Emanuel, G., unpublished analysis.

Problems

10.1 A shock tube has helium as its high-pressure gas (region 4) and N_2 as its low-pressure gas (region 1). Both gases are at room temperature. A pressure ratio $(p_2/p_1) = 3$ is measured. Determine p_5/p_1, p_4/p_1, M_2, and M_5.

10.2 For a monatomic gas, where $\gamma_1 = \gamma_4 = 5/3$, $W_1 = W_4$, $T_1 = T_4 = 300$ K, and $(p_4/p_1) = 10$, compute T_2 and T_5 for a shock-tube flow.

10.3 A shock tube operates with $(T_4/T_1) = 1$ and $(p_2/p_1) = 10$. One time it has He as the high-pressure gas and Ar as the low-pressure gas, and then vice versa. Compute p_4/p_1 for both cases.

10.4 A combustion-driven shock tube has the initial conditions given in the table. Determine p_2/p_1, T_2/T_1, V_s, and M_s.

	Driver	Driven
p, atm	200	1
T, K	1500	300
W, kg/kmole	25	40
γ	1.3	5/3

10.5 For the reflected shock region of Problem 10.4, determine p_3/p_1, T_3/T_1, and the reflected shock speed V_r.

10.6 Derive the shock-tube times $a_1 t_a/\ell$ and $a_1 t_b/\ell$ as functions of γ_1 and p_2/p_1. The dashed line in the sketch is the contact surface, whereas line ab is the reflected shock wave. For argon as the low-pressure gas at 300 K, $\ell = 3$ m, and $(p_2/p_1) = 15$, determine t_a and t_b.

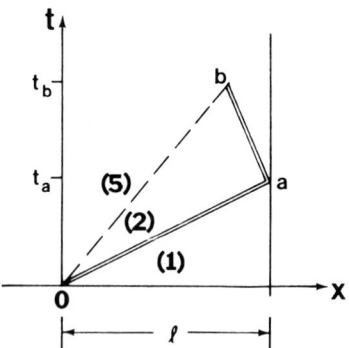

10.7 A shock tube has air at 300 K on both sides of the diaphragm. The contact surface has a speed of 800 m/s. Determine p_2/p_1, p_5/p_1, and p_4/p_1.

10.8 The initial conditions for a shock tube are listed in the table. Determine T_3 and p_3/p_1, when $(p_4/p_1) \to \infty$.

	Driver	Driven
W, kg/kmole	4	40
γ	5/3	5/3
T, K	300	300

PART II

11. GOVERNING EQUATIONS

In this chapter, we derive the governing equations for unsteady, three-dimensional gasdynamics. The usual assumptions of inviscid, adiabatic flow with no body forces are employed. The resulting equations are the basis for all subsequent analysis, including the applications dealt with in Chaps. 17–20.

For purposes of simplicity, a Cartesian coordinate system x_i is employed, in which the respective velocity components are u_i. (Other coordinate systems are of practical importance; these will be discussed in Chap. 13.) Our derivation of the equations of motion will closely parallel the one-dimensional derivation in Chap. 3. Hence, maximum use is made of the substantial derivative

$$\frac{D}{Dt} = \frac{\partial}{\partial t} + \sum_{i=1}^{3} u_i \frac{\partial}{\partial x_i}$$

This approach has the virtue of providing a simple and consistent derivation for the governing equations.

11.1 VOLUME DILATATION

We consider a particle of fluid with a differential volume

$$d\tau = dx_1 \, dx_2 \, dx_3$$

For the conservation of mass derivation, we will need to know how $d\tau$, for a fixed mass fluid particle, changes with time. The change is provided by the substantial derivative $D(d\tau)/Dt$.

As was shown in Chap. 3, for a one-dimensional flow, where

$$x = x_1, \qquad u = u_1, \qquad \frac{D}{Dt} = \frac{\partial}{\partial t} + u\frac{\partial}{\partial x}$$

we obtain [Eq. (3.5)]

$$\frac{1}{dx}\frac{D\,dx}{Dt} = \frac{D\,\ell n(dx)}{Dt} = \frac{\partial u}{\partial x}$$

This result is extended to the three-dimensional case by writing*

$$\frac{D\ell n(dx)_i}{Dt} = \frac{\partial u_i}{\partial x_i}, \quad i = 1, 2, 3$$

We sum over i to obtain

$$\sum_{i=1}^{3} \frac{D\ell n(dx)_i}{Dt} = \sum_{i=1}^{3} \frac{\partial u_i}{\partial x_i}$$

$$\frac{D\ell n(dx_1 dx_2 dx_3)}{Dt} = \sum_{i=1}^{3} \frac{\partial u_i}{\partial x_i}$$

$$\frac{1}{d\tau} \frac{D d\tau}{Dt} = \sum_{i=1}^{3} \frac{\partial u_i}{\partial x_i} \quad (11.1)$$

Equation (11.1) is the desired result, where the right side is the divergence of the velocity.

11.2 CONSERVATION EQUATIONS

As was done in Chap. 3, we derive the conservation equations for mass, momentum, and energy.

Mass

The mass of a fluid particle $\rho\, d\tau$ is a constant. Consequently,

$$\frac{D(\rho\, d\tau)}{Dt} = 0 \quad (11.2)$$

and we use Eq. (11.1) to obtain

$$d\tau \frac{D\rho}{Dt} + \rho \frac{D d\tau}{Dt} = 0$$

$$d\tau \frac{D\rho}{Dt} + \rho\, d\tau \sum_{i=1}^{3} \frac{\partial u_i}{\partial x_i} = 0$$

$$\frac{D\rho}{Dt} + \rho \sum_{i=1}^{3} \frac{\partial u_i}{\partial x_i} = 0 \quad (11.3)$$

---

*A repeated index usually implies summation over that index. This convention is not used because of the frequent exceptions, such as the term on the right side of the following equation.

GOVERNING EQUATIONS

Equation (11.3) is the equation for conservation of mass, or continuity. A flow is incompressible if the density of a fluid particle remains constant with time, in which case the divergence of the velocity is zero. A more useful form is obtained by utilizing the substantial derivative for $D\rho/Dt$, to obtain

$$\frac{\partial \rho}{\partial t} + \sum_{i=1}^{3} \frac{\partial}{\partial x_i}(\rho u_i) = 0 \qquad (11.4)$$

Momentum

Consider again a fluid particle at a fixed instant of time. Since no body or viscous forces occur, the particle experiences only a normal, surface force caused by the pressure. The sketch in Fig. 11.1 shows the forces that act on the fluid particle in the x_1 direction. Since x_1 increases to the right, the applied force in the positive direction is

$$dF_1 = p\,dx_2\,dx_3 - \left(p + \frac{\partial p}{\partial x_1}dx_1\right)dx_2\,dx_3 = -\frac{\partial p}{\partial x_1}d\tau$$

Newton's second law of motion for the fluid particle is the governing principle. It states that the vector sum of the applied forces equals the time rate of change of the momentum of the particle. For the x_1 component, we have the scalar equation

$$\frac{D(\rho\,d\tau\,u_1)}{Dt} = -\frac{\partial p}{\partial x_1}d\tau$$

where the right side is the applied force. The left side is expanded and Eq. (11.2) is utilized to obtain

$$u_1\frac{D(\rho\,d\tau)}{Dt} + \rho\,d\tau\frac{Du_1}{Dt} = -\frac{\partial p}{\partial x_1}d\tau$$

or

$$\rho\frac{Du_1}{Dt} = -\frac{\partial p}{\partial x_1}$$

This relation represents, in the x_1 direction, mass × acceleration = applied force. Here, Du_1/Dt is the acceleration, and the negative of the pressure

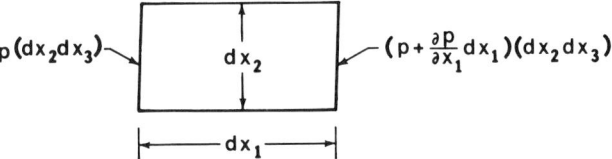

Fig. 11.1 Fluid element showing the pressure forces in the x_1 direction.

gradient represents the applied force. Consequently, for a flow to accelerate, the pressure gradient must be negative. If the gradient is zero, then either u_1 is a constant or the density is zero. Since the x_1 direction is in no way special, we have

$$\rho \frac{Du_i}{Dt} = -\frac{\partial p}{\partial x_i}, \quad i = 1, 2, 3 \qquad (11.5)$$

for the momentum equation.

Energy

For a particle of fluid in an inviscid, adiabatic flow, Eq. (3.10b),

$$\frac{Dh_0}{Dt} = \frac{1}{\rho}\frac{\partial p}{\partial t} \qquad (11.6)$$

still provides the energy equation. The stagnation enthalpy is given by

$$h_0 = e + \frac{p}{\rho} + \frac{1}{2}\sum_{i=1}^{3} u_i^2 = h + \frac{1}{2}\sum_{i=1}^{3} u_i^2 \qquad (11.7)$$

Observe that the stagnation enthalpy of a fluid particle changes with time when there is an unsteady pressure field. Of course, this result requires an inviscid adiabatic flow.

As in Sec. 3.4, one can show that

$$\frac{Ds}{Dt} = 0$$

and the flow is isentropic. This holds for all regions in which the primary variables (p, ρ, e, u_i) are continuous. However, discontinuities may occur, across which these variables, and thus s, are discontinuous.

11.3 CONSERVATIVE FORM

Equations (11.4)–(11.6) constitute the governing equations. In scalar form, they are five in number. An alternate form for these equations, which is important in computational fluid dynamics, is called the conservative form. It has a divergencelike structure[1]

$$\frac{\partial q_0}{\partial t} + \sum_{i=1}^{3}\frac{\partial q_i}{\partial x_i} + q_4 = 0 \qquad (11.8)$$

This equation represents the five scalar equations of motion. Consequently, each q_i has five elements, which will be given shortly. Suffice it to say, the q_i are functions only of the various dependent variables. Generally, the elements of the inhomogeneous term q_4 are zero; hereafter, we set $q_4 = 0$, except when noted otherwise.

Equation (11.4) for continuity is already in the conservative form. To transform momentum, first change the summation index in Eq. (11.4) to j and then multiply by u_i. Finally, add the result to Eq. (11.5) to obtain

$$u_i \frac{\partial \rho}{\partial t} + \sum_{j=1}^{3} u_i \frac{\partial}{\partial x_j}(\rho u_j) + \rho \frac{\partial u_i}{\partial t} + \sum_{j=1}^{3} \rho u_j \frac{\partial u_i}{\partial x_j} = -\frac{\partial p}{\partial x_i}$$

Upon simplification, this becomes

$$\frac{\partial (\rho u_i)}{\partial t} + \sum_{j=1}^{3} \frac{\partial}{\partial x_j}(\rho u_j u_i) = -\frac{\partial p}{\partial x_i}$$

or

$$\frac{\partial (\rho u_i)}{\partial t} + \sum_{j=1}^{3} \frac{\partial}{\partial x_j}(p \delta_{ij} + \rho u_i u_j) = 0, \qquad i = 1, 2, 3 \qquad (11.9)$$

where the Kronecker delta is

$$\delta_{ij} = 0, \qquad i \neq j$$

$$= 1, \qquad i = j$$

Equation (11.9) is the conservative form for the momentum equation.

To transform the energy equation, first multiply continuity by h_0, and add the result to Eq. (11.6), to obtain

$$h_0 \frac{\partial \rho}{\partial t} + \sum_{i=1}^{3} h_0 \frac{\partial}{\partial x_i}(\rho u_i) + \rho \frac{\partial h_0}{\partial t} + \sum_{i=1}^{3} \rho u_i \frac{\partial h_0}{\partial x_i} = \frac{\partial p}{\partial t}$$

This becomes the conservative form for the energy equation

$$\frac{\partial (-p + \rho h_0)}{\partial t} + \sum_{i=1}^{3} \frac{\partial}{\partial x_i}(\rho u_i h_0) = 0 \qquad (11.10)$$

We see that the q_i in Eq. (11.8) can be written as

$$q_0 = \begin{pmatrix} \rho \\ \rho u_1 \\ \rho u_2 \\ \rho u_3 \\ -p + \rho h_0 \end{pmatrix}, \quad q_1 = \begin{pmatrix} \rho u_1 \\ p + \rho u_1^2 \\ \rho u_1 u_2 \\ \rho u_1 u_3 \\ \rho u_1 h_0 \end{pmatrix}, \quad q_2 = \begin{pmatrix} \rho u_2 \\ \rho u_1 u_2 \\ p + \rho u_2^2 \\ \rho u_2 u_3 \\ \rho u_2 h_0 \end{pmatrix}, \quad q_3 = \begin{pmatrix} \rho u_3 \\ \rho u_1 u_3 \\ \rho u_2 u_3 \\ p + \rho u_3^2 \\ \rho u_3 h_0 \end{pmatrix}$$

$$(11.11)$$

where h_0 is given by Eqs. (11.7). The dependent variables are p, ρ, e, and the three u_i, making a total of six. The system of five equations is closed by appending two thermodynamics state equations. For these, we use Eqs. (2.12) with $v = \rho^{-1}$. Consequently, there are seven equations, where the dependent variables are the foregoing six plus the temperature.

11.4 BOUNDARY AND INITIAL CONDITIONS

For unsteady, three-dimensional flow, sufficient initial conditions are necessary to determine fully the $\partial q_i/\partial x_i$ at $t = 0$. Once these derivatives are known, Eq. (11.8) determines $\partial q_0/\partial t$, thereby allowing the computation to proceed in time. Thus, initial values are needed for five independent variables, say, p, T, and the u_i. The state equations then determine ρ and e; after which h_0 is found.

Boundary conditions for a steady or an unsteady flow are also needed. Since neither heat transfer nor viscous forces are considered, the only impermeable wall condition allowed is a velocity tangency condition. Upstream, or inlet, conditions are generally required and most often consist of a uniform flow. Outlet conditions are not required when the outlet flow is supersonic. If some, or all, of the outlet flow is subsonic, then outlet conditions are required. Fortunately, much of the time, the outlet flow is supersonic and outlet conditions are of no concern. This is the case for the nozzle, aerodynamic window, and waverider applications discussed in Chaps. 17, 18, and 20.

Reference

[1] Lax, P. D., "Weak Solutions of Nonlinear Hyperbolic Equations and Their Numerical Computation," *Communications on Pure and Applied Mathematics*, Vol. VII, 1954, pp. 159–193.

Problems

11.1 Derive the relation

$$\frac{D(\rho J)}{Dt} = 0$$

where the Jacobian J is (see Appendix N)

$$J = \frac{\partial(x_1, x_2, x_3)}{\partial(\xi_1, \xi_2, \xi_3)}$$

The new coordinates ξ_i are arbitrary.

12. SHOCK WAVES

In this chapter, we reexamine normal and oblique shock waves (first considered in Chap. 5) from a more sophisticated viewpoint. Our primary objective is to provide a formulation for a nonplanar shock wave in an unsteady three-dimensional flow. In this circumstance, the shock wave itself may be unsteady, i.e., it may be accelerating or decelerating. This formulation will prove useful in subsequent chapters.

12.1 ONE-DIMENSIONAL FLOW

We first consider a one-dimensional flow, in which the equations of motion are [see Eqs. (11.8) and (11.11)]:

$$\frac{\partial \rho}{\partial t} + \frac{\partial}{\partial x}(\rho u) = 0$$

$$\frac{\partial(\rho u)}{\partial t} + \frac{\partial}{\partial x}(p + \rho u^2) = 0 \qquad (12.1)$$

$$\frac{\partial(-p + \rho h_0)}{\partial t} + \frac{\partial}{\partial x}(\rho u h_0) = 0$$

These equations can be written as

$$\frac{\partial q_0}{\partial t} + \frac{\partial q_1}{\partial x} = 0 \qquad (12.2)$$

Let us examine the possibility of a discontinuous solution of Eq. (12.2). In the x-t plane, let the curve

$$x_s = x_s(t) \qquad (12.3)$$

represent the path of a discontinuity in the solution. Opposite sides of the curve are denoted by a ( )' and ( )'', respectively, as shown in Fig. 12.1. Equation (12.2) is satisfied on each side of the curve; hence, we have

$$\frac{\partial q_0'}{\partial t} + \frac{\partial q_1'}{\partial x} = 0$$

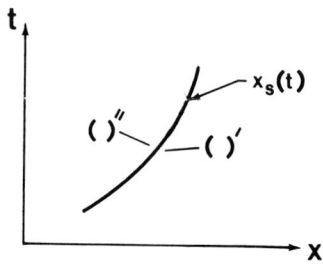

Fig. 12.1 Curve along which a solution may be discontinuous.

and

$$\frac{\partial q_0''}{\partial t} + \frac{\partial q_1''}{\partial x} = 0$$

By taking the difference, we obtain

$$\frac{\partial (q_0'' - q_0')}{\partial t} + \frac{\partial (q_1'' - q_1')}{\partial x} = 0 \tag{12.4}$$

However, x and t are no longer independent of each other; they are related by Eq. (12.3). In Eq. (12.4), we must use

$$\partial t \to dt, \qquad \partial x \to u_s \, dt$$

where

$$u_s = \frac{dx_s}{dt}$$

Equation (12.4) thus becomes

$$u_s \, d(q_0'' - q_0') + d(q_1'' - q_1') = 0$$

We integrate across the x_s curve at a fixed time, so that u_s is a constant, to obtain

$$u_s q_0'' + q_1'' = u_s q_0' + q_1' \tag{12.5}$$

For a discontinuous solution to exist, the quantity $(u_s q_0 + q_1)$ must be conserved across the discontinuity.

Since Eq. (12.2) represents three scalar equations, three $u_s q_0 + q_1$ quantities are conserved. From Eqs. (12.1), these are

$$u_s \rho'' + \rho'' u'' = (u_s \rho + \rho u)'$$

$$u_s \rho'' u'' + p'' + \rho'' u''^2 = (u_s \rho u + p + \rho u^2)' \tag{12.6}$$

$$u_s(-p'' + \rho'' h_0'') + \rho'' u'' h_0'' = [u_s(-p + \rho h_0) + \rho u h_0]'$$

where

$$h'_0 = h' + \tfrac{1}{2}u'^2, \qquad h''_0 = h'' + \tfrac{1}{2}u''^2 \qquad (12.7)$$

A coordinate system is introduced that moves with the discontinuity. This is accomplished with the transformation

$$u_1 = u_s + u', \qquad u_2 = u_s + u'' \qquad (12.8)$$

where the speeds u_1 and u_2 are measured relative to the x_s curve. Equations (12.8) are used for eliminating u' and u'', but not necessarily u_s, from Eqs. (12.6) and (12.7). The transformation encompassed by Eqs. (12.8) leaves all thermodynamic static variables unchanged. However, for consistency in terminology, we alter the notation

$$(p', \rho', \ldots) \to (p_1, \rho_1, \ldots)$$

$$(p'', \rho'', \ldots) \to (p_2, \rho_2, \ldots)$$

for all such variables. After a modest amount of algebra, Eqs. (12.6) are replaced with

$$\rho_2 u_2 = \rho_1 u_1$$
$$p_2 + \rho_2 u_2^2 = p_1 + \rho_1 u_1^2 \qquad (12.9)$$
$$h_{02} = h_{01}$$

where all terms containing u_s, in fact, have canceled. These equations are just the shock wave jump conditions, Eqs. (5.2). As with the conservation equations, two thermodynamic state equations are required for closure of the system.

Equations (12.9) are the Rankine-Hugoniot equations for a normal shock wave in a coordinate system fixed with the shock. They represent conservation of mass, momentum, and energy across a one-dimensional discontinuity. These equations can also be derived directly from Eqs. (12.1) by simply setting the unsteady terms equal to zero. Our more involved derivation shows that these equations hold even in an unsteady flow because u_s does not appear in Eqs. (12.9). The physical reason for this is that a discontinuity contains no mass. Hence, the time rate of change of mass, energy, or momentum within the shock is identically zero. If this were not the case, the jump conditions would be differential equations instead of algebraic ones that explicitly do not depend on time. Insofar as the above derivation is concerned, the variable speed u_s of the shock is arbitrary. To determine u_s, boundary conditions are needed.

12.2 OBLIQUE SHOCK WAVES

Jump Conditions

We next consider the general case of an unsteady three-dimensional surface,

$$F(x_1, x_2, x_3, t) = 0 \tag{12.10}$$

using Cartesian coordinates, across which a discontinuous solution of the conservation equations may exist. Since the surface $F = 0$ contains no mass, the jump conditions across it will be algebraic equations that are independent of time in a coordinate system fixed with the discontinuity. This is the case even when $(\partial F / \partial t) \neq 0$. As we will show, the jump conditions are for an oblique shock wave.

The upstream side, where $\mathbf{V}_1$ is into the surface, is denoted as side 1. The downstream side is side 2. Consider any point P of the surface, and an orthonormal vector basis centered at this point, as shown in Fig. 12.2. Let $\hat{\mathbf{n}}$ be a unit normal vector to the surface whose direction satisfies $\mathbf{V}_1 \cdot \hat{\mathbf{n}} \geq 0$. A surface unit tangent vector $\hat{\mathbf{t}}$ is chosen that is in the plane of $\mathbf{V}_1$ and $\hat{\mathbf{n}}$ and where $\mathbf{V}_1 \cdot \hat{\mathbf{t}} \geq 0$. A second tangent vector $\hat{\mathbf{b}}$ is orthogonal to $\hat{\mathbf{n}}$ and $\hat{\mathbf{t}}$. If $\mathbf{V}_1$ and $\hat{\mathbf{n}}$ are collinear, then the vectors $\hat{\mathbf{t}}$ and $\hat{\mathbf{b}}$ are unnecessary because the jump conditions are for a normal shock, as given in the preceding section.

Conservation of momentum across the discontinuity shows that $\mathbf{V}_2$ must be in the same plane as $\mathbf{V}_1$ and $\hat{\mathbf{n}}$. (This is demonstrated as part of Problem 12.2.) Thus, the three-dimensional unsteady flow across the discontinuity has been reduced locally, in a shock fixed coordinate system, to a steady two-dimensional flow, as shown in Fig. 12.3. In the figure, $\hat{\mathbf{b}}$ is perpendicular to the plane of the page, whereas all vectors shown are in the plane of

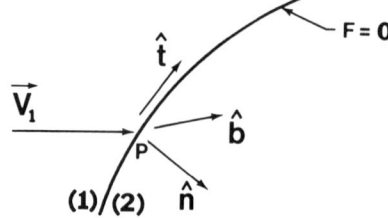

Fig. 12.2 Orthonormal basis at a point P of the surface $F = 0$.

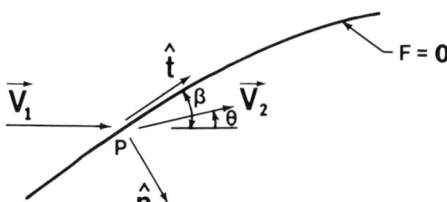

Fig. 12.3 Velocities and angles in a plane perpendicular to $\hat{\mathbf{b}}$.

the page. However, this two-dimensional flow was analyzed in Sec. 5.2, and we will utilize results from this earlier section. Hence, θ is the velocity turn angle and β is the shock wave angle, as shown in the figure.

The velocity on opposite sides of the shock can be written as

$$\mathbf{V}_i = u_{in}\hat{\mathbf{n}} + u_{it}\hat{\mathbf{t}}, \qquad i = 1, 2 \tag{12.11}$$

where the n and t subscripts denote normal and tangential components relative to the shock. The jump conditions, equivalent to Eqs. (12.9), are

$$\begin{aligned}\rho_2 u_{2n} &= \rho_1 u_{1n} \\ p_2 + \rho_2 u_{2n}^2 &= p_1 + \rho_1 u_{1n}^2 \\ u_{2t} &= u_{1t} \\ h_{02} &= h_{01}\end{aligned} \tag{12.12}$$

In view of Fig. 12.3, the velocity components u_{in} and u_{it}, in terms of θ, β, and the speed V_1, are given by

$$\begin{aligned} u_{1n} &= V_1 \sin \beta \\ u_{1t} &= u_{2t} = V_1 \cos \beta \\ u_{2n} &= V_1 \cos \beta \tan(\beta - \theta) \end{aligned} \tag{12.13}$$

Except for $u_{2t} = u_{1t}$, the jump conditions can be transformed into a form compatible with Eq. (12.10). The resulting equations are

$$\rho_1 \mathbf{V}_1 \cdot \nabla F = \rho_2 \mathbf{V}_2 \cdot \nabla F$$

$$\rho_1(\mathbf{V}_1 \cdot \nabla F)^2 + p_1(\nabla F)^2 = \rho_2(\mathbf{V}_2 \cdot \nabla F)^2 + p_2(\nabla F)^2 \tag{12.14}$$

$$h_1(\nabla F)^2 + \tfrac{1}{2}(\mathbf{V}_1 \cdot \nabla F)^2 = h_2(\nabla F)^2 + \tfrac{1}{2}(\mathbf{V}_2 \cdot \nabla F)^2$$

where ∇F is the spatial gradient of F. Equations (12.14) represent conservation of mass, momentum, and energy, respectively, across $F = 0$, without any restriction on the thermodynamic state relations. If a perfect gas is assumed, then the enthalpy h is replaced by $\gamma p / [(\gamma - 1)\rho]$. The derivation of Eqs. (12.14) is the subject of Problem 12.2. As with Eqs. (12.12), Eqs. (12.14) hold for steady and unsteady flows. They are general vector equations that also hold in any coordinate system, providing F is expressed in the same system.

Other types of discontinuities are possible, although shock waves are the most important. A slipstream or contact surface also involves a discontinuity in the variables. A third type of discontinuity can occur in the derivatives of the dependent variables across a Mach line. These discontinuities are discussed in later chapters.

Geometrical Analysis

A firm connection between the $\hat{\mathbf{t}}, \hat{\mathbf{n}}, \beta, \theta, \ldots$ variables and the shock geometry, Eq. (12.10), has not yet been made. The following analysis rectifies this situation. Without loss of generality, we assume $\mathbf{V}_1$ is given by

$$\mathbf{V}_1 = V_1 \hat{\mathbf{l}}_1 \tag{12.15}$$

where the Cartesian normal basis $\hat{\mathbf{l}}_i$ corresponds to the coordinates x_i of Eq. (12.10). Thus, the x_i now represent a local coordinate system fixed to the shock wave at point P with $\mathbf{V}_1$ parallel to x_1. This complexity is easily resolved for the vast majority of applications where the freestream flow ahead of the shock is uniform. In this circumstance, choose x_1 to be along the freestream velocity $\mathbf{V}_1$, and the local coordinate restriction is no longer necessary.

Using vector analysis, relations are established between $\hat{\mathbf{n}}, \hat{\mathbf{t}}, \hat{\mathbf{b}}$, and the $\hat{\mathbf{l}}_i$. The surface normal $\hat{\mathbf{n}}$ is given by

$$\hat{\mathbf{n}} = \frac{\nabla F}{|\nabla F|} = \frac{\sum\limits_{i=1}^{3} (\partial F/\partial x_i)\hat{\mathbf{l}}_i}{\left[\sum\limits_{i=1}^{3} (\partial F/\partial x_i)^2\right]^{\frac{1}{2}}} \tag{12.16}$$

where the arbitrary sign of F is chosen such that $(\partial F/\partial x_1) \geq 0$, thereby yielding $\mathbf{V}_1 \cdot \hat{\mathbf{n}} \geq 0$, as previously required. Since $\hat{\mathbf{t}}$ is in the $\hat{\mathbf{n}}, \mathbf{V}_1$ plane, we can write

$$\hat{\mathbf{t}} = a_0 \hat{\mathbf{n}} + a_1 \frac{\mathbf{V}_1}{V_1} = a_0 \hat{\mathbf{n}} + a_1 \hat{\mathbf{l}}_1$$

In addition, $\hat{\mathbf{t}}$ satisfies

$$\hat{\mathbf{t}} \cdot \hat{\mathbf{n}} = 0, \qquad \hat{\mathbf{t}} \cdot \hat{\mathbf{t}} = 1, \qquad a_1 \geq 0$$

where the a_1 inequality stems from the earlier requirement that $\mathbf{V}_1 \cdot \hat{\mathbf{t}} \geq 0$. The above relations yield

$$a_0 = -\frac{\partial F/\partial x_1}{|\nabla F'|}, \qquad a_1 = \frac{|\nabla F|}{|\nabla F'|}$$

Hence, $\hat{\mathbf{t}}$ is given by

$$\hat{\mathbf{t}} = \frac{|\nabla F'|}{|\nabla F|} \hat{\mathbf{l}}_1 - \frac{\partial F/\partial x_1}{|\nabla F||\nabla F'|} \sum_{i=2}^{3} \frac{\partial F}{\partial x_i} \hat{\mathbf{l}}_i \tag{12.17}$$

where

$$|\nabla F'| = \left[\sum_{i=2}^{3}\left(\frac{\partial F}{\partial x_i}\right)^2\right]^{\frac{1}{2}}$$

To obtain $\hat{\mathbf{b}}$, set it equal to

$$\hat{\mathbf{b}} = \hat{\mathbf{n}} \times \hat{\mathbf{t}}$$

With Eqs. (12.16) and (12.17), this becomes

$$\hat{\mathbf{b}} = \begin{vmatrix} \hat{\mathbf{i}}_1 & \hat{\mathbf{i}}_2 & \hat{\mathbf{i}}_3 \\ \dfrac{1}{|\nabla F|}\dfrac{\partial F}{\partial x_1} & \dfrac{1}{|\nabla F|}\dfrac{\partial F}{\partial x_2} & \dfrac{1}{|\nabla F|}\dfrac{\partial F}{\partial x_3} \\ \dfrac{|\nabla F'|}{|\nabla F|} & -\dfrac{(\partial F/\partial x_1)(\partial F/\partial x_2)}{|\nabla F||\nabla F'|} & -\dfrac{(\partial F/\partial x_1)(\partial F/\partial x_3)}{|\nabla F||\nabla F'|} \end{vmatrix}$$

which expands to

$$\hat{\mathbf{b}} = \frac{1}{|\nabla F'|}\frac{\partial F}{\partial x_3}\hat{\mathbf{i}}_2 - \frac{1}{|\nabla F'|}\frac{\partial F}{\partial x_2}\hat{\mathbf{i}}_3 \qquad (12.18)$$

Clearly, we have

$$\hat{\mathbf{b}} \cdot \hat{\mathbf{b}} = 1, \qquad \hat{\mathbf{b}} \cdot \mathbf{V}_1 = 0$$

Equations (12.16)–(12.18) are the desired relations for the $\hat{\mathbf{n}}, \hat{\mathbf{t}}, \hat{\mathbf{b}}$ basis set in terms of the shock shape, Eq. (12.10), and Eq. (12.15) for $\mathbf{V}_1$.

From Eqs. (12.11) for $\mathbf{V}_2$, and the jump condition, Eqs. (12.12),

$$\mathbf{V}_1 \cdot \hat{\mathbf{t}} = \mathbf{V}_2 \cdot \hat{\mathbf{t}}$$

we easily obtain

$$u_{2t} = \frac{|\nabla F'|}{|\nabla F|} V_1$$

By comparison with Eqs. (12.13), the shock angle β is given by

$$\cos\beta = \frac{|\nabla F'|}{|\nabla F|} \qquad (12.19)$$

We determine u_{2n} by first eliminating p_2 and ρ_2 from Eqs. (12.12), as follows:

$$\rho_2 = \rho_1 \frac{u_{1n}}{u_{2n}}, \qquad p_2 = p_1 + \rho_1 u_{1n}^2 - \rho_1 u_{1n} u_{2n}$$

The remaining relations in Eqs. (12.12) are used and a perfect gas is assumed, to obtain a quadratic equation for u_{2n}

$$u_{2n}^2 - \frac{2\gamma}{\gamma+1}\left(\frac{p_1}{\rho_1} + u_{1n}^2\right)\frac{u_{2n}}{u_{1n}} + \frac{2\gamma}{\gamma+1}\frac{p_1}{\rho_1} + \frac{\gamma-1}{\gamma+1}u_{1n}^2 = 0$$

Equations (12.13) provide u_{1n}, and we also use

$$\frac{p_1}{\rho_1} = \frac{a_1^2}{\gamma}, \quad V_1 = a_1 M_1$$

The above quadratic equation for u_{2n} becomes

$$\frac{\gamma+1}{2} M_1 \sin\beta \left(\frac{u_{2n}}{a_1}\right)^2 - (1 + \gamma M_1^2 \sin^2\beta)\left(\frac{u_{2n}}{a_1}\right)$$

$$+ M_1 \sin\beta \left(1 + \frac{\gamma-1}{2} M_1^2 \sin^2\beta\right) = 0$$

One root yields the trivial result $u_{2n} = u_{1n}$. The other root provides

$$\frac{u_{2n}}{a_1} = \frac{2}{\gamma+1}\frac{1 + [(\gamma-1)/2] M_1^2 \sin^2\beta}{M_1 \sin\beta}$$

Hence, $\mathbf{V}_2$ can be written as

$$\frac{\mathbf{V}_2}{a_1} = M_1(\cos\beta)\hat{\mathbf{t}} + \frac{2}{\gamma+1}\frac{1 + [(\gamma-1)/2] M_1^2 \sin^2\beta}{M_1 \sin\beta}\hat{\mathbf{n}} \quad (12.20)$$

where Eq. (12.19) provides β in terms of F.

By comparing the above u_{2n} equation with the one in Eqs. (12.13), we have

$$M_1 \cos\beta \tan(\beta - \theta) = \frac{2}{\gamma+1}\frac{1 + [(\gamma-1)/2] M_1^2 \sin^2\beta}{M_1 \sin\beta}$$

or

$$\tan(\beta - \theta) = \frac{2}{\gamma+1}\frac{1 + [(\gamma-1)/2] M_1^2 \sin^2\beta}{M_1^2 \sin\beta \cos\beta} \quad (12.21a)$$

Upon solving for θ, we obtain Eq. (5.23),

$$\tan\theta = \frac{1}{\tan\beta}\frac{M_1^2 \sin^2\beta - 1}{1 + \{[(\gamma+1)/2] - \sin^2\beta\} M_1^2} \quad (12.21b)$$

Either of Eqs. (12.21) can be used to determine θ in conjunction with Eq. (12.19).

The magnitude of $\mathbf{V}_2$ can now be determined as follows:

$$V_2 = \left(u_{2t}^2 + u_{2n}^2\right)^{\frac{1}{2}} = \frac{a_1 M_1 \cos\beta}{\cos(\beta - \theta)}$$

$$= \frac{2}{\gamma+1} \frac{a_1}{M_1 \sin\beta} \left\{1 + (\gamma-1)M_1^2\sin^2\beta + \left[\left(\frac{\gamma+1}{2}\right)^2 - \gamma\sin^2\beta\right]M_1^4\sin^2\beta\right\}^{\frac{1}{2}}$$

(12.22)

The corresponding relation for M_2, in the form $M_2 = M_2(\gamma, \beta, \theta)$, is the subject of Problem 12.1.

The foregoing relations, along with those from Chap. 5, are sufficient for determining any variable on the downstream side of the shock, providing Eqs. (12.1) and upstream conditions are known.

Other quantities, such as the vorticity and entropy on the downstream side of the shock, are examined in later chapters. It is worth noting that the equations have been formulated in a manner that might assist in their computer implementation.

Problems

12.1 Use the notation of Fig. 5.5 to derive the relation

$$M_2 = M_2(\gamma, \beta, \theta)$$

Be sure to simplify your results as much as possible. The solution to this problem is often utilized in subsequent chapters.

12.2 Start with the integral form of the conservation equations and derive Eqs. (12.14).

12.3 In the simplest possible form, determine the equation for $\mathbf{V}_2$ in terms of $\hat{1}_{i}$, M_1, and derivatives of $F(x_1, x_2, x_3, t) = 0$. Use your results to determine $\mathbf{V}_2$ when

(a) $\qquad F = ax_1 + bx_2 + c$

(b) $\qquad F = ax_1 + bx_2^2 + c$

where a, b, and c are constants, and x_2 is in the $\mathbf{V}_1, \mathbf{V}_2$ plane.

12.4 Use the result from Problem 12.1 to determine the β, θ equation when $M_2 = 1$, in the form

$$\cos^2\theta = f(\gamma, \cos^2\beta)$$

If there is more than one root, determine the correct one.

12.5 Consider parabolic coordinates ξ_j given by

$$x_1 = \tfrac{1}{2}(\xi_1^2 - \xi_2^2), \qquad x_2 = \xi_1\xi_2\sin\xi_3, \qquad x_3 = \xi_1\xi_2\cos\xi_3$$

A shock wave occurs at

$$F = 1 - \xi_2 = 0$$

and the freestream velocity is

$$\mathbf{V}_\infty = V_\infty \hat{\mathbf{i}}_1$$

where V_∞ is a constant. Assume a perfect gas and denote conditions just downstream of the shock with a d subscript. Determine the shock jump conditions in the form p_d/p_∞, T_d/T_∞, and ρ_d/ρ_∞ as functions of γ, M_∞, and ξ_1.

12.6 Assume explicit relations for the transformation

$$x_i = x_i(\xi_1, \xi_2, \xi_3), \qquad i = 1, 2, 3$$

and the shock shape

$$F(\xi_1, \xi_2, \xi_3, t) = 0$$

are known. Start with Eqs. (12.14), and derive a relation for ρ_d/ρ_∞ in terms of the new coordinates, where the ∞ and d subscripts denote upstream and downstream shock conditions, respectively. The freestream velocity is $V_\infty \hat{\mathbf{i}}_1$. Assume that the ξ_j are orthogonal and the gas is perfect, and simplify your results.

12.7 Extend Problem 12.6 to obtain p_d/p_∞, T_d/T_∞, and M_d.

12.8 Repeat Problems 12.6 and 12.7 for a van der Waals gas whose thermal equation of state is

$$p = \frac{RT}{v - \beta} - \frac{\alpha}{v^2}$$

where R, α, and β are constants. Also obtain for the speed of sound, a_d/a_∞. Use the results of Problems 2.8, 2.9, and 2.11.

12.9 Extend Problem 12.8 by obtaining for the entropy s_d/s_∞. Use the results of Problem 2.10.

13. TRANSFORMATION OF THE CONSERVATION EQUATIONS

In practice, many different coordinate systems have been found to be useful. The choice of a coordinate system is important and is usually chosen on the basis of the boundary and initial conditions. If a flow possesses some geometrical symmetry, it is natural to choose a coordinate system that takes advantage of the symmetry.

Not all coordinate systems, however, are based on the boundary conditions, e.g., characteristic coordinates. Most systems are orthogonal, but there are important exceptions, such as characteristic coordinates.

In Cartesian coordinates, the dependent variables are chosen as p, ρ, e, and u_i. The thermodynamic variables require no alteration when transforming to new coordinates. However, one invariably needs the velocity components in the new system. We perform this vector operation in the first subsection of Sec. 13.1. In the second subsection, the governing equations are transformed using Jacobian theory. A concise review of this theory is provided in Appendix N.

As is already apparent, we make frequent use of the equations of motion when written in their conservative form, as given by Eqs. (11.8) and (11.11). In fact, this form, with Cartesian coordinates, is consistently our starting point for further derivations. In view of this, a general transformation is provided in the subsection headed "Invariance of the Conservative Form."

The theory of Sec. 13.1 is put to use in Secs. 13.2 and 13.3, which both deal with steady two-dimensional or axisymmetric flow. In the first of these sections, the transformation to the new coordinate system is known, whereas in Sec. 13.3 it is not known. In Sec. 13.3 the equations of motion are obtained in a general, orthogonal, curvilinear coordinate system. These later equations are then used to formulate the equations of motion in natural coordinates (Sec. 13.4), and with the hodograph transformation (Sec. 13.5). They are also the starting point for the characteristic theory analysis in Sec. 16.1.

13.1 GENERAL THEORY

Velocity Components

We start with a Cartesian coordinate system x_i, in terms of which the position and velocity vectors are

$$\mathbf{r} = \sum_{i=1}^{3} x_i \hat{\mathbf{1}}_i$$

and

$$\mathbf{V} = \frac{D\mathbf{r}}{Dt} = \sum_{i=1}^{3} u_i \hat{1}_i \qquad (13.1)$$

where the $\hat{1}_i$ denotes an orthonormal Cartesian basis. We transform to coordinates

$$\xi_j = \xi_j(x_1, x_2, x_3), \qquad j = 1, 2, 3 \qquad (13.2)$$

where the new system is not necessarily orthogonal but is steady. The flow itself may still be unsteady. We assume that the Jacobian

$$\frac{\partial(\xi_1, \xi_2, \xi_3)}{\partial(x_1, x_2, x_3)}$$

is not zero except at isolated points, so that the inverse transformation

$$x_i = x_i(\xi_1, \xi_2, \xi_3), \qquad i = 1, 2, 3 \qquad (13.3)$$

exists. It is the inverse form, Eq. (13.3), that is needed for the transformation rather than Eq. (13.2).

We denote by A the matrix

$$A = \left(\frac{\partial x_i}{\partial \xi_j}\right) = \begin{vmatrix} \frac{\partial x_1}{\partial \xi_1} & \frac{\partial x_1}{\partial \xi_2} & \frac{\partial x_1}{\partial \xi_3} \\ \frac{\partial x_2}{\partial \xi_1} & \frac{\partial x_2}{\partial \xi_2} & \frac{\partial x_2}{\partial \xi_3} \\ \frac{\partial x_3}{\partial \xi_1} & \frac{\partial x_3}{\partial \xi_2} & \frac{\partial x_3}{\partial \xi_3} \end{vmatrix} \qquad (13.4)$$

For later reference, it is worth noting that the determinant of A is the Jacobian

$$|A| = \frac{\partial(x_1, x_2, x_3)}{\partial(\xi_1, \xi_2, \xi_3)} = J(x_1, x_2, x_3)$$

of the transformation Eq. (13.3).

Let the unnormalized vector $\mathbf{e}_j$ be tangent to ξ_j and point in the direction of increasing ξ_j. From vector analysis, the transformation between $\mathbf{e}_j$ and the $\hat{1}_i$ is

$$(\mathbf{e})_i = A_t(\hat{1})_i \qquad (13.5a)$$

where

$$(\mathbf{e}) = (\mathbf{e}_1, \mathbf{e}_2, \mathbf{e}_3), \quad (\hat{\mathbf{I}}) = (\hat{\mathbf{I}}_1, \hat{\mathbf{I}}_2, \hat{\mathbf{I}}_3)$$

and the subscript t denotes the transposed matrix. The inverse of Eq. (13.5a) then is

$$(\hat{\mathbf{I}})_t = A_t^{-1}(\mathbf{e})_t \tag{13.5b}$$

It is, of course, convenient to use a normalized basis $(\hat{\mathbf{e}})$. This is obtained from

$$\mathbf{e}_j = \sum_{i=1}^{3} \frac{\partial x_i}{\partial \xi_j} \hat{\mathbf{I}}_i$$

$$|\mathbf{e}_j| = \left[\sum_{i=1}^{3} \left(\frac{\partial x_i}{\partial \xi_j} \right)^2 \right]^{\frac{1}{2}} = h_j \tag{13.6}$$

where the h_j are the scale factors and should not be confused with enthalpy. [The term scale factor is usually used with an orthogonal transformation. Equation (13.6), however, also holds for nonorthogonal transformations.] Hence, the $\hat{\mathbf{e}}_j$ are given by

$$\hat{\mathbf{e}}_j = \frac{1}{h_j} \sum_{i=1}^{3} \frac{\partial x_i}{\partial \xi_j} \hat{\mathbf{I}}_i \tag{13.7}$$

One can show for an orthogonal transformation that

$$A^{-1} = \begin{pmatrix} \frac{1}{h_1^2} & 0 & 0 \\ 0 & \frac{1}{h_2^2} & 0 \\ 0 & 0 & \frac{1}{h_3^2} \end{pmatrix} A_t \tag{13.8}$$

which readily provides A^{-1} in this case.

We denote by v_j the velocity components in the transformed coordinate system, i.e.,

$$\mathbf{V} = \sum_{j=1}^{3} v_j \hat{\mathbf{e}}_j$$

where $v_j \hat{\mathbf{e}}_j$ is the component tangent to ξ_j. In view of Eq. (13.7), this becomes

$$\mathbf{V} = \sum_{j=1}^{3} v_j \frac{1}{h_j} \sum_{i=1}^{3} \frac{\partial x_i}{\partial \xi_j} \hat{\mathbf{I}}_i = \sum_{i=1}^{3} \hat{\mathbf{I}}_i \sum_{j=1}^{3} \frac{\partial x_i}{\partial \xi_j} \frac{v_j}{h_j}$$

A comparison with Eq. (13.1) yields the transformation for the velocity components

$$u_i = \sum_{j=1}^{3} \frac{\partial x_i}{\partial \xi_j} \frac{v_j}{h_j}, \qquad i = 1, 2, 3$$

In matrix form this is written as

$$(u)_t = A(v/h)_t \qquad (13.9a)$$

where, for instance,

$$(v/h)_t = \begin{pmatrix} v_1/h_1 \\ v_2/h_2 \\ v_3/h_3 \end{pmatrix}$$

Occasionally, the inverse of Eq. (13.9a)

$$(v/h)_t = A^{-1}(u)_t \qquad (13.9b)$$

is needed. Equation (13.9a) shows that only A and the scale factors are needed for the desired velocity component transformation.

Coordinate Transformation

For notational convenience, set $x_0 = t$, so that Eq. (11.8) can be written as

$$\sum_{i=0}^{3} \frac{\partial q_i}{\partial x_i} = 0 \qquad (13.10)$$

In Jacobian form this becomes

$$\frac{\partial(q_0, x_1, x_2, x_3)}{\partial(x_0, x_1, x_2, x_3)} + \frac{\partial(q_1, x_0, x_2, x_3)}{\partial(x_1, x_0, x_2, x_3)}$$

$$+ \frac{\partial(q_2, x_0, x_1, x_3)}{\partial(x_2, x_0, x_1, x_3)} + \frac{\partial(q_3, x_0, x_1, x_2)}{\partial(x_3, x_0, x_1, x_2)} = 0$$

where each term in Eq. (13.10) is expanded into a 4×4 Jacobian in accord with rule (2) in Appendix N. This relation is revised (see Appendix N) to the symmetric form

$$\frac{\partial(q_0, x_1, x_2, x_3)}{\partial(x_0, \ldots, x_3)} + \frac{\partial(x_0, q_1, x_2, x_3)}{\partial(x_0, \ldots, x_3)}$$

$$+ \frac{\partial(x_0, x_1, q_2, x_3)}{\partial(x_0, \ldots, x_3)} + \frac{\partial(x_0, x_1, x_2, q_3)}{\partial(x_0, \ldots, x_3)} = 0$$

With the transformation

$$x_i = x_i(\xi_0, \ldots, \xi_3), \qquad i = 0, \ldots, 3 \tag{13.11}$$

the above becomes

$$\frac{\partial(q_0, x_1, x_2, x_3)}{\partial(\xi_0, \ldots, \xi_3)} + \frac{\partial(x_0, q_1, x_2, x_3)}{\partial(\xi_0, \ldots, \xi_3)}$$

$$+ \frac{\partial(x_0, x_1, q_2, x_3)}{\partial(\xi_0, \ldots, \xi_3)} + \frac{\partial(x_0, x_1, x_2, q_3)}{\partial(\xi_0, \ldots, \xi_3)} = 0 \tag{13.12}$$

The transformation, Eq. (13.11), is not necessarily orthogonal or steady.

The transformed equations are Eq. (13.12), where the q_i are given by Eq. (11.11), and, therefore, still depend on the u_i. A new set, $\tilde{q}_k(p, \rho, v_j) = q_k(p, \rho, u_j)$, is then obtained by utilizing Eq. (13.9a) to eliminate the u_j.

Upon completion of the coordinate transformation and the change to $\tilde{q}_k$, one or more terms in Eq. (13.12) may become zero. A term, say, the second Jacobian, is zero if each of the elements $(x_0, \tilde{q}_1, x_2, x_3)$ in the numerator is independent of, at least, one of the ξ_j ($j = 0, \ldots, 3$). The other nonzero, fourth-order Jacobians may then contract down to a lower order. A contraction occurs when the $\tilde{q}_k$ are independent of one ξ_j coordinate. If, for instance, the $\tilde{q}_k$ depend on only ξ_1 and ξ_2, then the nonzero Jacobians become second order. Each spatial contraction results in a reduction by one in the number of nontrivial momentum equations. This is accomplished by combining the original momentum equations in a manner that eliminates the irrelevant space variable, which is ξ_3 in our discussion.

Although Eq. (13.10) is in the conservative form, this is generally not true of Eq. (13.12). We demonstrate this assertion in Secs. 13.2 and 13.3, which illustrate the transformation procedures of this section. However, we first derive a general transformation under which the conservative form is invariant.

Invariance of the Conservative Form

We again start with Eq. (13.10) and use Eq. (13.11), whose Jacobian is written as

$$J = J(x_0, \ldots, x_3) = \frac{\partial(x_0, \ldots, x_3)}{\partial(\xi_0, \ldots, \xi_3)}$$

The transformation represented by Eq. (13.11) is a general one in that the functional form of the $x_i(\xi_0,\ldots,\xi_3)$ is not specified. Nevertheless, the q_i can be replaced with new variables Q_j, such that Eq. (13.10) holds for the Q_j when these are functions of the ξ_j. These new variables will explicitly depend on the q_i and on the coordinate transformation.

Since there are four q_i, there will be four Q_j, each of which has five elements. In the derivation, the q_i and Q_j are treated as vectors, thus the transformation actually determines the elements of Q_j in terms of those of q_i. A generalization of Eqs. (N.6) in Appendix N is first obtained, which will be needed shortly. Jacobian theory is then used to replace the x_i with ξ_j. New variables Q_j are defined, and the derivative $\partial Q_j/\partial \xi_j$ is evaluated. We conclude by showing that the sum over j of these derivative terms equals zero.

We begin the above generalization by considering the derivative

$$\frac{\partial \xi_0}{\partial x_0} = \frac{\partial(\xi_0, x_1, x_2, x_3)/\partial(\xi_0,\ldots,\xi_3)}{\partial(x_0,\ldots,x_3)/\partial(\xi_0,\ldots,\xi_3)} = \frac{\partial(x_1, x_2, x_3)/\partial(\xi_1,\xi_2,\xi_3)}{J}$$

Equation (13.11) determines on the right side both the numerator and the J denominator. In a similar manner, we obtain

$$\frac{\partial \xi_0}{\partial x_1} = \frac{\partial(\xi_0, x_0, x_2, x_3)/\partial(\xi_0,\ldots,\xi_3)}{\partial(x_1, x_0, x_2, x_3)/\partial(\xi_0,\ldots,\xi_3)} = -\frac{\partial(x_0, x_2, x_3)/\partial(\xi_1,\xi_2,\xi_3)}{J}$$

$$\frac{\partial \xi_1}{\partial x_0} = \frac{\partial(\xi_1, x_1, x_2, x_3)/\partial(\xi_0,\ldots,\xi_3)}{\partial(x_0,\ldots,x_3)/\partial(\xi_0,\ldots,\xi_3)}$$

$$= -\frac{\partial(x_1, x_2, x_3)/\partial(\xi_0,\xi_2,\xi_3)}{J},\ldots$$

The foregoing special cases confirm the general result

$$J(x_0,\ldots,x_3)\frac{\partial \xi_j}{\partial x_i} = (-1)^{i+j}\frac{\partial(x_0,\ldots,x_{i-1},x_{i+1},\ldots,x_3)}{\partial(\xi_0,\ldots,\xi_{j-1},\xi_{j+1},\ldots,\xi_3)} \quad (13.13a)$$

If we interchange the roles of x and ξ and recall that the inverse of J is $1/J$, a second relation is obtained:

$$\frac{\partial x_k}{\partial \xi_j} = (-1)^{j+k}J(x_0,\ldots,x_3)\frac{\partial(\xi_0,\ldots,\xi_{j-1},\xi_{j+1},\ldots,\xi_3)}{\partial(x_0,\ldots,x_{k-1},x_{k+1},\ldots,x_3)} \quad (13.13b)$$

TRANSFORMATION OF THE CONSERVATION EQUATIONS

With the aid of Eq. (13.13b), we compute by the chain rule

$$\frac{\partial q_i}{\partial \xi_j} = \sum_{k=0}^{3} \frac{\partial q_i}{\partial x_k} \frac{\partial x_k}{\partial \xi_j}$$

$$= J \sum_{k=0}^{3} (-1)^{k+j} \frac{\partial(\xi_0, \ldots, \xi_{j-1}, \xi_{j+1}, \ldots, \xi_3)}{\partial(x_0, \ldots, x_{k-1}, x_{k+1}, \ldots, x_3)} \frac{\partial q_i}{\partial x_k} \quad (13.14)$$

New variables Q_j ($j = 0, \ldots, 3$) are now defined by

$$Q_j = J \sum_{i=0}^{3} \frac{\partial \xi_j}{\partial x_i} q_i$$

$$= \sum_{i=0}^{3} (-1)^{i+j} \frac{\partial(x_0, \ldots, x_{i-1}, x_{i+1}, \ldots, x_3)}{\partial(\xi_0, \ldots, \xi_{j-1}, \xi_{j+1}, \ldots, \xi_3)} q_i \quad (13.15)$$

where Eq. (13.13a) is used. The ξ_j derivative of Q_j then is

$$\frac{\partial Q_j}{\partial \xi_j} = \sum_{i=0}^{3} (-1)^{i+j} \left\{ q_i \frac{\partial}{\partial \xi_j} \frac{\partial(x_0, \ldots, x_{i-1}, x_{i+1}, \ldots, x_3)}{\partial(\xi_0, \ldots, \xi_{j-1}, \xi_{j+1}, \ldots, \xi_3)} \right.$$

$$\left. + \frac{\partial(x_0, \ldots, x_{i-1}, x_{i+1}, \ldots, x_3)}{\partial(\xi_0, \ldots, \xi_{j-1}, \xi_{j+1}, \ldots, \xi_3)} \frac{\partial q_i}{\partial \xi_j} \right\} \quad (13.16)$$

The rightmost term is evaluated by utilizing Eq. (13.14), to obtain

$$\sum_{i=0}^{3} (-1)^{i+j} \frac{\partial(x_0, \ldots, x_{i-1}, x_{i+1}, \ldots, x_3)}{\partial(\xi_0, \ldots, \xi_{j-1}, \xi_{j+1}, \ldots, \xi_3)} \frac{\partial q_i}{\partial \xi_j}$$

$$= J \sum_{i=0}^{3} \sum_{k=0}^{3} (-1)^{i+k} \frac{\partial(x_0, \ldots, x_{i-1}, x_{i+1}, \ldots, x_3)}{\partial(x_0, \ldots, x_{k-1}, x_{k+1}, \ldots, x_3)} \frac{\partial q_i}{\partial x_k}$$

where $(-1)^{2j} = 1$. However, all the x_i are independent variables, so that

$$\frac{\partial(x_0, \ldots, x_{i-1}, x_{i+1}, \ldots, x_3)}{\partial(x_0, \ldots, x_{k-1}, x_{k+1}, \ldots, x_3)} = \delta_{ik}$$

The rightmost term in Eq. (13.16) equals

$$J \sum_{i=0}^{3} \frac{\partial q_i}{\partial x_i} = 0$$

by virtue of Eq. (13.10).

We now sum Eq. (13.16) over j and interchange on the right side the i and j summations to obtain

$$\sum_{j=0}^{3} \frac{\partial Q_j}{\partial \xi_j} = \sum_{i=0}^{3} (-1)^i q_i \sum_{j=0}^{3} (-1)^j \frac{\partial}{\partial \xi_j} \frac{\partial(x_0, \ldots, x_{i-1}, x_{i+1}, \ldots, x_3)}{\partial(\xi_0, \ldots, \xi_{j-1}, \xi_{j+1}, \ldots, \xi_3)}$$

Consider the j summation on the right side with i fixed at zero. This quantity is

$$\sum_{j=0}^{3} = \frac{\partial}{\partial \xi_0} \frac{\partial(x_1, x_2, x_3)}{\partial(\xi_1, \xi_2, \xi_3)} - \frac{\partial}{\partial \xi_1} \frac{\partial(x_1, x_2, x_3)}{\partial(\xi_0, \xi_2, \xi_3)} + \cdots$$

We readily evaluate the sum when the upper limit is 1 or 2:

$$\sum_{j=0}^{1} = \frac{\partial}{\partial \xi_0} \frac{\partial x_1}{\partial \xi_1} - \frac{\partial}{\partial \xi_1} \frac{\partial x_1}{\partial \xi_0} = 0$$

$$\sum_{j=0}^{2} = \frac{\partial}{\partial \xi_0} \frac{\partial(x_1, x_2)}{\partial(\xi_1, \xi_2)} - \frac{\partial}{\partial \xi_1} \frac{\partial(x_1, x_2)}{\partial(\xi_0, \xi_2)} + \frac{\partial}{\partial \xi_2} \frac{\partial(x_1, x_2)}{\partial(\xi_0, \xi_1)}$$

$$= \frac{\partial}{\partial \xi_0} \left(\frac{\partial x_1}{\partial \xi_1} \frac{\partial x_2}{\partial \xi_2} - \frac{\partial x_1}{\partial \xi_2} \frac{\partial x_2}{\partial \xi_1} \right) - \cdots$$

Fully expanded, the right side of the last j summation shows that the resulting 12 terms cancel each other; hence,

$$\sum_{j=0}^{2} = 0$$

It is left as an exercise (see Problem 13.9) for the reader to show that

$$\sum_{j=0}^{n} (-1)^j \frac{\partial}{\partial \xi_j} \frac{\partial(x_0, \ldots, x_{i-1}, x_{i+1}, \ldots, x_n)}{\partial(\xi_0, \ldots, \xi_{j-1}, \xi_{j+1}, \ldots, \xi_n)} = 0$$

for an arbitrary i and any integer $n \geq 1$. We conclude that

$$\sum_{j=0}^{3} \frac{\partial Q_j}{\partial \xi_j} = 0 \qquad (13.17)$$

Equation (13.17) is the desired result. Thus, an arbitrary transformation, Eq. (13.11), will result in a conservative form providing new Q are defined in accord with Eq. (13.15).

The derivation of conservative form invariance is quite general in several aspects:

(1) The restriction on i or j going from 0 to 3 is arbitrary. We could have used 0 to n with equal ease.

(2) The transformation need not be steady or orthogonal. We require only that Eq. (13.11) exist and the Jacobian be nonzero.

(3) Since the elements of the q_i have not been used, our results are not specific to gasdynamics. They hold for any equation, or set of equations, with the form of Eq. (13.10). The transformation of the velocity components is unaffected by requiring conservative form invariance. Thus, velocity components tangential to the new coordinates can be introduced.

(4) For simplicity, we stated that the starting point, Eq. (13.10), utilized Cartesian coordinates. However, the coordinates x_i in Eq. (13.10) need not be Cartesian or orthogonal.

(5) Suppose an assumption on spatial symmetry is invoked after the new coordinates and tangential velocity components are introduced. Such an assumption would be that of an axially symmetric flow. Implementation of the assumption may lead to equations that are not in conservative form. Homework problems illustrate this point and item (3).

13.2 STEADY TWO-DIMENSIONAL OR AXISYMMETRIC FLOW—I

In this section, a simple illustration is provided of the foregoing transformation theory. Nevertheless, it is useful to state the assumptions not made. We do not assume subsonic or supersonic flow, a perfect gas, irrotational flow (defined later), or isentropic, homentropic, or isoenergetic (constant h_0) flow.

Since the flow is steady, $\tilde{q}_0$, x_1, x_2, and x_3 are independent of ξ_0 ($=t$). Hence,

$$\frac{\partial(\tilde{q}_0, x_1, x_2, x_3)}{\partial(\xi_0, \ldots, \xi_3)} = 0$$

and the Jacobians in Eq. (13.12) contract to third order, i.e.,

$$\frac{\partial(\tilde{q}_1, x_2, x_3)}{\partial(\xi_1, \xi_2, \xi_3)} + \frac{\partial(x_1, \tilde{q}_2, x_3)}{\partial(\xi_1, \xi_2, \xi_3)} + \frac{\partial(x_1, x_2, \tilde{q}_3)}{\partial(\xi_1, \xi_2, \xi_3)} = 0 \quad (13.18)$$

For planar two-dimensional flow, replace the ξ, v, and $\tilde{q}$ with

$$\xi_i = x_i, \qquad v_i = u_i, \qquad \tilde{q}_i = q_i, \qquad i = 1, 2$$

Next, delete the q_3 term since, by assumption, the elements in the numerator of this term do not depend on x_3. As a consequence, Eq. (13.18) reduces to

$$\frac{\partial q_1}{\partial x_1} + \frac{\partial q_2}{\partial x_2} = 0 \quad (13.19)$$

where

$$q_1 = \begin{pmatrix} \rho u_1 \\ p + \rho u_1^2 \\ \rho u_1 u_2 \\ \rho u_1 h_0 \end{pmatrix}, \quad q_2 = \begin{pmatrix} \rho u_2 \\ \rho u_1 u_2 \\ p + \rho u_2^2 \\ \rho u_2 h_0 \end{pmatrix} \quad (13.20)$$

Written out, the first two components of Eq. (13.19) are

$$\frac{\partial(\rho u_1)}{\partial x_1} + \frac{\partial(\rho u_2)}{\partial x_2} = 0$$

$$\frac{\partial(p + \rho u_1^2)}{\partial x_1} + \frac{\partial(\rho u_1 u_2)}{\partial x_2} = 0$$

For this case, Eqs. (13.19) and (13.20) are the final form for the equations of motion.

A less trivial illustration is the axisymmetric case, whose cylindrical polar coordinates are shown in Fig. 13.1. We thus have

$$\xi_1 = x_1, \quad \xi_2 = (x_2^2 + x_3^2)^{\frac{1}{2}}, \quad \xi_3 = \tan^{-1}(x_3/x_2)$$

or, upon inversion,

$$x_1 = \xi_1, \quad x_2 = \xi_2 \cos \xi_3, \quad x_3 = \xi_2 \sin \xi_3 \quad (13.21)$$

The A matrix and the scale factors are obtained from Eqs. (13.21):

$$A = \begin{pmatrix} 1 & 0 & 0 \\ 0 & \cos \xi_3 & -\xi_2 \sin \xi_3 \\ 0 & \sin \xi_3 & \xi_2 \cos \xi_3 \end{pmatrix} \quad (13.22)$$

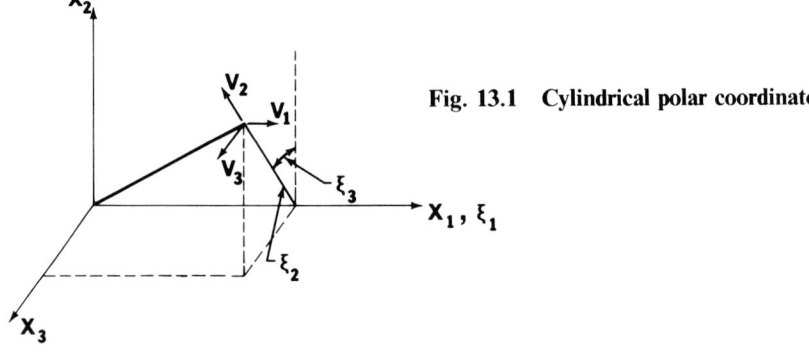

Fig. 13.1 Cylindrical polar coordinates.

TRANSFORMATION OF THE CONSERVATION EQUATIONS

$$h_1 = h_2 = 1, \qquad h_3 = \xi_2 \tag{13.23}$$

Equations (13.9) then yield

$$u_1 = v_1$$
$$u_2 = v_2 \cos \xi_3 - v_3 \sin \xi_3 \tag{13.24a}$$
$$u_3 = v_2 \sin \xi_3 + v_3 \cos \xi_3$$

and

$$v_1 = u_1$$
$$v_2 = u_2 \cos \xi_3 + u_3 \sin \xi_3 \tag{13.24b}$$
$$v_3 = -u_2 \sin \xi_3 + u_3 \cos \xi_3$$

Next, we determine the three terms in Eq. (13.18) as follows:

$$\frac{\partial(q_1, x_2, x_3)}{\partial(\xi_1, \xi_2, \xi_3)} = \begin{vmatrix} \frac{\partial q_1}{\partial \xi_1} & \frac{\partial q_1}{\partial \xi_2} & \frac{\partial q_1}{\partial \xi_3} \\ 0 & \cos \xi_3 & -\xi_2 \sin \xi_3 \\ 0 & \sin \xi_3 & \xi_2 \cos \xi_3 \end{vmatrix} = \xi_2 \frac{\partial q_1}{\partial \xi_1} \tag{13.25a}$$

$$\frac{\partial(x_1, q_2, x_3)}{\partial(\xi_1, \xi_2, \xi_3)} = \xi_2 \frac{\partial q_2}{\partial \xi_2} \cos \xi_3 - \frac{\partial q_2}{\partial \xi_3} \sin \xi_3 \tag{13.25b}$$

$$\frac{\partial(x_1, x_2, q_3)}{\partial(\xi_1, \xi_2, \xi_3)} = \xi_2 \frac{\partial q_3}{\partial \xi_2} \sin \xi_3 + \frac{\partial q_3}{\partial \xi_3} \cos \xi_3 \tag{13.25c}$$

In these Jacobians, we write q_i rather than $\tilde{q}_i$, since the $u \to v$ transformation, Eq. (13.24a), has not yet been imposed. Furthermore, it is wrong to set

$$\frac{\partial q_i}{\partial \xi_3} = 0$$

because the q_i depend on the u_i, and the derivatives $\partial u_i / \partial \xi_3$ are not zero. In place of Eq. (13.18), we now have

$$\frac{\partial q_1}{\partial \xi_1} + \frac{\partial}{\partial \xi_2}(q_2 \cos \xi_3 + q_3 \sin \xi_3) - \frac{\sin \xi_3}{\xi_2} \frac{\partial q_2}{\partial \xi_3} + \frac{\cos \xi_3}{\xi_2} \frac{\partial q_3}{\partial \xi_3} = 0$$

$$\tag{13.26}$$

In view of Eqs. (13.24a), the q_i are

$$q_1 = \begin{pmatrix} \rho v_1 \\ p + \rho v_1^2 \\ \rho v_1 u_2 \\ \rho v_1 u_3 \\ \rho v_1 h_0 \end{pmatrix}, \quad q_2 = \begin{pmatrix} \rho u_2 \\ \rho v_1 u_2 \\ p + \rho u_2^2 \\ \rho u_2 u_3 \\ \rho u_2 h_0 \end{pmatrix}, \quad q_3 = \begin{pmatrix} \rho u_3 \\ \rho v_1 u_3 \\ \rho u_2 u_3 \\ p + \rho u_3^2 \\ \rho u_3 h_0 \end{pmatrix}$$

and, for convenience,

$$q_2 \cos \xi_3 + q_3 \sin \xi_3 = \begin{pmatrix} \rho v_2 \\ \rho v_1 v_2 \\ p \cos \xi_3 + \rho v_2 u_2 \\ p \sin \xi_3 + \rho v_2 u_3 \\ \rho v_2 h_0 \end{pmatrix}$$

In the above, a partial velocity component changeover has been accomplished. We now utilize the symmetry conditions

$$\frac{\partial \rho}{\partial \xi_3} = \frac{\partial p}{\partial \xi_3} = \frac{\partial h_0}{\partial \xi_3} = \frac{\partial v_j}{\partial \xi_3} = 0$$

and complete the velocity component transformation. As a consequence, the two rightmost terms in Eq. (13.26) reduce to

$$-\frac{\sin \xi_3}{\xi_2} \frac{\partial q_2}{\partial \xi_3} + \frac{\cos \xi_3}{\xi_2} \frac{\partial q_3}{\partial \xi_3} = \begin{pmatrix} \frac{\rho v_2}{\xi_2} \\ \frac{\rho v_1 v_2}{\xi_2} \\ \frac{\rho}{\xi_2} \left[(v_2^2 - v_3^2) \cos \xi_3 - 2 v_2 v_3 \sin \xi_3 \right] \\ \frac{\rho}{\xi_2} \left[(v_2^2 - v_3^2) \sin \xi_3 + 2 v_2 v_3 \cos \xi_3 \right] \\ \frac{\rho v_2 h_0}{\xi_2} \end{pmatrix}$$

With this replacement in Eq. (13.26), only the second and third momentum equations contain ξ_3. This coordinate is eliminated by multiplying the third equation in Eq. (13.26) by $\cos \xi_3$ and the fourth by $\sin \xi_3$, and then adding the two together. The resulting momentum equation

$$\frac{\partial}{\partial \xi_1}(\rho v_1 v_2) + \frac{\partial}{\partial \xi_2}(p + \rho v_2^2) + \frac{\rho (v_2^2 - v_3^2)}{\xi_2} = 0$$

is free of ξ_3.

The foregoing results can be summarized as follows:

$$\frac{\partial \tilde{q}_1}{\partial \xi_1} + \frac{\partial \tilde{q}_2}{\partial \xi_2} + \sigma \tilde{q}_4 = 0 \quad (13.27)$$

where

$$\tilde{q}_1 = \begin{pmatrix} \rho v_1 \\ p + \rho v_1^2 \\ \rho v_1 v_2 \\ \rho v_1 h_0 \end{pmatrix}, \quad \tilde{q}_2 = \begin{pmatrix} \rho v_2 \\ \rho v_1 v_2 \\ p + \rho v_2^2 \\ \rho v_2 h_0 \end{pmatrix}, \quad \tilde{q}_4 = \begin{pmatrix} \rho v_2/\xi_2 \\ \rho v_1 v_2/\xi_2 \\ \rho(v_2^2 - v_3^2)/\xi_2 \\ \rho v_2 h_0/\xi_2 \end{pmatrix} \quad (13.28)$$

$$\sigma = 0, \text{ two-dimensional flow}$$
$$= 1, \text{ axisymmetric flow} \quad (13.29)$$

and

$$h_0 = h + \frac{1}{2} \sum_{j=1}^{3} v_j^2$$

The parameter σ enables us to include the earlier two-dimensional result. Observe that the equations for continuity and energy become redundant when $h_0 = $ const.

By subtracting out continuity from the momentum and energy equations, we obtain from Eq. (13.27) the following more familiar form for the conservation equations:

$$\frac{\partial(\rho v_1)}{\partial \xi_1} + \frac{\partial(\rho v_2)}{\partial \xi_2} + \sigma \frac{\rho v_2}{\xi_2} = 0$$

$$v_1 \frac{\partial v_1}{\partial \xi_1} + v_2 \frac{\partial v_1}{\partial \xi_2} + \frac{1}{\rho} \frac{\partial p}{\partial \xi_1} = 0$$

$$v_1 \frac{\partial v_2}{\partial \xi_1} + v_2 \frac{\partial v_2}{\partial \xi_2} - \frac{\sigma v_3^2}{\xi_2} + \frac{1}{\rho} \frac{\partial p}{\partial \xi_2} = 0 \quad (13.30)$$

$$v_1 \frac{\partial h_0}{\partial \xi_1} + v_2 \frac{\partial h_0}{\partial \xi_2} = 0$$

Although

$$\frac{\partial v_3}{\partial \xi_3} = 0$$

v_3 itself need not be zero. In fact, planar vortical flow is given by

$$v_1 = v_2 = 0, \quad v_3 = v_3(\xi_2) \quad (13.31a)$$

which yields

$$\frac{\rho v_3^2}{\xi_2} = \frac{\partial p}{\partial \xi_2} \qquad (13.31b)$$

$$h + \tfrac{1}{2} v_3^2 = h_0 = \text{const} \qquad (13.31c)$$

There is a second momentum equation

$$\xi_2 v_3 = \text{const} \qquad (13.31d)$$

which was the one previously eliminated.

13.3 STEADY TWO-DIMENSIONAL OR AXISYMMETRIC FLOW—II

In Sec. 13.2, the $x_i = x_i(\xi_j)$ equations are known. We now suppose that this transformation is not explicitly known. We only assume that it exists, that it is nonsingular, and that the ξ_j coordinates are orthogonal. Because of the generality involved, it will be necessary to augment substantially the theory introduced in Sec. 13.1. As in Sec. 13.2, we do not assume subsonic or supersonic flow, a perfect gas, or irrotational, isentropic, or isoenergetic flow. Some of the material in this section is based on an article by Meyer.[1]

There are several reasons for deriving the equations of motion in a curvilinear orthogonal coordinate system. Foremost among these is the utility of the derived results for establishing the theory of characteristics given in Sec. 16.1. However, there are other reasons, e.g., the ready application of the equations to natural coordinates, which is the subject of Sec. 13.4.

Geometric Considerations

The starting point will be Eqs. (13.30), with $v_3 = 0$. For notational simplicity, we replace ξ_i and v_i in these equations with x_i and u_i, respectively. For convenience, the equations are repeated in the altered notation:

$$\frac{\partial(\rho u_1)}{\partial x_1} + \frac{\partial(\rho u_2)}{\partial x_2} + \sigma \frac{\rho u_2}{x_2} = 0$$

$$u_1 \frac{\partial u_1}{\partial x_1} + u_2 \frac{\partial u_1}{\partial x_2} + \frac{1}{\rho} \frac{\partial p}{\partial x_1} = 0$$

$$u_1 \frac{\partial u_2}{\partial x_1} + u_2 \frac{\partial u_2}{\partial x_2} + \frac{1}{\rho} \frac{\partial p}{\partial x_2} = 0 \qquad (13.32)$$

$$u_1 \frac{\partial h_0}{\partial x_1} + u_2 \frac{\partial h_0}{\partial x_2} = 0$$

These equations are to be transformed to new, arbitrary, orthogonal coordinates ξ_1 and ξ_2, whose respective velocity components are v_1 and v_2.

As is apparent from Eqs. (13.32), we have chosen x_3 to be perpendicular to the plane of the flow in the two-dimensional case. In the axisymmetric case, x_1 is the symmetry axis, x_2 the radial coordinate, and x_3 the azimuthal angle. Figure 13.2 shows a sketch of the old and new coordinates. The quantities $h_j d\xi_j$ are the differential length elements, θ is the angle between ξ_1 and x_1, and κ_j are the local curvatures of the ξ_j coordinates, respectively. Figure 13.3 is used to relate θ and the curvatures κ_j to the h_j scale factors. The infinitesimal length $h_2 d\xi_2$ is a circular arc whose origin is point 0, providing the ξ coordinates are orthogonal. Since arc length equals the product of the radius with the included angle, we obtain from the figure

$$\frac{1}{\kappa_2} d\theta \bigg|_{\xi_1} = \frac{1}{\kappa_2}\left(\frac{\partial \theta}{\partial \xi_2} d\xi_2\right) = h_2 d\xi_2$$

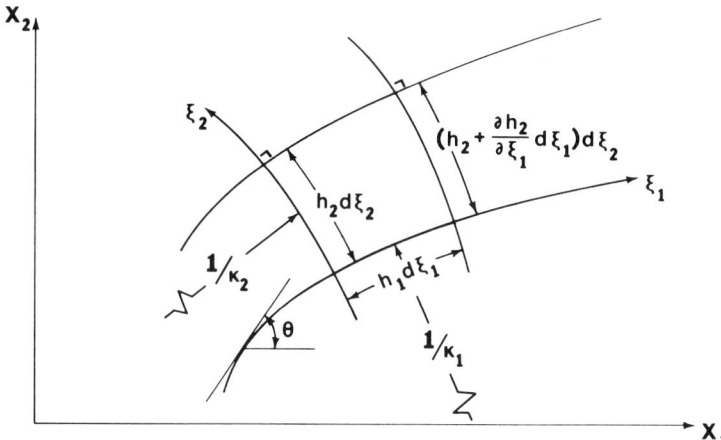

Fig. 13.2 Original two-dimensional or axisymmetric coordinate system and new curvilinear orthogonal coordinate system.

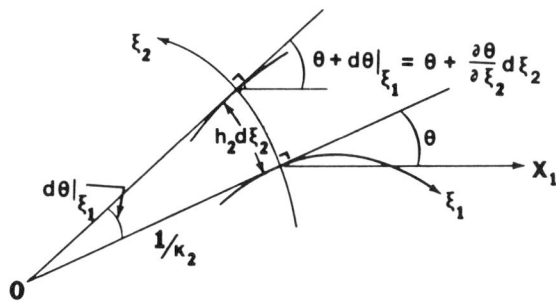

Fig. 13.3 Geometrical basis for Eqs. (13.33).

or

$$\kappa_2 = \frac{1}{h_2} \frac{\partial \theta}{\partial \xi_2} \qquad (13.33a)$$

In a similar manner, we obtain

$$\kappa_1 = -\frac{1}{h_1} \frac{\partial \theta}{\partial \xi_1} \qquad (13.33b)$$

Figure 13.3 is extended to include two $\xi_1 = $ const arcs, as shown in Fig. 13.4. For the outermost arc, we have

$$\left(\frac{1}{\kappa_2} + h_1 \, d\xi_1 \right) d\theta \bigg|_{\xi_1} = \left(h_2 + \frac{\partial h_2}{\partial \xi_1} d\xi_1 \right) d\xi_2$$

which simplifies to

$$h_1 \, d\xi_1 \, d\theta \big|_{\xi_1} = \frac{\partial h_2}{\partial \xi_1} d\xi_1 \, d\xi_2$$

$$h_1 \left(\frac{\partial \theta}{\partial \xi_2} d\xi_2 \right) = \frac{\partial h_2}{\partial \xi_1} d\xi_2$$

$$\frac{\partial \theta}{\partial \xi_2} = \frac{1}{h_1} \frac{\partial h_2}{\partial \xi_1} \qquad (13.34a)$$

In a similar manner, a second relation is obtained for θ:

$$\frac{\partial \theta}{\partial \xi_1} = -\frac{1}{h_2} \frac{\partial h_1}{\partial \xi_2} \qquad (13.34b)$$

Eliminating θ from Eqs. (13.33) and (13.34) yields for the curvatures

$$\kappa_1 = \frac{1}{h_1 h_2} \frac{\partial h_1}{\partial \xi_2}, \qquad \kappa_2 = \frac{1}{h_1 h_2} \frac{\partial h_2}{\partial \xi_1} \qquad (13.35)$$

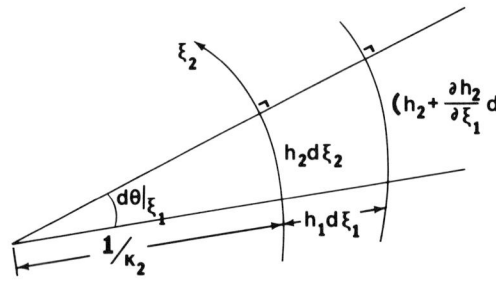

Fig. 13.4 Extension of Fig. 13.3 showing two $\xi_1 = $ const arcs.

TRANSFORMATION OF THE CONSERVATION EQUATIONS   229

Thus, the scale factors fully determine θ and the κ_j. Since

$$\frac{\partial^2 \theta}{\partial \xi_1 \partial \xi_2} = \frac{\partial^2 \theta}{\partial \xi_2 \partial \xi_1}$$

Eqs. (13.34) result in the scale-factor condition

$$\frac{\partial}{\partial \xi_1}\left(\frac{1}{h_1}\frac{\partial h_2}{\partial \xi_1}\right) + \frac{\partial}{\partial \xi_2}\left(\frac{1}{h_2}\frac{\partial h_1}{\partial \xi_2}\right) = 0 \qquad (13.36a)$$

which can also be written as a curvature condition

$$\frac{1}{h_2}\frac{\partial \kappa_1}{\partial \xi_2} + \frac{1}{h_1}\frac{\partial \kappa_2}{\partial \xi_1} + \kappa_1^2 + \kappa_2^2 = 0 \qquad (13.36b)$$

Equation (13.6) actually determines the scale factors, i.e.,

$$h_j = \left[\left(\frac{\partial x_1}{\partial \xi_j}\right)^2 + \left(\frac{\partial x_2}{\partial \xi_j}\right)^2\right]^{\frac{1}{2}} \qquad (13.37)$$

An orthogonality condition for the ξ_j is derived by noting that the vectors $\mathbf{e}_j$, which are tangential to the ξ_j, are given by

$$\mathbf{e}_j = \sum_{i=1}^{2} \frac{\partial x_i}{\partial \xi_j}\hat{1}_i$$

Orthogonality requires that $\mathbf{e}_1 \cdot \mathbf{e}_2 = 0$ or

$$\frac{\partial x_1}{\partial \xi_1}\frac{\partial x_1}{\partial \xi_2} + \frac{\partial x_2}{\partial \xi_1}\frac{\partial x_2}{\partial \xi_2} = 0 \qquad (13.38)$$

There are other orthogonality relations that are equivalent to this equation. The Jacobian of the transformation is given by

$$J(x_1, x_2) = \frac{\partial(x_1, x_2)}{\partial(\xi_1, \xi_2)} = \frac{\partial x_1}{\partial \xi_1}\frac{\partial x_2}{\partial \xi_2} - \frac{\partial x_1}{\partial \xi_2}\frac{\partial x_2}{\partial \xi_1} \qquad (13.39a)$$

[It is convenient here to use Eq. (13.39a) for the Jacobian rather than its inverse.] By means of Eqs. (13.37) and (13.38), one can show that

$$J = h_1 h_2 \qquad (13.39b)$$

which is interpreted to mean that a unit element of area in the x plane transforms into an element with area $h_1 h_2$ in the ξ plane.

Transformation of the Conservation Equations

As before, we begin by transforming the velocity components u_i in Eqs. (13.32) to v_j. The A matrix is

$$A = \begin{pmatrix} \dfrac{\partial x_1}{\partial \xi_1} & \dfrac{\partial x_1}{\partial \xi_2} \\ \dfrac{\partial x_2}{\partial \xi_1} & \dfrac{\partial x_2}{\partial \xi_2} \end{pmatrix}$$

and Eq. (13.9a) is given by

$$u_1 = \frac{\partial x_1}{\partial \xi_1}\frac{v_1}{h_1} + \frac{\partial x_1}{\partial \xi_2}\frac{v_2}{h_2}$$
$$u_2 = \frac{\partial x_2}{\partial \xi_1}\frac{v_1}{h_1} + \frac{\partial x_2}{\partial \xi_2}\frac{v_2}{h_2} \tag{13.40}$$

Another set of useful geometric relations can now be developed, based on Fig. 13.5. From the figure, we have*

$$u_1 = v_1 \cos\theta - v_2 \sin\theta$$
$$u_2 = v_1 \sin\theta + v_2 \cos\theta \tag{13.41}$$

and by comparison with Eqs. (13.40),

$$\frac{\partial x_1}{\partial \xi_1} = h_1 \cos\theta, \qquad \frac{\partial x_1}{\partial \xi_2} = -h_2 \sin\theta$$
$$\frac{\partial x_2}{\partial \xi_1} = h_1 \sin\theta, \qquad \frac{\partial x_2}{\partial \xi_2} = h_2 \cos\theta \tag{13.42}$$

These relations identically satisfy Eqs. (13.37) and (13.38), and Eq. (13.39b) is easily obtained from Eq. (13.39a).

The continuity equation is expanded to

$$\frac{D\rho}{Dt} + \rho\left(\frac{\partial u_1}{\partial x_1} + \frac{\partial u_2}{\partial x_2}\right) + \sigma\frac{\rho u_2}{x_2} = 0 \tag{13.43}$$

---

*The inverse of Eqs. (13.41) is

$$v_1 = u_1 \cos\theta + u_2 \sin\theta$$
$$v_2 = -u_1 \sin\theta + u_2 \cos\theta$$

TRANSFORMATION OF THE CONSERVATION EQUATIONS   231

where the substantial derivative of any function ϕ is given by

$$\frac{D\phi}{Dt} = u_1 \frac{\partial \phi}{\partial x_1} + u_2 \frac{\partial \phi}{\partial x_2} \tag{13.44a}$$

In view of the fact that this derivative appears in each of the equations of motion, we transform it to the new coordinates. This transformation is given by

$$\frac{D\phi}{Dt} = u_1 \frac{\partial(\phi, x_2)}{\partial(\xi_1, \xi_2)} \frac{1}{J} + u_2 \frac{\partial(x_1, \phi)}{\partial(\xi_1, \xi_2)} \frac{1}{J}$$

$$J\frac{D\phi}{Dt} = \left(\frac{\partial x_1}{\partial \xi_1}\frac{v_1}{h_1} + \frac{\partial x_1}{\partial \xi_2}\frac{v_2}{h_2}\right)\left(\frac{\partial \phi}{\partial \xi_1}\frac{\partial x_2}{\partial \xi_2} - \frac{\partial \phi}{\partial \xi_2}\frac{\partial x_2}{\partial \xi_1}\right)$$

$$+ \left(\frac{\partial x_2}{\partial \xi_1}\frac{v_1}{h_1} + \frac{\partial x_2}{\partial \xi_2}\frac{v_2}{h_2}\right)\left(\frac{\partial \phi}{\partial \xi_2}\frac{\partial x_1}{\partial \xi_1} - \frac{\partial \phi}{\partial \xi_1}\frac{\partial x_1}{\partial \xi_2}\right)$$

$$= J\frac{v_1}{h_1}\frac{\partial \phi}{\partial \xi_1} + J\frac{v_2}{h_2}\frac{\partial \phi}{\partial \xi_2}$$

or

$$\frac{D\phi}{Dt} = \frac{v_1}{h_1}\frac{\partial \phi}{\partial \xi_1} + \frac{v_2}{h_2}\frac{\partial \phi}{\partial \xi_2} \tag{13.44b}$$

With the derivatives

$$\frac{\partial J}{\partial \xi_1} = \frac{\partial x_2}{\partial \xi_2}\frac{\partial^2 x_1}{\partial \xi_1^2} + \frac{\partial x_1}{\partial \xi_1}\frac{\partial^2 x_2}{\partial \xi_1 \partial \xi_2} - \frac{\partial x_2}{\partial \xi_1}\frac{\partial^2 x_1}{\partial \xi_1 \partial \xi_2} - \frac{\partial x_1}{\partial \xi_2}\frac{\partial^2 x_1}{\partial \xi_1^2}$$

$$\frac{\partial J}{\partial \xi_2} = \frac{\partial x_2}{\partial \xi_2}\frac{\partial^2 x_1}{\partial \xi_1 \partial \xi_2} + \frac{\partial x_1}{\partial \xi_1}\frac{\partial^2 x_2}{\partial \xi_2^2} - \frac{\partial x_2}{\partial \xi_1}\frac{\partial^2 x_1}{\partial \xi_2^2} - \frac{\partial x_1}{\partial \xi_2}\frac{\partial^2 x_2}{\partial \xi_1 \partial \xi_2}$$

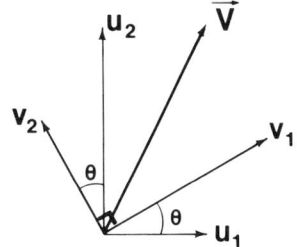

Fig. 13.5 Velocity components in the Cartesian coordinate system and in the curvilinear system.

the divergence of velocity transforms as follows:

$$\frac{\partial u_1}{\partial x_1} + \frac{\partial u_2}{\partial x_2} = \frac{1}{J}\left[\frac{\partial(u_1, x_2)}{\partial(\xi_1, \xi_2)} + \frac{\partial(x_1, u_2)}{\partial(\xi_1, \xi_2)}\right]$$

$$= \frac{1}{J}\left(\frac{\partial u_1}{\partial \xi_1}\frac{\partial x_2}{\partial \xi_2} - \frac{\partial u_1}{\partial \xi_2}\frac{\partial x_2}{\partial \xi_1} + \frac{\partial x_1}{\partial \xi_1}\frac{\partial u_2}{\partial \xi_2} - \frac{\partial x_1}{\partial \xi_2}\frac{\partial u_2}{\partial \xi_1}\right)$$

$$= \frac{1}{J}\left[\frac{v_1}{h_1}\frac{\partial J}{\partial \xi_1} + J\frac{\partial(v_1/h_1)}{\partial \xi_1} + \frac{v_2}{h_2}\frac{\partial J}{\partial \xi_2} + J\frac{\partial(v_2/h_2)}{\partial \xi_2}\right]$$

$$= \frac{1}{J}\left[\frac{\partial}{\partial \xi_1}\left(\frac{v_1 J}{h_1}\right) + \frac{\partial}{\partial \xi_2}\left(\frac{v_2 J}{h_2}\right)\right]$$

$$= \frac{1}{h_1 h_2}\left[\frac{\partial}{\partial \xi_1}(h_2 v_1) + \frac{\partial}{\partial \xi_2}(h_1 v_2)\right]$$

$$= \frac{1}{h_1}\frac{\partial v_1}{\partial \xi_1} + \frac{1}{h_2}\frac{\partial v_2}{\partial \xi_2} + \frac{v_1}{h_1 h_2}\frac{\partial h_2}{\partial \xi_1} + \frac{v_2}{h_1 h_2}\frac{\partial h_1}{\partial \xi_2}$$

The algebraic manipulations between the second and third steps, which use Eqs. (13.40), have been omitted, and Eq. (13.39b) is used to replace J.

The foregoing results enable us to write Eq. (13.43) for continuity as

$$\frac{v_1}{h_1}\frac{\partial \rho}{\partial \xi_1} + \frac{v_2}{h_2}\frac{\partial \rho}{\partial \xi_2} + \frac{\rho}{h_1}\frac{\partial v_1}{\partial \xi_1} + \frac{\rho}{h_2}\frac{\partial v_2}{\partial \xi_2}$$

$$+ \frac{\rho}{h_1 h_2}\left(v_1\frac{\partial h_2}{\partial \xi_1} + v_2\frac{\partial h_1}{\partial \xi_2}\right) + \sigma\frac{\rho}{x_2}\left(\frac{\partial x_1}{\partial \xi_1}\frac{v_1}{h_1} + \frac{\partial x_2}{\partial \xi_2}\frac{v_2}{h_2}\right) = 0$$

By introducing Eqs. (13.35) and (13.42), we obtain

$$\frac{1}{h_1}\frac{\partial(\rho v_1)}{\partial \xi_1} + \frac{1}{h_2}\frac{\partial(\rho v_2)}{\partial \xi_2} + \rho(\kappa_2 v_1 + \kappa_1 v_2) + \frac{\sigma \rho}{x_2}(v_1 \sin\theta + v_2 \cos\theta) = 0 \quad (13.45)$$

where x_2 is a function of ξ_j. This is the desired form for continuity.

We next transform the momentum equations in Eqs. (13.32), starting with the pressure gradients. With the aid of Eqs. (13.42), we obtain

$$\frac{\partial p}{\partial x_1} = \frac{\partial(p, x_2)}{\partial(x_1, x_2)} = \frac{\partial(p, x_2)}{\partial(\xi_1, \xi_2)}\frac{1}{J}$$

$$= \frac{1}{h_1 h_2}\left(\frac{\partial p}{\partial \xi_1}\frac{\partial x_2}{\partial \xi_2} - \frac{\partial p}{\partial \xi_2}\frac{\partial x_2}{\partial \xi_1}\right) = \frac{\cos\theta}{h_1}\frac{\partial p}{\partial \xi_1} - \frac{\sin\theta}{h_2}\frac{\partial p}{\partial \xi_2} \quad (13.46a)$$

TRANSFORMATION OF THE CONSERVATION EQUATIONS 233

and similarly

$$\frac{\partial p}{\partial x_2} = \frac{\sin\theta}{h_1}\frac{\partial p}{\partial \xi_1} + \frac{\cos\theta}{h_2}\frac{\partial p}{\partial \xi_2} \tag{13.46b}$$

By utilizing Eqs. (13.44b) and (13.46), we have for the momentum equations:

$$\frac{v_1}{h_1}\frac{\partial u_1}{\partial \xi_1} + \frac{v_2}{h_2}\frac{\partial u_1}{\partial \xi_2} + \frac{1}{\rho}\left(\frac{\cos\theta}{h_1}\frac{\partial p}{\partial \xi_1} - \frac{\sin\theta}{h_2}\frac{\partial p}{\partial \xi_2}\right) = 0$$

$$\frac{v_1}{h_1}\frac{\partial u_2}{\partial \xi_1} + \frac{v_2}{h_2}\frac{\partial u_2}{\partial \xi_2} + \frac{1}{\rho}\left(\frac{\sin\theta}{h_1}\frac{\partial p}{\partial \xi_1} + \frac{\cos\theta}{h_2}\frac{\partial p}{\partial \xi_2}\right) = 0$$

The u_i are replaced with Eqs. (13.34) and (13.41) to yield

$$\sin\theta\left(\frac{v_1^2}{h_1 h_2}\frac{\partial h_1}{\partial \xi_2} - \frac{v_1}{h_1}\frac{\partial v_2}{\partial \xi_1} - \frac{v_1 v_2}{h_1 h_2}\frac{\partial h_2}{\partial \xi_1} - \frac{v_2}{h_2}\frac{\partial v_2}{\partial \xi_2} - \frac{1}{\rho h_2}\frac{\partial p}{\partial \xi_2}\right)$$

$$+ \cos\theta\left(\frac{v_1}{h_1}\frac{\partial v_1}{\partial \xi_1} + \frac{v_1 v_2}{h_1 h_2}\frac{\partial h_1}{\partial \xi_2} + \frac{v_2}{h_2}\frac{\partial v_1}{\partial \xi_2} - \frac{v_2^2}{h_1 h_2}\frac{\partial h_2}{\partial \xi_1} + \frac{1}{\rho h_1}\frac{\partial p}{\partial \xi_1}\right) = 0$$

$$\sin\theta\left(\frac{v_1}{h_1}\frac{\partial v_1}{\partial \xi_1} + \frac{v_1 v_2}{h_1 h_2}\frac{\partial h_1}{\partial \xi_2} + \frac{v_2}{h_2}\frac{\partial v_1}{\partial \xi_2} - \frac{v_2^2}{h_1 h_2}\frac{\partial h_2}{\partial \xi_1} + \frac{1}{\rho h_1}\frac{\partial p}{\partial \xi_1}\right)$$

$$- \cos\theta\left(\frac{v_1^2}{h_1 h_2}\frac{\partial h_1}{\partial \xi_2} - \frac{v_1}{h_1}\frac{\partial v_2}{\partial \xi_1} - \frac{v_1 v_2}{h_1 h_2}\frac{\partial h_2}{\partial \xi_1} - \frac{v_2}{h_2}\frac{\partial v_2}{\partial \xi_2} - \frac{1}{\rho h_2}\frac{\partial p}{\partial \xi_2}\right) = 0$$

These equations have the form

$$A\sin\theta + B\cos\theta = 0$$
$$B\sin\theta - A\cos\theta = 0$$

which results in

$$A = B = 0$$

The $\partial h_i/\partial \xi_j$ derivatives are eliminated by using Eqs. (13.35), to obtain the final form for conservation of momentum:

$$\frac{v_1}{h_1}\frac{\partial v_1}{\partial \xi_1} + \frac{v_2}{h_2}\frac{\partial v_1}{\partial \xi_2} + \kappa_1 v_1 v_2 - \kappa_2 v_2^2 + \frac{1}{\rho h_1}\frac{\partial p}{\partial \xi_1} = 0 \tag{13.47a}$$

$$\frac{v_1}{h_1}\frac{\partial v_2}{\partial \xi_1} + \frac{v_2}{h_2}\frac{\partial v_2}{\partial \xi_2} - \kappa_1 v_1^2 + \kappa_2 v_1 v_2 + \frac{1}{\rho h_2}\frac{\partial p}{\partial \xi_2} = 0 \tag{13.47b}$$

Conservation of energy readily reduces to

$$\frac{v_1}{h_1}\frac{\partial h_0}{\partial \xi_1} + \frac{v_2}{h_2}\frac{\partial h_0}{\partial \xi_2} = 0 \qquad (13.48)$$

where

$$h_0 = e + \frac{p}{\rho} + \frac{1}{2}(v_1^2 + v_2^2) \qquad (13.49)$$

In their new form, the conservation equations are Eqs. (13.45), (13.47), and (13.48). Assume that h_0 is replaced by Eq. (13.49) and that two thermodynamic state equations are available for the elimination of e and p. The unknowns then are v_1, v_2, ρ, and T, which the four conservation equations fully determine. However, these equations also contain κ_1, κ_2, h_1, h_2, θ, and x_2, which are functions of ξ_1 and ξ_2. If the transformation $x_i = x_i(\xi_j)$ is explicitly known, then the h_j are determined by Eq. (13.37), κ_j by Eqs. (13.35), and θ by Eqs. (13.42). The only restrictions on the transformation are that $J \neq 0$ and Eq. (13.38) must hold for orthogonality.

In general, however, the transformation is not known a priori, but is found as part of the solution. The unknowns now include κ_1, κ_2, h_1, h_2, x_1, x_2, and θ. The conservation equations are supplemented by enough of Eqs. (13.34), (13.35), (13.37), and (13.42) to cover these variables. In addition, boundary conditions are needed for variables such as θ and h_1. Equation (13.38) for orthogonality is not needed because Eqs. (13.42) identically satisfy this condition. Despite its complexity, this formulation of the conservation equations will provide the theoretical framework for much of the subsequent analysis.

Shock Wave Equations

Of course, a coordinate transformation may affect the shock wave conditions, as well as the governing equations and velocity components. We thus reexamine the shock relations of Sec. 12.2 in a manner consistent with this section. Since any shock is a surface in three dimensions, the theory of surfaces[3] will be used. For purposes of simplicity, however, the discussion is limited to a two-dimensional or axisymmetric shock, whose equation is given by

$$F = x_1 - f(x_2) = 0 \qquad (13.50)$$

where x_1 is the symmetry axis and x_2 is the radial coordinate when the flow is axisymmetric.

The shock may be an attached or detached wave. For a surface of revolution, we have

$$dl^2 = (1 + f'^2)(dx_2)^2 + \sigma x_2^2(dx_3)^2 \qquad (13.51)$$

TRANSFORMATION OF THE CONSERVATION EQUATIONS   235

where l is the arc length along the surface, x_3 is the transverse coordinate when $\sigma = 0$ or the azimuthal angle when $\sigma = 1$, and a prime denotes the derivative

$$f' = \frac{df}{dx_2}$$

Following Ref. 3, one can obtain the coefficients

$$E = 1 + f'^2, \qquad F = 0, \qquad G = \sigma x_2^2$$

$$e = \frac{f''}{\left(1 + f'^2\right)^{\frac{1}{2}}}, \qquad f = 0, \qquad g = \frac{\sigma x_2 f'}{\left(1 + f'^2\right)^{\frac{1}{2}}}$$

for the first (E, F, G) and second (e, f, g) fundamental differential forms of a surface. [The coefficients $F = f = 0$ should not be confused with the F and f in Eq. (13.50). These zero coefficients are not used in the rest of this subsection.]

Consider any curve on a surface in three dimensions. At a point P on the curve, a circle can be constructed that is tangent to the curve at this point and has the same curvature as the curve. The inverse of the radius of the (osculating) circle is a curvature of the surface at P. One can show that any point on the surface has two extremum (a minimum and a maximum) curvatures, which are associated with two orthogonal curves on the surface. These are the principal curvatures.

A meridian plane, $x_3 = $ const, will intersect the $F = 0$ surface in a curve that passes through the origin, as shown by the dashed line in Fig. 13.6. The

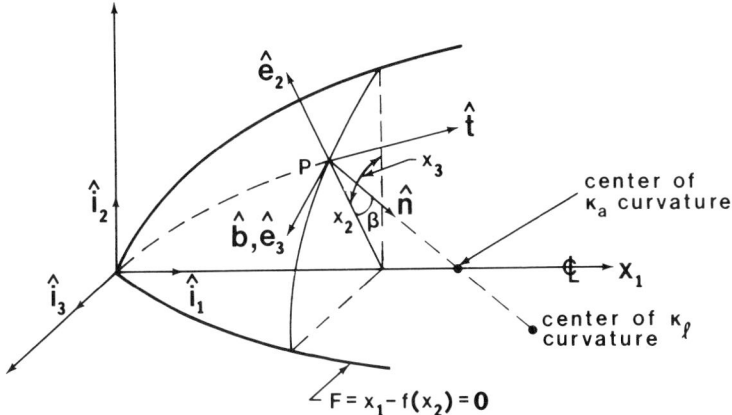

Fig. 13.6 Coordinate systems and center of the principal curvatures for an axisymmetric surface.

longitudinal curvature associated with this curve is given by[3]

$$\kappa_\ell = \frac{e}{E} = \frac{f''}{(1+f'^2)^{\frac{3}{2}}} \qquad (13.52a)$$

A second curve is obtained by cutting the surface with a plane that is perpendicular to the meridian plane. (As will be evident, this plane is generally not perpendicular to the x_1 axis.) The resulting azimuthal curvature is[3]

$$\kappa_a = \frac{g}{G} = \frac{\sigma f'}{x_2(1+f'^2)^{\frac{1}{2}}} \qquad (13.52b)$$

where $\kappa_a = 0$ in the two-dimensional case. These are the principal curvatures that characterize the surface. For instance, consider any other surface curve in Fig. 13.6 that has an angle α relative to the vector $\hat{\mathbf{t}}$. Euler's theorem,

$$\kappa = \kappa_\ell \cos^2\alpha + \kappa_a \sin^2\alpha$$

then provides the corresponding curvature κ.

The total, or Gaussian, curvature is defined by

$$K = \kappa_\ell \kappa_a = \frac{\sigma f' f''}{x_2(1+f'^2)^2}$$

A surface is called elliptic if, at each point, $K > 0$ or, equivalently, $\sigma f' f'' > 0$. Similarly, if $K = 0$, the surface is called parabolic. The surface sketched in Fig. 13.6 is elliptic, and our subsequent discussion assumes an elliptic or parabolic surface. Planar or conical shock waves are examples of parabolic surfaces. The figure shows the coordinates x_i and the corresponding orthonormal basis $\hat{\mathbf{e}}_i$, where $\hat{\mathbf{e}}_i$ is parallel to $\hat{\mathbf{l}}_i$. Also shown is the orthonormal basis $\hat{\mathbf{n}}$, $\hat{\mathbf{t}}$, and $\hat{\mathbf{b}}$ introduced in Sec. 12.2. Observe that $\hat{\mathbf{b}}$ and $\hat{\mathbf{e}}_3$ coincide and that $\hat{\mathbf{n}}$ is an inward-pointing vector.

As in Chap. 12, we assume a uniform freestream with a constant velocity

$$\mathbf{V}_1 = V_1 \hat{\mathbf{e}}_1 \qquad (13.53)$$

In line with Fig. 12.3, we have

$$\hat{\mathbf{t}} \cdot \hat{\mathbf{e}}_1 = \cos\beta$$

where β is the usual shock angle measured relative to the freestream velocity.

The various gradients required in Sec. 12.2 are given by

$$\nabla F = \hat{\mathbf{e}}_1 - f' \hat{\mathbf{e}}_2$$

$$|\nabla F| = \left(1 + f'^2\right)^{\frac{1}{2}}$$

$$|\nabla F'| = f'$$

From Eq. (12.19), we have

$$\cos \beta = \frac{|\nabla F'|}{|\nabla F|} = \frac{f'}{\left(1 + f'^2\right)^{\frac{1}{2}}} \qquad (13.54a)$$

Hence,

$$\sin \beta = \frac{1}{\left(1 + f'^2\right)^{\frac{1}{2}}} \qquad (13.54b)$$

$$\tan \beta = \frac{1}{f'} \qquad (13.54c)$$

and Eq. (13.54c) yields

$$f'' = -\frac{1}{\sin^2 \beta} \frac{d\beta}{dx_2} \qquad (13.55)$$

For axisymmetric flow, an elliptic surface is equivalent to a negative value for $d\beta/dx_2$. It is a simple matter to show that Eqs. (12.16)–(12.18) yield

$$\hat{\mathbf{n}} = \hat{\mathbf{e}}_1 \sin \beta - \hat{\mathbf{e}}_2 \cos \beta$$

$$\hat{\mathbf{t}} = \hat{\mathbf{e}}_1 \cos \beta + \hat{\mathbf{e}}_2 \sin \beta$$

$$\hat{\mathbf{b}} = \hat{\mathbf{e}}_3$$

Equations (13.52) thus become

$$\kappa_\ell = -\sin \beta \frac{d\beta}{dx_2} \qquad (13.56a)$$

$$\kappa_a = \sigma \frac{\cos \beta}{x_2} \qquad (13.56b)$$

These relations enable us to show that $\hat{\mathbf{n}}$ is at an angle β relative to the x_2 coordinate in Fig. 13.6 and that both centers of curvature fall on the line defined by $\hat{\mathbf{n}}$. In fact, the azimuthal center of curvature is located on the symmetry axis.

The velocity on the downstream side of the shock $\mathbf{V}_2$ is given by

$$\mathbf{V}_2 = v_{x_1}\hat{\mathbf{e}}_1 + v_{x_2}\hat{\mathbf{e}}_2$$

where

$$\frac{v_{x_1}}{V_1} = 1 - \frac{2}{\gamma+1}\frac{M_1^2\sin^2\beta - 1}{M_1^2}$$

$$\frac{v_{x_2}}{V_1} = \frac{2}{\gamma+1}\frac{M_1^2\sin^2\beta - 1}{M_1^2\tan\beta}$$

The velocity turn angle θ is given by Eq. (12.21b), or as

$$\tan\theta = \frac{v_{x_2}}{v_{x_1}}$$

Equations (12.14) then reduce to

$$\frac{\rho_2}{\rho_1} = \frac{[(\gamma+1)/2]\,M_1^2\sin^2\beta}{1 + [(\gamma-1)/2]\,M_1^2\sin^2\beta}$$

$$\frac{p_2}{p_1} = \frac{2}{\gamma+1}\left(\gamma M_1^2\sin^2\beta - \frac{\gamma-1}{2}\right) \tag{13.57}$$

$$\frac{T_2}{T_1} = \left(\frac{2}{\gamma+1}\right)^2 \frac{\{1 + [(\gamma-1)/2]\,M_1^2\sin^2\beta\}\{\gamma M_1^2\sin^2\beta - [(\gamma-1)/2]\}}{M_1^2\sin^2\beta}$$

The derivation has resulted in the usual oblique shock relations first considered in Chap. 5.

13.4 NATURAL COORDINATES

The transformed conservation equations of Sec. 13.3 have a particularly elegant form when natural coordinates are used. In this system, the ξ_2 = const lines are chosen as the streamlines, and ξ_1 increases in the direction of positive v_1. Thus, the ξ_j coordinate system is coincident with and perpendicular to the streamlines. As a consequence,

$$v_1 = v, \qquad v_2 = 0 \tag{13.58}$$

and the equations of motion simplify to

$$\frac{1}{h_1}\frac{\partial(\rho v)}{\partial \xi_1} + \kappa_2 \rho v + \sigma \frac{\rho v \sin\theta}{x_2} = 0$$

$$\rho \frac{v}{h_1}\frac{\partial v}{\partial \xi_1} + \frac{1}{h_1}\frac{\partial p}{\partial \xi_1} = 0$$

$$\kappa_1 \rho v^2 - \frac{1}{h_2}\frac{\partial p}{\partial \xi_2} = 0 \tag{13.59a}$$

$$\frac{1}{h_1}\frac{\partial h_0}{\partial \xi_1} = 0$$

Actual arc lengths in the physical plane s and n, along and transverse to the streamlines, are introduced by means of

$$\frac{\partial}{\partial s} = \frac{1}{h_1}\frac{\partial}{\partial \xi_1}, \qquad \frac{\partial}{\partial n} = \frac{1}{h_2}\frac{\partial}{\partial \xi_2}$$

We thereby obtain

$$\frac{\partial(\rho v)}{\partial s} + \rho v \frac{\partial \theta}{\partial n} + \sigma \frac{\rho v \sin\theta}{x_2} = 0$$

$$\rho v \frac{\partial v}{\partial s} + \frac{\partial p}{\partial s} = 0 \tag{13.59b}$$

$$\rho v^2 \frac{\partial \theta}{\partial s} + \frac{\partial p}{\partial n} = 0$$

and

$$e + \frac{p}{\rho} + \frac{1}{2}v^2 = h_0(n) \tag{13.60}$$

where the κ_j are replaced by

$$\kappa_1 = -\frac{1}{h_1}\frac{\partial \theta}{\partial \xi_1} = -\frac{\partial \theta}{\partial s}, \qquad \kappa_2 = \frac{1}{h_2}\frac{\partial \theta}{\partial \xi_2} = \frac{\partial \theta}{\partial n}$$

The unknowns are p, ρ, v, and the slope θ of the streamlines relative to the x_1 axis. The stagnation enthalpy $h_0(n)$ is considered a known function that is determined by upstream boundary conditions.

In practice, natural coordinates are not used for an axisymmetric flow because the $1/x_2$ factor in the continuity equation is an unknown function of s and n. Natural coordinates are used for two-dimensional flows, since the governing equations have a particularly simple form. For instance, the characteristic and compatibility equations (see Chap. 16) are readily derived,[2] starting with these equations. Despite the complexity of inverting the transformation to Cartesian coordinates, two-dimensional natural coordinates are occasionally used in computational fluid dynamics.

13.5 HODOGRAPH TRANSFORMATION

The hodograph plane has either u_1, u_2 or θ, v as its coordinates. These are the independent variables, rather than the x_i, or s and n. The reason for this interchange will be apparent later. We pursue this transformation by starting with Eqs. (13.59b) and by introducing additional independent assumptions:

(1) Isoenergetic flow, $h_0 = $ const.
(2) Two-dimensional flow, $\sigma = 0$.
(3) A thermally and calorically perfect gas.
(4) Homentropic flow, entropy is a constant.

No Mach number restriction is imposed. As a result of items (3) and (4), the pressure is given by

$$\frac{p}{p_0} = \left(\frac{\rho}{\rho_0}\right)^\gamma \tag{13.61}$$

where p_0 and ρ_0 are the constant stagnation pressure and density. Equations (13.59b) become

$$\frac{1}{\rho}\frac{\partial \rho}{\partial s} + \frac{1}{v}\frac{\partial v}{\partial s} + \frac{\partial \theta}{\partial n} = 0$$

$$v\frac{\partial v}{\partial s} + \gamma\frac{p_0}{\rho_0^\gamma}\rho^{\gamma-2}\frac{\partial \rho}{\partial s} = 0 \tag{13.62}$$

$$v^2\frac{\partial \theta}{\partial s} + \gamma\frac{p_0}{\rho_0^\gamma}\rho^{\gamma-2}\frac{\partial \rho}{\partial n} = 0$$

Let X and Y be the new independent variables. In Jacobian form, Eqs. (13.62) are written as

$$\frac{1}{\rho}\frac{\partial(\rho, n)}{\partial(X, Y)} + \frac{1}{v}\frac{\partial(v, n)}{\partial(X, Y)} + \frac{\partial(s, \theta)}{\partial(X, Y)} = 0$$

$$v\frac{\partial(v, n)}{\partial(X, Y)} + \gamma\frac{p_0}{\rho_0^\gamma}\rho^{\gamma-2}\frac{\partial(\rho, n)}{\partial(X, Y)} = 0$$

$$v^2\frac{\partial(\theta, n)}{\partial(X, Y)} + \gamma\frac{p_0}{\rho_0^\gamma}\rho^{\gamma-2}\frac{\partial(s, \rho)}{\partial(X, Y)} = 0$$

TRANSFORMATION OF THE CONSERVATION EQUATIONS   241

We now choose for X and Y

$$X = v, \qquad Y = \theta$$

The Jacobians in the continuity equation are given by

$$\frac{\partial(\rho, n)}{\partial(v, \theta)} = \begin{vmatrix} \frac{\partial \rho}{\partial v} & \frac{\partial \rho}{\partial \theta} \\ \frac{\partial n}{\partial v} & \frac{\partial n}{\partial \theta} \end{vmatrix} = \frac{\partial \rho}{\partial v}\frac{\partial n}{\partial \theta} - \frac{\partial \rho}{\partial \theta}\frac{\partial n}{\partial v}$$

$$\frac{\partial(v, n)}{\partial(v, \theta)} = \frac{\partial n}{\partial \theta}, \qquad \frac{\partial(s, \theta)}{\partial(v, \theta)} = \frac{\partial s}{\partial v}$$

with similar results for the Jacobians in the other equations. Thus, the conservation equations become

$$\frac{1}{\rho}\left(\frac{\partial \rho}{\partial v}\frac{\partial n}{\partial \theta} - \frac{\partial \rho}{\partial \theta}\frac{\partial n}{\partial v}\right) + \frac{1}{v}\frac{\partial n}{\partial \theta} + \frac{\partial s}{\partial v} = 0$$

$$v\frac{\partial n}{\partial \theta} + \gamma\frac{p_0}{\rho_0^\gamma}\rho^{\gamma-2}\left(\frac{\partial \rho}{\partial v}\frac{\partial n}{\partial \theta} - \frac{\partial \rho}{\partial \theta}\frac{\partial n}{\partial v}\right) = 0$$

$$v^2\frac{\partial n}{\partial v} + \gamma\frac{p_0}{\rho_0^\gamma}\rho^{\gamma-2}\left(\frac{\partial \rho}{\partial v}\frac{\partial s}{\partial \theta} - \frac{\partial \rho}{\partial \theta}\frac{\partial s}{\partial v}\right) = 0$$

Assumptions (1) and (3) are used to obtain for Eq. (13.60) the following energy equation:

$$\frac{\gamma}{\gamma-1}\frac{p_0}{\rho_0^\gamma}\rho^{\gamma-1} + \frac{1}{2}v^2 = h_0$$

from which we have

$$\frac{\partial \rho}{\partial \theta} = 0$$

$$\frac{\partial \rho}{\partial v} = -\frac{v}{\gamma(p_0/\rho_0^\gamma)\rho^{\gamma-2}}$$

The conservation equations become

$$-\frac{v}{\gamma(p_0/\rho_0^\gamma)\rho^{\gamma-2}}\frac{\partial n}{\partial \theta} + \frac{1}{v}\frac{\partial n}{\partial \theta} + \frac{\partial s}{\partial v} = 0, \qquad \frac{\partial n}{\partial \theta} - \frac{\partial n}{\partial \theta} = 0$$

$$v\frac{\partial n}{\partial v} - \frac{\partial s}{\partial \theta} = 0 \qquad\qquad (13.63a)$$

where the second equation is an identity. Thus, the number of governing equations is reduced by one. The first equation simplifies to

$$\frac{h_0 - \frac{1}{2}[(\gamma+1)/(\gamma-1)]v^2}{h_0 - \frac{1}{2}v^2} \frac{1}{v}\frac{\partial n}{\partial \theta} + \frac{\partial s}{\partial v} = 0 \qquad (13.63b)$$

By cross-differentiation, we can eliminate s from Eqs. (13.63) to obtain

$$\frac{h_0 - \frac{1}{2}[(\gamma+1)/(\gamma-1)]v^2}{h_0 - \frac{1}{2}v^2} \frac{\partial^2 n}{\partial \theta^2} + v^2\frac{\partial^2 n}{\partial v^2} + v\frac{\partial n}{\partial v} = 0 \qquad (13.64)$$

In contrast to Eqs. (13.62), which are nonlinear, Eqs. (13.63) and (13.64) are linear equations. Within the stated assumptions, we have obtained a linear system that is exact, and for which superposition of solutions can be used. As illustrated by Problem 13.13, the transformation also yields linear equations when applied to unsteady one-dimensional flow.

The difficulty with the hodograph formulation is usually in the complexity of the boundary conditions. Nevertheless, it is used for studying the flow of a free jet. With a constant ambient pressure, the flow speed v at the edge of the jet is a known constant, while on the centerline, $n(v, \theta = 0) = 0$. In addition, a boundary condition is needed at the start of the jet where it leaves a duct or reservoir. See Chap. VII in Ref. 1 for an extensive discussion of the hodograph transformation.

As discussed in Ref. 2, the approximate equation that describes transonic flow is inherently nonlinear. Again, the hodograph transformation results in a linear equation. For this reason, the bulk of transonic flow analysis is done in the hodograph plane.

Aside from the previously stated assumptions, the Jacobian of the transformation cannot be zero except at isolated points. Thus, v and θ must be independent of each other. As a consequence, the hodograph transformation cannot yield similarity solutions, such as for Prandtl-Meyer flow (Chap. 6). These are referred to as "missing" solutions.

Since the gas is perfect, we have

$$h_0 = \frac{a^2}{\gamma - 1} + \frac{1}{2}v^2$$

where

$$a = \text{speed of sound} = (\gamma RT)^{\frac{1}{2}}$$

With $M = (v/a)$, Eq. (13.64) becomes

$$(1 - M^2)\frac{\partial^2 n}{\partial \theta^2} + v^2\frac{\partial^2 n}{\partial v^2} + v\frac{\partial n}{\partial v} = 0 \qquad (13.65)$$

Equation (13.65) is a second-order partial differential equation (PDE) of the classic form

$$A\phi_{xx} + 2B\phi_{xy} + C\phi_{yy} = D \qquad (13.66)$$

which was discussed in Sec. 9.3. As we know, Eq. (13.66) falls into one of the three categories according to the sign of $B^2 - AC$. For Eq. (13.65) we have

$$B^2 - AC = -v^2(1 - M^2)$$

Hence, Eqs. (13.64) and (13.65) are elliptic when $M < 1$ and hyperbolic when $M > 1$.

In the v, θ plane, the solution of Eq. (13.65) is restricted to a semi-infinite strip ($0 \le v$, $-\pi \le \theta \le \pi$). The flow is supersonic when $v > v^*$, where the sonic speed is

$$v^* = \left(2\frac{\gamma - 1}{\gamma + 1} h_0\right)^{\frac{1}{2}}$$

and stagnates at the left boundary when $v = 0$.

References

[1] Meyers, R. E., "The Method of Characteristics," *Modern Developments in Fluid Dynamics*, edited by L. Howarth, Oxford University Press, New York, 1953.

[2] Liepmann, H. W. and Roshko, A., *Elements of Gasdynamics*, J. Wiley & Sons, New York, 1957.

[3] Struik, D. J., *Differential Geometry*, Chap. 2, Addison-Wesley Publishing Co., Reading, MA, 1950.

Problems

13.1 Consider unsteady two-dimensional flow starting with Eq. (13.10) and a four-element set for the q_i to be obtained from Eq. (11.11). (Delete q_3 and set $u_3 = 0$.) For the transformation

$$x_0 = t, \qquad \xi_0 = t$$
$$x_1 = x_1(\xi_0, \xi_1, \xi_2)$$
$$x_2 = x_2(\xi_0, \xi_1, \xi_2)$$

determine explicit results for the conservative form Q_j. It is not necessary to determine the elements of Q_j. Just determine the simplified relations between Q_j and q_i.

13.2 Consider steady two-dimensional flow, and utilize Problem 13.1 and the material in the first two subsections of Sec. 13.3. For the conservative form

$$\frac{\partial Q_1}{\partial \xi_1} + \frac{\partial Q_2}{\partial \xi_2} = 0$$

determine the elements of Q_1 and Q_2 in terms of ρ, p, h_0, v_j, h_j, and θ, where $j = 1, 2$. Next, specialize your results to natural coordinates, retaining at all times the conservative form. How are the h_j determined? Show that your results are consistent with Eqs. (13.59a) when $\sigma = 0$.

13.3 Consider two-dimensional or axisymmetric steady flow, with $v_3 = 0$, about a cone or a wedge with semivertex angle θ_b, as shown in the sketch below. Transform Eqs. (13.27) to body-oriented x, y coordinates. Your final result should have a form similar to Eqs. (13.30).

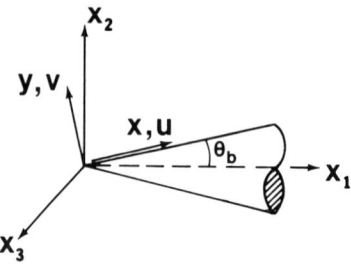

13.4 Show that

$$\left| \frac{\partial(\xi_1, \ldots, \xi_n)}{\partial(\eta_1, \ldots, \eta_n)} \right| = \prod_{j=1}^{n} \frac{k_j}{h_j}$$

where the space dimensionality n is a positive integer, and the η_j and ξ_j coordinate systems are each orthogonal. The h_j and k_j are the scale factors associated with the ξ and η coordinate systems relative to a Cartesian system, respectively. This relation is a generalization of Eq. (13.39b).

13.5 Start with Eqs. (13.59a) for the two-dimensional or axisymmetric flow. Further, assume homentropic and isoenergetic flow of a perfect gas. Without using the theory in the subsection headed "Invariance of the Conservative Form" in Sec. 13.1, write the governing PDEs in conservative form. Simplify your results as much as possible.

13.6 Prove Eq. (13.8).

TRANSFORMATION OF THE CONSERVATION EQUATIONS   245

13.7 Suppose that x_i are Cartesian coordinates and ξ_j are curvilinear orthogonal coordinates, where

$$x_i = x_i(\xi_1, \xi_2, \xi_3), \qquad i = 1, 2, 3$$

Show that

$$\sum_{j=1}^{3} \frac{1}{h_j^2} \frac{\partial x_i}{\partial \xi_j} \frac{\partial x_k}{\partial \xi_j} = \delta_{ik}, \qquad i, k = 1, 2, 3$$

13.8 (a) Use the theory in the subsection headed "Invariance of the Conservative Form" in Sec. 13.1 to derive the conservative form

$$\sum_{j=1}^{3} \frac{\partial Q_j}{\partial \xi_j} = 0$$

for steady flow in cylindrical polar coordinates, defined in Fig. 13.1. Your answer should be the Q_j, which should depend on the new velocity components v_j.

(b) With $v_3 = 0$ and with p, ρ, and v_j independent of ξ_3, write the governing equations in (a) in conservative form without a q_4 term [see Eq. (11.8)], except for the radial momentum equation.

13.9 Prove that

$$\sum_{j=0}^{n} (-1)^j \frac{\partial}{\partial \xi_j} \frac{\partial(x_0, \ldots, x_{i-1}, x_{i+1}, \ldots, x_n)}{\partial(\xi_0, \ldots, \xi_{j-1}, \xi_{j+1}, \ldots, \xi_n)} = 0$$

for any integer $n \geq 1$. This equation appears just before Eq. (13.17).

13.10 Consider steady, two-dimensional isentropic flow of a perfect gas. Use natural coordinates and let θ and

$$\phi = \ln(p/p_r)^{1/\gamma}$$

be the dependent variables, where p_r is a constant reference pressure. Determine the equations of motion with the Mach number M appearing in the coefficients of the derivatives. Determine an algebraic relation for M in terms of θ, ϕ, and h_0 when the flow is homentropic.

13.11 Use the hodograph transformation to transform the nonlinear equation

$$\phi_{tt} + F(\phi_t, \phi_x)\phi_{xx} = 0$$

into a linear second-order PDE with x as the dependent variable. Determine the character (hyperbolic, elliptic, or parabolic) of the new PDE. For the new coordinates use

$$u = \phi_t, \qquad v = \phi_x$$

13.12 As in Problem 13.11, use the hodograph transformation to transform the equation

$$\phi_{tt} + F(\phi_t, \phi_x)\phi_{xx} = G(\phi_t, \phi_x)$$

into a second-order PDE with x as the dependent variable. Determine the character of the new PDE.

13.13 Consider unsteady, one-dimensional homentropic motion of a perfect gas, given by

$$\frac{\partial a}{\partial t} + u\frac{\partial a}{\partial x} + \frac{\gamma - 1}{2}a\frac{\partial u}{\partial x} = 0$$

$$\frac{\partial u}{\partial t} + \frac{2}{\gamma - 1}a\frac{\partial a}{\partial x} + u\frac{\partial u}{\partial x} = 0$$

where a is the speed of sound and u is the flow speed (see Sec. 9.3).

(a) Apply the hodograph transformation such that the new dependent variables are x and t.

(b) Again, apply the hodograph transformation to the results of (a), such that a and u become the new dependent variables.

(c) Use the results of (a) to obtain a second-order PDE for t. Classify this equation.

(d) Suppose you have a solution, $t = t(a, u)$, of the (c) equation for t. Based on this solution, explain how you would, in principle, obtain algebraic equations for p, ρ, u, a, and M as functions of x and t.

(e) Consider (1) uniform flow and (2) a centered rarefaction wave as in shock-tube flow. Do the equations derived in (a) apply? Explain your answer.

13.14 Consider the transformation

$$x_1 = \xi_1 \xi_2, \qquad x_2 = \tfrac{1}{2}(\xi_1^2 - \xi_2^2)$$

Determine

$$J = \frac{\partial(x_1, x_2)}{\partial(\xi_1, \xi_2)}$$

as a function of ξ_1 and ξ_2, and

$$J^{-1} = \frac{\partial(\xi_1, \xi_2)}{\partial(x_1, x_2)}$$

as a function of x_1 and x_2. Verify Eqs. (N.6) in Appendix N, and show that

$$\left.\frac{\partial x_1}{\partial \xi_1}\right|_{\xi_2} \neq \frac{1}{\left.\dfrac{\partial \xi_1}{\partial x_1}\right|_{x_2}}$$

13.15 Determine the vorticity, defined by

$$\boldsymbol{\omega} = \nabla \times \mathbf{V}$$

for the flowfield of Sec. 13.3. Use this result to simplify Eqs. (13.47) by eliminating the curvatures. Finally, determine ω when natural coordinates are used.

13.16 Derive Eq. (13.56a) for κ_ℓ with the same approach used to derive Eqs. (13.33).

14. DEFINITIONS AND THEOREMS

Since the pioneering work of Euler and the Bernoullis, a sizable number of simplifications or assumptions have been introduced to assist in solving the governing equations. The most prominent of these are:
(1) Steady flow.
(2) Two-dimensional flow.
(3) Axisymmetric flow.
(4) Isentropic flow.
(5) Homentropic flow.
(6) Isoenergetic flow.
(7) Irrotational flow.
(8) Perfect gas.
(9) Incompressible flow.

Many practical flows are steady and of restricted dimensionality. In addition, the isoenergetic and perfect gas assumptions are frequently warranted. The only assumption we will not consider is the incompressible one.

A principal objective of this chapter is to define and discuss many of the foregoing assumptions. Some of these have already been invoked in earlier chapters. The irrotational flow assumption is one exception. Section 14.1 treats this assumption, along with such related topics as vorticity and Crocco's equation. As we will see, not all the above assumptions are independent of each other.

The most common upstream condition for a steady flow is a uniform one. This flow is also homentropic, isoenergetic, and irrotational. If it is supersonic, then shock waves are generally encountered. These are usually curved, and the flow downstream of them is rotational and isentropic but not homentropic. However, the isoenergetic property is preserved, which is why it is of greater importance than some of the other conditions. The principal concern of Sec. 14.2 is therefore with steady, rotational, isoenergetic flow that is two-dimensional or axisymmetric. Extensive use is made of Crocco's equation, and several different stream-function formulations are developed. Finally, in Sec. 14.3 the homogeneity property of the conservation equations is demonstrated. This property is important for computational fluid dynamics.

Since the amount of material that might be covered in this chapter is voluminous, selectivity and brevity are necessary. Furthermore, not all the above simplifications are of practical importance in a high-speed flow. A number of theorems that are primarily relevant to incompressible flow are not considered here, among them the Helmholtz vortex theorems, which can be found in many textbooks. (One textbook that is also a good reference for the next section is by Owczarek.[1])

14.1 BASIC CONCEPTS

We begin with the vorticity ω which, by definition, is the curl of the velocity

$$\omega = \nabla \times \mathbf{V} \qquad (14.1)$$

A related quantity is the circulation Γ, defined by

$$\Gamma = \int_C \mathbf{V} \cdot d\mathbf{r} \qquad (14.2)$$

where the integral is taken along a closed curve C in the counterclockwise direction. One can show that

$$\frac{D\Gamma}{Dt} = \int_C \frac{D\mathbf{V}}{Dt} \cdot d\mathbf{r} \qquad (14.3)$$

which is called Kelvin's equation. The vorticity and circulation are related by the surface integral

$$\Gamma = \int_S \boldsymbol{\omega} \cdot d\mathbf{A} = \int_S \boldsymbol{\omega} \cdot \hat{\mathbf{n}} \, dA$$

or

$$\frac{d\Gamma}{dA} = \hat{\mathbf{n}} \cdot \boldsymbol{\omega} \qquad (14.4)$$

where dA, $\hat{\mathbf{n}}$, and C are illustrated in Fig. 14.1. The unit vector $\hat{\mathbf{n}}$ is normal to the area dA, where $\hat{\mathbf{n}}$ and the curve C form a right-handed system.

We now consider a solid body rotation with a constant angular speed ω_b. Choose $\hat{\mathbf{n}}$ to lie along the axis of rotation. Let C be a circle of radius r, centered about the axis and lying in a plane perpendicular to $\hat{\mathbf{n}}$. Then,

$$\mathbf{V} = \omega_b r \hat{\mathbf{t}}$$

where $\hat{\mathbf{t}}$ is a unit vector tangent to C as shown in Fig. 14.1. To simplify the computation, choose a Cartesian coordinate system with $\hat{\mathbf{n}}$ along the

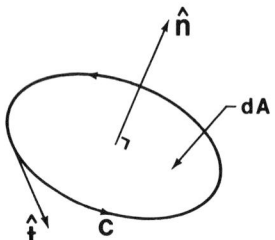

Fig. 14.1 Contour for the circulation integral.

positive x_3 axis. We have

$$\hat{t} = -\frac{x_2}{r}\hat{i}_1 + \frac{x_1}{r}\hat{i}_2$$

so that

$$\mathbf{V} = \omega_b\left(-x_2\hat{i}_1 + x_1\hat{i}_2\right)$$

and the curl of $\mathbf{V}$ is

$$\nabla \times \mathbf{V} = \begin{vmatrix} \hat{i}_1 & \hat{i}_2 & \hat{i}_3 \\ \dfrac{\partial}{\partial x_1} & \dfrac{\partial}{\partial x_2} & \dfrac{\partial}{\partial x_3} \\ -\omega_b x_2 & \omega_b x_1 & 0 \end{vmatrix} = 2\omega_b \hat{i}_3$$

Hence, Eq. (14.1) becomes

$$\boldsymbol{\omega} = \nabla \times \mathbf{V} = 2\omega_b \hat{n}$$

Thus, the vorticity vector is along the axis of rotation and equals, in magnitude, twice the angular speed ω_b. The circulation, given by Eq. (14.2), is computed as the product of $|\mathbf{V}|$ times the path length

$$\Gamma = (\omega_b r)2\pi r = 2\pi r^2 \omega_b = 2A\omega_b$$

where A is the area of the circle whose perimeter is C. Hence, by differentiation, we obtain

$$\frac{d\Gamma}{dA} = 2\omega_b = \hat{n} \cdot \boldsymbol{\omega}$$

in accord with Eq. (14.4). Also note that

$$\frac{D\Gamma}{Dt} = 0$$

since Γ is a constant, for a given r. Alternatively, the acceleration $D\mathbf{V}/Dt$, which is radially directed, is therefore perpendicular to $d\mathbf{r}$ [see Eq. (14.3)].

A motion is called irrotational if

$$\boldsymbol{\omega} = \nabla \times \mathbf{V} = 0 \tag{14.5}$$

in which case a single-valued velocity potential, to within a constant, can be introduced by

$$\mathbf{V} = \nabla \phi \tag{14.6}$$

providing the region is simply connected. The reason for Eq. (14.6) is that

$$\nabla \times \nabla \phi = 0$$

is a vector identity. In view of the integral definition of Γ, an irrotational motion in a simply connected region has zero circulation.

The most common example of an irrotational motion is the uniform flow that generally exists upstream of a body in a wind tunnel. In flight, the upstream condition is presumed to be quiescent, which is also irrotational.

We now derive Crocco's equation by first writing for momentum the vector equation

$$\frac{D\mathbf{V}}{Dt} = -\frac{1}{\rho}\nabla p \qquad (14.7a)$$

The substantial derivative of any scalar function ϕ, in general, is given by

$$\frac{D\phi}{Dt} = \frac{\partial \phi}{\partial t} + \mathbf{V} \cdot \nabla \phi$$

However, this relation is not always appropriate for the substantial derivative of $\mathbf{V}$, whose general expansion is given by*

$$\frac{D\mathbf{V}}{Dt} = \frac{\partial \mathbf{V}}{\partial t} + \boldsymbol{\omega} \times \mathbf{V} + \frac{1}{2}\nabla V^2$$

Consequently, Eq. (14.7a) can be written as

$$-\frac{1}{\rho}\nabla p = \frac{\partial \mathbf{V}}{\partial t} - \mathbf{V} \times \boldsymbol{\omega} + \frac{1}{2}\nabla V^2 \qquad (14.7b)$$

Equation (2.11),

$$de = T\,ds - p\,dv \qquad (14.8a)$$

is a general thermodynamic relation that holds in any arbitrary direction at a given instant of time. As a consequence, Eq. (14.8a) can be written as

$$\nabla e = T\nabla s - p\nabla v$$

We now replace e and v by

$$e = h - \frac{p}{\rho} = h_0 - \frac{1}{2}V^2 - \frac{p}{\rho}, \qquad v = \frac{1}{\rho}$$

---

*In Cartesian coordinates $\mathbf{V} \cdot \nabla \mathbf{V}$ is correct. The difficulty occurs in curvilinear coordinates, where $\mathbf{V} \cdot \nabla \mathbf{V}$ is not well defined. The transition from $\mathbf{V} \cdot \nabla \mathbf{V}$ is by means of the vector identity $\nabla(\mathbf{A} \cdot \mathbf{B}) = (\mathbf{A} \cdot \nabla)\mathbf{B} + (\mathbf{B} \cdot \nabla)\mathbf{A} + \mathbf{A} \times (\nabla \times \mathbf{B}) + \mathbf{B} \times (\nabla \times \mathbf{A})$ with $\mathbf{A} = \mathbf{B} = \mathbf{V}$.

to obtain

$$-\frac{1}{\rho}\nabla p - \frac{1}{2}\nabla V^2 = T\nabla s - \nabla h_0 \qquad (14.8b)$$

The pressure gradient term is eliminated from Eqs. (14.7b) and (14.8b), to obtain Crocco's equation

$$\frac{\partial \mathbf{V}}{\partial t} - \mathbf{V} \times \boldsymbol{\omega} = T\nabla s - \nabla h_0 \qquad (14.9)$$

The only significant assumptions are that the flow is inviscid and that Eq. (14.8a) holds for the thermodynamic variables. Equation (14.9) is a vector equation that is useful for establishing fundamental properties of a flowfield. Along a streamline, Crocco's equation becomes

$$\frac{1}{2}\frac{\partial V^2}{\partial t} = T\mathbf{V} \cdot \nabla s - \mathbf{V} \cdot \nabla h_0 \qquad (14.10)$$

since the triple product $\mathbf{V} \cdot \mathbf{V} \times \boldsymbol{\omega}$ is identically zero.

As discussed in Chap. 9, in an unsteady flow h_0 is generally not constant. Hence, even if the flow is adiabatic, $\mathbf{V} \cdot \nabla h_0$ is not necessarily zero. If the flow is steady and adiabatic, then $\mathbf{V} \cdot \nabla h_0 = 0$ as shown by Eq. (11.6). As it stands, however, Crocco's equation does not require a steady or an adiabatic flow.

Many of the assumptions listed at the beginning of the chapter directly impact Eqs. (14.9) and (14.10), as is evident from Table 14.1. Any two of the last three assumptions in the table imply the remaining one for a steady flow. The isentropic equation is for a steady flow. If the flow is unsteady, the entropy of a fluid particle is constant, i.e., $Ds/Dt = 0$ instead of $\mathbf{V} \cdot \nabla s = 0$. One can also distinguish between a constant stagnation enthalpy for a fluid particle and that for a region of flow. However, we will not make this later distinction.

The most common type of flow is a steady isoenergetic one, in which case Eq. (14.9) becomes

$$-\mathbf{V} \times \boldsymbol{\omega} = T\nabla s$$

Table 14.1
Assumptions Affecting Crocco's Equation

Assumption	Equation
Steady flow	$\frac{\partial}{\partial t} = 0$
Isentropic flow	$\mathbf{V} \cdot \nabla s = 0$
Homentropic flow	$\nabla s = 0$
Isoenergetic flow	$\nabla h_0 = 0$
Irrotational flow	$\boldsymbol{\omega} = 0$

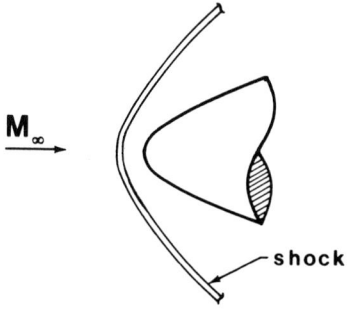

Fig. 14.2 Curved bow shock.

If the flow is uniform, then $\omega = 0$. Hence, we have $\nabla s = 0$, and the flow is also homentropic. One can show, by means of Eqs. (14.3), (14.7a), and (14.8b), that an irrotational flow will remain irrotational. This result, however, does not hold across a discontinuity. In particular, a curved shock, Fig. 14.2, is such a discontinuity. The jump in entropy across the shock depends on the inclination angle of the shock. Consequently, the entropy on the downstream side of the shock varies. As a result, the downstream flow is isentropic but not homotropic. In this flow region $\mathbf{V} \cdot \nabla s = 0$, but $\nabla s \neq 0$. By Crocco's equation, the flow must be rotational. Curved shock waves therefore are a source of vorticity. However, flow downstream of a conical, axisymmetric shock wave that curves only in the azimuthal direction is homentropic and therefore also irrotational (see Sec. 15.3).

Thus, a curved shock wave introduces vorticity. This shock-produced vorticity, however, does not remain constant. As shown by Problem 14.15, along any streamline downstream of the shock, $\partial \omega / \partial s$ is generally nonzero.

As an example, consider unsteady one-dimensional flow, which occurs in a shock tube. It is easy to see that the flow is irrotational, since

$$\nabla \times \mathbf{V} = \begin{vmatrix} \hat{\mathbf{i}}_1 & \hat{\mathbf{i}}_2 & \hat{\mathbf{i}}_3 \\ \dfrac{\partial}{\partial x_1} & \dfrac{\partial}{\partial x_2} & \dfrac{\partial}{\partial x_3} \\ u_1(x_1) & 0 & 0 \end{vmatrix} = 0$$

Hence, Crocco's equation becomes a scalar equation

$$\frac{\partial u}{\partial t} = T \frac{\partial s}{\partial x} - \frac{\partial h_0}{\partial x}$$

where $x = x_1$, and $u = u_1(x, t)$. Except across a shock wave or a contact surface, the flow is homentropic, i.e., $(\partial s / \partial x) = 0$. We thus have

$$\frac{\partial u}{\partial t} = - \frac{\partial h_0}{\partial x}$$

which becomes

$$\frac{Du}{Dt} = -\frac{\partial h}{\partial x} \tag{14.11}$$

With the perfect-gas relations

$$h = \frac{\gamma}{\gamma - 1}\frac{p}{\rho}, \qquad p \sim \rho^\gamma$$

Eq. (14.11) reduces to the momentum equation.

14.2 STREAM AND POTENTIAL FUNCTIONS

When appropriate, the use of a stream or potential function as the single dependent variable provides a considerable simplification of the equations of motion. Their use, however, is generally limited to a two-dimensional or axisymmetric flow. We discuss two different stream functions in this section. Less attention is paid to the potential function, since it requires an irrotational flow.

For steady two-dimensional flow, the conservation equations have the form [see Eq. (13.27)]

$$\frac{\partial q_1}{\partial x_1} + \frac{\partial q_2}{\partial x_2} = 0 \tag{14.12}$$

where each q consists of four elements. Except for the radial momentum equation, the other three axially symmetric flow equations also have this form [see Problem 13.8(b)]. Any one of these scalar equations can be satisfied by the substitution

$$q_{1j} = \frac{\partial \psi}{\partial x_2}, \qquad q_{2j} = -\frac{\partial \psi}{\partial x_1} \tag{14.13}$$

where the subscript j denotes the particular equation to be satisfied. Historically, this substitution has been used only for continuity, and ψ is then called the stream function.

Continuity is given by the first of Eqs. (13.32), and can be written as

$$\frac{\partial(\rho u_1 x_2^\sigma)}{\partial x_1} + \frac{\partial(\rho u_2 x_2^\sigma)}{\partial x_2} = 0 \tag{14.14}$$

where the x_i and u_i have the same meaning as introduced early in Sec. 13.3. The stream function is then defined by

$$\frac{\partial \psi}{\partial x_1} = -\rho u_2 x_2^\sigma, \qquad \frac{\partial \psi}{\partial x_2} = \rho u_1 x_2^\sigma \tag{14.15}$$

and Eq. (14.14) is identically satisfied. An important property of the stream function in a steady flow is that lines of constant ψ are streamlines that no material crosses.

We now assume an isoenergetic flow of a perfect gas. One of our objectives in this section is to develop equations that hold behind a curved shock wave. Therefore, we avoid the homentropic and irrotational assumptions. In place of a differential energy equation, we use

$$h_0 = \frac{\gamma}{\gamma-1}\frac{p}{\rho} + \frac{1}{2}(u_1^2 + u_2^2) = \text{const} \tag{14.16a}$$

which can be solved for p

$$p = \frac{\gamma-1}{\gamma}\left[h_0 - \frac{1}{2}(u_1^2 + u_2^2)\right]\rho \tag{14.16b}$$

Equations (13.32) for momentum can be written as

$$\frac{1}{2}\frac{\partial}{\partial x_1}[u_1^2 - (\gamma-1)u_2^2] + \gamma u_2\frac{\partial u_1}{\partial x_2} + \gamma\frac{p}{\rho^2}\frac{\partial \rho}{\partial x_1} = 0$$

$$\frac{1}{2}\frac{\partial}{\partial x_2}[-(\gamma-1)u_1^2 + u_2^2] + \gamma u_1\frac{\partial u_2}{\partial x_1} + \gamma\frac{p}{\rho^2}\frac{\partial \rho}{\partial x_2} = 0$$
(14.17)

Equations (14.15) allow u_1 and u_2 to be replaced by ψ, and Eq. (14.16b) is used to eliminate p. Thus, Eqs. (14.17) are two equations for ψ and ρ.

Crocco introduced a different stream-function approach, which we now present. We assume steady isoenergetic flow of a perfect gas. We do not assume two-dimensional or axisymmetric flow until much later. Crocco's equation is

$$-\mathbf{V} \times \boldsymbol{\omega} = T\nabla s \tag{14.18}$$

which we multiply by $\mathbf{V} \times$ to obtain

$$\mathbf{V} \times \mathbf{V} \times \boldsymbol{\omega} = -T\mathbf{V} \times \nabla s$$

The vector identity

$$\mathbf{A} \times (\mathbf{B} \times \mathbf{C}) = \mathbf{B}(\mathbf{A} \cdot \mathbf{C}) - \mathbf{C}(\mathbf{A} \cdot \mathbf{B})$$

with $\mathbf{A} = \mathbf{B} = \mathbf{V}$ yields $-V^2\boldsymbol{\omega}$ for the left side. In addition, Eq. (14.16a) can be written as

$$T = \frac{\gamma-1}{2\gamma R}V_m^2\left[1 - \left(\frac{V}{V_m}\right)^2\right]$$

DEFINITIONS AND THEOREMS

where the maximum possible speed V_m is

$$V_m = (2h_0)^{\frac{1}{2}}$$

Consequently, the vorticity is given by

$$\omega = \frac{\gamma - 1}{2\gamma R}\left[\left(\frac{V_m}{V}\right)^2 - 1\right]\mathbf{V} \times \nabla s \qquad (14.19)$$

This relation provides the vorticity, anywhere in the flow, in terms of the velocity components and the entropy gradient.

Next, multiply Eq. (14.18) by $\mathbf{V} \cdot$ to obtain

$$\mathbf{V} \cdot \nabla s = 0 \qquad (14.20)$$

which means that the flow is isentropic.

For a two-dimensional or axisymmetric flow, a single stream function ψ can be introduced. From Eq. (14.20) we have

$$s = s(\psi)$$

and, consequently,

$$\nabla s = \frac{ds}{d\psi}\nabla\psi$$

$$\mathbf{V} \times \nabla s = \frac{ds}{d\psi}\mathbf{V} \times \nabla\psi \qquad (14.21a)$$

In the three-dimensional case, two stream functions are necessary, and we have[2]

$$s = s(\psi_1, \psi_2)$$

$$\mathbf{V} \times \nabla s = \frac{\partial s}{\partial \psi_1}\mathbf{V} \times \nabla\psi_1 + \frac{\partial s}{\partial \psi_2}\mathbf{V} \times \nabla\psi_2 \qquad (14.21b)$$

Thus, one of Eqs. (14.21) is used for the right side of Eq. (14.19). For a steady flow, continuity can be written as

$$\nabla \cdot (\rho \mathbf{V}) = 0 \qquad (14.22)$$

Using vector identities, one can show that Eq. (14.22) is satisfied by

$$\rho \mathbf{V} = \nabla\psi_1 \times \nabla\psi_2 \qquad (14.23)$$

It is worth noting that the line of intersection of the two surfaces

$$\psi_i = c_i, \qquad i = 1, 2$$

is a streamline, since

$$\mathbf{V} \cdot \nabla \psi_i = 0, \qquad i = 1, 2$$

For a perfect gas, we have

$$a^2 = \gamma \frac{p}{\rho}$$

and

$$s = \text{const} + \frac{R}{\gamma - 1} \ell n (p/\rho^\gamma)$$

where the entropy relation can be modified to

$$s = \text{const} + \frac{1}{\gamma - 1} \ell n\, a^2 - R\, \ell n\, \rho$$

(The relation $p \sim \rho^\gamma$ is a homentropic relation, not an isentropic one. Since the flow is only isentropic, it cannot be used. Our analysis is designed to avoid this limitation.) By taking $\mathbf{V} \cdot \nabla$ of this last equation, and setting the result equal to zero, we obtain

$$\mathbf{V} \cdot \nabla \ell n\, \rho = \frac{1}{\gamma - 1} \mathbf{V} \cdot \nabla \ell n\, a^2$$

However, Eq. (14.16a) can be written as

$$a^2 = \frac{\gamma - 1}{2} V_m^2 \left[1 - \left(\frac{V}{V_m}\right)^2\right]$$

Eliminating a^2 from the above two relations, yields

$$\mathbf{V} \cdot \nabla \ell n\, \rho = \mathbf{V} \cdot \nabla \left\{ \ell n \left[1 - \left(\frac{V}{V_m}\right)^2\right]^{1/(\gamma - 1)} \right\}$$

Equation (14.22) can be expanded to

$$\nabla \cdot \mathbf{V} + \mathbf{V} \cdot \nabla \ell n\, \rho = 0$$

Combining the foregoing equations results in

$$\nabla \cdot \mathbf{V} + \mathbf{V} \cdot \nabla \left\{ \ell n \left[1 - \left(\frac{V}{V_m} \right)^2 \right]^{1/(\gamma-1)} \right\} = 0$$

or

$$\nabla \cdot \left\{ \left[1 - \left(\frac{V}{V_m} \right)^2 \right]^{1/(\gamma-1)} \mathbf{V} \right\} = 0 \qquad (14.24)$$

Equation (14.24), derived by Crocco, is equivalent to Eq. (14.22). Both hold in a three-dimensional steady flow.

Equations (14.19), (14.21b), and (14.24) hold for a three-dimensional flow. However, further progress will require the two-dimensional ($\sigma = 0$) or axisymmetric ($\sigma = 1$) assumption. In this case, Eq. (14.24) becomes

$$\frac{\partial}{\partial x_1} \left\{ x_2^\sigma u_1 \left[1 - \left(\frac{V}{V_m} \right)^2 \right]^{1/(\gamma-1)} \right\} + \frac{\partial}{\partial x_2} \left\{ x_2^\sigma u_2 \left[1 - \left(\frac{V}{V_m} \right)^2 \right]^{1/(\gamma-1)} \right\} = 0$$

where x_1 is the symmetry axis and x_2 is either the Cartesian vertical or the radial coordinate. Hence, a stream function ψ that does not depend on the density can be defined as

$$\frac{\partial \psi}{\partial x_1} = -u_2 \left[1 - \left(\frac{V}{V_m} \right)^2 \right]^{1/(\gamma-1)} x_2^\sigma$$

$$\frac{\partial \psi}{\partial x_2} = u_1 \left[1 - \left(\frac{V}{V_m} \right)^2 \right]^{1/(\gamma-1)} x_2^\sigma \qquad (14.25)$$

These express Crocco's stream-function equations. In this case, the scale factors and velocity are [see Eqs. (13.23)]

$$h_1 = h_2 = 1, \qquad h_3 = x_2^\sigma$$

$$\mathbf{V} = u_1 \hat{\mathbf{e}}_1 + u_2 \hat{\mathbf{e}}_2$$

so that

$$\mathbf{V} \times \nabla \psi = x_2^\sigma V^2 \left[1 - \left(\frac{V}{V_m} \right)^2 \right]^{1/(\gamma-1)} \hat{\mathbf{e}}_3$$

Finally, Eq. (14.19) becomes

$$\boldsymbol{\omega} = \frac{\gamma-1}{2\gamma R} V_m^2 \left[1 - \left(\frac{V}{V_m} \right)^2 \right]^{\gamma/(\gamma-1)} \frac{ds}{d\psi} x_2^\sigma \hat{\mathbf{e}}_3 \qquad (14.26)$$

where ψ is the Crocco stream function. Since $\boldsymbol{\omega} = \nabla \times \mathbf{V}$, one can readily show that

$$\boldsymbol{\omega} = \left(\frac{\partial u_2}{\partial x_1} - \frac{\partial u_1}{\partial x_2}\right)\hat{\mathbf{e}}_3 \qquad (14.27)$$

We then can eliminate $\boldsymbol{\omega}$, u_1, and u_2 from Eqs. (14.25)–(14.27) to obtain a second-order, quasi-linear PDE for ψ, as shown in Problem 14.8.

Let us momentarily drop the isoenergetic condition in favor of an irrotational one. We immediately have

$$\boldsymbol{\omega} = 0$$

which from Eq. (14.27) becomes

$$\frac{\partial u_2}{\partial x_1} = \frac{\partial u_1}{\partial x_2}$$

for both two-dimensional and axisymmetric flow. This equation is identically satisfied by a potential function given by Eq. (14.6). If the flow is also isoenergetic, it is then homentropic according to Crocco's equation. In this irrotational situation, the governing equations reduce to the form

$$\left(a^2 - \phi_{x_1}^2\right)\phi_{x_1 x_1} - 2\phi_{x_1}\phi_{x_2}\phi_{x_1 x_2} + \left(a^2 - \phi_{x_2}^2\right)\phi_{x_2 x_2} + \sigma\frac{a^2}{x_2}\phi_{x_2} = 0 \quad (14.28)$$

where the speed of sound a is given by the energy equation

$$a^2 = a_0^2 - \frac{\gamma - 1}{2}\left(\phi_{x_1}^2 + \phi_{x_2}^2\right) \qquad (14.29)$$

Equations (14.28) and (14.29) also result in a second-order, quasi-linear PDE.

It is instructive to transform Eq. (14.28) by means of the hodograph transformation. The velocity components u_1 and u_2 thereby become the new independent variables. Later, we will also change the dependent variable. The resulting equation will be a relatively simple linear PDE when $\sigma = 0$, but a nonlinear (not quasi-linear) PDE when σ is unity.

We first replace a_0 with the maximum possible speed

$$a_0^2 = \frac{\gamma - 1}{2}V_m^2$$

and combine Eqs. (14.28) and (14.29) to obtain

$$\left(V_m^2 - \frac{\gamma+1}{\gamma-1}u_1^2 - u_2^2\right)\phi_{x_1x_1} - \frac{4}{\gamma-1}u_1u_2\phi_{x_1x_2} + \left(V_m^2 - u_1^2 - \frac{\gamma+1}{\gamma-1}u_2^2\right)\phi_{x_2x_2}$$

$$+ \sigma\left(V_m^2 - u_1^2 - u_2^2\right)\frac{u_2}{x_2} = 0$$

For the derivatives, new independent variables are introduced

$$u_1 = \phi_{x_1} = X, \qquad u_2 = \phi_{x_2} = Y$$

and the derivatives are written in Jacobian form

$$\left(V_m^2 - \frac{\gamma+1}{\gamma-1}u_1^2 - u_2^2\right)\frac{\partial(X, x_2)}{\partial(x_1, x_2)} + \frac{4}{\gamma-1}u_1u_2\frac{\partial(X, x_1)}{\partial(x_1, x_2)}$$

$$- \left(V_m^2 - u_1^2 - \frac{\gamma+1}{\gamma-1}u_2^2\right)\frac{\partial(Y, x_1)}{\partial(x_1, x_2)} + \sigma\left(V_m^2 - u_1^2 - u_2^2\right)\frac{u_2}{x_2} = 0$$

This equation is multiplied by $\partial(x_1, x_2)/\partial(X, Y)$ to yield

$$\left(V_m^2 - \frac{\gamma+1}{\gamma-1}u_1^2 - u_2^2\right)\frac{\partial x_2}{\partial Y} + \frac{4}{\gamma-1}u_1u_2\frac{\partial x_1}{\partial Y} + \left(V_m^2 - u_1^2 - \frac{\gamma+1}{\gamma-1}u_2^2\right)\frac{\partial x_1}{\partial X}$$

$$+ \sigma\left(V_m^2 - u_1^2 - u_2^2\right)\frac{u_2}{x_2}\left(\frac{\partial x_1}{\partial X}\frac{\partial x_2}{\partial Y} - \frac{\partial x_1}{\partial Y}\frac{\partial x_2}{\partial X}\right) = 0 \qquad (14.30)$$

since

$$\frac{\partial(X, x_2)}{\partial(X, Y)} = \frac{\partial x_2}{\partial Y}, \ldots$$

The four derivatives in Eq. (14.30) require evaluation. For this, a new dependent variable χ is introduced by means of a Legendre transformation

$$\chi = \phi - x_1u_1 - x_2u_2 = \phi - x_1X - x_2Y$$

The chain rule is used to evaluate $\partial \chi/\partial X$, as follows:

$$\chi_X = \phi_{x_1}\frac{\partial x_1}{\partial X} + \phi_{x_2}\frac{\partial x_2}{\partial X} - X\frac{\partial x_1}{\partial X} - x_1 - Y\frac{\partial x_2}{\partial X}$$

$$= X\frac{\partial x_1}{\partial X} + Y\frac{\partial x_2}{\partial X} - X\frac{\partial x_1}{\partial X} - x_1 - Y\frac{\partial x_2}{\partial X} = -x_1$$

and similarly

$$\chi_Y = -x_2$$

Hence, the derivatives are given by

$$\frac{\partial x_2}{\partial Y} = -\chi_{YY} = -\chi_{u_2 u_2}$$

$$\frac{\partial x_1}{\partial Y} = -\chi_{XY} = -\chi_{u_1 u_2}$$

$$\frac{\partial x_1}{\partial X} = -\chi_{XX} = -\chi_{u_1 u_1}$$

$$\frac{\partial x_2}{\partial X} = -\chi_{XY} = -\chi_{u_1 u_2}$$

Equation (14.30) thus becomes

$$\left(V_m^2 - \frac{\gamma+1}{\gamma-1}u_1^2 - u_2^2\right)\chi_{u_2 u_2} + \frac{4}{\gamma-1}u_1 u_2 \chi_{u_1 u_2} + \left(V_m^2 - u_1^2 - \frac{\gamma+1}{\gamma-1}u_2^2\right)\chi_{u_1 u_1}$$

$$+ \sigma\left(V_m^2 - u_1^2 - u_2^2\right)\frac{u_2}{\chi_{u_2}}\left(\chi_{u_1 u_1}\chi_{u_2 u_2} - \chi_{u_1 u_2}^2\right) = 0$$

which is the desired result. As previously stated, the equation is linear when the flow is two-dimensional. This result should be compared with the one at the end of Sec. 13.5.

14.3 HOMOGENEITY OF THE CONSERVATION EQUATIONS

A function $\phi(x_1, \ldots, x_n)$ is called homogeneous of degree m if

$$\phi(\lambda x_1, \ldots, \lambda x_n) = \lambda^m \phi(x_1, \ldots, x_n) \tag{14.31}$$

for all values of the parameter λ. A necessary and sufficient condition for ϕ to be homogeneous of degree m is that Euler's relation

$$\sum_{k=1}^n \frac{\partial \phi}{\partial x_k} x_k = m\phi \tag{14.32}$$

hold. Also, one can show that if ϕ is homogeneous of degree m, then the ℓth derivative of ϕ is homogeneous of degree $m - \ell$.

The conservation equations in conservative form, Eqs. (11.8) and (11.11), are not homogeneous. It is computationally advantageous, for reasons mentioned at the end of this section, to have an unsteady form for the gasdynamic equations that is both conservative and homogeneous. We will do this for the three-dimensional unsteady equations; however, the approach is applicable to either the steady or unsteady equations in any dimensionality. We utilize Cartesian coordinates, although their use is not essential.

For notational simplicity, we will write the jth element of q_i ($i = 0, \ldots, 3$) as q_{ij}. The elements q_{0j} are of special significance and will be denoted by θ_j. Thus, we have

$$q_0 = \begin{pmatrix} \theta_1 \\ \theta_2 \\ \theta_3 \\ \theta_4 \\ \theta_5 \end{pmatrix} = \begin{pmatrix} \rho \\ \rho u_1 \\ \rho u_2 \\ \rho u_3 \\ \rho h_0 - p \end{pmatrix} \tag{14.33}$$

A thermally and calorically perfect gas is assumed; hence,

$$\theta_5 = \rho h_0 - p = \rho \left[e + \frac{1}{2} \sum_{i=1}^{3} u_i^2 \right] = \frac{p}{\gamma - 1} + \frac{1}{2} \sum_{i=1}^{3} \rho u_i^2$$

This relation can be solved for p, with the result that p and ρh_0 are given by

$$p = (\gamma - 1) \left[\theta_5 - \frac{1}{2\theta_1} \sum_{j=2}^{4} \theta_j^2 \right] \tag{14.34a}$$

and

$$\rho h_0 = \gamma \theta_5 - \frac{\gamma - 1}{2\theta_1} \sum_{j=2}^{4} \theta_j^2 \tag{14.34b}$$

When the θ_j are known, the original variables are easily found, since

$$\rho = \theta_1, \quad u_1 = \frac{\theta_2}{\theta_1}, \quad u_2 = \frac{\theta_3}{\theta_1}, \ldots$$

Equations (14.33) and (14.34) enable us to write the remaining q_i as

$$q_1 = \begin{pmatrix} \theta_2 \\ (\gamma-1)\theta_5 + \dfrac{3-\gamma}{2}\dfrac{\theta_2^2}{\theta_1} - \dfrac{\gamma-1}{2\theta_1}(\theta_3^2 + \theta_4^2) \\ \theta_2\theta_3/\theta_1 \\ \theta_2\theta_4/\theta_1 \\ \dfrac{\theta_2}{\theta_1}\left(\gamma\theta_5 - \dfrac{\gamma-1}{2\theta_1}\sum_{j=2}^{4}\theta_j^2\right) \end{pmatrix} \qquad (14.35a)$$

$$q_2 = \begin{pmatrix} \theta_3 \\ \theta_2\theta_3/\theta_1 \\ (\gamma-1)\theta_5 + \dfrac{3-\gamma}{2}\dfrac{\theta_3^2}{\theta_1} - \dfrac{\gamma-1}{2\theta_1}(\theta_2^2 + \theta_4^2) \\ \theta_3\theta_4/\theta_1 \\ \dfrac{\theta_3}{\theta_1}\left(\gamma\theta_5 - \dfrac{\gamma-1}{2\theta_1}\sum_{j=2}^{4}\theta_j^2\right) \end{pmatrix} \qquad (14.35b)$$

$$q_3 = \begin{pmatrix} \theta_4 \\ \theta_2\theta_4/\theta_1 \\ \theta_3\theta_4/\theta_1 \\ (\gamma-1)\theta_5 + \dfrac{3-\gamma}{2}\dfrac{\theta_4^2}{\theta_1} - \dfrac{\gamma-1}{2\theta_1}(\theta_2^2 + \theta_3^2) \\ \dfrac{\theta_4}{\theta_1}\left(\gamma\theta_5 - \dfrac{\gamma-1}{2\theta_1}\sum_{j=2}^{4}\theta_j^2\right) \end{pmatrix} \qquad (14.35c)$$

It is easy to show that

$$q_{ij}(\lambda\theta_1,\ldots,\lambda\theta_5) = \lambda q_{ij}(\theta_1,\ldots,\theta_5), \qquad i = 0,\ldots,3 \qquad (14.36)$$

This is evident in Eq. (14.33) for the elements of q_0. Consider next the element q_{23} in Eq. (14.35b). For this element, we have

$$q_{23}(\lambda\theta_1,\ldots,\lambda\theta_5) = (\gamma-1)\lambda\theta_5 + \frac{3-\gamma}{2}\frac{\lambda^2\theta_3^2}{\lambda\theta_1} - \frac{\gamma-1}{2\lambda\theta_1}(\lambda^2\theta_2^2 + \lambda^2\theta_4^2)$$

$$= \lambda q_{23}(\theta_1,\ldots,\theta_5)$$

and q_{23} is homogeneous of degree one. Similarly, one can show that the other q_{ij} are also homogeneous of degree one.

The Euler equation corresponding to Eq. (14.36) is given by

$$\sum_{k=1}^{5} \frac{\partial q_{ij}}{\partial \theta_k}\theta_k = q_{ij}, \qquad i=0,\ldots,3, \qquad j=1,\ldots,5 \qquad (14.37)$$

This relation is actually an identity. To see this, we write Eq. (14.37) for the q_{23} element

$$q_{23} = \sum_{k=1}^{5} \frac{\partial q_{23}}{\partial \theta_k}\theta_k = \left[-\frac{3-\gamma}{2}\frac{\theta_3^2}{\theta_1^2} + \frac{\gamma-1}{2\theta_1^2}(\theta_2^2+\theta_4^2)\right]\theta_1 + \left[-\frac{(\gamma-1)}{\theta_1}\theta_2\right]\theta_2$$

$$+ \left[(3-\gamma)\frac{\theta_3}{\theta_1}\right]\theta_3 + \left[-\frac{(\gamma-1)}{\theta_1}\theta_4\right]\theta_4 + [\gamma-1]\theta_5 \qquad (14.38)$$

After simplification, the right side merely equals q_{23} as given in Eq. (14.35b).

Since q_{ij} is homogeneous of degree one, its derivative with respect to time is homogeneous of degree zero. Thus, $\partial q_{ij}/\partial t$ satisfies the Euler equation

$$\sum_{k=1}^{5} \frac{\partial^2 q_{ij}}{\partial t\, \partial\theta_k}\theta_k = 0, \qquad i=1,2,3, \qquad j=1,\ldots,5 \qquad (14.39)$$

where the order of differentiation is immaterial. There are $3 \times 5 = 15$ nontrivial scalar equations of this form, since the $i=0$ equations are identically zero.

Equation (14.37) is useful when introducing matrix methods, since the q_i can be written as

$$\begin{pmatrix} q_{01} \\ q_{02} \\ \vdots \\ q_{05} \end{pmatrix} = \begin{pmatrix} 1 & 0 & \cdots & 0 \\ 0 & 1 & \cdots & 0 \\ \vdots & \vdots & \ddots & \vdots \\ 0 & \cdot & \cdots & 1 \end{pmatrix} \begin{pmatrix} \theta_1 \\ \theta_2 \\ \vdots \\ \theta_5 \end{pmatrix}$$

$$\begin{pmatrix} q_{11} \\ q_{12} \\ \vdots \\ q_{15} \end{pmatrix} = \begin{pmatrix} \dfrac{\partial q_{11}}{\partial \theta_1} & \dfrac{\partial q_{11}}{\partial \theta_2} & \cdots & \dfrac{\partial q_{11}}{\partial \theta_5} \\ \dfrac{\partial q_{12}}{\partial \theta_1} & \dfrac{\partial q_{12}}{\partial \theta_2} & \cdots & \vdots \\ \vdots & & & \\ \dfrac{\partial q_{15}}{\partial \theta_1} & \cdots & \cdots & \dfrac{\partial q_{15}}{\partial \theta_5} \end{pmatrix} \begin{pmatrix} \theta_1 \\ \theta_2 \\ \vdots \\ \theta_5 \end{pmatrix}$$

$$\vdots$$

Let θ and q_i be the vectors $(\theta_1, \ldots, \theta_5)$ and $(q_{i1}, \ldots, q_{i5})$, respectively. Then

$$q_0 = I\theta, \qquad q_i = A_i \theta, \qquad i = 1, 2, 3 \tag{14.40}$$

where I is a 5×5 identity matrix, the A_i are the 5×5 matrices $(\partial q_i / \partial \theta)$, and where A_1 is shown above. Note that the determinant of A_i is the Jacobian

$$|A_i| = \frac{\partial(q_{i1}, \ldots, q_{i5})}{\partial(\theta_1, \ldots, \theta_5)}$$

and that the elements of the A_i depend only on γ and θ. This is evident from Eq. (14.38), where the third row of A_2 consists of the five terms within the square brackets.

Equations (14.40) are particularly useful when computing an unsteady flow. The q_0 vector contains the unknown dependent variables. Aside from constants, the other q_i depend only on the components of θ, thus minimizing and simplifying the required number of algebraic computations.

For a calculation to move forward in time from $t = n\,\Delta t$ to $(n+1)\,\Delta t$, it is necessary to compute q_i^{n+1}, $i \neq 0$, where the superscript indicates the time. The following derivation for q_i^{n+1} illustrates the usefulness of Eqs. (14.39) and (14.40). From Eqs. (14.40), we have

$$q_i^n = A_i^n \theta^n$$

and

$$q_i^{n+1} = A_i^{n+1} \theta^{n+1}$$

where

$$A_i^{n+1} = A_i^n + \left(\frac{\partial A_i}{\partial t}\right)^n \Delta t + \mathcal{O}(\Delta t^2)$$

Consequently, the time increment for q_i is

$$q_i^{n+1} - q_i^n = A_i^n \theta^{n+1} + \left(\frac{\partial A_i}{\partial t}\right)^n \theta^{n+1} \Delta t - A_i^n \theta^n + \mathcal{O}(\Delta t^2) \quad (14.41)$$

Next, we show that the term $(\partial A_i/\partial t)^n \theta^{n+1} \Delta t$ in Eq. (14.41) is of order $(\Delta t)^2$ and can be neglected. From Eq. (14.39) we have the matrix product

$$\left(\frac{\partial A_i}{\partial t}\right)^n \theta^n = 0 \quad (14.42)$$

We write a Taylor series expansion for θ^{n+1} as

$$\theta^{n+1} = \theta^n + \left(\frac{\partial \theta}{\partial t}\right)^n \Delta t + \mathcal{O}(\Delta t^2)$$

to obtain, by matrix multiplication,

$$\left(\frac{\partial A_i}{\partial t}\right)^n \theta^{n+1} \Delta t = \left(\frac{\partial A_i}{\partial t}\right)^n \left(\frac{\partial \theta}{\partial t}\right)^n (\Delta t)^2 + \mathcal{O}(\Delta t^3)$$

where the Eq. (14.42) term has been set equal to zero. With this result, Eq. (14.41) becomes

$$q_i^{n+1} = q_i^n + A_i^n(\theta^{n+1} - \theta^n) + \mathcal{O}(\Delta t^2), \quad i = 1, 2, 3 \quad (14.43)$$

which is a basic differencing formula in several CFD algorithms. As a consequence of using a homogeneous formulation, Eq. (14.43) has the important advantage of not containing any derivatives of A_i. Nevertheless, the time differencing of the q is second-order accurate and is provided by matrix multiplications, with one multiplication per i. Equation (14.43) is implicit since θ^{n+1} appears on the right side. However, an implicit formula for the temporal differencing is often required for numerical stability. Our discussion here is far from complete. We merely indicate the usefulness of the homogeneity property.

References

[1] Owczarek, J. A., *Fundamentals of Gas Dynamics*, International Textbook Co., Scranton, PA, 1964.

[2] Karamcheti, K., *Principles of Ideal Fluid Aerodynamics*, John Wiley & Sons, New York, 1966, Sec. 4.9.

Problems

14.1 Consider a steady, two-dimensional free shear layer as shown in the sketch below. Determine ω inside the layer and the circulation about the path a- $\cdots$ -a. As $\Delta x_2 \to 0$, the shear layer is called a vortex sheet. In this limit, determine the circulation per unit distance in the flow direction, and the surface vorticity ω_s, defined by

$$\omega_s = \hat{1}_2 \times [\mathbf{u}_1(d) - \mathbf{u}_1(a)]$$

where the velocities outside the layer are labeled in the sketch.

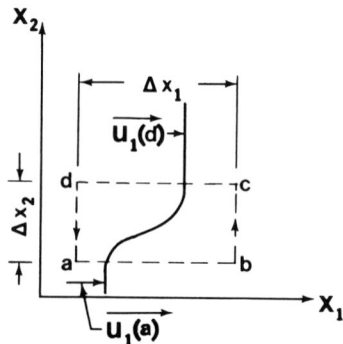

14.2 Derive for the shock wave $F(x_1, x_2, x_3, t) = 0$ an explicit vector formula for the vorticity just downstream of the shock ω_d in terms of derivatives of F. Assume $\mathbf{V}_1$ constant and a perfect gas. Use your result to determine ω_d when

(a) $F = ax_1 + bx_2 + c$

(b) $F = ax_1 + bx_2^2 + c$

where a, b, and c are constants. Simplify your results as much as possible. (See Problem 12.3.)

14.3 With $\mathbf{V}_2$ constant with respect to time and the shock given by Eqs. (a) and (b) in Problem 14.2, determine the entropy gradient $(\nabla s)_d$ in terms of conditions upstream of the shock, and the coordinates x_1 and x_2. Assume a perfect gas and isoenergetic flow. Do not attempt to develop a single general formula for $(\nabla s)_d$.

14.4 Write the homogeneous conservative equations for the unsteady one-dimensional flow of a perfect gas. Write specific equations for q_i, A_i, and θ_i. Verify that $\partial q_i / \partial t$ is homogeneous of degree zero.

DEFINITIONS AND THEOREMS

14.5 Reconsider Eqs. (12.14) with $F(x_1, x_2, x_3, t) = 0$ for the shock surface. Convert the shock jump conditions into a form compatible with the conservative, homogeneous equations of Sec. 14.3. In other words, replace ρ, p, and $\mathbf{V}$ with θ_i.

14.6 Derive Eqs. (14.17). Then, use Eqs. (14.15) to transform the x_1 momentum equation so that ψ and ρ are the dependent variables.

14.7 Start with the Crocco stream function, and assume a perfect gas, axisymmetric flow, and irrotational flow. Derive two first-order PDEs for u_1 and u_2. Simplify your results.

14.8 Consider a curvilinear orthogonal coordinate system ξ_j, where $\xi_i = x_i$ in a planar case, $\xi_1 = x_1 =$ coordinate along the symmetry axis, and $\xi_2 = x_2 =$ radial coordinate in the axisymmetric case. In both cases, set $\partial(\ )/\partial\xi_3 = 0$, and let the velocity be

$$\mathbf{V} = \sum_{i=1}^{2} u_i \hat{\mathbf{e}}_i$$

Determine the second-order PDE for the Crocco stream function ψ. The coefficients of the second derivatives $\psi_{x_i x_j}$ will involve u_1 and u_2.

14.9 Start with conditions given in Problem 12.5. Replace V_∞ and V_m with M_∞. Determine $ds/d\psi$ anywhere downstream of the shock, where ψ is the Crocco stream function. Next, determine ω in terms of M and ψ anywhere in the flow downstream of the shock. Simplify your answers as much as possible.

14.10 Consider steady, inviscid, adiabatic, two-dimensional or axisymmetric flow of a perfect gas in which the streamlines are straight lines parallel to a planar wall or a circular cylinder. Determine the most general form of the solution in this case. Is the flow isoenergetic, homentropic, or irrotational?

14.11 In conjunction with Fig. 6.4, use a Cartesian coordinate system x_1, x_2 centered at the bend in the wall and with x_1 oriented in the direction of the upstream velocity. Thus, the angle η is given by $\tan^{-1}(x_2/x_1)$. For the flow inside the expansion, determine u_1/a_∞, u_2/a_∞, p/p_0, and ρ/ρ_0 explicitly in terms of η with γ and M_∞ as parameters. The subscript infinity refers to the uniform flow upstream of the expansion. Hint: First show that the Mach number is given by

$$M^2 = 1 + \frac{\gamma+1}{\gamma-1} \tan^2 \left\{ \left(\frac{\gamma-1}{\gamma+1} \right)^{\frac{1}{2}} \left[\nu(M_\infty) + \frac{\pi}{2} - \eta \right] \right\}$$

where ν is the Prandtl-Meyer function.

14.12 Assume that the flow in Problem 14.10 is upstream of a centered, two-dimensional expansion with straight Mach lines, as in Fig. 6.4. Start with Eqs. (13.32), and transform the equations to η and a stream function ψ, as given by Eqs. (14.15). Assume that upstream of the expansion $h_0(\psi)$ is a known arbitrary function. Determine the explicit solution for M, p, ρ, u_1, u_2, and ω inside the expansion in terms of η and ψ. The solution represents a generalization of that for Problem 14.11.

14.13 Consider steady, irrotational, axisymmetric or two-dimensional flow of a perfect gas. Do not assume isoenergetic flow or that the swirl velocity component u_3 is zero. Let x_1 be the symmetry axis, x_2 the transverse or radial coordinate, and

$$\mathbf{V} = \sum_{i=1}^{3} u_i \hat{\mathbf{e}}_i$$

(a) Determine the consequences of the irrotationality condition and of Crocco's equation.
(b) Introduce a velocity potential $\hat{\phi}$. Determine the governing equations for $\hat{\phi}$ starting with Eqs. (13.32). What type of decoupling occurs, and what can be used to replace the energy equation?
(c) Use your previous results to derive a generalization of Eqs. (14.28) and (14.29) that allows for $u_3 = \text{const}$ when $\sigma = 0$ and for swirl when $\sigma = 1$.
(d) Consider the incompressible limit. Determine the governing equation for the velocity potential, and determine the pressure in terms of this potential by integrating the momentum equation.

14.14 For a two-dimensional or axisymmetric flow with a uniform freestream, show that the vorticity ω_2, just downstream of the shock, is given by

$$\omega_2 = -\frac{4}{\gamma + 1} V_1 \kappa_\ell \cos \beta \hat{\mathbf{e}}_3$$

where κ_ℓ is the longitudinal curvature of the shock. Compare this with Eq. (14.26) of Sec. 14.2 to obtain $(ds/d\psi)_2$, which can depend on the shock angle β. Determine the equation for $(d\psi/d\beta)_2$ as a function of β.

14.15 Use the result

$$\boldsymbol{\omega} = \left(\frac{1}{h_1} \frac{\partial v_2}{\partial \xi_1} - \frac{1}{h_2} \frac{\partial v_1}{\partial \xi_2} + v_2 \kappa_2 - v_1 \kappa_1 \right) \hat{\mathbf{e}}_3$$

of Problem 13.15 to determine $\partial S/\partial s$ and $\partial S/\partial n$ in a steady, isoenergetic flow (S is the entropy, and s and n are natural coordinates). Determine an equation for the change in vorticity along a streamline $\partial \omega/\partial s$ in terms of the streamline pressure gradient and Mach number (assume a perfect gas). Show that $\partial \omega/\partial s$ cannot be zero unless $(\partial S/\partial n) \times (\partial v/\partial s)$ is zero.

14.16 Show that unsteady isoenergetic flow cannot be isentropic. For a perfect gas, derive the streamline relation

$$\frac{1}{\rho}\mathbf{V}\cdot\nabla\rho = -\frac{1}{a^2}\left(\gamma\frac{\partial V^2}{\partial t} + \frac{1}{2}\mathbf{V}\cdot\nabla V^2\right)$$

14.17 Use the thermodynamic relation for a fluid mixture with compositional changes

$$de = T\,ds - p\,dv + \sum \mu_i\,dn_i$$

to obtain a generalization of Crocco's equation. In the equation, μ_i is the chemical potential of a mole of species i, and n_i is the number of moles of species i per unit mass of fluid.

15. EXACT SOLUTIONS OF STEADY HOMENTROPIC FLOW OF A PERFECT GAS

15.1 PRELIMINARY REMARKS

Exact algebraic solutions for two-dimensional or three-dimensional flows are always important. They lead to physical insight, often correspond to significant real flows, and are useful for checking out computer codes or experimental apparatus. In this chapter, we exclude steady or unsteady one-dimensional flows, such as quasi-one-dimensional nozzle flow or shock-tube flow, as well as the theory of characteristics, which can be used to generate an exact solution. One-dimensional flows have been treated in Part I, and the theory of characteristics will be examined in Chap. 16. Because of its considerable importance, we also discuss supersonic flow past a cone where only ordinary differential equations occur. These are solved numerically. The reduction of the steady or unsteady partial differential equations to ordinary differential equations results in a similarity solution. Flow past a cone is only one such solution.

Reference 1 provides the following list of exact solutions:

(1) Vortex flow.

(2) Radial flow.

(3) Spiral flow.

(4) Prandtl-Meyer flow, including flow past a convex corner.

(5) Transition from radial to uniform flow.

(6) Flow past a wedge.

(7) Flat plate at incidence in a supersonic stream.

(8) Double-wedge airfoil.

(9) Supersonic flow past a cone.

Vortex and radial flow are special cases of spiral flow. Thus, we treat the first three cases simultaneously as a spiral flow. Prandtl-Meyer flow, treated in Chap. 6, is not reexamined. The same is true of topics (6)–(8), which were referred to as oblique shock theory and shock-expansion theory (Chaps. 5 and 6). Topic (5) is not treated here since it will be examined in an application context in Chap. 17. As a consequence, only spiral flow and flow past a cone are treated. (Spherical source or sink flow is the subject of Problem 15.3.)

15.2 SPIRAL FLOW

This section considers a two-dimensional flow for which an exact solution is available. In this flow, a particle of fluid moves, either inward or outward, in a spiral about the origin. There are two limiting cases. In the first, the particle moves in a circle about the origin. The flow is thus an inviscid vortex motion. In the second, the particle moves radially inward or outward and corresponds to a source or sink flow. Both limiting cases are important in Chaps. 17 and 18, which deal with several applications.

As shown in Fig. 15.1, polar coordinates are used, where u and v denote the radial and transverse velocity components. Both components are independent of the azimuthal angle θ. The governing equations are written as

$$2\pi r \rho u = \dot{m} \tag{15.1}$$

$$2\pi r v = \Gamma \tag{15.2}$$

$$V_m^2 \left[1 - \left(\frac{\rho}{\rho_0}\right)^{\gamma-1}\right] = u^2 + v^2 \tag{15.3}$$

where the mass flow rate $\dot{m}$, circulation Γ, and stagnation density ρ_0 are constants. Equations (15.1)–(15.3) are for continuity, circulation, and energy, respectively.

If $\dot{m}$ is zero, the streamlines are circles, and we have a vortex flow. If $\dot{m}$ is not zero but Γ is zero, then the streamlines are straight and the flow is a source or sink flow. The flow is radially inward when u and $\dot{m}$ are negative.

From these equations we deduce that

$$r^2 = \left(\frac{\Gamma}{2\pi V_m}\right)^2 \frac{(\rho/\rho_0)^2 + k^2}{(\rho/\rho_0)^2 \left[1 - (\rho/\rho_0)^{\gamma-1}\right]} \tag{15.4}$$

where

$$k = \frac{\dot{m}}{\rho_0 \Gamma}$$

We observe that as $r \to \infty$, ρ tends to zero or to ρ_0. Furthermore, there is a minimum value for r at an intermediate value of ρ. Any quantity at this

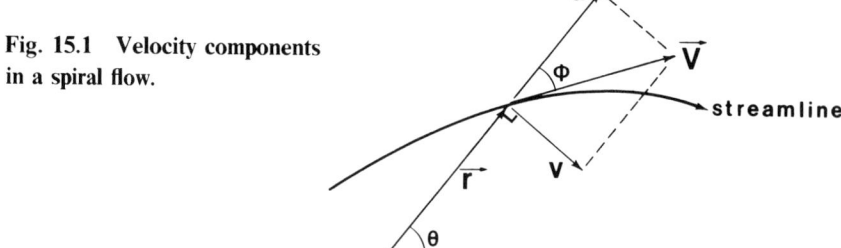

Fig. 15.1 Velocity components in a spiral flow.

minimum r location is denoted by a superscript asterisk (*). By logarithmic differentiation of Eq. (15.4), we obtain

$$(\gamma - 1)\left(\frac{\rho^*}{\rho_0}\right)^{\gamma+1} + (\gamma + 1)k^2\left(\frac{\rho^*}{\rho_0}\right)^{\gamma-1} - 2k^2 = 0 \qquad (15.5a)$$

and

$$r^* = \left(\frac{\Gamma}{2\pi V_m}\right)\left[1 - \frac{\gamma+1}{2}\left(\frac{\rho^*}{\rho_0}\right)^{\gamma-1}\right]^{-\frac{1}{2}} \qquad (15.5b)$$

for ρ^* and r^*. The radial component u at the minimum radius is given by Eq. (15.1) as

$$|u^*| = \frac{|\dot{m}|/\rho_0}{2\pi r^*(\rho^*/\rho_0)} = \left(\frac{\gamma-1}{2}\right)^{\frac{1}{2}} V_m \left(\frac{\rho^*}{\rho_0}\right)^{(\gamma-1)/2} = a^* \qquad (15.6)$$

and is sonic providing $\dot{m} \neq 0$. This is the basis for the asterisk notation. The circle where $u = u^*$, or the origin when $\Gamma = 0$, is either a source or a sink for the flow, depending on whether u is positive or negative.

For each sense of the flow, inward or outward, two possibilities exist. In the first,

$$\rho \to 0 \text{ as } r \to \infty$$

and the flow is everywhere supersonic. In the second,

$$\rho \to \rho_0 \text{ as } r \to \infty$$

and the radial component of the velocity corresponds to a subsonic Mach number. However, there is an annular region about r^* where the flow itself is supersonic.

The circle $r = r^*$ is called a limit line and falls outside any physically possible flowfield. For example, one can show that the acceleration $u(du/dr)$ is infinite at r^*. The occurrence of a limit line and of a multiple-valued solution is not restricted to spiral flow but occurs in other flows. One flow that is also double-valued and conceptually similar to spiral flow is nozzle/diffuser flow.

The reason for calling the flow a spiral one stems, in part, from Fig. 15.1. We have

$$\tan \phi = \frac{v}{u} = \frac{1}{k}\frac{\rho}{\rho_0}$$

and the angle that $\mathbf{V}$ makes with $\mathbf{r}$ varies with the density. A streamline thus has a spiral-like configuration, the orientation depending on whether $\rho \to 0$ or increases to ρ_0 as $r \to \infty$.

As mentioned earlier, a purely inviscid vortex motion occurs when $\dot{m} = 0$. In this case, the circular limit line $r = r^*$ occurs at a finite radius. When $\Gamma = 0$ the velocity is directed radially outward or inward. Equation (15.5b) is indeterminate in this limit, but one may show that

$$\frac{\rho^*}{\rho_0} = \left(\frac{2}{\gamma+1}\right)^{1/(\gamma-1)}, \qquad r^* = \left(\frac{2}{\gamma-1}\right)^{\frac{1}{2}}\left(\frac{\gamma+1}{2}\right)^{(\gamma+1)/2(\gamma-1)} \frac{\dot{m}}{2\pi\rho_0 V_m}$$

for conditions on the circular sonic line.

15.3 SUPERSONIC FLOW PAST A CONE

Often called Taylor-Maccoll flow, supersonic flow about a cone at zero angle of attack was first examined theoretically by Busemann in 1929 and experimentally by Taylor and Maccoll in the 1930s. Two examples of its frequent occurrence are supersonic engine inlets and flow about sharp-nosed missiles.

If the semivertex angle θ_b is not too large, there will be a conical shock wave attached to the apex of the body, as in Fig. 15.2. The streamline angle θ is measured relative to the centerline. Just downstream of the shock, θ equals θ_2. With increasing distance downstream of the shock, the streamlines become parallel to the body, except for the one that wets the body. For these streamlines, θ increases from θ_2 and far downstream becomes asymptotic to θ_b. Since $\theta_2 \neq \theta_b$, the solution is singular at the tip of the cone.

The conical shock wave is of uniform strength, and the streamline turn angle θ_2 at the shock is given by Eq. (12.21b). The Mach number M_2 just downstream of the shock is given by Problem 12.1. Thus, γ, M_1, and β are sufficient to establish conditions at location 2.

Downstream of the shock wave, the flow is homentropic and irrotational. An inviscid tangency condition for the velocity holds at the wall.

As in the oblique shock case, there is no intrinsic length scale. Consequently, the solution for the flow between the shock and the body depends only on the angular coordinate η, shown in Fig. 15.3. Thus, **V** and its angle relative to the surface of the body ϕ are constant on a conical surface with a

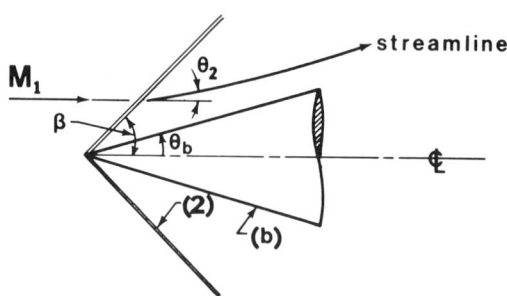

Fig. 15.2 Supersonic flowfield for a cone at zero incidence.

EXACT SOLUTIONS OF STEADY HOMENTROPIC FLOW

semivertex angle $\eta + \theta_b$. (For convenience, the velocity component perpendicular to the body v is shown in the positive direction. In the flowfield, v and ϕ are actually negative.) The orientation of the velocity vector is restricted to the range

$$\theta_2 - \theta_b \leq \phi \leq 0$$

In terms of the body-oriented x, y coordinate system shown in Fig. 15.3, the equations of motion are given by Problem 13.3, except for energy, which is replaced by $(Ds/Dt) = 0$ for a perfect gas. The equations are:

$$\frac{\partial(\rho u)}{\partial x} + \frac{\partial(\rho v)}{\partial y} + \sigma \rho \frac{u \sin\theta_b + v \cos\theta_b}{x \sin\theta_b + y \cos\theta_b} = 0$$

$$u\frac{\partial u}{\partial x} + v\frac{\partial u}{\partial y} + \frac{1}{\rho}\frac{\partial p}{\partial x} = 0$$

$$u\frac{\partial v}{\partial x} + v\frac{\partial v}{\partial y} + \frac{1}{\rho}\frac{\partial p}{\partial y} = 0 \qquad (15.7)$$

$$u\frac{\partial p}{\partial x} + v\frac{\partial p}{\partial y} = \gamma\frac{p}{\rho}\left(u\frac{\partial \rho}{\partial x} + v\frac{\partial \rho}{\partial y}\right)$$

In view of the foregoing discussion, new dependent and independent (similarity) variables are introduced:

$$\xi = x, \qquad \eta = \tan^{-1}(y/x)$$

$$p = p_b P(\eta), \qquad \rho = \rho_b R(\eta) \qquad (15.8)$$

$$u = u_b Q(\eta)\cos\phi, \qquad v = u_b Q(\eta)\sin\phi$$

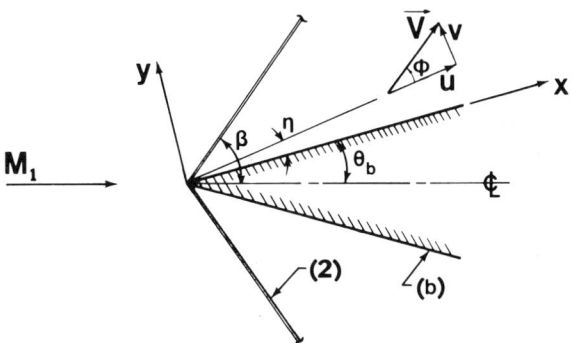

Fig. 15.3   Coordinates and nomenclature for flow about a cone.

where the b subscript denotes a body value. On the body, where $\eta = 0$,

$$P(0) = R(0) = Q(0) = 1 \tag{15.9}$$

and the tangency condition is automatically satisfied. As usual, it is convenient to introduce the Mach number

$$M^2 = \frac{V^2}{\gamma(p/\rho)} = M_b^2 \frac{RQ^2}{P} \tag{15.10}$$

After some algebra, Eqs. (15.7) can be shown[2] to yield two integrals of the motion and two, first-order, ordinary differential equations:

$$P = \left\{ \frac{1 + [(\gamma-1)/2] M_b^2}{1 + [(\gamma-1)/2] M^2} \right\}^{\gamma/(\gamma-1)} \tag{15.11a}$$

$$Q = \frac{M}{M_b} \left\{ \frac{1 + [(\gamma-1)/2] M_b^2}{1 + [(\gamma-1)/2] M^2} \right\}^{\frac{1}{2}} \tag{15.11b}$$

$$\frac{d\phi}{d\eta} = \sigma \frac{\sin(\phi + \theta_b)}{\sin(\eta + \theta_b)} \frac{\cos(\phi - \eta)}{M^2 \sin^2(\phi - \eta) - 1} \tag{15.12a}$$

$$\frac{dM}{d\eta} = \sigma \frac{\sin(\phi + \theta_b)}{\sin(\eta + \theta_b)} \frac{M\{1 + [(\gamma-1)/2] M^2\} \sin(\phi - \eta)}{M^2 \sin^2(\phi - \eta) - 1} \tag{15.12b}$$

Equations (15.11) will be recognized as the homentropic relations for pressure and velocity, respectively. The derivation[2] given here is substantially different from what can be found in other texts, where a complicated single ordinary differential equation (ODE) of second order is obtained.

The dependent variables in Eqs. (15.12) are ϕ and M. The equations can be integrated numerically, starting with body values

$$\phi(0) = 0, \quad M(0) = M_b \tag{15.13}$$

It is worth noting that the shock angle β and the wall values ρ_b, p_b, and u_b do not enter into Eqs. (15.11)–(15.13). The only required parameters are σ, γ, θ_b, and M_b. Integration of Eqs. (15.12) terminates at the shock when

$$\frac{[M^2 \sin^2(\phi - \eta) - 1] \sin(\eta + \theta_b)}{\{1 + [(\gamma-1)/2] M^2 \sin^2(\phi - \eta)\} \sin(\phi + \theta_b)} \cos(\phi - \eta) + 1 = 0 \tag{15.14}$$

EXACT SOLUTIONS OF STEADY HOMENTROPIC FLOW

is satisfied. This relation stems directly from the shock wave discussion earlier in this section. When Eq. (15.14) is satisfied, we have

$$\eta_2 = \eta, \qquad M_2 = M, \qquad \theta_2 = \phi + \theta_b, \qquad \beta = \eta + \theta_b \qquad (15.15)$$

and the freestream Mach number and pressure are given by

$$M_1 = \left[\sin^2\beta - \frac{\gamma+1}{2} \frac{\sin\theta_2 \sin\beta}{\cos(\theta_2 - \beta)} \right]^{-\frac{1}{2}} \qquad (15.16a)$$

$$\frac{p_1}{p_b} = \frac{P(\eta_2)}{[2/(\gamma+1)] M_1^2 \sin^2\beta - (\gamma-1)/(\gamma+1)} \qquad (15.16b)$$

With σ, γ, and θ_b fixed, it is possible for M_b to be too large or too small. If M_b is too large, then M_1^2 is negative. In other words, a real value for M_1 does not exist for the chosen M_b value. If M_b is too small, then the shock is not attached and Eq. (15.14) cannot be satisfied for any η in the range

$$0 < \eta < \frac{\pi}{2} - \theta_b$$

Usually both M_1 and θ_b are prescribed, resulting in a two-point boundary-value problem. In the foregoing inverse approach, θ_b is fixed as the simpler of the two choices, thus requiring an iteration on M_b until M_1 is matched. Once this has occurred, matching p_1 is simple, since p_b only appears in Eqs. (15.8) and (15.16b).

For the two-dimensional case where $\sigma = 0$, ϕ and M retain their wall values everywhere downstream of the shock. Thus, the oblique shock solution is recovered. This result applies when θ_b is positive, since the wall compresses the flow.

When θ_b is negative, however, the flow must expand around the turn such that no characteristic length appears and the wave nature of the flow is preserved. As we know, this adjustment occurs by means of a centered Prandtl-Meyer expansion. Because there is no intrinsic length, Eqs. (15.11) and (15.12) should still hold. To counter the $\sigma = 0$ in Eqs. (15.12), the equations are made indeterminate by setting the denominator equal to zero

$$M \sin(\phi - \eta) = 1$$

which becomes

$$\tan(\phi - \eta) = \frac{1}{(M^2 - 1)^{\frac{1}{2}}}$$

The indeterminancy is removed by dividing Eq. (15.12b) by Eq. (15.12a), to obtain

$$\frac{dM}{d\phi} = M\left[1 + \frac{\gamma-1}{2} M^2\right] \tan(\phi - \eta)$$

By eliminating the tangent factor in the two preceding equations, we have

$$\int_0^\nu d\phi = \int_1^M \frac{(M^2 - 1)^{\frac{1}{2}}}{M\{1 + [(\gamma - 1)/2]M^2\}} \, dM$$

which integrates to the Prandtl-Meyer function

$$\nu = \left(\frac{\gamma + 1}{\gamma - 1}\right)^{\frac{1}{2}} \tan^{-1}\left[\frac{\gamma - 1}{\gamma + 1}(M^2 - 1)\right]^{\frac{1}{2}} - \tan^{-1}(M^2 - 1)^{\frac{1}{2}} \quad (15.17)$$

Aside from conical flow, we have shown that Eqs. (15.11) and (15.12) also yield centered Prandtl-Meyer flow and the flow behind an oblique planar shock wave.

There is considerable similarity between the planar and axisymmetric solutions. For instance, with a given value of θ_b, there is a minimum value of M_1 for an attached shock wave. When the shock is attached for a given θ_b, both weak and strong solutions exist. As in the planar case, only the weak solution is expected to occur, and the flowfield downstream of the strong solution shock is everywhere subsonic. This aspect is further discussed in Sec. 19.2.

While the similarities are well known, there are differences between the planar and axisymmetric solutions, which are now discussed. See also the discussion near the end of Sec. 16.2.

Except for the streamline that wets the body, a particle of fluid experiences an isentropic compression as it travels downstream from the shock wave. It is, therefore, possible to have a sonic condition on a conical surface that is situated between the body and the shock, where the flow is supersonic upstream and subsonic downstream of the surface. Such flowfields have been experimentally observed.[3] They demonstrate the ability of a supersonic flow to negotiate a smooth, shock-free transition to subsonic flow. These experiments[3] further verify the conical flow theory even when the Mach number on the surface of the cone is subsonic. In this instance, the theory holds asymptotically in the vicinity of the apex. Thus, the influence of the downstream pressure decreases as the apex is approached. The rate at which it decays presumably depends on the extent, in terms of η, of the subsonic region.

Various figures relating to the weak solution for $\gamma = 1.405$ can be found in Ref. 4. Rasmussen[5] provides accurate approximate relations for the flowfield that can replace Eqs. (15.12). These relations can be used to establish initial estimates for unknown boundary conditions or for hypersonic flow studies, as discussed in Sec. 20.4.

References

[1] Bickley, W. G., "Some Exact Solutions of the Equations of Steady Homentropic Flow of an Inviscid Gas," *Modern Developments in Fluid Dynamics*, Vol. I, edited by L. Howarth, Oxford University Press, New York, 1953, Chap. V.

[2]Emanuel, G., "Blowing from a Porous Cone with an Embedded Shock Wave," *AIAA Journal*, Vol. 8, Feb. 1970, pp. 283–286.

[3]Solomon, G. E., "Transonic Flow Past Cone-Cylinders," Ph.D. Dissertation, California Institute of Technology, Pasadena, 1953.

[4]Ames Research Staff, *Equations, Tables and Charts for Compressible Flow*, NACA Rept. 1135.

[5]Rasmussen, M., "On Hypersonic Flow Past an Unyawed Cone," *AIAA Journal*, Vol. 5, Aug. 1970, pp. 1495–1497.

Problems

15.1 For spiral flow, determine a general relation for a conventional stream function, in the form

$$\psi = \psi(\theta, \rho/\rho_0; \gamma, k, \Gamma)$$

based on polar coordinates. This relation, in conjunction with Eq. (15.4), provides the shape of a $\psi = $ const streamline.

15.2 Determine $\tilde{r}/r^*$ for a spiral flow, where $\tilde{r}$ is the radius at which $M = 1$. Simplify your results and investigate the $k = 0$ limit.

15.3 Determine the solution for spherically symmetric source or sink flow.

15.4 Determine the radius r^* of the limiting sphere for the spherically symmetric flow of Problem 15.3. Determine the acceleration and Mach number on the limiting sphere.

15.5 Prove or derive Eq. (15.14).

Computational Problems:

15.6 Integrate numerically Eqs. (15.12) for $\sigma = 1$. Prepare curves for p_b/p_1 and M_b in terms of θ_b and M_1 for $\gamma = 1.4$ and $5/3$. Compare your results with the approximate formulas in Ref. 5 and with the exact results in Fig. 4 as well as charts 4 and 5 of Ref. 4.

16. THEORY OF CHARACTERISTICS

We distinguish between the theory of characteristics and the method of characteristics (MOC). The theory investigates fundamental properties of the equations of motion, in contrast to applications for obtaining solutions to specific problems. The MOC has been a major aerodynamic tool for solving high-speed flows since the late 1920s because the numerical calculations were simple enough to be done without computers. Today, of course, computers are used.

There are a variety of ways to introduce the concepts behind the theory of characteristics. A simplified approach was used in Sec. 9.3 for unsteady one-dimensional flow. An alternate approach[1] can be based on Sec. 13.3, as is done in Sec. 16.1. This approach is useful for directly establishing the gasdynamic concepts inherent in a hyperbolic system of equations. These concepts, summarized in Sec. 16.1 under the heading "Theory," are the bases for the applications described in Chaps. 17, 18, and part of 19.

Section 16.2 contains a third, more mathematical approach, which is tailor-made for two important flows. The theory is briefly applied to unsteady one-dimensional flow, which was analyzed in Secs. 9.3 and 9.4. Then, a more extensive discussion is provided of steady two-dimensional or axisymmetric flow where natural coordinates are used.

An excellent introductory presentation is contained in Chap. 12 of Ref. 2. You may also want to consult the standard work by Courant and Friedrichs.[3]

16.1 STEADY TWO-DIMENSIONAL OR AXISYMMETRIC FLOW

Preliminary Analysis

We start with the equations of motion derived in Sec. 13.3 for steady two-dimensional or axisymmetric flow. These are Eqs. (13.45), (13.47), and (13.48), which constitute four first-order, coupled, nonlinear PDEs. The independent variables are the orthogonal, curvilinear coordinates ξ_1 and ξ_2, while ρ, T, v_1, and v_2 are the dependent variables. For a closed system, the PDEs are supplemented with three algebraic equations. These are Eq. (13.49), which defines h_0, and two thermodynamic relations, i.e.,

$$p = p(\rho, T), \qquad e = e(\rho, T) \tag{16.1}$$

A third thermodynamic equation

$$p = p(\rho, s) \tag{16.2}$$

is actually more useful than Eqs. (16.1). This stems from its differential form

$$dp = \left.\frac{\partial p}{\partial \rho}\right|_s d\rho + \left.\frac{\partial p}{\partial s}\right|_\rho ds = a^2 d\rho + \left.\frac{\partial p}{\partial s}\right|_\rho ds \tag{16.3}$$

where a is the speed of sound.

We now assume the flow to be isentropic; hence, s is constant along a streamline, except where it crosses a discontinuity. Therefore, Crocco's equation, Eq. (14.10), becomes

$$\mathbf{V} \cdot \nabla h_0 = 0 \tag{16.4}$$

However, this is Eq. (13.48) for conservation of energy. We, therefore, can replace Eq. (16.4) with the isentropic relation

$$\mathbf{V} \cdot \nabla s = 0 \tag{16.5}$$

as was done in Sec. 15.3.

Equation (16.3) holds in any direction and can be written as

$$\nabla p = a^2 \nabla \rho + \left.\frac{\partial p}{\partial s}\right|_\rho \nabla s$$

Taking the dot product with $\mathbf{V}$ and using Eq. (16.5) leads to

$$\mathbf{V} \cdot \nabla p = a^2 \mathbf{V} \cdot \nabla \rho$$

or

$$\frac{v_1}{h_1}\frac{\partial p}{\partial \xi_1} + \frac{v_2}{h_2}\frac{\partial p}{\partial \xi_2} = a^2 \left[\frac{v_1}{h_1}\frac{\partial \rho}{\partial \xi_1} + \frac{v_2}{h_2}\frac{\partial \rho}{\partial \xi_2}\right] \tag{16.6}$$

Equation (16.6) thus replaces Eq. (16.4) or (16.5). Our system, therefore, consists of Eqs. (13.45) and (13.47), as well as Eq. (16.6). Now, the dependent variables are p, ρ, v_1, and v_2. The only additional thermodynamic equation needed for closure is $a = a(p, \rho)$, although the actual functional form for $a(p, \rho)$ is not required.

Before proceeding, the assumptions not invoked are listed. These are
(1) Subsonic or supersonic flow.
(2) Irrotational flow.
(3) Homentropic flow.
(4) Isoenergetic flow.
(5) Perfect gas.

The only new assumption, relative to Sec. 13.3, is the isentropic one.

For the subsequent analysis, we need to manipulate the governing equations into a more suitable form. We first eliminate $\mathbf{V}\cdot\nabla p$ from Eq. (16.6) and Eq. (13.45) to obtain

$$a^2\left(\frac{1}{h_1}\frac{\partial v_1}{\partial \xi_1} + \frac{1}{h_2}\frac{\partial v_2}{\partial \xi_2} + \kappa_2 v_1 + \kappa_1 v_2 + R\right) + \frac{v_1}{\rho h_1}\frac{\partial p}{\partial \xi_1} + \frac{v_2}{\rho h_2}\frac{\partial p}{\partial \xi_2} = 0 \tag{16.7}$$

where

$$R = \frac{\sigma}{x_2}(v_1 \sin\theta + v_2 \cos\theta)$$

Multiply Eq. (13.47b) by v_2, and then subtract the result from Eq. (16.7), to obtain

$$\frac{a^2}{h_1}\frac{\partial v_1}{\partial \xi_1} - \frac{v_1 v_2}{h_1}\frac{\partial v_2}{\partial \xi_1} + (a^2 + v_1^2)v_2 \kappa_1 + a^2 R + \frac{v_1}{\rho h_1}\frac{\partial p}{\partial \xi_1}$$

$$= (v_2^2 - a^2)\left(\frac{1}{h_2}\frac{\partial v_2}{\partial \xi_2} + v_1 \kappa_2\right) \tag{16.8}$$

Now multiply Eq. (16.7) by v_2, and subtract Eq. (13.47b) after it is multiplied by a^2. This yields

$$a^2\left[\frac{v_2}{h_1}\frac{\partial v_1}{\partial \xi_1} - \frac{v_1}{h_1}\frac{\partial v_2}{\partial \xi_1} + (v_1^2 + v_2^2)\kappa_1 + v_2 R\right] + \frac{v_1 v_2}{\rho h_1}\frac{\partial p}{\partial \xi_1}$$

$$= -(v_2^2 - a^2)\frac{1}{\rho h_2}\frac{\partial p}{\partial \xi_2} \tag{16.9}$$

Next, multiply Eq. (13.45) by $(v_2^2 - a^2)/\rho$, and add the result to Eq. (16.8), to obtain

$$\frac{v_2^2}{h_1}\frac{\partial v_1}{\partial \xi_1} - \frac{v_1 v_2}{h_1}\frac{\partial v_2}{\partial \xi_1} + \frac{v_1}{\rho h_1}\frac{\partial p}{\partial \xi_1} + \frac{v_1(v_2^2 - a^2)}{\rho h_1}\frac{\partial \rho}{\partial \xi_1}$$

$$+ v_2(v_1^2 + v_2^2)\kappa_1 + v_2^2 R = -\frac{v_2(v_2^2 - a^2)}{\rho h_2}\frac{\partial \rho}{\partial \xi_2} \tag{16.10}$$

Finally, eliminate κ_2 from Eq. (16.8) and Eq. (13.47a), to yield

$$a^2\left[\frac{v_1^2 + v_2^2}{h_1}\frac{\partial v_1}{\partial \xi_1} + \frac{v_1}{\rho h_1}\frac{\partial p}{\partial \xi_1} + v_2(v_1^2 + v_2^2)\kappa_1 + v_2^2 R\right] - \frac{v_1 v_2^2}{2h_1}\frac{\partial(v_1^2 + v_2^2)}{\partial \xi_1}$$

$$= \frac{v_2(v_2^2 - a^2)}{2h_2}\frac{\partial(v_1^2 + v_2^2)}{\partial \xi_2} \tag{16.11}$$

Equations (16.8)–(16.11) replace the former equations of motion. These new equations have one common feature: The derivative terms with respect to ξ_2 only appear on the right side and, in every equation, are multiplied by $(v_2^2 - a^2)$.

Theory

Evidently, the derivative terms with respect to ξ_2 disappear in Eqs. (16.8)–(16.11) whenever

$$v_2 = \pm a \tag{16.12}$$

On the lines $\xi_2 =$ const where Eq. (16.12) is satisfied, derivatives normal to these lines do not appear in the equations. However, this statement presumes that these lines exist. Clearly, if the Mach number is less than unity, then no component of the velocity can exceed a, and Eq. (16.12) cannot be satisfied. On the other hand, if $M \geq 1$, there are as many as two directions for which Eq. (16.12) holds, as shown in Fig. 16.1. At point P, Eq. (16.12) holds on the two dashed lines, one of which is the $\xi_2 =$ const line. In general, the ξ_2 coordinate, oriented 90 deg counterclockwise from the ξ_1 coordinate, does not correspond to the second Eq. (16.12) direction. The lines in Fig. 16.1 on which Eq. (16.12) holds are called Mach lines. They are a special case, limited to a steady flow, of what are generally called characteristic lines. The derivation of Eqs. (16.8)–(16.11) has thus selected $\xi_2 =$ const as one set of the Mach lines. The two sets form a nonorthogonal coordinate system in a two-dimensional or axisymmetric geometry. In three dimensions, the Mach lines become conical surfaces called Mach cones. The existence of Mach lines or cones is possible only when the equations are hyperbolic.

In the above discussion, we have established the important result that a steady flow must be sonic or supersonic for the existence of real Mach lines.

With Eq. (16.12), Eqs. (16.8)–(16.11) reduce to two equations

$$\frac{a}{h_1}\frac{\partial v_1}{\partial \xi_1} - \frac{v_1}{h_1}\frac{\partial a}{\partial \xi_1} + \frac{v_1}{\rho a h_1}\frac{\partial p}{\partial \xi_1} + aR \pm (a^2 + v_1^2)\kappa_1 = 0 \tag{16.13}$$

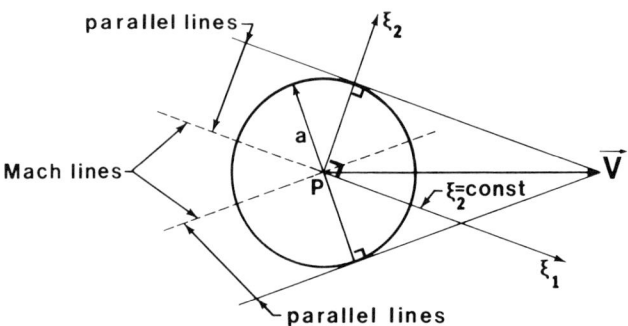

Fig. 16.1 Mach lines in a steady supersonic flow.

where the $\pm$ signs in Eqs. (16.12) and (16.13) coincide. Since only derivatives with respect to ξ_1 appear, Eq. (16.13) constitutes two ODEs, one per Mach line direction. (There should be a $\pm$ subscript on ξ_1 to denote the two different directions.) Thus, the four PDEs are redundant on these lines. Equations (16.13) are called the compatibility equations or conditions. Overall, there still are four unknowns to be determined by four equations. These consist of Eqs. (16.13) plus two equations, given later, for the two characteristic directions.

Let L be a smooth line segment in the ξ plane that is nowhere coincident with a Mach line (that is, the Mach lines cannot be tangent to L). Arbitrary values for p, ρ, v_1, and v_2 can be specified on L, which we choose to be a ξ_1 (noncharacteristic) coordinate. We require only that the variables be continuous and that their first derivatives along L exist. If there is a point of discontinuity on the line, simply divide L into two segments, each of which then satisfies the continuity condition. Since L is nowhere coincident with a Mach line, none of the coefficients on the right side of Eqs. (16.8)–(16.11) are zero. Therefore, these equations provide unique values for the normal derivatives ($\partial p/\partial \xi_2$, $\partial \rho/\partial \xi_2$, $\partial v_1/\partial \xi_2$, and $\partial v_2/\partial \xi_2$) on L, inasmuch as the left sides of the equations are fully determined along L. Consequently, the solution can be extended to a new line lying near L. In this manner the solution can be extended over a finite region of flow on both sides of L. This analytical construction, it should be noted, is not restricted to a supersonic flow.

When the flow is supersonic, we can demonstrate a uniqueness theorem for the solution. Consider prescribed values on L from point A to point B where, again, L is not coincident with a Mach line. Construct the Mach lines that pass through these two points, as shown in Fig. 16.2. Along the four Mach lines $AC, BC, \ldots$, the normal derivatives are undetermined. Consequently, the solution from A to B on L cannot be extended to the region outside the $ACBD$ quadrangle. (This is in sharp contrast to the elliptic, or subsonic, case, where no such restriction occurs.) However, inside the quadrangle the solution is uniquely determined by a Taylor series expansion about L that utilizes Eqs. (16.8)–(16.11), and derivatives of these equations, to evaluate the Taylor series coefficients to any order.

On the data line L the dependent variables are arbitrary. However, there are restrictions on L, aside from not coinciding with a Mach line. Roughly,

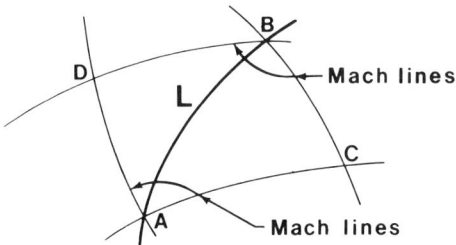

Fig. 16.2 Mach lines that pass through points A and B on a line L that nowhere coincides with a Mach line.

these can be summarized by saying that L must result in a physically acceptable solution. Two nonphysical examples for the data line are sketched in Fig. 16.3. A proper L is one that provides a unique real solution in a region encompassing L. (See Ref. 3, Chap. 2, for further discussion of permissible data.) Given such an L, a portion of the quadrangle can be terminated by one or more of the following obstacles:
(1) Wall.
(2) Symmetry axis.
(3) Free-surface boundary.
(4) Sonic line.
(5) Shock wave.
(6) Slipstream or contact surface.

The first five items are self-evident. The last item will be discussed later.

Two important consequences of the uniqueness theorem are the concepts of a region of influence and a domain of dependence. Consider a data line L that is nowhere coincident with a Mach line, as shown in Fig. 16.4. The downstream domain of dependence of the data from A to B on L is then the triangular region ABC. On the other hand, the data at point A on L influence only the region downstream of A that lies between the two Mach lines AC and AD.

A data line L may also coincide with a Mach line. However, fewer dependent variables can be specified in order to ensure that Eq. (16.13) still holds along the data line.

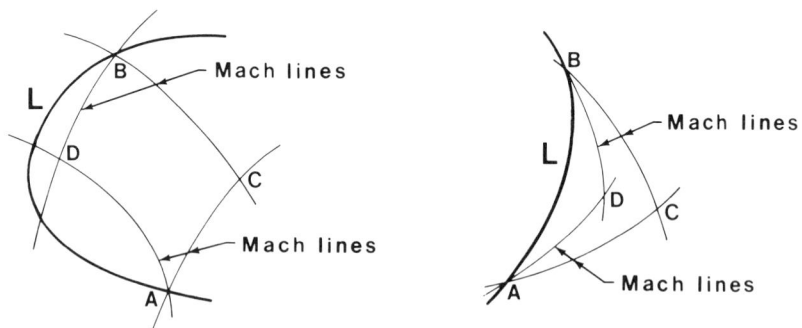

Fig. 16.3 Two nonphysical examples for the data line L.

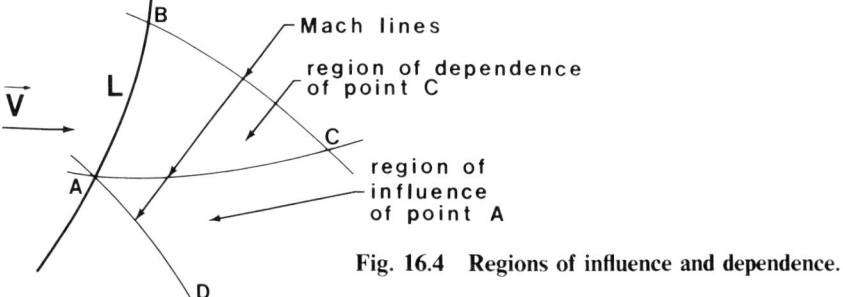

Fig. 16.4 Regions of influence and dependence.

The concept of Mach lines is familiar to us from our earlier analysis of expansion waves in steady and unsteady flows. However, Mach lines are not the only lines along which the equations of motion admit discontinuous normal derivatives. Equations (13.59a), which use natural coordinates, contain three equations that involve only the derivative $\partial(\ )/\partial \xi_1$, which is along a streamline. Only one transverse derivative, $\partial p/\partial \xi_2$, appears, and this derivative is not required to be continuous across streamlines. In the earlier discussion, we tacitly required the dependent variables to be continuous both along and across a Mach line. This requirement must be dropped with respect to certain streamlines, which are called slipstreams. A slipstream allows the dependent variables (except pressure) to be discontinuous across it. As a consequence, $\partial p/\partial \xi_2$ can be discontinuous because ρv^2 can be discontinuous across a streamline. Thus, streamlines are characteristic lines that differ from Mach lines in two major respects. First, the dependent variables (except pressure), as well as their normal derivatives including the derivative of the pressure, may be discontinuous. Second, these lines or surfaces are not restricted to a supersonic flow.

Consider a region of uniform flow adjacent to one with nonuniform flow. The boundary between the two regions must be a Mach line, a streamline, or a shock wave. Our subsequent discussion, however, is limited to Mach lines. If it isn't a Mach line, then it is an ordinary line, and the uniqueness theorem would result in uniform flow in the supposedly nonuniform flow region. The presence of a Mach line as a boundary enables all the derivatives normal to it to be discontinuous. This property, in fact, is the most fundamental difference between steady supersonic and subsonic flows, where such discontinuities are not permissible.

In a region of uniform flow, the streamlines and Mach lines are straight. Consequently, the border between uniform and nonuniform flow regions is a straight line. We now examine the question: Are any Mach lines straight in a nonuniform flow region? To answer this question, we need to reconsider the hodograph transformation of Sec. 13.5. As shown in that section and in the associated problems, the transformation results in linear equations, providing the original PDEs are homogeneous, along with other generally satisfied conditions. In this situation, the PDEs are called reducible (see the last paragraph in this section or Sec. 29 of Ref. 3), and a nonuniform flow region that borders one with uniform flow is called a simple wave region. In a simple wave region, one set of Mach lines, or characteristics, are straight, and the dependent variables are constant along them. The value of any dependent variable, however, changes from line to line. These straight Mach lines are of the same family as the one on the border. The second family of Mach lines are curved, and the dependent variables change continuously along them as they cross the nonuniform flow region.

A region where both Mach line families are curved is called a nonsimple wave region. If the equations are reducible, one can show that a nonsimple wave region can only be bordered by a simple wave region, except at discrete points, or at one or more of the previously mentioned obstacles, such as a sonic line.

We now address the question: When are the equations of motion reducible? One can show that they are reducible for an unsteady, one-dimensional flow or a steady, supersonic two-dimensional flow. In the latter case, we have $\sigma = R = 0$. Thus, the concept of a simple wave region occurs for a two-dimensional flow, i.e., a Prandtl-Meyer flow, but does not occur for an axisymmetric or three-dimensional flow.

16.2 COMPATIBILITY AND CHARACTERISTIC EQUATIONS

It was useful, in Sec. 16.1, to use orthogonal, curvilinear coordinates for deriving a variety of results specific to gasdynamics. However, the derivation of Eqs. (16.8)–(16.11) lacks generality. To provide balance, we present a general procedure for the analysis of a system of first-order equations. Although the procedure is general, for purposes of simplicity, only two equations in two independent variables are considered. Of course, these two equations are equivalent to one second-order PDE, such as the wave equation. It should be noted that the general procedure and the specific system of equations examined are not restricted to gasdynamics alone. The results are then applied in the second subsection to unsteady one-dimensional flow and in the third to steady two-dimensional or axisymmetric flow.

Theory for Two Independent Variables

We consider two PDEs with unknowns ϕ_1 and ϕ_2:

$$a_1^{(1)} \frac{\partial \phi_1}{\partial x_1} + a_2^{(1)} \frac{\partial \phi_2}{\partial x_1} + b_1^{(1)} \frac{\partial \phi_1}{\partial x_2} + b_2^{(1)} \frac{\partial \phi_2}{\partial x_2} + c^{(1)} = 0$$

$$a_1^{(2)} \frac{\partial \phi_1}{\partial x_1} + a_2^{(2)} \frac{\partial \phi_2}{\partial x_1} + b_1^{(2)} \frac{\partial \phi_1}{\partial x_2} + b_2^{(2)} \frac{\partial \phi_2}{\partial x_2} + c^{(2)} = 0$$

which can be written as

$$\sum_{i=1}^{2} a_i^{(j)} \frac{\partial \phi_i}{\partial x_1} + \sum_{i=1}^{2} b_i^{(j)} \frac{\partial \phi_i}{\partial x_2} + c^{(j)} = 0, \qquad j = 1, 2 \qquad (16.14)$$

The a, b, and c coefficients are restricted to be functions only of x_1, x_2, ϕ_1, and ϕ_2. Thus, Eqs. (16.14) are either linear or quasi-linear. (An equation is quasi-linear when it is linear with respect to its highest-order derivatives.) We first determine under what conditions these equations are hyperbolic. The characteristic equations and compatibility conditions are then derived. Furthermore, our final results for these equations are to be obtained in a convenient form that allows for their ready utilization.

With two independent variables, there are two characteristic directions, providing these directions exist. Let us assume they exist. Then, by a transformation to characteristic coordinates, Eqs. (16.14) become ODEs, one per direction. For this transformation, Jacobian theory is used. The new nonorthogonal characteristic coordinates are designated as ξ_1 and ξ_2. One of the early results of the analysis will be the condition for the existence of real characteristics.

THEORY OF CHARACTERISTICS

We proceed by writing Eqs. (16.14) in Jacobian form

$$\sum_{i=1}^{2} a_i^{(j)} \frac{\partial(\phi_i, x_2)}{\partial(x_1, x_2)} - \sum_{i=1}^{2} b_i^{(j)} \frac{\partial(\phi_i, x_1)}{\partial(x_1, x_2)} + c^{(j)} = 0$$

This relation is multiplied by J, where J is

$$J = \frac{\partial(x_1, x_2)}{\partial(\xi_1, \xi_2)} \tag{16.15}$$

to obtain

$$\sum_{i=1}^{2} a_i^{(j)} \frac{\partial(\phi_i, x_2)}{\partial(\xi_1, \xi_2)} - \sum_{i=1}^{2} b_i^{(j)} \frac{\partial(\phi_i, x_1)}{\partial(\xi_1, \xi_2)} + Jc^{(j)} = 0, \qquad j = 1, 2$$

The second of these equations is multiplied by λ, which need not be a constant, and the result is added to the first equation, to obtain

$$\sum_{i=1}^{2} \left(a_i^{(1)} + \lambda a_i^{(2)}\right) \frac{\partial(\phi_i, x_2)}{\partial(\xi_1, \xi_2)} - \sum_{i=1}^{2} \left(b_i^{(1)} + \lambda b_i^{(2)}\right) \frac{\partial(\phi_i, x_1)}{\partial(\xi_1, \xi_2)}$$

$$+ J\left(c^{(1)} + \lambda c^{(2)}\right) = 0 \tag{16.16a}$$

The Jacobians inside the summations are written out as

$$\frac{\partial(\phi_i, x_2)}{\partial(\xi_1, \xi_2)} = \frac{\partial x_2}{\partial \xi_2} \frac{\partial \phi_i}{\partial \xi_1} - \frac{\partial x_2}{\partial \xi_1} \frac{\partial \phi_i}{\partial \xi_2}, \quad \frac{\partial(\phi_i, x_1)}{\partial(\xi_1, \xi_2)} = \frac{\partial x_1}{\partial \xi_2} \frac{\partial \phi_i}{\partial \xi_1} - \frac{\partial x_1}{\partial \xi_1} \frac{\partial \phi_i}{\partial \xi_2}$$

$$\tag{16.17}$$

We substitute Eqs. (16.17) into (16.16a) to obtain

$$\sum_{i=1}^{2} A_i \frac{\partial \phi_i}{\partial \xi_1} - \sum_{i=1}^{2} B_i \frac{\partial \phi_i}{\partial \xi_2} + J\left(c^{(1)} + \lambda c^{(2)}\right) = 0 \tag{16.16b}$$

where

$$A_i = \left(a_i^{(1)} + \lambda a_i^{(2)}\right) \frac{\partial x_2}{\partial \xi_2} - \left(b_i^{(1)} + \lambda b_i^{(2)}\right) \frac{\partial x_1}{\partial \xi_2}$$

$$\tag{16.18}$$

$$B_i = \left(a_i^{(1)} + \lambda a_i^{(2)}\right) \frac{\partial x_2}{\partial \xi_1} - \left(b_i^{(1)} + \lambda b_i^{(2)}\right) \frac{\partial x_1}{\partial \xi_1}$$

If there are two distinct values for λ, then Eq. (16.16b) represents two equations that can replace Eqs. (16.14). We now discuss how we arrive at these values.

By hypothesis, ξ_1 is a characteristic coordinate, i.e., the lines $\xi_2 =$ const are characteristic lines. Hence, λ must be chosen such that

$$B_1 = B_2 = 0 \tag{16.19}$$

In this case, Eq. (16.16b) becomes a compatibility equation, which is an ODE, on the $\xi_2 =$ const lines. Equations (16.19) can be written as

$$\frac{(\partial x_2/\partial \xi_1)}{(\partial x_1/\partial \xi_1)} = \frac{b_1^{(1)} + \lambda b_1^{(2)}}{a_1^{(1)} + \lambda a_1^{(2)}} = \frac{b_2^{(1)} + \lambda b_2^{(2)}}{a_2^{(1)} + \lambda a_2^{(2)}} \tag{16.20}$$

The rightmost equality yields a quadratic for λ:

$$C_2 \lambda^2 + C_1 \lambda + C_0 = 0 \tag{16.21}$$

where

$$C_0 = a_2^{(1)} b_1^{(1)} - a_1^{(1)} b_2^{(1)}$$

$$C_1 = a_2^{(1)} b_1^{(2)} + a_2^{(2)} b_1^{(1)} - a_1^{(1)} b_2^{(2)} - a_1^{(2)} b_2^{(1)} \tag{16.22}$$

$$C_2 = a_2^{(2)} b_1^{(2)} - a_1^{(2)} b_2^{(2)}$$

The two roots of Eq. (16.21) are given by

$$\lambda_\pm = \frac{1}{2C_2} \left[-C_1 \pm \sqrt{C_1^2 - 4C_0 C_2} \right] \tag{16.23}$$

(This step requires that $C_2 \neq 0$.) In general, the λ are functions of x_1, x_2, ϕ_1, and ϕ_2 through their dependence on the a and b. In fact, λ cannot depend on any derivatives of ϕ_1 or ϕ_2, if the process of changing Eq. (16.16b) into an ordinary differential equation is to be valid. The requirement that Eqs. (16.14) be quasi-linear, however, ensures that λ does not depend on these derivatives.

The first of our goals can now be satisfied. If

$$C_1^2 > 4C_0 C_2$$

then $\lambda_\pm$ are real and distinct and Eqs. (16.14) are hyperbolic.

The coordinate ξ_2 is also a characteristic coordinate. If we set

$$A_1 = A_2 = 0$$

we obtain, in place of Eqs. (16.20),

$$\frac{(\partial x_2/\partial \xi_2)}{(\partial x_1/\partial \xi_2)} = \frac{b_1^{(1)} + \lambda b_1^{(2)}}{a_1^{(1)} + \lambda a_1^{(2)}} = \frac{b_2^{(1)} + \lambda b_2^{(2)}}{a_2^{(1)} + \lambda a_2^{(2)}} \tag{16.24}$$

Since the two rightmost terms are the same as in Eqs. (16.20), Eqs. (16.21)–(16.23) are unchanged. Hence, we can associate λ_+ with the ξ_2 = const characteristic lines and λ_- with the ξ_1 = const lines, or vice versa.

The equations for the characteristic lines in terms of x_1 and x_2, are obtained from Eqs. (16.20) and (16.24), as follows:

$$\frac{(\partial x_2/\partial \xi_1)}{(\partial x_1/\partial \xi_1)} = \frac{dx_2}{dx_1}\bigg|_{\xi_2} = \frac{b_1^{(1)} + \lambda_+ b_1^{(2)}}{a_1^{(1)} + \lambda_+ a_1^{(2)}}, \quad C_+ \quad (16.25a)$$

$$\frac{(\partial x_2/\partial \xi_2)}{(\partial x_1/\partial \xi_2)} = \frac{dx_2}{dx_1}\bigg|_{\xi_1} = \frac{b_1^{(1)} + \lambda_- b_1^{(2)}}{a_1^{(1)} + \lambda_- a_1^{(2)}}, \quad C_- \quad (16.25b)$$

The characteristic lines ξ_2 = const (ξ_1 = const) associated with λ_+ (λ_-) are designated, by convention, as the C_+ (C_-) lines. The quantity

$$\frac{dx_2}{dx_1}\bigg|_{\xi_2}$$

means that ξ_2 is constant on the curve given by Eq. (16.25a) in the x plane.

The compatibility equation for the C_+ characteristics is determined by Eqs. (16.16) and (16.19). The resulting equation can be appreciably simplified by using the following relations:

$$\lambda_+ - \lambda_- = \frac{1}{C_2}(C_1^2 - 4C_0C_2)^{\frac{1}{2}}$$

$$\lambda_+ \lambda_- = \frac{C_0}{C_2}$$

$$D_i = a_i^{(2)}b_1^{(1)} - a_1^{(1)}b_i^{(2)} - a_i^{(1)}b_1^{(2)} + a_1^{(2)}b_i^{(1)}, \quad i = 1, 2$$

$$A_i = \left(a_i^{(1)} + \lambda_+ a_i^{(2)}\right)\frac{\partial x_1}{\partial \xi_2}\frac{b_1^{(1)} + \lambda_- b_1^{(2)}}{a_1^{(1)} + \lambda_- a_1^{(2)}} - \left(b_i^{(1)} + \lambda_+ b_i^{(2)}\right)\frac{\partial x_1}{\partial \xi_2}$$

$$= \frac{(C_1^2 - 4C_0C_2)^{\frac{1}{2}}}{2C_2\left(a_1^{(1)} + \lambda_- a_1^{(2)}\right)}\frac{\partial x_1}{\partial \xi_2}\left[D_i - \delta_{i2}(C_1^2 - 4C_0C_2)^{\frac{1}{2}}\right] \quad (16.26)$$

$$\delta_{i2} = 0, \quad i = 1$$
$$= 1, \quad i = 2$$

$$J = \frac{\partial x_1}{\partial \xi_1} \frac{\partial x_2}{\partial \xi_2} - \frac{\partial x_1}{\partial \xi_2} \frac{\partial x_2}{\partial \xi_1}$$

$$= \frac{\partial x_1}{\partial \xi_1} \frac{\partial x_1}{\partial \xi_2} \frac{b_1^{(1)} + \lambda_- b_1^{(2)}}{a_1^{(1)} + \lambda_- a_1^{(2)}} + \frac{\partial x_1}{\partial \xi_2} \frac{\partial x_1}{\partial \xi_1} \frac{b_1^{(1)} + \lambda_+ b_1^{(2)}}{a_1^{(1)} + \lambda_+ a_1^{(2)}}$$

$$= \frac{D_1 \left(C_1^2 - 4 C_0 C_2\right)^{\frac{1}{2}}}{2 C_2 \left(a_1^{(1)} + \lambda_- a_1^{(2)}\right) \left(a_1^{(1)} + \lambda_+ a_1^{(2)}\right)} \frac{\partial x_1}{\partial \xi_1} \frac{\partial x_1}{\partial \xi_2} \qquad (16.27)$$

In Eq. (16.18) for A_i, the $\partial x_2/\partial \xi_2$ is replaced with Eq. (16.25b), which uses λ_- whereas λ_+ is used in the rest of the equation, since $B_i = 0$, i.e., $\xi_2 = $ const. With the foregoing relations, the C_+ compatibility equation reduces to

$$\frac{\partial \phi_1}{\partial \xi_1} + \frac{D_2 - \left(C_1^2 - 4 C_0 C_2\right)^{\frac{1}{2}}}{D_1} \frac{\partial \phi_2}{\partial \xi_1} + \frac{c^{(1)} + \lambda_+ c^{(2)}}{a_1^{(1)} + \lambda_+ a_1^{(2)}} \frac{\partial x_1}{\partial \xi_1} = 0, \quad C_+ \quad (16.28)$$

In a similar manner, we obtain, for the C_- compatibility equation,

$$B_i = -\frac{\left(C_1^2 - 4 C_0 C_2\right)^{\frac{1}{2}}}{2 C_2 \left(a_1^{(1)} + \lambda_+ a_1^{(2)}\right)} \frac{\partial x_1}{\partial \xi_1} \left[D_i + \delta_{i2} \left(C_1^2 - 4 C_0 C_2\right)^{\frac{1}{2}}\right] \quad (16.29)$$

$$\frac{\partial \phi_1}{\partial \xi_2} + \frac{D_2 + \left(C_1^2 - 4 C_0 C_2\right)^{\frac{1}{2}}}{D_1} \frac{\partial \phi_2}{\partial \xi_2} + \frac{c^{(1)} + \lambda_- c^{(2)}}{a_1^{(1)} + \lambda_- a_1^{(2)}} \frac{\partial x_1}{\partial \xi_2} = 0, \quad C_- \quad (16.30)$$

After much algebra, we have achieved our goal of obtaining relatively simple relations for the compatibility and characteristic equations. These are summarized in Table 16.1. The ease of application will be demonstrated in the remaining subsections.

Example: Unsteady One-Dimensional Motion

Let us check the foregoing theory by applying it to a flow whose solution we already know. We consider the isentropic motion of a perfect gas. As shown in Sec. 9.3, the pertinent equations, written in the form of Eqs. (16.14), are

$$\frac{\partial a}{\partial t} + u \frac{\partial a}{\partial x} + \frac{\gamma - 1}{2} a \frac{\partial u}{\partial x} = 0$$

$$\frac{\partial u}{\partial t} + \frac{2}{\gamma - 1} a \frac{\partial a}{\partial x} + u \frac{\partial u}{\partial x} = 0$$
(16.31)

THEORY OF CHARACTERISTICS

where a is the speed of sound and u is the flow speed. Many quantities of interest are provided in Table 16.2. From this table, we see that $(C_1^2 - C_0 C_2)^{\frac{1}{2}}$ is always real. Hence, unsteady, compressible one-dimensional flow is always hyperbolic, even when the flow is subsonic. (This result also holds in two or three dimensions.)

In accord with our previous notation, we refer to the $\xi_2 = \text{const}$ lines as the C_+ characteristics. These are given by Eq. (16.25a) as

$$\left.\frac{dx_2}{dx_1}\right|_{\xi_2} = \frac{dx_+}{dt} = u + a, \qquad C_+ \tag{16.32a}$$

Similarly, the C_- characteristics are given by Eq. (16.25b) as

$$\left.\frac{dx_2}{dx_1}\right|_{\xi_1} = \frac{dx_-}{dt} = u - a, \qquad C_- \tag{16.32b}$$

Table 16.1 Summary of Characteristic Results

Characteristic Lines	λ	dx_2/dx_1 Equation	Compatibility Equation
C_+, $\xi_2 = \text{const}$	λ_+	(16.25a)	(16.28)
C_-, $\xi_1 = \text{const}$	λ_-	(16.25b)	(16.30)

Table 16.2 Unsteady, One-Dimensional Flow Parameters

Quantity	Value			
Independent variables	$x_1 = t$, $x_2 = x$			
Dependent variables	$\phi_1 = a$, $\phi_2 = u$			
$a_1^{(1)}, \ldots, c^{(2)}$	$a_1^{(1)} = 1$,	$a_2^{(1)} = 0$,	$b_1^{(1)} = u$,	$b_2^{(1)} = \frac{\gamma-1}{2}a$, $c^{(1)} = 0$
	$a_1^{(2)} = 0$,	$a_2^{(2)} = 1$,	$b_1^{(2)} = \frac{2}{\gamma-1}a$,	$b_2^{(2)} = u$, $c^{(2)} = 0$
C_i	$C_0 = -\frac{\gamma-1}{2}a$, $C_1 = 0$, $C_2 = \frac{2}{\gamma-1}a$			
$(C_1^2 - 4C_0C_2)^{\frac{1}{2}}$	$2a$			
$\lambda_\pm$	$\pm\frac{\gamma-1}{2}$			
D_i	$D_1 = -\frac{4}{\gamma-1}a$, $D_2 = 0$			

Equations (16.28) and (16.30) are the corresponding compatibility equations, which can be written as

$$\frac{\partial a}{\partial \xi_1} + \frac{\gamma - 1}{2} \frac{\partial u}{\partial \xi_1} = 0, \quad C_+$$

$$\frac{\partial a}{\partial \xi_2} - \frac{\gamma - 1}{2} \frac{\partial u}{\partial \xi_2} = 0, \quad C_-$$

Since these are ODEs, the functions of integration are actually constants of integration. The equations are integrated, to yield

$$J_\pm = u \pm \frac{2}{\gamma - 1} a \qquad (16.33)$$

The $J_\pm$ are the constants of integration, which in Sec. 9.3 are called the Riemann invariants. This illustrates the general result that the Riemann invariants are simply the constants (or functions) of integration associated with the compatibility equations.

Example: Natural Coordinates

Steady two-dimensional or axisymmetric flow is considered using natural coordinates. We further assume isoenergetic and homentropic flow of a perfect gas. (These assumptions, for example, are consistent with nozzle flow. For rotational flow, such as that behind a curved shock, the theory in

Table 16.3 Two-Dimensional or Axisymmetric Flow Parameters in Natural Coordinates

Quantity	Value
Independent variables	$x_1 = s, \quad x_2 = n$
Dependent variables	$\phi_1 = M, \quad \phi_2 = \theta$
$a_1^{(1)}, \ldots, c^{(2)}$	$a_1^{(1)} = 0, \quad a_2^{(1)} = 1, \quad b_1^{(1)} = -\frac{1}{MX}, \quad b_2^{(1)} = 0, \quad c^{(1)} = 0$
	$a_1^{(2)} = -\frac{M^2 - 1}{MX}, \quad a_2^{(2)} = 0, \quad b_1^{(2)} = 0, \quad b_2^{(2)} = 1, \quad c^{(2)} = \frac{\sigma \sin \theta}{r}$
C_i	$C_0 = -\frac{1}{MX}, \quad C_1 = 0, \quad C_2 = \frac{M^2 - 1}{MX}$
$(C_1^2 - 4C_0C_2)^{\frac{1}{2}}$	$\frac{2}{MX}(M^2 - 1)^{\frac{1}{2}}$
$\lambda_\pm$	$\pm(M^2 - 1)^{-\frac{1}{2}}$
D_i	$D_1 = \frac{2(M^2 - 1)}{M^2 X^2}, \quad D_2 = 0$

Sec. 16.1 or in Ref. 2 should be used. See also Problem 16.10.) Equation (13.60) becomes

$$\frac{a}{a_0} = X^{-\frac{1}{2}} \tag{16.34a}$$

where, for notational convenience, we introduce

$$X \equiv 1 + \frac{\gamma - 1}{2} M^2 \tag{16.34b}$$

and all stagnation conditions are constant. The density, pressure, and speed v also can be written in terms of the Mach number

$$\frac{\rho}{\rho_0} = X^{-1/(\gamma-1)}, \qquad \frac{p}{p_0} = X^{-\gamma/(\gamma-1)}, \qquad v = aM = a_0 M X^{-\frac{1}{2}} \tag{16.35}$$

Equations (13.59b) become

$$\frac{\partial \ln[MX^{-(\gamma+1)/2(\gamma-1)}]}{\partial s} + \frac{\partial \theta}{\partial n} + \sigma \frac{\sin\theta}{r} = 0$$

$$\gamma M X^{-(\gamma+1)/2(\gamma-1)} \frac{\partial (MX^{-\frac{1}{2}})}{\partial s} + \frac{\partial X^{-\gamma/(\gamma-1)}}{\partial s} = 0$$

$$\gamma M^2 X^{-\gamma/(\gamma-1)} \frac{\partial \theta}{\partial s} + \frac{\partial X^{-\gamma/(\gamma-1)}}{\partial n} = 0$$

where θ and M are the dependent variables, and $\rho_0 a_0^2 = \gamma p_0$ is used. For notational convenience, the radial distance to point s, n is changed from x_2 to r, and x_1 is written as x. Since there are three equations for two unknowns, one of the equations must be redundant or an identity. This equation is probably the second one, since both derivatives are with respect to s. By substituting Eq. (16.34b) into this relation, we obtain an identity. The remaining two equations can be written as

$$\frac{\partial \theta}{\partial s} - \frac{1}{MX} \frac{\partial M}{\partial n} = 0$$

$$\frac{1 - M^2}{MX} \frac{\partial M}{\partial s} + \frac{\partial \theta}{\partial n} + \sigma \frac{\sin\theta}{r} = 0 \tag{16.36}$$

which are in the form of Eqs. (16.14).

Quantities of interest are provided in Table 16.3. Two aspects are different from the unsteady case. First, the inhomogeneous term $c^{(2)}$ is not zero. Second, real characteristics exist only if $M \geq 1$. It is useful to note that if Eqs. (16.36) are interchanged, $a_1^{(1)}$ interchanges with $a_1^{(2)}$, etc. In this case, C_0 interchanges with C_2, and the λ and D are altered. However, the main result is that the C_+ characteristics change to C_-, and vice versa.

We again refer to the λ_+ (λ_-), or $\xi_2 = $ const ($\xi_1 = $ const), characteristics as the C_+ (C_-) characteristics. From Eqs. (16.25) we obtain

$$\frac{dn_+}{ds} = \frac{1}{(M^2-1)^{\frac{1}{2}}}, \quad C_+$$

$$\frac{dn_-}{ds} = -\frac{1}{(M^2-1)^{\frac{1}{2}}}, \quad C_-$$

From Fig. 6.2, we have

$$\tan \mu = \frac{1}{(M^2-1)^{\frac{1}{2}}} \tag{16.37}$$

Hence, the characteristics or Mach lines are curves in the (x, r) plane that are given by

$$\frac{dn_\pm}{ds} = \pm \tan \mu \tag{16.38}$$

This result enables us to sketch the Mach lines as shown in Fig. 16.5. The C_- (C_+) characteristics are referred to as right-running (left-running) because they appear to be moving in that direction when the flow is viewed in the downstream direction.

With the use of Table 16.3, Eqs. (16.28) and (16.30) are readily obtained as

$$\frac{(M^2-1)^{\frac{1}{2}}}{MX}\frac{\partial M}{\partial \xi_1} - \frac{\partial \theta}{\partial \xi_1} - \frac{\sigma \sin \theta}{(M^2-1)^{\frac{1}{2}}r}\frac{\partial s}{\partial \xi_1} = 0, \quad C_+$$

$$\frac{(M^2-1)^{\frac{1}{2}}}{MX}\frac{\partial M}{\partial \xi_2} + \frac{\partial \theta}{\partial \xi_2} - \frac{\sigma \sin \theta}{(M^2-1)^{\frac{1}{2}}r}\frac{\partial s}{\partial \xi_2} = 0, \quad C_-$$

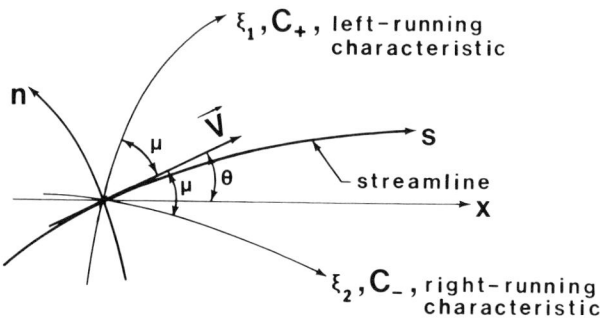

Fig. 16.5 Steady supersonic flow showing the left-running and right-running characteristics.

THEORY OF CHARACTERISTICS

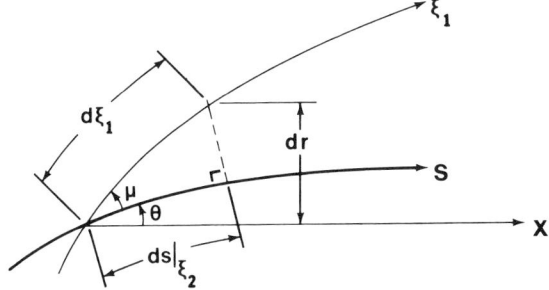

Fig. 16.6 Relationship between differential-length elements.

To integrate along the respective characteristics, multiply the first equation by $d\xi_1$ and the second equation by $d\xi_2$ to obtain

$$\int \frac{(M^2-1)^{\frac{1}{2}}}{MX} dM - \theta - \sigma \int \frac{\sin\theta}{(M^2-1)^{\frac{1}{2}}r} \frac{\partial s}{\partial \xi_1} d\xi_1 = J_+, \qquad C_+$$

$$\int \frac{(M^2-1)^{\frac{1}{2}}}{MX} dM + \theta - \sigma \int \frac{\sin\theta}{(M^2-1)^{\frac{1}{2}}r} \frac{\partial s}{\partial \xi_2} d\xi_2 = J_-, \qquad C_-$$

where $J_\pm$ are the Riemann invariants. In the two integrals, we have used

$$dM = \frac{\partial M}{\partial \xi_1} d\xi_1, \qquad dM = \frac{\partial M}{\partial \xi_2} d\xi_2$$

where ξ_2 is constant in the $d\xi_1$ integral, since the integral is restricted to a C_+ characteristic. We have, from Eq. (6.7),

$$d\nu = \frac{(M^2-1)^{\frac{1}{2}}}{MX} dM$$

where ν is the Prandtl-Meyer function given by Eq. (6.8). To evaluate the $d\xi_1$ integral, we use Fig. 16.6. From this figure* and Fig. 6.2, we obtain two relations:

$$\frac{ds|_{\xi_2}}{d\xi_1} = \frac{\partial s}{\partial \xi_1} = \cos\mu = \frac{(M^2-1)^{\frac{1}{2}}}{M}$$

---

*$ds|_{\xi_2}$ is the part of ds contributed by $d\xi_1$. There is an equal contribution to ds from $d\xi_2$.

and

$$d\xi_1 = \frac{dr}{\sin(\theta+\mu)} = \frac{1}{\sin\theta}\frac{M\,dr}{(M^2-1)^{\frac{1}{2}}+\cot\theta}$$

A similar sketch for the $d\xi_2$ integral would yield

$$\frac{\partial s}{\partial \xi_2} = \cos\mu = \frac{(M^2-1)^{\frac{1}{2}}}{M}$$

and

$$d\xi_2 = \frac{dr}{\sin(\theta-\mu)} = \frac{1}{\sin\theta}\frac{M\,dr}{(M^2-1)^{\frac{1}{2}}-\cot\theta}$$

We therefore have

$$\nu - \theta - \sigma\int \frac{dr/r}{(M^2-1)^{\frac{1}{2}}+\cot\theta} = J_+, \qquad C_+$$

$$\nu + \theta - \sigma\int \frac{dr/r}{(M^2-1)^{\frac{1}{2}}-\cot\theta} = J_-, \qquad C_- \tag{16.39}$$

for the compatibility relations.

In the axisymmetric case, the integrals can only be evaluated numerically. In this circumstance, a more convenient form is a differential one, given by

$$d\nu - d\theta - \frac{1}{(M-1)^{\frac{1}{2}}+\cot\theta}\frac{dr}{r} = 0, \qquad C_+$$

$$d\nu + d\theta - \frac{1}{(M-1)^{\frac{1}{2}}-\cot\theta}\frac{dr}{r} = 0, \qquad C_- \tag{16.40a}$$

In place of Eq. (16.38) for the characteristics, it is more convenient to use

$$\frac{dr}{dx} = \tan(\theta+\mu), \qquad C_+$$

$$\frac{dr}{dx} = \tan(\theta-\mu), \qquad C_- \tag{16.40b}$$

However, Eqs. (16.40) are still awkward since they require both $\nu(M)$ and $\mu(M)$. These two functions, however, are directly related by

$$\nu = \left(\frac{\gamma+1}{\gamma-1}\right)^{\frac{1}{2}}\tan^{-1}\left[\left(\frac{\gamma-1}{\gamma+1}\right)^{\frac{1}{2}}\cot\mu\right] + \mu - \frac{\pi}{2}$$

As a consequence, a computationally more satisfactory system of equations (except near a sonic line) can be obtained:

$$\left. \begin{array}{c} \dfrac{1}{[(\gamma-1)/2]+[(\gamma+1)/2]\tan^2\mu}\,d\mu + d\theta + \dfrac{\tan\theta\tan\mu}{1-\tan\theta\tan\mu}\dfrac{dx}{r} = 0 \\ \dfrac{dr}{dx} = \dfrac{\tan\theta+\tan\mu}{1-\tan\theta\tan\mu} \end{array} \right\} C_+$$

(16.41)

$$\left. \begin{array}{c} \dfrac{1}{[(\gamma-1)/2]+[(\gamma+1)/2]\tan^2\mu}\,d\mu - d\theta + \dfrac{\tan\theta\tan\mu}{1+\tan\theta\tan\mu}\dfrac{dx}{r} = 0 \\ \dfrac{dr}{dx} = \dfrac{\tan\theta-\tan\mu}{1+\tan\theta\tan\mu} \end{array} \right\} C_-$$

Here, the coefficients only involve $\tan\theta$ and $\tan\mu$. No intermediary functions or parameters are required.

The dx/r terms in the compatibility equations are indeterminant on a symmetry axis, where $\theta = 0$ and $r = 0$. These terms are evaluated as follows:

$$\dfrac{\tan\theta\tan\mu}{1\mp\tan\theta\tan\mu}\dfrac{dx}{r} \to (\tan\mu)\dfrac{\theta}{r}\,dx \to \tan\mu\dfrac{d\theta}{dr}\,dx$$

$$\to (\tan\mu)\,d\theta\,\dfrac{1}{\tan(\theta\pm\mu)} \to \mp d\theta$$

In going from θ/r to $d\theta/dr$, L'Hospital's rule is used. We thus obtain

$$\dfrac{1}{[(\gamma-1)/2]+[(\gamma+1)/2]\tan^2\mu}\,d\mu + 2\,d\theta = 0, \qquad C_+$$

$$\dfrac{1}{[(\gamma-1)/2]+[(\gamma+1)/2]\tan^2\mu}\,d\mu - 2\,d\theta = 0, \qquad C_-$$

in the vicinity of the axis.

As a brief illustration, let us examine the conical flowfield of Sec. 15.3. Equations (16.41) yield, for the C_+ characteristic,

$$\dfrac{dr}{dx} = \dfrac{\tan\theta+\tan\mu}{1-\tan\theta\tan\mu} = \dfrac{1+(M^2-1)^{\frac{1}{2}}\tan(\phi+\theta_b)}{(M^2-1)^{\frac{1}{2}}-\tan(\phi+\theta_b)}, \quad C_+ \quad (16.42a)$$

where we use Eq. (16.37) and $\theta = \phi + \theta_b$. Similarly, for the C_- characteristic, we obtain

$$\frac{dr}{dx} = \frac{(M^2-1)^{\frac{1}{2}}\tan(\phi + \theta_b) - 1}{(M^2-1)^{\frac{1}{2}} + \tan(\phi + \theta_b)}, \qquad C_- \qquad (16.42b)$$

Equations (16.42) are numerically integrated outward from the body along with the compatibility equations. Of course, this integration is possible only when $M_b \geq 1$. It is worth noting that the rays $\eta = $ const are not characteristic lines. For this to be the case, one of the dr/dx derivatives would have to be a constant. Furthermore, the flow downstream of the shock is a nonsimple wave region where both characteristic families are curved.

In the two-dimensional case, where $\sigma = 0$, Eqs. (16.39) yield an integrated result

$$\begin{aligned} \nu - \theta &= J_+, & C_+ \\ \nu + \theta &= J_-, & C_- \end{aligned} \qquad (16.43)$$

Suppose we have uniform flow over a flat wall that has a convex turn between points A and B, as shown in Fig 16.7. In the region upstream of the turn, Eqs. (16.38) yield straight lines for the characteristics. In the region of the bend, some of the C_- characteristics cross the expansion from region 1 to 2. Conditions in region 2 are determined by these C_- characteristics, i.e.,

$$J_- = \nu(M_1) + \theta_1 = \nu(M_2) + \theta_2$$

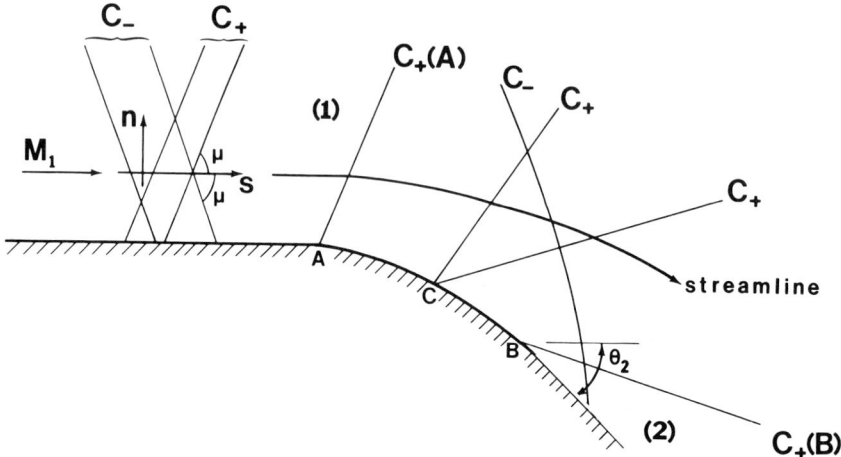

Fig. 16.7 Planar wall with a convex turn between points A and B. There is a discontinuity in wall slope at points A and C. Regions 1 and 2 have uniform supersonic flow.

With $\theta_1 = 0$, and noting that $\theta_2 < 0$, we obtain

$$\nu(M_2) = \nu(M_2) + |\theta_2|$$

which coincides with results in Chap. 6.

With M and θ continuous, Eqs. (16.36) show that a discontinuity of a derivative of order m will cause a discontinuity in a derivative of the other dependent variable of the same order. Thus, if along some line the derivative $\partial^2\theta/\partial s^2$ is discontinuous, then $\partial^2 M/\partial n\,\partial s$ is discontinuous along the same line. These lines are confined to the Mach lines and streamlines. If the wall in Fig. 16.7 has a discontinuity in slope at point A, then the normal derivatives of $\rho, p, v, M,\ldots$, will have a discontinuity along $C_+(A)$. On the other hand, if the wall shape at point B is first discontinuous in the second derivative, then the second normal derivatives of the variables will be discontinuous along $C_+(B)$.

Most frequently these discontinuities, particularly of the first derivative, occur on the leading and trailing edges of an expansion. This is the case for both steady and unsteady expansions because these Mach lines are the demarcation between uniform and nonuniform regions.

However, discontinuities in normal derivatives are not restricted to the leading and trailing edges, as shown in Fig. 16.7. The wall is curved from point A to B, but at an interior point C there is a discontinuity in the wall slope. This discontinuity results in a centered expansion that is interior to the larger expansion. Along the leading and trailing edges of the centered expansion, the normal derivatives are discontinuous.

References

[1] Meyer, R. E., "The Method of Characteristics," *Modern Developments in Fluid Dynamics*, Vol. I, edited by L. Howarth, Oxford University Press, New York, 1953, Chap. III.

[2] Liepmann, H. W. and Roshko, A., *Elements of Gasdynamics*, John Wiley & Sons, New York, 1957, Chap. 12.

[3] Courant, R. and Friedrichs, K. O., *Supersonic Flow and Shock Waves*, Interscience Publishers, New York, 1948.

Problems

16.1 Apply Crocco's equation to steady two-dimensional or axisymmetric flow, without swirl ($v_3 = 0$), using the curvilinear coordinates of Sec. 16.1. Your result should be two scalar equations. Introduce natural coordinates (s, n), and simplify your results. Finally, assume a perfect gas, and derive a relation for $\partial p/\partial n$. (Denote entropy by S.)

16.2 Determine the characteristic and compatibility equations for the equations used in Problem 13.12:

$$\phi_{tt} + F(\phi_t, \phi_x)\phi_{xx} = G(\phi_t, \phi_x)$$

What is the condition for real characteristics? With

$$F = -\phi_x^2, \quad G = \text{const}$$

integrate the compatibility equations.

16.3 A piston has the trajectory shown in the sketch below. Region 1 is quiescent, and for $t \geq t_b$ the piston speed u_p is constant. Determine J_- at the piston's surface for any characteristic that originates between $t = 0$ and t_b. Assume now that

$$\frac{p(0,t)}{p_1} = \frac{1+t}{1+\alpha t}$$

where $\alpha > 1$. Determine an explicit equation for $x_p = x_p(t_p)$, and state any conditions or restrictions that apply.

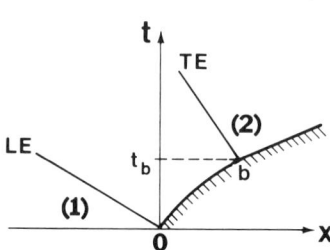

16.4 Assume the gas in Problem 16.3 is helium and that

$$\frac{p_2}{p_1} = \frac{1}{3}, \quad T_1 = 300 \text{ K}, \quad p_1 = 10^5 \text{ N/m}^2$$

Determine $u_p(t_b)$, M_2, the slope dt/dx of the trailing-edge characteristic, p_2, and p_{02}.

16.5 Assume that the Mach number is supersonic on the downstream side of an oblique shock wave. In contrast to Fig. 16.5, show that both the C_+ and C_- Mach lines are located on the same side of the downstream velocity V_2. Consequently, the C_- Mach lines are the reflection from the shock wave of the C_+ lines.

16.6 Consider a steady, two-dimensional supersonic flow that contains a slipstream. Determine the change in the slope (a labeled sketch is adequate) and in the Riemann invariants for the $C_\pm$ characteristics that cross the slipstream.

16.7 A piston is withdrawn from a quiescent gas in accord with

$$x_p = \alpha \exp(\beta t_p) - \alpha$$

where α and β are positive constants. Derive an equation for the Mach number $M(x,t)$ at any point inside the expansion.

16.8 Consider the unsteady, one-dimensional isentropic motion of a perfect gas that has $\gamma = 1$. (Note that the limit $\gamma \to 1$ is a singular limit.) Determine the governing equations with ρ and u as the dependent variables. Next, determine $\lambda_\pm$, $J_\pm$, and $H_\pm$, where the Riemann invariants are given by

$$\frac{\partial J_\pm}{\partial t} + H_\pm \frac{\partial J_\pm}{\partial x} = 0$$

Set $J = J_+$ and determine a second-order PDE for J. Determine if this PDE is elliptic, parabolic, or hyperbolic.

16.9 Equation (6.4) is based on Fig. 6.3 and does not hold for axisymmetric flow. Derive the counterpart to this equation that holds for both two-dimensional and axisymmetric homentropic flow.

16.10 Show that for supersonic, isoenergetic, two-dimensional or axisymmetric rotational flow, the characteristic equations can be written as

$$\left. \begin{aligned} d\theta - d\nu + \frac{(M^2-1)^{\frac{1}{2}}}{\gamma M^2} d\ln p_0 + \frac{\sigma \sin\theta}{(M^2-1)^{\frac{1}{2}}\cos\theta - \sin\theta} \frac{dx_1}{x_2} &= 0 \\ \frac{dx_2}{dx_1} = \frac{(M^2-1)^{\frac{1}{2}}\sin\theta + \cos\theta}{(M^2-1)^{\frac{1}{2}}\cos\theta - \sin\theta} & \end{aligned} \right\} C_+$$

$$\left. \begin{aligned} \ln p_0 &= \text{const} \\ \frac{dx_2}{dx_1} &= \tan\theta \end{aligned} \right\} C_0$$

$$\left. \begin{aligned} d\theta + d\nu - \frac{(M^2-1)^{\frac{1}{2}}}{\gamma M^2} d\ln p_0 - \frac{\sigma \sin\theta}{\sin\theta + (M^2-1)^{\frac{1}{2}}\cos\theta} \frac{dx_1}{x_2} &= 0 \\ \frac{dx_2}{dx_1} = \frac{(M^2-1)^{\frac{1}{2}}\sin\theta - \cos\theta}{\sin\theta + (M^2-1)^{\frac{1}{2}}\cos\theta} & \end{aligned} \right\} C_-$$

where C_0 denotes the streamline characteristic. As a starting point for the derivation, you may want to use the equations of motion in the following form:

$$\cos\theta \frac{\partial p_0}{\partial x_1} + \sin\theta \frac{\partial p_0}{\partial x_2} = 0$$

$$\sin\theta \frac{\partial \theta}{\partial x_1} - \cos\theta \frac{\partial \theta}{\partial x_2} + \frac{M^2-1}{MX}\left(\cos\theta \frac{\partial M}{\partial x_1} + \sin\theta \frac{\partial M}{\partial x_2}\right) - \frac{\sigma \sin\theta}{x_2} = 0$$

$$\cos\theta \frac{\partial \theta}{\partial x_1} + \sin\theta \frac{\partial \theta}{\partial x_2} + \frac{1}{MX}\left(\sin\theta \frac{\partial M}{\partial x_1} - \cos\theta \frac{\partial M}{\partial x_2}\right)$$

$$- \frac{1}{\gamma M^2 p_0 \sin\theta} \frac{\partial p_0}{\partial x_1} = 0$$

where

$$X = 1 + \frac{\gamma-1}{2}M^2$$

16.11 Consider a centered Prandtl-Meyer expansion as shown in Fig. 6.4. Use an r, η coordinate system, where η is shown in the figure and r is the radial distance from the corner. Let r_{LE} be an arbitrary distance along the leading edge of the expansion. Derive the equation

$$\frac{r}{r_{LE}} = f(\eta; \gamma, M_\infty)$$

for a right-running characteristic line that crosses the expansion. Under what conditions will the characteristic line first approach the origin when entering the expansion? (Hint: Use the results of Problem 14.11.)

17. MINIMUM-LENGTH NOZZLES

A supersonic nozzle with a minimum throat-to-exit length and a uniform flow at the exit is called a minimum-length nozzle (MLN). The concept is applicable to both two-dimensional and axisymmetric flows. A distinctive feature of an MLN is that the wall has a sharp corner at the throat. Most nozzles are not of minimum length and have a smoothly contoured wall at the throat. The practical importance of MLNs is discussed in the next section. One reason why MLNs have not been widely used is the fragmentary status of their theory. This chapter partly rectifies this deficiency.

17.1 PRELIMINARY REMARKS

Shapiro[1] presents an analysis for a two-dimensional MLN with a straight sonic line at the throat. However, there are two types of MLNs. One type assumes a straight sonic line for the flow at the throat. In the other, the sonic line is assumed to be an arc of a circle. Both types hold for two-dimensional or axisymmetric flow. Four different configurations are thus possible. Interestingly, the literature for one type does not reference the other.[2] As a consequence, a systematic comparison between configurations has heretofore not been done.

All configurations require a sharp corner at the throat. If the sonic line is straight, the corner generates a centered Prandtl-Meyer expansion. For convenience, this nozzle type will be referred to as the straight sonic line MLN.[1-5] In the second type, the sonic line is an arc of a circle, and no centered expansion occurs. This type will be called a curved sonic line MLN.[6-9]

Our discussion indicates that the published material is fragmentary and sometimes in error. Furthermore, parametric results that depend, for example, on the ratio of specific heats, γ, do not exist for either type of MLN. Indeed, Fig. 15.37 for $\gamma = 1.4$ in Ref. 1 represents the only previously published parametric result. Analysis and parametric results will be presented in this chapter.

For a given exit Mach number, a principal feature of an MLN nozzle is its ability to minimize the boundary-layer thicknesses at the exit plane. This is a direct result of imposing the throat-to-exit-pressure change over the shortest possible distance. This aspect should prove advantageous for a hypersonic wind tunnel that normally has a thick boundary layer in the test section. Inviscid results will be shown to indicate the possibility of a significant reduction in nozzle length that can be obtained with an axisymmetric MLN.

The practical importance of an MLN is not limited to wind-tunnel design. Any application in which a "clean" supersonic flow is needed and physical size and weight are to be minimized should benefit from an MLN. Thus, axial impulse turbine and thrust rocket nozzle design as well as high-power laser technology are several possibilities.

One laser example would involve high-energy, high-repetitive-rate pulsed lasers. For example, excimer lasers are short-pulsed gas lasers whose optical wavelength is in the ultraviolet. The wavelength, however, can be changed to a different value by a process called Raman shifting. This is done in a separate device whose conversion medium is a gas, such as hydrogen. Neither the shifting nor the lasing process is overly efficient. Thus, a large amount of waste energy is deposited in a few nanoseconds in the gas inside the Raman device. The energy deposition is nonuniform and results in a density perturbation that can severely degrade the optical, or beam, quality of the subsequent laser pulse. When a high repetitive laser pulse rate is desired, a uniform supersonic flow should be used for the gas in the Raman device. A supersonic flow has the advantage of a short clearing time[10] for the disturbance created by the energy-transfer process.

An MLN may become important in a number of technologies. These include:

(1) The gasdynamic laser, where this type of nozzle is called a rapid expansion nozzle.[4] (Only the two-dimensional straight sonic line nozzle has been used in this application.) It maximizes the fraction of nitrogen vibrational energy that is frozen.

(2) Continuous-wave chemical lasers that generally operate at a very low Reynolds number and thus contain thick laminar boundary layers in the nozzles. The MLN concept minimizes the rate of boundary-layer growth, and would be applicable to both the HF/DF laser and the more recent supersonic oxygen-iodine laser.

(3) The aerodynamic window, including the free-vortex window discussed in the next chapter. Again, the MLN minimizes boundary-layer growth.

(4) Laser isotope separation or laser-induced chemical processes, where a low static temperature is needed for spectral simplification. An MLN would delay condensation of the highly supercooled condensible species.

Although the importance of the MLN has been recognized for the gasdynamic laser, this is not the case for any of the other applications.

The next section provides a comprehensive analytical theory for inviscid flow in an MLN with a curved sonic line. Section 17.3 considers the straight sonic line MLN, while Sec. 17.4 provides a limited comparison of the different geometries and types of MLNs.

17.2 CURVED SONIC LINE MLN

We consider steady two-dimensional or axisymmetric flow in an MLN where the sonic line is curved. The gas is perfect and the flow is assumed to be isoenergetic and homentropic. Since the nozzle is an MLN, it has a sharp corner at the throat, and the flow at the exit is supersonic and uniform with a Mach number M_f. If the throat has a smooth contour, the initial data line can still be a circular arc at a Mach number of unity. This nozzle, however, is not of minimum length.

Figure 17.1 shows a sketch of the nozzle, the coordinate system, and some of the nomenclature. A conventional x, y Cartesian coordinate system with velocity components u, v is utilized. In the axisymmetric case, y and v are the radial coordinate and radial velocity component, respectively. The sonic line, denoted with an asterisk, is an arc of a circle of radius R^* that extends from point A to A'. The center of the arc is the origin of the coordinate system. The section of wall from A' to B' has a constant slope θ^*. This section is a frustum of a cone when the nozzle is axisymmetric. The triangular region, $B-C'-x_f$, is a uniform flow region with Mach number M_f. According to characteristic theory, BC' is a straight Mach line with Mach angle μ_f, and the region upstream of this line must be a simple wave region when the flow is two-dimensional. This region, it will turn out, is $B-B'-C'$. The Mach lines in it, which are of the same family as BC', are also straight lines (when the flow is two-dimensional). One such Mach line is PP', which will play an important role in the subsequent analysis.

Between AA' and BB' the flow is a two-dimensional or a spherical source flow; nevertheless, this is a nonsimple wave region. (The spherical source flow solution for this region is considered in Problem 15.3.) All variables in the source flow region are independent of θ; they vary only with R.

The location of point B' is arbitrary. It is generally chosen to coincide with point A', since this is thought to yield an MLN.[8] As will be shown, this presumption is incorrect. Therefore, in the subsequent derivation, the location of B' is kept arbitrary. Its location is later determined by the minimum-length condition.

We start the analysis by first considering the source flow region. When the flow is two-dimensional, the cross-sectional area A between the two walls is a section of a cylinder with radius R and unit depth. It is a spherical cap, with radius R, in the axisymmetric case. We thus have

$$A = 2\pi^\sigma \theta^{*1-\sigma}(1 - \cos\theta^*)^\sigma R^{1+\sigma} \qquad (17.1)$$

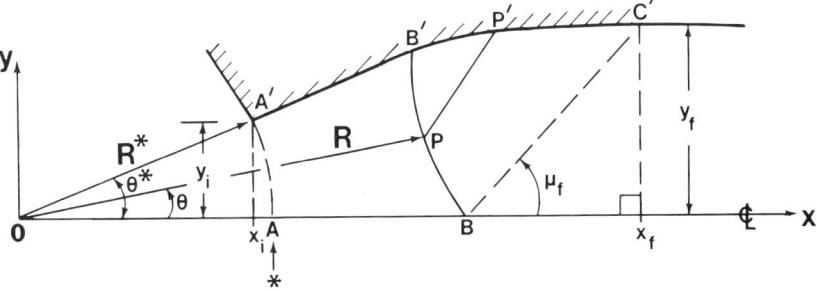

Fig. 17.1 Coordinate system and nomenclature for a two-dimensional or axisymmetric MLN with a curved sonic line.

This relation holds at the sonic line where A and R become A^* and R^*, respectively. The area ratio in the source flow region is

$$\frac{A}{A^*} = \left(\frac{R}{R^*}\right)^{1+\sigma}$$

This ratio is related to the Mach number by the usual homentropic relation

$$\left(\frac{R}{R^*}\right)^{1+\sigma} = \left[\left(\frac{2}{\gamma+1}\right)X\right]^{(\gamma+1)/[2(\gamma-1)]} \frac{1}{M} \quad (17.2)$$

where again we use

$$X \equiv 1 + \frac{\gamma-1}{2}M^2$$

We shall shortly need the logarithmic derivative of Eq. (17.2), which is

$$\frac{dR}{R} = \frac{1}{1+\sigma} \frac{M^2-1}{MX} dM \quad (17.3)$$

The mass flow rate is also needed. It can be evaluated at the sonic line as

$$\dot{m} = (\rho A V)^*$$

$$= 2\pi^\sigma \left(\frac{2}{\gamma+1}\right)^{(\gamma+1)/[2(\gamma-1)]} (\theta^*)^{1-\sigma}(1-\cos\theta^*)^\sigma (R^*)^{1+\sigma} \left(\frac{\gamma}{R_g T_0}\right)^{\frac{1}{2}} p_0$$

$$(17.4)$$

where, in this chapter only, R_g is the gas constant.

We next determine the shape of the right-running borderline Mach line BB'. Actually, the derivation can be used for any right-running or left-running Mach line in the source flow region. A section of the BB' Mach line is

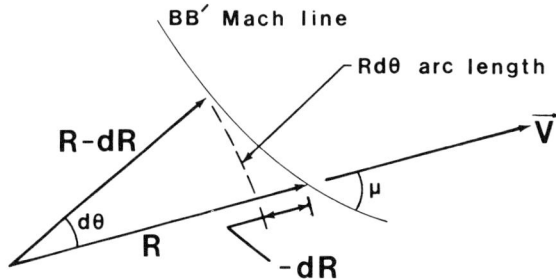

Fig. 17.2 Various differentials associated with a right-running Mach line in the source flow region.

shown in Fig. 17.2. From the figure, we obtain

$$\frac{R\,d\theta}{-dR} = \tan\mu = \frac{1}{(M^2-1)^{\frac{1}{2}}}$$

or

$$\frac{dR}{R} = -(M^2-1)^{\frac{1}{2}}\,d\theta \qquad (17.5)$$

(For a left-running Mach line, change the minus sign to a plus in Eq. (17.5). The resulting two differential equations are the equations for the Mach lines in R, θ coordinates.) We eliminate dR/R from Eqs. (17.3) and (17.5) to obtain

$$-\int_0^\theta d\theta = \frac{1}{1+\sigma}\int_{M_f}^{M}\frac{(M^2-1)^{\frac{1}{2}}}{MX}\,dM$$

As indicated in the integrals, at point B we have $\theta = 0$ and $M = M_f$. We thus obtain

$$\theta = \frac{1}{1+\sigma}\left[\nu(M_f) - \nu(M)\right] \qquad (17.6)$$

where $\nu(M)$ is the Prandtl-Meyer function. Therefore, the shape of the BB' Mach line

$$R/R^* = f(\theta; \gamma, M_f, \sigma) \qquad (17.7)$$

is given by Eqs. (17.2) and (17.6) with M as a parameter.

Before continuing with the analysis, we observe that this type of nozzle can be viewed as if it were an optical lens that takes and transforms a source flow into a parallel one. The flow for this nozzle is reversible; that is, if the direction is reversed, a parallel flow is transformed into a converging flow. Of course, the adverse pressure gradient may cause boundary-layer separation. In accord with the lens analogy, the theory can be used to transform one uniform supersonic flow with Mach number M_1 into a uniform supersonic flow with Mach number M_2, where M_2 may be greater or less than M_1. The type of duct required is shown in Fig. 17.3. The sketch is for the $M_2 > M_1$ case, which has a favorable pressure gradient. By simply reversing the flow direction, one obtains the compressive case, where the pressure gradient is unfavorable. When the flow is two-dimensional, the nonsimple source flow region is bordered on both sides by simple wave regions, except at two discrete points. An exact analytical solution is possible when the flow is two-dimensional, including the shape of the curved walls. (See Problem 17.4.)

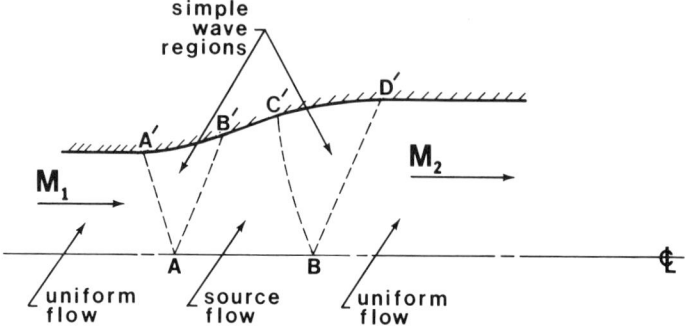

Fig. 17.3 Duct for transforming uniform supersonic flow from Mach number M_1 to M_2. In the axisymmetric case, the two simple wave regions are nonsimple wave regions.

When the flow is axisymmetric, simple wave regions do not exist. Regions $A-A'-B'$ and $B-C'-D'$ in Fig. 17.3 are nonsimple wave regions whose solution is obtained numerically with the MOC. For region $A-A'-B'$, we briefly indicate that sufficient data are known for the AA' and AB' Mach lines, including their location. Mach line AA' is straight and $\theta = 0$, $M = M_1$ along it. The analysis leading to Eqs. (17.6) and (17.7) is used for the left-running curved Mach line AB', with M_f replaced by M_1. Equations (17.1)–(17.3) for the source flow solution also hold on line AB'. The MOC solution would start just above point A and terminate when the wall from A' to B' is determined. A similar construction would be used for region $B-C'-D'$. The location of the origin for the source flow region is not arbitrary. It must be chosen such that the wall Mach numbers satisfy

$$M_{B'} \leq M_{C'}$$

The construction is somewhat simplified if points B' and C' coincide.

One possible difficulty is the convergence of Mach lines of a given family as they approach the centerline of an axially symmetric flow. Should overlap start to occur in the Mach lines, a shock wave forms instead. Overlap cannot occur in our construction, since it starts near points A and B. In turn, this is a result of using a source flow region between lines AB' and BC'.

Reference 11 discusses an alternate approach for transforming an axially symmetric uniform flow into one with a different Mach number. In this approach, however, Mach line overlap occurs and the flow is not shock-free. Except for the problems, we terminate the lens analogy discussion.

The shape of the wall from A' to B' in Fig. 17.1 is fully determined by θ^*. However, we need to determine the shape of the curved wall between B' and C'. In the two-dimensional case, $B-B'-C'$ is a simple wave region and the left-running Mach lines that impinge on the wall reflect with zero strength. This reflection cancellation condition is common in nozzle design work. To accomplish it, the wall must turn the flow the same amount as the

Mach line. In the subsequent derivation for the wall shape, it will not be necessary to utilize this reflection condition explicitly although it certainly holds.

Since simple wave regions do not exist in the axisymmetric case, the Mach line PP' is not straight, and flow conditions are not constant along it. An exact wall shape from B' to C' requires a numerical solution for the $B-B'-C'$ region for which sufficient M and θ data are known on the Mach lines BB' and BC'. An alternative to this approach is an approximate analytical solution. Fortunately, such a solution exists.[12] We utilize this approximate solution for the axisymmetric case because it provides exact results for the nozzle length $x_f - x_i$ and for the location of points B' and C'. Consequently, the nozzle's area ratio $(y_f/R^*)^2/[2(1 - \cos\theta^*)]$ is determined exactly. In this approach, PP' is considered straight and ρV is taken to be constant along it. The slope of PP' and the ρV product are exact on the line only at point P. For the two-dimensional nozzle, the slope of PP' and the ρV product are exact everywhere on the PP' Mach line. Thus, the approximate axisymmetric approach is exact in the two-dimensional case.

Explicit equations for the wall shape are obtained by a mass balance argument. We first determine the mass flow rate through the spherical annular source flow section located between points P and B'. This is evaluated first at point B' by using Eqs. (17.1) and (17.4). It is then evaluated at point P by setting θ^* equal to θ in these two equations. We take the difference to obtain for the flow rate through the section

$$\dot{m}_P = 2\pi^\sigma \left[(1+\sigma)(\theta^* - \theta) + \sigma(\cos\theta - \cos\theta^*) \right] R^{1+\sigma} (\rho V)_P \quad (17.8)$$

where a P subscript means the quantity is evaluated at point P.

The mass flow rate across the straight annular section between points P and P' is also equal to $\dot{m}_P$. It is given by

$$\dot{m}_P = \rho_P A_{fr} V_P \sin\mu \quad (17.9)$$

where $V_P \sin\mu$ is the component of $\mathbf{V}_P$ that is perpendicular to PP'. Let ℓ be the distance between points P and P'. As shown in Fig. 17.4, A_{fr} is the area $(2\ell)1$ for the two flat surfaces in the two-dimensional case, or the frustum of the PP' cone in the axisymmetric case. Hence,

$$A_{fr} = 2\ell, \qquad\qquad \sigma = 0$$

$$= \pi(r_1 + r_2)\left[(\Delta x)^2 + (r_2 - r_1)^2\right]^{\frac{1}{2}}, \qquad \sigma = 1$$

By simple geometry we have

$$r_1 = R\sin\theta, \qquad r_2 = R\sin\theta + \ell\sin(\theta + \mu), \qquad \Delta x = \ell\cos(\theta + \mu)$$

so that

$$A_{fr} = 2\pi\ell[R\sin\theta + \tfrac{1}{2}\ell\sin(\theta+\mu)], \quad \sigma = 1$$

or

$$A_{fr} = 2\pi^\sigma \ell[R\sin\theta + \tfrac{1}{2}\ell\sin(\theta+\mu)]^\sigma \quad (17.10)$$

for both geometries.

With A_{fr} given by Eq. (17.10), we eliminate $(\dot{m}/\rho V)_P$ from Eqs. (17.8) and (17.9) to obtain

$$\frac{\ell}{R} = \frac{\theta^* - \theta}{\sin\mu}, \quad \sigma = 0 \quad (17.11a)$$

and

$$\sin(\theta+\mu)\left(\frac{\ell}{R}\right)^2 + 2\sin\theta\left(\frac{\ell}{R}\right) + \frac{2(\cos\theta^* - \cos\theta)}{\sin\mu} = 0, \quad \sigma = 1$$

The appropriate solution of this quadratic is

$$\frac{\ell}{R} = \frac{\sin\theta}{\sin(\theta+\mu)}\left[1 + \frac{2\sin(\theta+\mu)(\cos\theta - \cos\theta^*)}{\sin\mu \sin^2\theta}\right]^{\frac{1}{2}} - 1, \quad \sigma = 1$$

(17.11b)

Equations (17.11) provide the functional result

$$\ell/R = g(\theta, M; \sigma, \theta^*)$$

where at point P, θ and M are related by Eq. (17.6), which introduces γ and M_f as parameters.

The arbitrary point P' on the wall is denoted by the coordinates x' and y'. For both geometries, they are given by

$$\frac{x'}{R} = \cos\theta + \frac{\ell}{R}\cos(\theta+\mu)$$

$$\frac{y'}{R} = \sin\theta + \frac{\ell}{R}\sin(\theta+\mu)$$

(17.12)

Note that R, θ denotes point P, not P' and that M is the same at both points and is given in terms of θ by Eq. (17.6). The coordinates R and θ are not independent of each other but are related by Eq. (17.7). Thus, ℓ/R, M, and R are determined by the angle θ and the independent fixed parameters: σ, γ, M_f, R^*, and θ^*.

On line $A'B'$, the Mach number is bounded according to $1 \leq M \leq M_{B'}$, where $M_{B'}$ is given by Eq. (17.6) as

$$\nu(M_{B'}) = \nu(M_f) - (1+\sigma)\theta^* \quad (17.13)$$

On the curve $B'C'$, we have $M_{B'} \leq M \leq M_f$. As $\theta^* \to 0$, we obtain $M_{B'} \to M_f$. This is not an interesting limit since the nozzle is quite long and since viscous effects dominate the flow. More interesting is the maximum value θ_m^* that θ^* can attain. This occurs when $B' \to A'$ and $M_{B'} \to 1$, so that Eq. (17.13) reduces to

$$\theta_m^* = \frac{\nu(M_f)}{1+\sigma} \tag{17.14}$$

It is important to note that θ_m^* does not necessarily provide a minimum length for the nozzle.

The sonic line radius R^* is determined by the mass flow rate, Eq. (17.4). However, it is necessary first to determine θ^* by Eq. (17.13) or (17.14) before using Eq. (17.4) to determine R^*.

Despite its algebraic complexity, we have an exact solution for a special case of a two-dimensional nozzle flow. This is the only closed-form, exact, inviscid nozzle flow solution we will find. Further aspects of the two-dimensional and axisymmetric solutions, such as the centerline or wall Mach number, are the subject of several problems.

The length of the nozzle is measured from x_i (Fig. 17.1) in order to facilitate a comparison with a straight sonic line MLN. We keep the sharp corner of the throat, with coordinates x_i, y_i, fixed for any geometrical comparisons. Thus, the nozzle length is given nondimensionally by

$$\frac{x_f - x_i}{y_i} = \frac{(x_f - x_B) + (x_B - R^*) + (R^* - x_i)}{y_i} \tag{17.15}$$

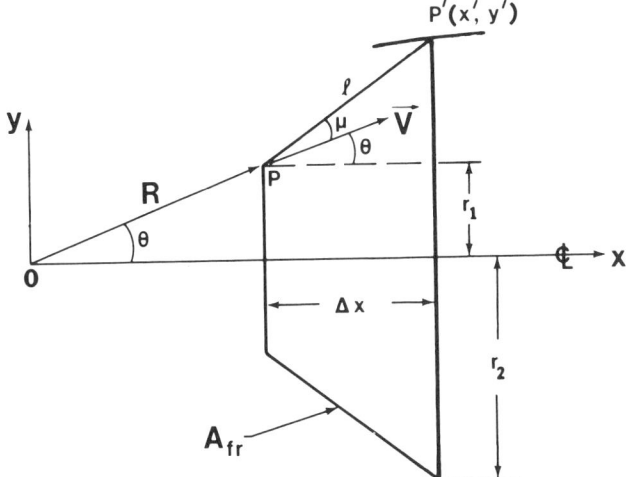

Fig. 17.4 Nomenclature associated with line PP'.

where $x_A = R^*$ is used. The ratio y_f/y_i, which is not the nozzle's area ratio, is determined first. From Eqs. (17.1) and (17.2), we have

$$\frac{y_f}{y_i} = \frac{1}{\sin\theta^*}\left[2^\sigma(\theta^*)^{1-\sigma}(1-\cos\theta^*)^\sigma \alpha_f\right]^{1/(1+\sigma)} \quad (17.16)$$

where

$$\alpha = \frac{1}{M}\left(\frac{2}{\gamma+1}X\right)^{(\gamma+1)/[2(\gamma-1)]}$$

From Fig. 17.1, we establish

$$\frac{(x_f - x_B)}{y_i} = \beta_f \frac{y_f}{y_i} \quad (17.17)$$

where

$$\beta = (M^2 - 1)^{\frac{1}{2}}$$

The last factor on the right in Eq. (17.15) is given by

$$\frac{R^* - x_i}{y_i} = \frac{R^* - R^*\cos\theta^*}{R^*\sin\theta^*} = \frac{1 - \cos\theta^*}{\sin\theta^*} \quad (17.18)$$

To obtain x_B, set $R_B = x_B$, and with Eq. (17.2), we have

$$x_B = R^*\alpha_f^{1/(1+\sigma)}$$

so that

$$\frac{x_B - R^*}{y_i} = \frac{1}{\sin\theta^*}\left(\alpha_f^{1/(1+\sigma)} - 1\right) \quad (17.19)$$

By combining Eqs. (17.15)–(17.19), we obtain the desired result

$$\frac{x_f - x_i}{y_i} = \frac{\alpha_f^{1/(1+\sigma)}}{\sin\theta^*}\left\{1 + \beta_f(\theta^*)^{1-\sigma}[2(1-\cos\theta^*)]^{\sigma/2}\right\} - \cot\theta^* \quad (17.20)$$

Recall that the location of point B' shown in Fig. 17.1 is arbitrary. We now choose this location so that $(x_f - x_i)/y_i$ is a minimum. However, this B' location is physically realistic only if the resulting value for θ^*, denoted as θ_M^*, does not exceed θ_m^*. (When θ^* equals θ_m^*, point B' coincides with point A' and θ^* is a maximum.)

MINIMUM-LENGTH NOZZLES

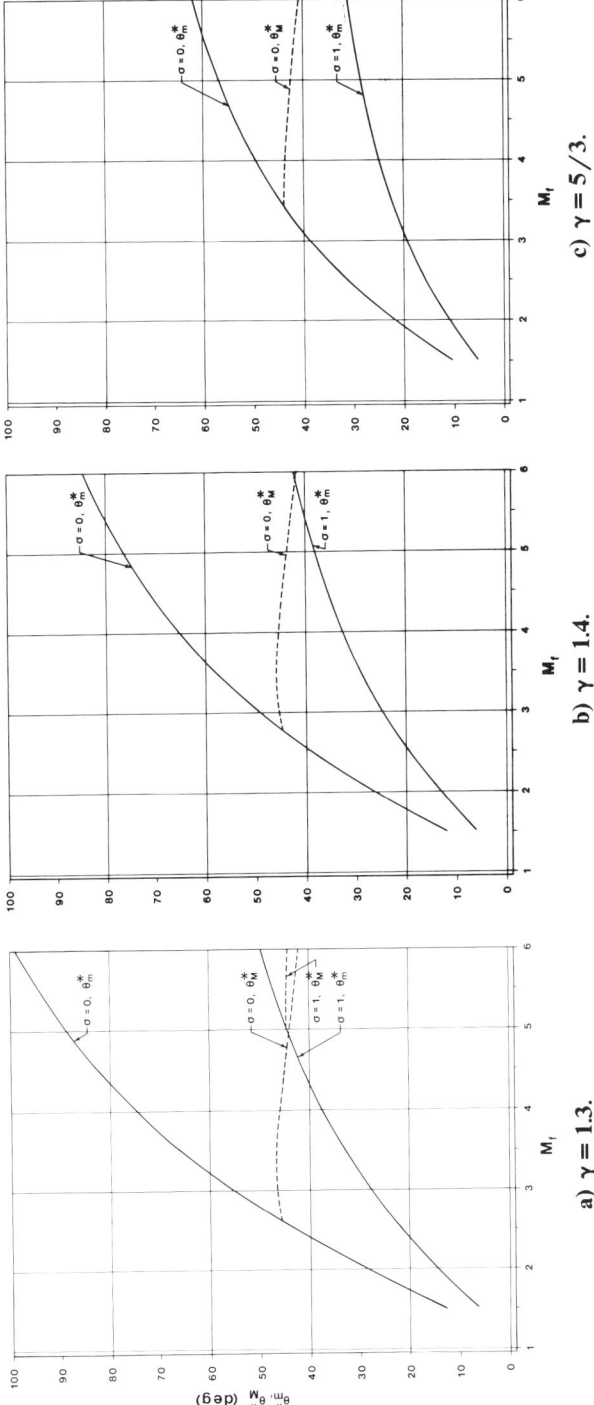

Fig. 17.5 Variation of θ_m^* and θ_M^* vs M_f for various γ for a curved sonic line MLN.

To facilitate the computation, separate the two σ cases. By differentiating Eq. (17.20), we obtain

$$\frac{\sin^2\theta^*}{y_i} \frac{\partial(x_f - x_i)}{\partial\theta^*} = 1 + \alpha_f\beta_f\sin\theta^* - \alpha_f(1 + \beta_f\theta^*)\cos\theta^*, \qquad \sigma = 0$$

$$= 1 + \left(\frac{\alpha_f}{2}\right)^{\frac{1}{2}}\left[\beta_f(1 - \cos\theta^*)^{\frac{3}{2}} - 2^{\frac{1}{2}}\cos\theta^*\right], \qquad \sigma = 1$$

(17.21)

The smallest positive value of θ^* that causes the right side of Eqs. (17.21) to be zero is θ_M^*. This value provides an MLN when $\theta_m^* \geq \theta_M^*$.

Figure 17.5 shows θ_m^* and θ_M^* vs M_f for $\gamma = 1.3$, 1.4, and 5/3. From the $\gamma = 1.3$ figure, we see that for an MLN, θ^* must be chosen in accord with

$$\theta^* = \theta_m^*, \qquad M_f < 2.6$$
$$\sigma = 0$$
$$= \theta_M^*, \qquad 2.6 \leq M_f$$

$$\theta^* = \theta_m^*, \qquad M_f < 5$$
$$\sigma = 1$$
$$= \theta_M^*, \qquad 5 \leq M_f$$

Whereas θ_m^* increases monotonically with M_f, θ_M^* does not. There is a sharp change in slope at the crossover point that depends on γ and σ. In fact, for $\gamma = 1.4$ and 5/3, the $\sigma = 1$, θ_M^* curve does not appear because the crossover point is beyond an M_f of 6. A very large value for the angle θ^* is not practical because the boundary layer may separate from the wall at the throat. In this regard, for the same γ and M_f, an axisymmetric nozzle is superior to a two-dimensional one. For the same reason, the use of θ_M^*, instead of θ_m^*, is imperative at high M_f values for a two-dimensional nozzle.

Figures 17.6 and 17.7 show $(x_B - x_i)/y_i$ and $(x_f - x_i)/y_i$ vs M_f, respectively, for the same σ and γ values of Fig. 17.5. All results are for an MLN; hence, θ^* is either θ_m^* or θ_M^*. A comparison of the figures shows that $(x_B - x_i)/(x_f - x_i)$ decreases as M_f increases. Consequently, at high M_f values, the source flow region occupies only a small percentage of the nozzle's throat-to-exit length. Since point B' is closer to the origin than is point B, the length of the straight section $A'B'$, which occurs only when $\theta^* = \theta_M^*$, is quite short relative to $x_f - x_i$.

A comparison of the different panels in Fig. 17.7 shows that the normalized nozzle length decreases rapidly with an increasing γ; that is, the rate of expansion is more rapid as γ increases.

The importance of using θ_M^* at the higher Mach numbers is shown graphically in Fig. 17.8 for two-dimensional nozzles. The quantity $(x_f - x_i)_{\theta_m^*}$

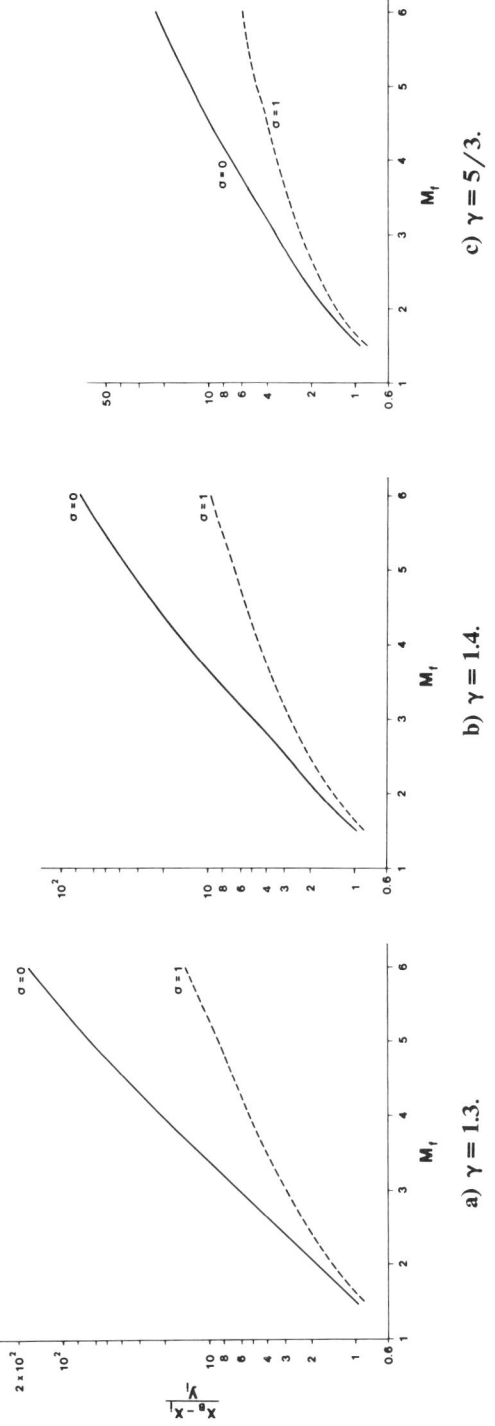

Fig. 17.6 Variation of $(x_B - x_i)/y_i$ vs M_f for various γ for a curved sonic line MLN.

a) $\gamma = 1.3$.   b) $\gamma = 1.4$.   c) $\gamma = 5/3$.

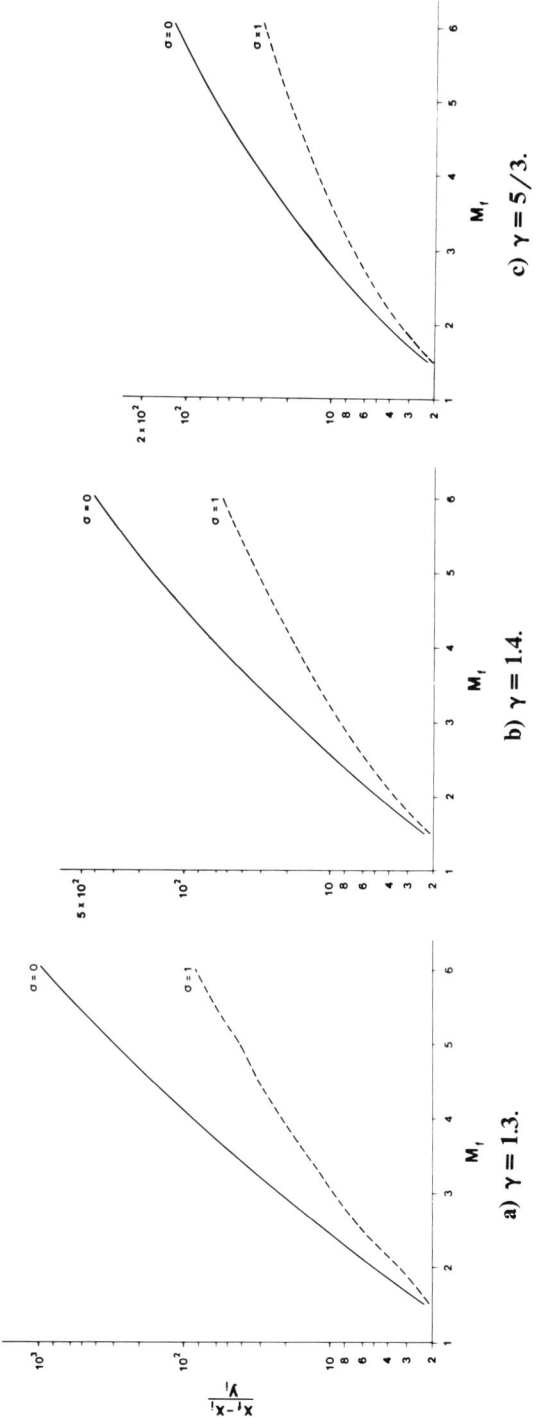

Fig. 17.7 Variation of $(x_f - x_i)/y_i$ vs M_f for various γ for a curved sonic line MLN.

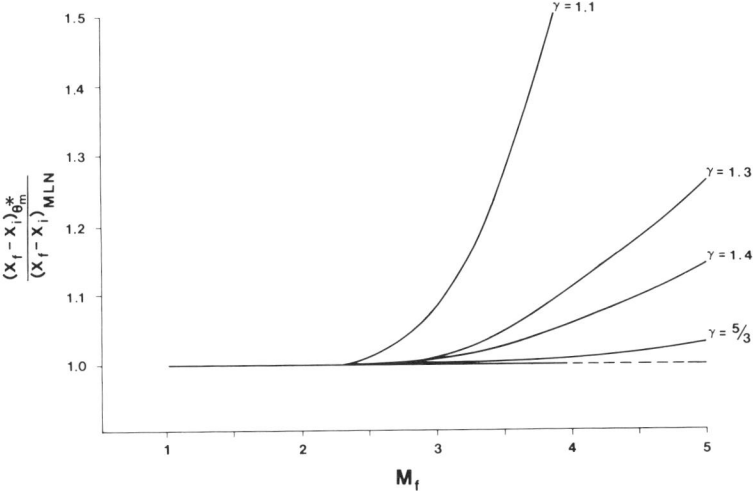

Fig. 17.8 Nozzle length based on θ_m^* divided by the correct length for a two-dimensional, curved sonic line nozzle.

denotes the nozzle length when $\theta^* = \theta_m^*$. Above the crossover point, where $\theta^* = \theta_M^*$, the additional length can be significant if θ_m^*, instead of θ_M^*, is used.

17.3 STRAIGHT SONIC LINE MLN

Figure 17.9 is a sketch of the nozzle showing some of the nomenclature. In both the two-dimensional and axisymmetric cases, a centered Prandtl-Meyer expansion originates at the corner of the throat. The expansion reflects from the centerline as left-running waves; or these left-running waves can be viewed as coming from the opposite corner. The region $A'-x_i-B$ is a nonsimple wave region called the kernel. The region $B-C'-x_f$ is a uniform flow region with Mach number M_f. When the flow is two-dimensional, the region $A'-B-C'$ is a simple wave region: it is a nonsimple wave region in the axisymmetric case.

In some situations, the sharp corner at the throat may be impractical. A nearly minimum-length nozzle, however, can be generated by using for the wall a streamline close to the $A'C'$ streamline. The wall now has a small but finite radius of curvature at the throat and an inflection point downstream of the throat.

The area ratio $(y_f/y_i)^{1+\sigma}$ simply equals A_f/A^*, as given by Eq. (17.2). At the throat, the flow is parallel to the x axis and the Mach number is unity. A streamline adjacent to the wall goes through a rapid Prandtl-Meyer expansion such that the Mach number $M_{A'}$ just downstream of the expansion is given by

$$\nu(M_{A'}) = \nu(M_i) + \theta^* = \theta^*$$

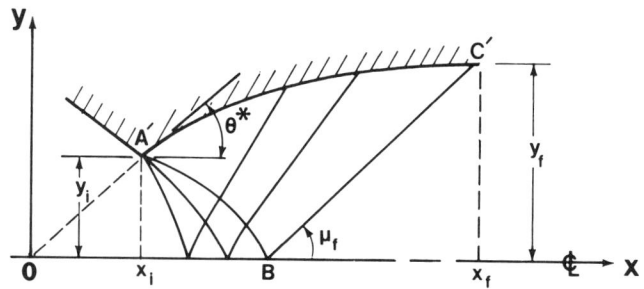

Fig. 17.9 Coordinate system and nomenclature for a two-dimensional or axisymmetric MLN with a straight sonic line.

The same streamline goes through a second expansion caused by the Prandtl-Meyer fan from the opposite corner when the flow is two-dimensional. This expansion increases M from $M_{A'}$ to M_f while decreasing θ from θ^* to zero. Consequently, we have

$$\nu(M_f) = \nu(M_{A'}) + \theta^*$$

By combining the above, we obtain for the two-dimensional case

$$\theta^* = \nu(M_{A'}) = \tfrac{1}{2}\nu_f \quad (17.22)$$

and the two expansions have equal strength.

The quantities $(y_f/y_i)^{1+\sigma}$ and θ^* (when $\sigma = 0$) are readily obtained analytically. All other parameters, however, can be obtained only by numerical solution of the MOC equations. In the curved sonic line case, the nonsimple wave region was analytically tractable because it consists of a source flow, but no such simplification occurs with a straight sonic line.

Saadat[13] provides the above numerical results in the two-dimensional case. Figures 17.10 and 17.11 show $(x_B - x_i)/y_i$ and $(x_f - x_i)/y_i$, respectively. As expected, these results are similar to those of Figs. 17.6 and 17.7. Figure 17.12 shows the kernel length divided by the overall length. This ratio is fairly insensitive to γ, and at a large M_f the kernel's length is only about 20% of the nozzle's length. The reason for both trends is that μ_f is independent of γ and becomes small with increasing M_f.

Figure 17.13 shows the Mach number M_e along the wall vs nondimensional arc length measured from the throat, given by

$$\tilde{s} = s/s_f \quad (17.23)$$

where s_f is the arc length from the throat to the exit. The Mach number at $\tilde{s} = 0$, previously denoted as $M_{A'}$, is not unity due to the Prandtl-Meyer expansion at the corner. Downstream of the corner, M_e smoothly increases

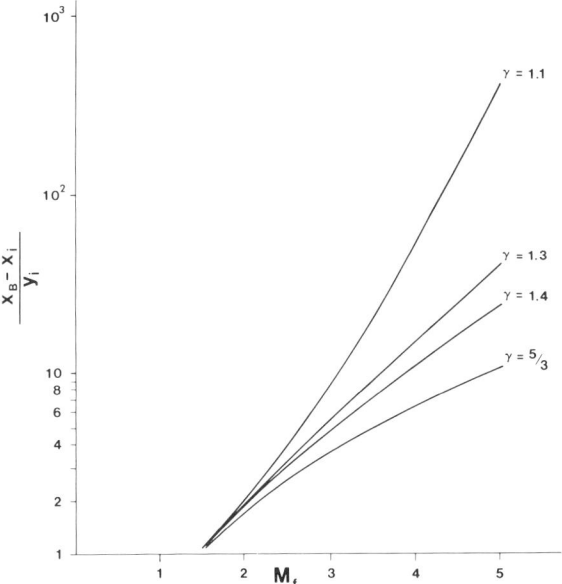

Fig. 17.10 $(x_B - x_i)/y_i$ vs M_f for various γ for a two-dimensional, straight sonic line MLN.

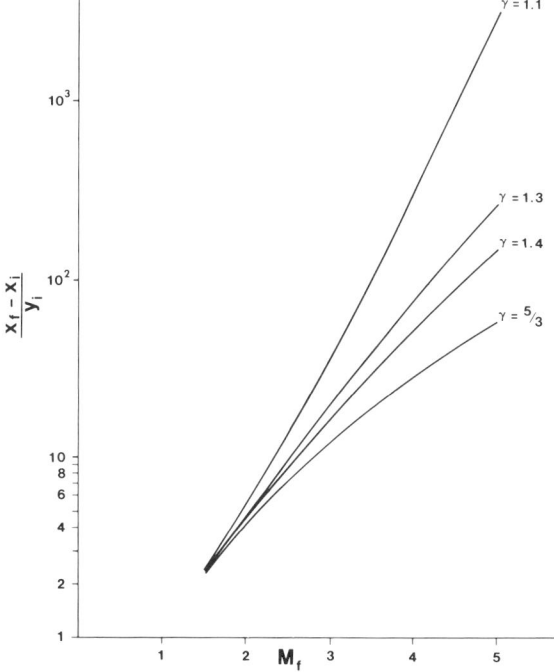

Fig. 17.11 $(x_f - x_i)/y_i$ vs M_f for various γ for a two-dimensional, straight sonic line MLN.

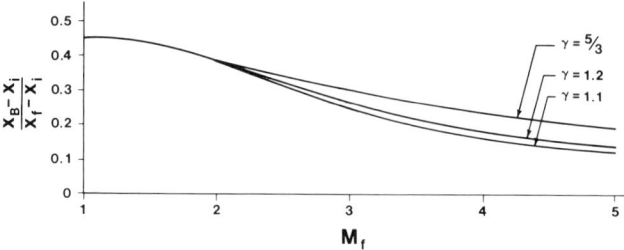

Fig. 17.12 $(x_B - x_i)/(x_f - x_i)$ vs M_f for various γ for a two-dimensional, straight sonic line MLN.

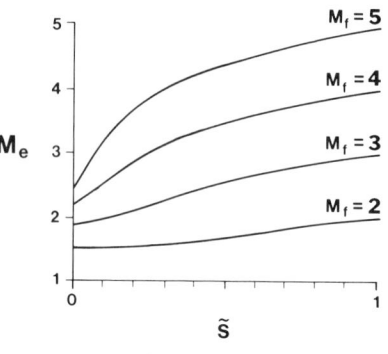

Fig. 17.13 Wall Mach number M_e vs normalized wall length $\tilde{s}$ for various M_f values for a two-dimensional, straight sonic line MLN when $\gamma = 1.4$.

to M_f. This variation contrasts with the curved sonic line nozzle in which M_e increases from unity to M_f.

This section has provided results for a two-dimensional flow only, because results for an axisymmetric flow do not yet exist. In the axisymmetric case, the Mach number and angle θ are known on the BC' Mach line (Fig. 17.9). By means of a numerical MOC calculation for the kernel, the solution for M and θ on the $A'B$ Mach line can be determined. Hence, a unique solution exists for the A'–B–C' nonsimple wave region.

17.4 MLN COMPARISONS

A comparison of Eq. (17.14) with Eq. (17.22) shows that θ^* is the same for the curved axisymmetric and straight two-dimensional nozzles. For the axisymmetric case, this result requires that points A' and B' coincide. Nevertheless, the axisymmetric nozzle is appreciably shorter than either of its two-dimensional counterparts. This is evident in Fig. 17.14, which compares the normalized length for three different nozzle configurations. Starting with an M_f of about 2.5, there is a substantial difference between the two-dimensional MLNs and the axisymmetric MLN with a curved sonic line. Despite the fact that θ^* is appreciably smaller for the axisymmetric nozzle with a curved sonic line (Fig. 17.5), its length is also substantially shorter. This result stems from the rapid three-dimensional expansion

MINIMUM-LENGTH NOZZLES

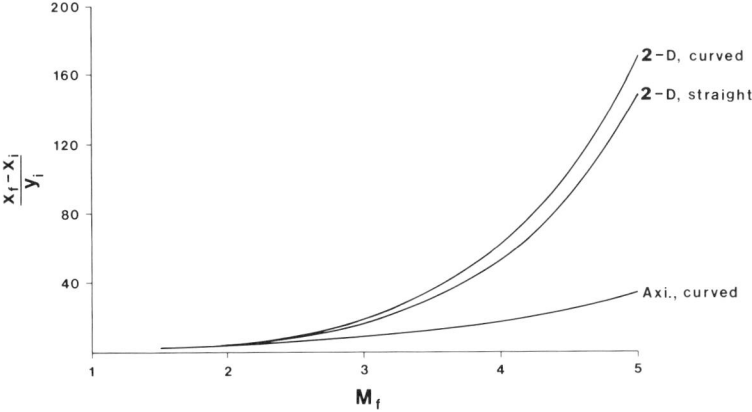

Fig. 17.14 Normalized length vs M_f when $\gamma = 1.4$ for three MLN configurations.

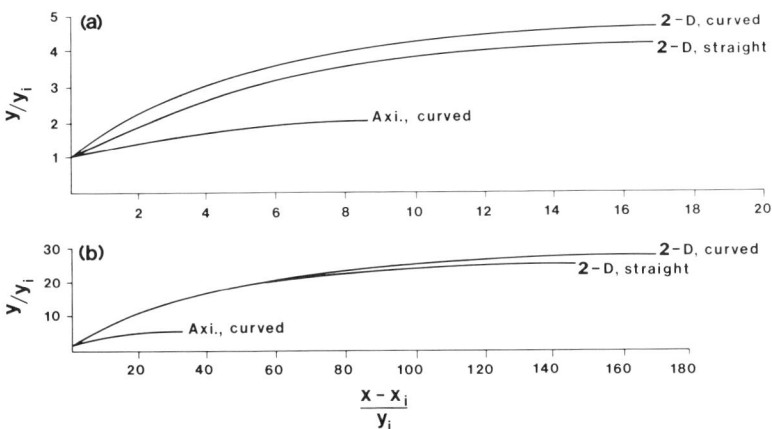

Fig. 17.15 Wall contour for the three MLN configurations of Fig. 17.14 when $\gamma = 1.4$ and (a) $M_f = 3$, (b) $M_f = 5$.

available to an axisymmetric flow. Thus, we verify the comment made in Sec. 17.1 that an MLN axisymmetric nozzle should be advantageous for hypersonic wind tunnels.

Figure 17.15 shows the normalized shape of nozzles with M_f equal to 3 and 5 when $\gamma = 1.4$. In both Figs. 17.15(a) and 17.15(b), the two-dimensional curved sonic line MLN has a straight wall with θ_M^* as the initial wall angle. However, point B' is so close to the throat that it is not readily discernible. Nevertheless, when $M_f = 5$, the overall nozzle length is reduced by 12% by the use of the straight segment. Both Figs. 17.15(a) and 17.15(b) show that the normalized length for the two-dimensional, straight sonic line MLN is shorter than its curved sonic line counterpart. This is also evident in Fig. 17.14. The disparity in Fig. 17.15 in the magnitude of y/y_i between the two-dimensional and axisymmetric nozzles is due to the difference between y_f and y_f^2.

The most interesting result in these figures is the substantial reduction in length afforded by an axisymmetric nozzle, especially when M_f becomes large. Even when $M_f = 3$, the length reduction is significant, and the initial wall angle is considerably reduced from its two-dimensional counterpart, as shown by Fig. 17.5.

References

[1] Shapiro, A., *Compressible Fluid Flow*, Vol. I, Ronald Press, New York, 1953, Chap. 15.

[2] A typical example is the excellent text by J. D. Anderson Jr., *Modern Compressible Flow*, McGraw-Hill Book Co., New York, 1982, Sec. 11.7.

[3] Shames, H. and Seashore, F. L., "Design Data for Graphical Construction of Two-Dimensional Sharp-Edge-Throat Supersonic Nozzles," NACA RM E8J12, 1948.

[4] Greenberg, R. A., Schneiderman, A. M., Ahouse, D. R., and Parmentier, E. M., "Rapid Expansion Nozzles for Gas Dynamic Lasers," *AIAA Journal*, Vol. 10, Nov. 1972, pp. 1494–1498.

[5] Vanco, M. R. and Goldman, L. J., "Computer Program for Design of Two-Dimensional Supersonic Nozzle with Sharp-Edged Throat," NASA TM X-1052, 1968.

[6] Bickley, W. G., "Some Exact Solutions of the Equations of Steady Homentropic Flow of an Inviscid Gas," *Modern Developments in Fluid Dynamics*, Vol. I, edited by L. Howarth, Oxford University Press, New York, 1953, Chap. V, Sec. 7.

[7] Redall, W. F. III, "Spherically-Symmetric Supersonic Source Flow: A New Use for the Prandtl-Meyer Function," *AIAA Journal*, Vol. 11, Dec. 1973, pp. 1787–1789.

[8] Dumitrescu, L. Z., "Minimum Length Axisymmetric Laval Nozzles," *AIAA Journal*, Vol. 13, April 1975, pp. 520–521.

[9] Foelsch, K., "The Analytical Design of an Axially Symmetric Laval Nozzle for a Parallel and Uniform Jet," *Journal of the Aeronautical Sciences*, Vol. 16, March 1949, pp. 161–166, 188.

[10] Emanuel, G., Cline, M. C., and Witte, K. H., "Laser-Induced Disturbance with Application to a Low Reynolds Number Flow," *AIAA Journal*, Vol. 19, Feb. 1981, pp. 226–231.

[11] Ferri, A., "Supersonic Flows with Shock Wave," *General Theory of High Speed Aerodynamics*, Vol. VI, edited by W. R. Sears, Princeton University Press, Princeton, NJ, 1954, Sec. H, 20, pp. 732–734.

[12] Beckwith, I. E., Ridyard, H. W., and Cromer, N., "The Aerodynamic Design of High Mach Number Nozzles Utilizing Axisymmetric Flow with Application to a Nozzle of Square Test Section," Langley Aeronautical Laboratory, NACA TN 2711, 1952.

[13] Saadat, A., "Analysis of Minimum Length Two-Dimensional Nozzles," M.S. Thesis, University of Oklahoma, Norman, 1982.

Problems

17.1 Assume that the following parameters are fixed:

$$\gamma, R_g, p_0, T_0, y_i$$

Maximize $\dot{m}$ when the sonic line is curved. Consider both two-dimensional and axisymmetric flow.

17.2 With $\theta^* = \theta_m^*$, $\sigma = 0$, and for a curved sonic line MLN, determine the equation for the differential arc length ds/R^* for the contour $A'C'$ as a function of γ and M. Simplify your results as much as possible.

17.3 For a two-dimensional MLN with a curved sonic line, show that Eqs. (17.12) satisfy the Mach wave wall cancellation condition.

17.4 Consider a two-dimensional duct with a uniform inlet flow that has a Mach number M_1. You are to design analytically a transition section such that the flow is shock-free, the given exit Mach number M_2 satisfies

$$M_2 > M_1 > 1$$

the exit flow is a uniform one, and the two simple wave regions coincide at one point. Determine the equations for the wall shape, and equations for all demarcation points, such as points A, A', etc., shown in Fig. 17.1. Normalize distances to the duct's half-height at the inlet, and simplify your results as much as possible.

Computational Problems:

17.5 For the parameter space:

$$\sigma = 0, 1$$
$$\gamma = 1.3, 1.4, 5/3$$
$$M_f = 1.5 \, (0.5) \, 6$$

determine the length $(x_B - x_i)/y_i$ and nozzle length $(x_f - x_i)/y_i$ for the curved sonic line MLN.

17.6 Use the curved sonic line theory for both two-dimensional and axisymmetric nozzle flow with points A' and B' in Fig. 17.1 coinciding. Use a caret to denote nondimensional lengths, where y_i is the reference length. Let $\hat{x}, \hat{y}$ be the coordinates of a point in either the source flow or B–B'–C' regions. (You will also need cylindrical coordinates R, ϕ, where ϕ is measured relative to the positive x axis.) Determine the conditions that point $\hat{x}, \hat{y}$ be in the source flow region. Next determine the conditions that the point falls in the B–B'–C' region. Computationally, determine the shape of the BB' Mach line for $\gamma = 1.4$, and

$$M_f = 1.5\,(0.5)\,2.5 \quad \text{when } \sigma = 0$$

$$= 1.5\,(0.5)\,5 \quad \text{when } \sigma = 1$$

This last result should be given in two figures, one for each σ, that show $\hat{y}_{BB'}$ vs $\hat{x}_{BB'} - \hat{x}_i$.

17.7 Extend your results to Problem 17.6 by determining the streamline angle θ, flow speed V, and pressure p at any point $\hat{x}, \hat{y}$ inside the source flow or B–B'–C' regions.

17.8 Continue with Problems 17.6 and 17.7. Determine the Mach number along the centerline as a function of $\hat{x}$. Determine for $\sigma = 0$ the Mach number M' along the wall in terms of $\hat{x}'$. Simplify your results as much as possible.

18. AERODYNAMIC WINDOW

18.1 PRELIMINARY REMARKS

The aerodynamic window, or aerowindow (AW) as it is usually called, is a natural extension of a minimum-length nozzle. The function of any AW is to provide a means of isolating a low-pressure laser cavity from the atmosphere while enabling a laser beam to pass through the gas flow. At low laser power levels, a transmitting material window suffices. However, the combination of thermal and structural stress with increasing power levels requires an AW. Most simply, an AW provides a gas flow, or jet, through which a laser beam passes, and which matches the (high) ambient pressure on the outside and the (low) laser cavity pressure on the inside. Desirable properties of any AW include:

(1) Low mass flow rate.
(2) Ability to support a large pressure ratio.
(3) Very low density disturbance level inside the jet, thus avoiding any laser beam distortion that might be caused by the jet.

Other factors can be considered, such as the leak rate, nozzle wall boundary layers, diffuser design, etc. However, the properties listed above are the most important factors.

The various types of supersonic AWs discussed in Refs. 1–3, and later in this section can be divided into two groups. In the first, an oblique shock or an expansion, or both, provide the pressure difference across the jet. In the second, an inviscid, supersonic vortex flow is utilized. This latter type is called a free-vortex AW. In terms of the above three items, the free-vortex AW is superior[3] to all other types. Consequently, this chapter focuses on the free-vortex AW.

The theory and results of a supporting experimental investigation for a free-vortex AW are found in Ref. 4. A simpler but more complete theory is presented in Sec. 18.2.

Items 1 and 3 are partly optimized by focusing the laser beam while it is still in the low-pressure cavity to a line focus in the vicinity of the AW. (A line focus is used instead of a point focus to avoid breakdown in the gas at the focal point where the beam intensity is extremely high.) As shown by the problems, this results in an AW nozzle with rather small dimensions. Furthermore, an AW designed for a large pressure ratio requires large Mach numbers. The combination of small dimensions and large Mach numbers results in thick boundary layers, which are always detrimental to any AW's operation. A significant reduction in the boundary-layer thickness can be obtained by an optimum design of the subsonic/supersonic nozzle that occurs upstream of the AW jet. An important part of this design would be a minimum-length nozzle.

In Sec. 18.3, a new approach is provided for the design of the downstream part of the supersonic nozzle that produces the free-vortex jet. This approach minimizes both the size of the nozzle and the boundary-layer thicknesses that occur in the exit plane of the nozzle. Furthermore, a free-vortex flow is provided without any unwanted disturbances, such as the weak shock waves that occurred in the experiments of Ref. 4.

Before embarking on the free-vortex window analysis in the next section, we briefly review the three simplest types of AWs, which are sketched in Fig. 18.1. In all three cases, the width of the opening for the beam is D. In

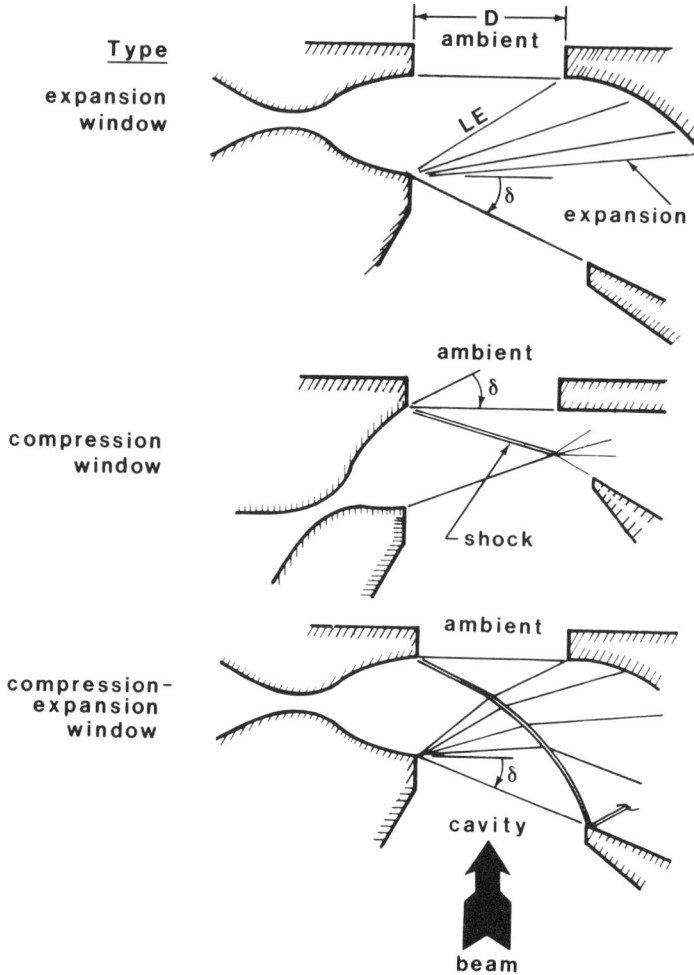

Fig. 18.1 Sketches of expansion, compression, and compression-expansion aerodynamic windows.

each sketch, the high (ambient) pressure is located on the top, and the low (cavity) pressure is located on the bottom. A two-dimensional nozzle on the left generates a uniform supersonic flow at its exit, while a diffuser on the right captures the flow and recovers as much of the initial stagnation pressure as possible.

With an expansion window, the pressure gradient in the direction of the laser beam is maintained by a centered Prandtl-Meyer expansion whose leading edge impinges on the wall near the entrance of the diffuser. When the leading edge impinges on the edge of the jet upstream of the diffuser, there is a massive leakage of ambient gas into the diffuser, unsteady jet flow, and optical distortion of the beam. Moreover, if the leading edge impinges well downstream on the diffuser's wall, then the mass flow rate is larger than need be.

With a compression window, an oblique shock sustains the pressure difference. The nozzle is canted so that the high-pressure side of the jet is perpendicular to the beam. By canting the nozzle, the contact area—and thus the mixing—between the jet and ambient gas is reduced. On the downstream side, the shock impinges on the low-pressure boundary of the jet at the entrance to the diffuser.

A compression-expansion window is a combination of the two. In this configuration, part of the pressure difference is provided by the shock and part by the expansion. The shock and expansion interact with the result that both are weaker downstream of the interaction.

Supersonic nozzles are far more common than AWs. However, an AW is worth discussing because several general internal aerodynamic design issues can be dealt with in a simple manner. These issues are not present when a nozzle is designed. Nevertheless, many design or operational aspects are not considered or are just briefly touched on. (Some of these issues, such as leakage, can only be experimentally investigated.) These include:

(1) Start-up procedures. This includes starting the nozzle and the diffuser and properly sequencing this procedure with the laser's operation.

(2) Nozzle wall boundary layers. For a small nozzle and high exit Mach numbers, these layers can be substantial in thickness.

(3) Mixing between the jet and the external gases.[5] The rate of mixing increases as the boundary-layer thickness increases at the nozzle exit.

(4) Leakage of jet gas into the laser cavity. This is important if the laser flow is to be recycled or if the laser cavity is to be a vacuum.

(5) The steadiness of the jet flow.

(6) Choosing the jet gas for refractive index matching with the ambient gas, and optimizing the overall optical quality of the transmitted laser beam.[6,7]

(7) Off-design performance as a result of a change in the ambient pressure. This is important if airborne operation is contemplated.

(8) Diffuser design and performance.

18.2 THEORY FOR A FREE-VORTEX AW

We assume steady, two-dimensional, homentropic, and isoenergetic flow of a perfect gas. The streamlines are circular. Consequently, the spiral flow solution of Sec. 15.2 holds with $\dot{m} = 0$. In Sec. 15.2, this $\dot{m}$ is for the radial flow component, which now is zero. Later, a different mass flow rate $\dot{m}$ for the v component of velocity is introduced.

With the foregoing assumptions, one can show that continuity is identically satisfied. The radial and azimuthal momentum equations reduce to

$$\frac{dp}{dr} = \frac{\rho v^2}{r} \tag{18.1}$$

and

$$vr = \text{const} \tag{18.2}$$

The homentropic equation

$$\frac{p}{p_0} = \left(\frac{\rho}{\rho_0}\right)^\gamma \tag{18.3}$$

satisfies conservation of energy and completes the list of conservation equations.

As usual, it is convenient to introduce the Mach number. The above equations plus the perfect gas equation of state reduce to

$$p = p_0 X^{-\gamma/(\gamma-1)} \tag{18.4}$$

$$\rho = \rho_0 X^{-1/(\gamma-1)} \tag{18.5}$$

$$v = aM = (\gamma R T_0)^{\frac{1}{2}} M X^{-\frac{1}{2}} \tag{18.6}$$

$$r = r_r X^{\frac{1}{2}}/M \tag{18.7}$$

where r_r is a reference length, and

$$X = 1 + \frac{\gamma - 1}{2} M^2 \tag{18.8}$$

Equations (18.4)–(18.8) identically satisfy Eq. (18.1), while Eq. (18.2) becomes Eq. (18.7). Thus, a free-vortex flowfield is fully described by Eqs. (18.4)–(18.8). As shown in Fig. 18.2, the outermost part of the flow is subsonic. From Eq. (18.7) we see that only the narrow annular region

$$\left(\frac{\gamma-1}{2}\right)^{\frac{1}{2}} r_r \leq r \leq \left(\frac{\gamma+1}{2}\right)^{\frac{1}{2}} r_r$$

is supersonic, and this is the region of interest. One reason for this interest is that the subsonic region, which is of infinite extent, can provide a pressure ratio of only 2. Generally, an AW requires a much larger pressure ratio.

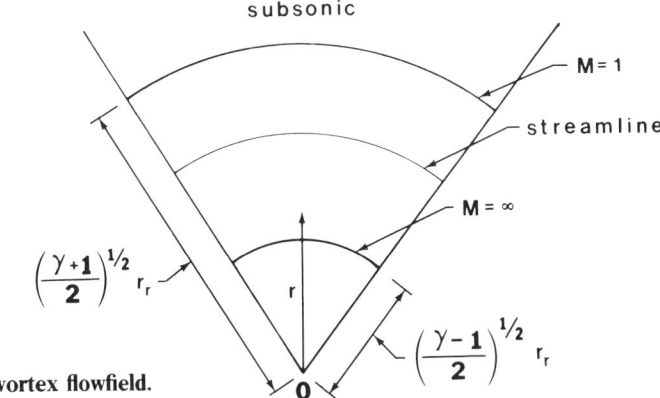

Fig. 18.2 Free-vortex flowfield.

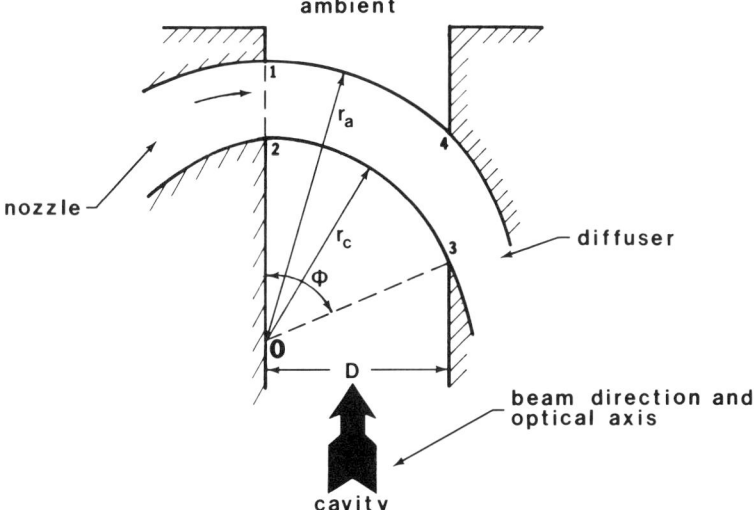

Fig. 18.3 Schematic of a free-vortex AW whose center is on the left wall.

A schematic of an AW is provided in Fig. 18.3. The flow is from left to right with the high pressure, p_a, along the 1–4 streamline, and the low pressure, p_c, along the 2–3 streamline. Consequently, the Mach number increases monotonically in the nozzle exit plane from points 1 to 2. The laser beam duct has cross-sectional dimensions of D and H, where we take $D \leq H$ without loss of generality. The laser beam usually has a line focus perpendicular to the plane of Fig. 18.3 near the AW. In this case, D is appreciably smaller than H, and the flow rate is minimized by choosing for D the orientation used in the figures.

The width of the laser beam duct is D, and the otherwise arbitrary inner jet radius r_c must exceed D. As will be shown, the jet mass flow rate $\dot{m}$ is minimized by minimizing r_c. The coordinate origin, point 0, is placed on the left wall. By design, the exit velocity is perpendicular to the nozzle exit plane. This choice thus simplifies the subsequent mass flow rate derivation and, at the same time, minimizes the mixing length between the jet and the ambient gas.

One can also locate the origin on the optical axis, as shown in Fig. 18.4. The mass flow rate is still determined along line 1–2, which is perpendicular to the velocity. To minimize mixing with the ambient gas, the outer wall can be extended to points 1' and 4'. The 1–1' and 4–4' extensions are circular arcs about the origin. In this configuration, the density gradient is symmetric about the optical axis.

The mass flow rate is evaluated at the nozzle exit plane by

$$\dot{m} = H \int_{r_c}^{r_a} \rho v \, dr$$

where the a and c subscripts, respectively, denote ambient and cavity conditions. With the aid of Eqs. (18.5)–(18.8) this becomes

$$\dot{m} = \frac{1}{2}\left(\frac{\gamma}{RT_0}\right)^{\frac{1}{2}} p_0 \frac{M_2}{X_2^{\frac{1}{2}}} r_c H \int_{X_1}^{X_2} X^{-[\gamma/(\gamma-1)]} \frac{dX}{X-1}$$

where

$$X_i = 1 + \frac{\gamma-1}{2} M_i^2, \qquad i=1,2$$

and the reference length can be chosen as

$$r_r = r_c \frac{M_2}{X_2^{\frac{1}{2}}}$$

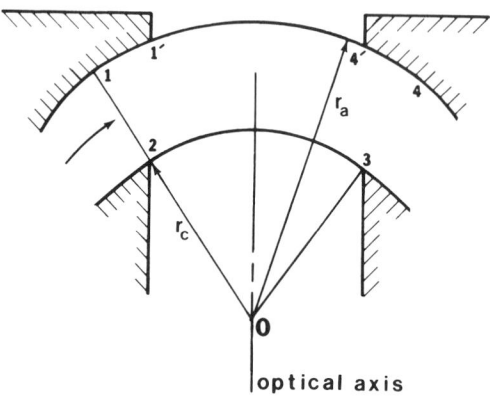

Fig. 18.4 Schematic of an AW whose center is on the optical axis.

AERODYNAMIC WINDOW

We introduce two pressure ratios

$$P_a = \frac{p_0}{p_a}, \qquad P_c = \frac{p_0}{p_c} \tag{18.9}$$

which satisfy

$$\left(\frac{\gamma+1}{2}\right)^{\gamma/(\gamma-1)} \leq P_a < P_c \tag{18.10}$$

in order that the flow be supersonic. In terms of these ratios, we have

$$M_1 = \left(\frac{2}{\gamma-1}\right)^{\frac{1}{2}} \left(P_a^{(\gamma-1)/\gamma} - 1\right)^{\frac{1}{2}}, \qquad X_1 = P_a^{(\gamma-1)/\gamma}$$

$$M_2 = \left(\frac{2}{\gamma-1}\right)^{\frac{1}{2}} \left(P_c^{(\gamma-1)/\gamma} - 1\right)^{\frac{1}{2}}, \qquad X_2 = P_c^{(\gamma-1)/\gamma}$$

With these equations, the mass flow rate becomes

$$\dot{m} = \left[\frac{\gamma}{2(\gamma-1)RT_0}\right]^{\frac{1}{2}} p_0 \left(1 - P_c^{-[(\gamma-1)/\gamma]}\right)^{\frac{1}{2}} \frac{DH}{\sin\phi} \int_{P_a^{(\gamma-1)/\gamma}}^{P_c^{(\gamma-1)/\gamma}} X^{-[\gamma/(\gamma-1)]} \frac{dX}{X-1} \tag{18.11a}$$

where

$$r_c = \frac{D}{\sin\phi}, \qquad \phi < \pi/2 \tag{18.12}$$

The integrand in Eq. (18.11a) is simplified by the introduction of

$$X = Z^2$$

In nondimensional form, Eq. (18.11a) becomes

$$\dot{\mathcal{M}} = \left(\frac{2\gamma}{\gamma-1}\right)^{\frac{1}{2}} \left[P_c^2 - P_c^{(\gamma+1)/\gamma}\right]^{\frac{1}{2}} I(\gamma) \tag{18.11b}$$

where

$$\dot{m} = \left[\frac{p_c}{(RT_0)^{\frac{1}{2}}} \frac{DH}{\sin\phi}\right] \dot{\mathcal{M}} \tag{18.13}$$

and

$$I(\gamma) = \int_{P_a^{(\gamma-1)/2\gamma}}^{P_c^{(\gamma-1)/2\gamma}} \frac{dZ}{Z^{(\gamma+1)/(\gamma-1)}(Z^2-1)} \tag{18.14}$$

The quantities $\dot{\mathcal{M}}$ and I depend on γ, P_a, and P_c. Only the coefficient of $\dot{\mathcal{M}}$ in Eq. (18.13) is then needed to obtain $\dot{m}$. From this scaling coefficient, we observe that $\dot{m}$ is proportional to DH and that the largest possible value for ϕ, subject to Eq. (18.12), minimizes $\dot{m}$. A large value for T_0 also minimizes $\dot{m}$, and may be needed to inhibit condensation.

The integral in Eq. (18.14) can be performed whenever

$$\gamma = (n+2)/n$$

where n is a positive integer. We thus obtain

$$I(5/3) = \frac{1}{3}\left(P_c^{-\frac{3}{5}} - P_a^{-\frac{3}{5}}\right) + P_c^{-\frac{1}{5}} - P_a^{-\frac{1}{5}} + \frac{1}{2}\ell n\left(\frac{P_a^{\frac{1}{5}}+1}{P_a^{\frac{1}{5}}-1}\frac{P_c^{\frac{1}{5}}-1}{P_c^{\frac{1}{5}}+1}\right) \quad (18.15a)$$

$$I(7/5) = \frac{1}{5}\left(P_c^{-\frac{5}{7}} - P_a^{-\frac{5}{7}}\right) + \frac{1}{3}\left(P_c^{-\frac{3}{7}} - P_a^{-\frac{3}{7}}\right)$$
$$+ P_c^{-\frac{1}{7}} - P_a^{-\frac{1}{7}} + \frac{1}{2}\ell n\left(\frac{P_a^{\frac{1}{7}}+1}{P_a^{\frac{1}{7}}-1}\frac{P_c^{\frac{1}{7}}-1}{P_c^{\frac{1}{7}}+1}\right) \quad (18.15b)$$

A pressure ratio p_a/p_c of about 40 is typical for a continuous-wave chemical laser operating at sea level. Other lasers can require a smaller or larger pressure ratio. For instance, the free-electron laser requires a hard vacuum in the cavity. It is therefore of interest to examine the foregoing result in the limit of

$$p_c \to 0 \text{ or } P_c \to \infty$$

A relatively noncondensible gas, helium, is chosen. In this limit, we obtain

$$\dot{m}_\infty = \left(\frac{5}{RT_0}\right)^{\frac{1}{2}} \frac{DH}{\sin\phi} p_0 I_\infty(5/3), \qquad P_c \to \infty$$

where

$$I_\infty(5/3) = -\frac{1}{3}P_a^{-\frac{3}{5}} - P_a^{-\frac{1}{5}} + \frac{1}{2}\ell n\left(\frac{P_a^{\frac{1}{5}}+1}{P_a^{\frac{1}{5}}-1}\right)$$

As a result, the flow rate changes very little, and an enormous p_a/p_c pressure ratio may be feasible. In fact, the free-vortex AW is the only type that is feasible in this limit. This pressure ratio and finite mass flow rate do not require a large value for r_a/r_c. The reason for this is that the inner part of the supersonic annulus, Fig. 18.2, is being used.

18.3 DESIGN PROCEDURE

Jet

In normal circumstances, the following parameters are known:

$$\gamma, R, T_0, p_a, p_c, D, H \qquad (18.16)$$

The principal unknowns to be determined are ϕ and p_0. The angle ϕ is chosen as large as possible but consistent with diffuser design requirements.

A diffuser will be used if the gas is to be recycled. A closed system for the AW gas, however, will require a heat exchanger and a compressor as well as a diffuser. In addition, a drying unit may also be required.

For an AW with a small mass flow rate, an open cycle is preferable. In this situation, a diffuser is unnecessary if p_0 is sufficiently large. A stable shock system, however, needs to be established in the jet downstream of the line from points 3 to 4' in Fig. 18.4.

Aside from pressure recovery considerations, the stagnation pressure is still to be determined. At this point, the only requirement on p_0 is

$$\left(\frac{\gamma+1}{2}\right)^{\gamma/(\gamma-1)} p_a \leq p_0$$

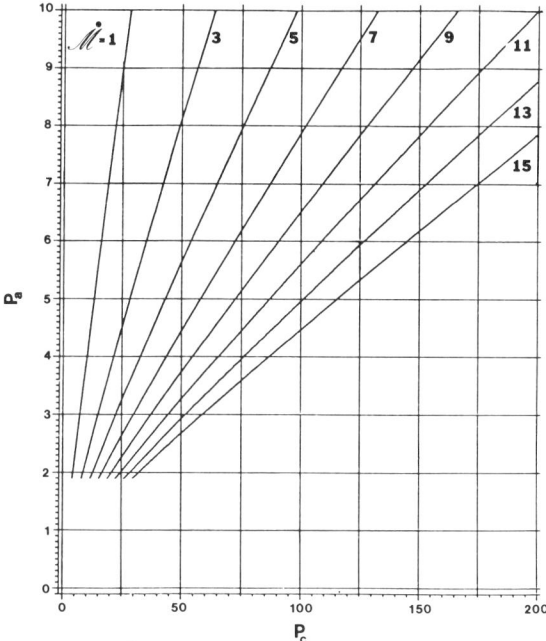

Fig. 18.5 $\dot{\mathcal{M}}$ as a function of P_a and P_c for $\gamma = 1.4$.

which is based on Eq. (18.10). We examine the possibility of choosing p_0 so as to minimize $\dot{m}$. For this, one can compute $(\partial \dot{m}/\partial p_0) = 0$ with the aid of Eqs. (18.9) and (18.13)–(18.15). A simpler approach is to plot lines of constant $\mathcal{M}$ values vs P_a and P_c. Figures 18.5 and 18.6 show such plots for γ values of 7/5 and 5/3, respectively.

With p_a and p_c known, but p_0 unknown, the operating point falls on a straight line through the origin

$$P_a = \frac{p_c}{p_a} P_c$$

When this relation is superimposed on either figure, it is apparent that $\dot{m}$ is a minimum only in the limit $p_0 \to \infty$. It is also apparent that for large P_a values, the rate of decrease in $\dot{m}$ with increasing p_0 is very slow. Depending on the gas supply method, a very high value for p_0 might not be warranted. For instance, if the gas is stored in a cylinder, then a high value for p_0 means a low utilization factor for the stored gas.

The reason for $\dot{m}$ decreasing with increasing p_0 is that the nozzle exit dimension $r_a - r_c$ also decreases with p_0. This is evident from the radius-ratio equation

$$\frac{r_a}{r_c} = \left[\frac{1 - P_c^{-(\gamma-1)/\gamma}}{1 - P_a^{-(\gamma-1)/\gamma}} \right]^{\frac{1}{2}}$$

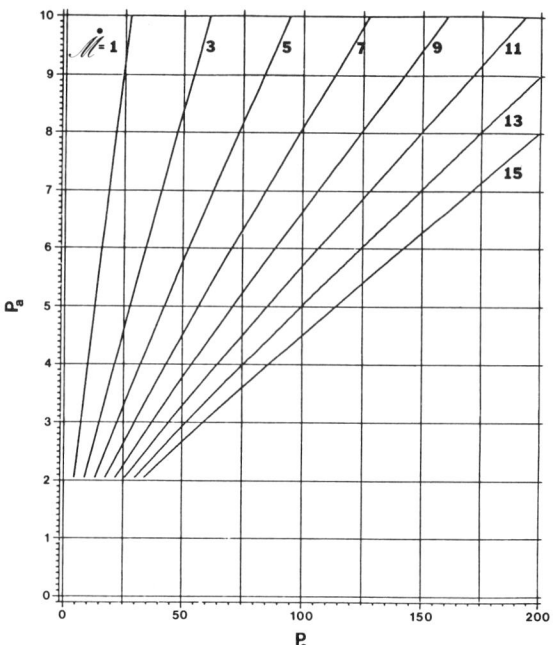

Fig. 18.6 $\mathcal{M}$ as a function of P_a and P_c for $\gamma = 5/3$.

AERODYNAMIC WINDOW

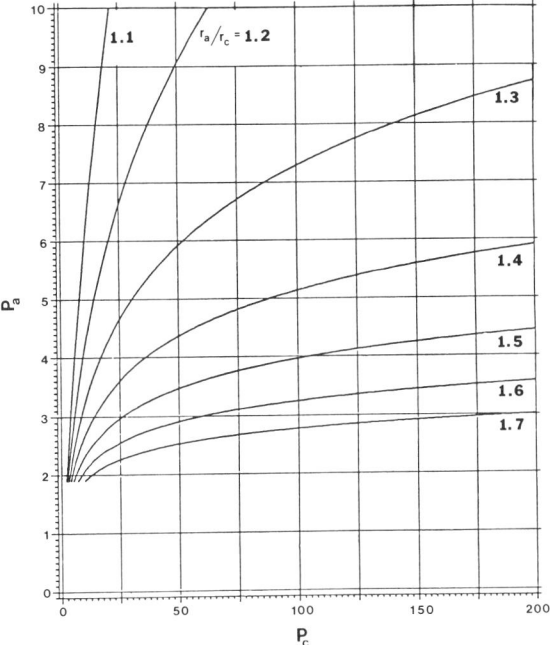

Fig. 18.7 r_a/r_c as a function of P_a and P_c for $\gamma = 1.4$.

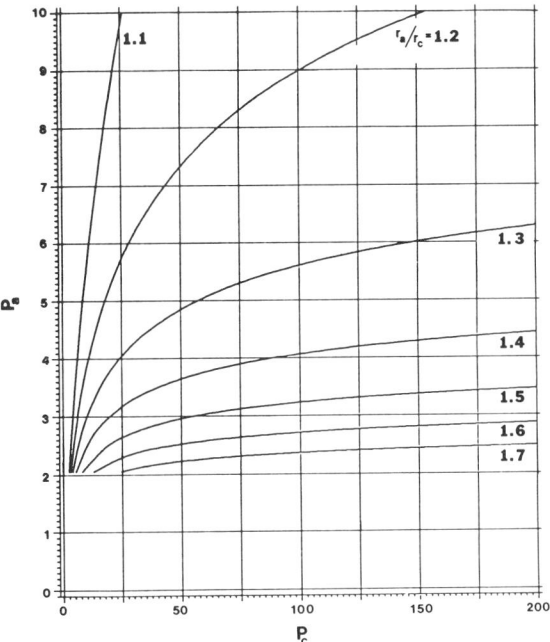

Fig. 18.8 r_a/r_c as a function of P_a and P_c for $\gamma = 5/3$.

which is shown in Figs. 18.7 and 18.8.

Based on an assumed knowledge of the items in (18.16), the jet design procedure is summarized as follows:

(1) Select the largest possible value for ϕ consistent with flow requirements downstream of point 3. Determine r_c.

(2) In conjunction with Figs. 18.5 and 18.6, select a value for p_0.

(3) Determine p_a, p_c, r_a, and the Mach number distribution in the nozzle exit plane. Check that condensation is not a problem for the given T_0 value.

Nozzle Design Procedure

We assume that the following parameters are known:

$$\gamma, R, T_0, p_0, r_a, r_c$$

along with the Mach number distribution $M(y)$ in the exit plane of the nozzle. This distribution stems from Eq. (18.7) and is given by

$$M(y) = \left[-\frac{\gamma - 1}{2} + \left(1 + \frac{y}{r_c}\right)^2 \frac{X_2}{M_2^2} \right]^{-\frac{1}{2}} \qquad (18.17)$$

where the coordinate y is along the nozzle exit plane and is measured from point 2, as shown in Fig. 18.9. With the origin on the left wall, we have in

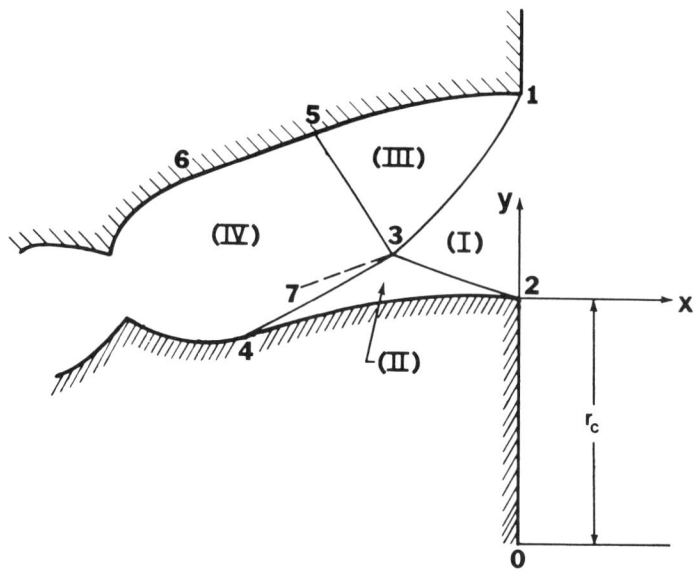

Fig. 18.9 Aerowindow nozzle schematic with two simple wave regions.

the nozzle exit plane,
$$\nu = \nu[M(y)], \qquad \theta = 0$$

where θ is the slope of the velocity relative to the x axis. Thus, there are sufficient data for determining the flow in region I, which is upstream of the nozzle's exit plane.

One possible design approach is shown in Fig. 18.9. Region I is a nonsimple wave region, which is also a free-vortex flow. Regions II and III are simple wave regions, while region IV is a uniform flow region. The boundaries 1–3–4 and 2–3–5 are Mach lines, where lines 3–4 and 3–5 are straight. From the results in Sec. 16.2, we obtain

$$\nu(M_{IV}) = \tfrac{1}{2}[\nu(M_2) + \nu(M_1)]$$

$$\theta_{IV} = \tfrac{1}{2}[\nu(M_2) - \nu(M_1)]$$

for the Mach number and velocity slope, shown as line 7–3, in region IV. Note that M_{IV} is given by a Prandtl-Meyer average of M_1 and M_2. Consequently, region II is an expansion, while region III is a compression. Further, the supersonic boundary layer on the wall between points 5 and 1 experiences an adverse pressure gradient. As a result, the boundary layer will thicken and may separate.

The foregoing approach is characterized by initial data prescribed at the nozzle exit plane that nowhere coincides with a characteristic direction, and by two simple wave regions, one of which is a compression. It turns out that a modification of the above construction exists that will take a uniform flow into the nonsimple wave region, with just a single intervening simple wave region. There are two possibilities in which the simple wave region is a compression or an expansion. Of course, only the expansion case is of interest. For its construction, the initial data must be prescribed on a Mach line. This fact is the key to the following approach.

Figure 18.10 is a schematic of the single simple wave region approach. Region I is a nonsimple wave region and is located between the nozzle exit plane and the right-running Mach line 5–3–2 that becomes the initial data line mentioned above. The flow in region I is a free-vortex flow, and consequently $M_5 = M_1$. The wall between points 5 and 1 merely inhibits mixing with the ambient gas. It is a circular arc of radius r_a, with its center located in the nozzle exit plane. As discussed in the second subsection of Sec. 16.1, data specified on a Mach line must be consistent with the corresponding compatibility equation. This is not a problem for the data specified on the 5–3–2 Mach line, since the data stem from an exact region I solution to the governing equations.

In regions I and II, it is useful to distinguish the flowfield above the left-running Mach line, which is located between points 4–3–1 in Fig. 18.10, from the flowfield below it. For the left-running Mach lines that are below 4–3–1, we have

$$\nu - \theta = \nu(M(y))$$

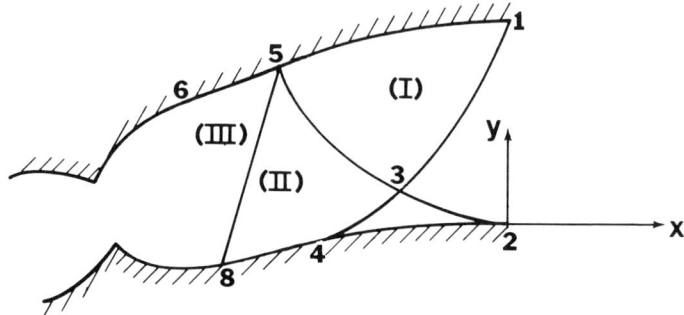

Fig. 18.10 Aerowindow nozzle schematic for a flow with a single simple wave region that is an expansion.

while for those above

$$\nu - \theta = \nu_1 - \theta_w$$

where θ_w is the known wall slope for any location between points 1 and 5.

Region II is a simple wave region, consisting of a Prandtl-Meyer expansion, and is bordered by a straight Mach line, 5–8, on the upstream side. Region III is a uniform flow region with Mach number M_1. A straight wall section between points 5 and 6 terminates when point 6 is opposite point 8. A minimum-length nozzle with exit Mach number M_1 should be utilized upstream of region III.

The shape of the wall, 8–4–2, can be found analytically. The first step would be to find the shape of the 5–3–2 Mach line that borders the free-vortex flow. It is most easily found using r, θ coordinates, shown in Fig. 18.11. The position vector **r** has the angle θ with respect to the y axis since **V** is perpendicular to **r**. From this figure, we have

$$\frac{dr}{r\,d\theta} = \tan \mu = (M^2 - 1)^{-\frac{1}{2}}$$

Next, we logarithmically differentiate Eq. (18.7), to obtain

$$\frac{dr}{r} = -\frac{dM}{MX}$$

Upon elimination of dr/r, we have

$$-d\theta = \frac{(M^2 - 1)^{\frac{1}{2}}}{MX}\,dM = d\nu$$

which yields

$$\theta = \nu(M_2) - \nu(M) \tag{18.18}$$

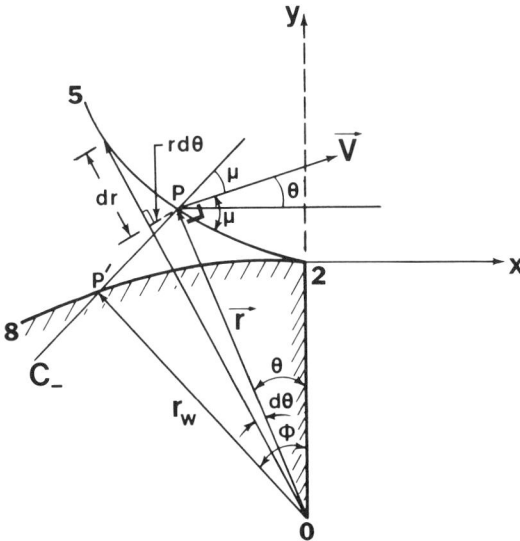

Fig. 18.11 Schematic for establishing the equations for the Mach line 2–3–5 and for the wall between points 2, 4, and 8.

along the 2–3–5 Mach line. Equation (18.7) is solved for β as

$$\beta = (M^2 - 1)^{\frac{1}{2}} = \left\{ \frac{[(\gamma + 1)/2] - (r/r_r)^2}{(r/r_r)^2 - [(\gamma - 1)/2]} \right\}^{\frac{1}{2}} \qquad (18.19)$$

After elimination of β, Eqs. (18.18) and (18.19) yield an explicit r, θ equation for the 2–3–5 Mach line.

As shown in Fig. 18.11, the C_- Mach line from P to P' is straight, and V and θ are constant along it. An explicit equation involving r_w and ϕ can be obtained for the wall by equating the mass flow rate through the 2–3–5 Mach line with the mass flow rate through the P–P' line. The analytical approach is essentially identical to that used in Sec. 17.2 for a two-dimensional MLN.

Since a mixing-layer spread rate tends to be proportional to the initial momentum thickness, the Fig. 18.10 procedure should result in a minimum rate of mixing on both the laser and ambient gas sides. In turn, this will result in improved beam quality and in less difficulty at the jet reattachment point(s) located on the downstream side of the optical duct.

As indicated at the end of Sec. 18.2, the free-vortex AW is feasible even when the cavity pressure is below 1 torr and the ambient pressure is 1 atm. However, the boundary layer in a low-density flow tends to grow rapidly. The best way to reduce this rate of growth is to use an MLN followed by the AW design outlined in this section. The low-density boundary layer is only on the lower wall in Fig. 18.10. However, a favorable pressure gradient is present all along this wall and should minimize the boundary-layer growth rate.

References

[1] Parmentier, E. M. and Greenberg, R. A., "Supersonic Flow Aerodynamic Windows of High-Power Lasers," *AIAA Journal*, Vol. 11, July 1973, pp. 943–949.

[2] Guile, R. N. and Mapes, S. N. Jr., "Aerodynamic Window Investigations, Final Technical Report," Vol. I, UARL Rept. L911204-12, April 1972.

[3] Guile, R. N., Decker, R. O., and Nomiyama, N. T., "Aerodynamic Window for a Focused DF Laser," UARL Rept. R75-911966-9, March 1975.

[4] Guile, R. N. and Hilding, W. E. "Investigation of a Free-Vortex Aerodynamic Window," AIAA Paper 75-122, Jan. 1975.

[5] Avidor, J. M., "Improved Free-Vortex, Subsonic Aerodynamic Window," *AIAA Journal*, Vol. 17, Nov. 1979, pp. 1267–1268.

[6] Hertzberg, A., "High-Power Gas Lasers: Applications and Future Developments," *Journal of Energy*, Vol. 1, Nov.–Dec. 1977, pp. 331–346.

[7] Fuhs, A. E., "Overview of Aero-Optical Phenomena," in *Wavefront Distortions in Power Optics*, Vol. 293, edited by A. E. Fuhs and S. E. Fuhs, Society of Photo-Optical Instrumentation Engineers, Bellingham, WA, 1981, pp. 36–55.

Problems

18.1 Starting with Eq. (18.14), derive Eqs. (18.15a) and (18.15b).

18.2 Derive the equations for $\dot{m}/HD$ in the expansion window case, assuming the following parameters are known:

helium gas, $T_0 = 300$ K, $p_0 = 10$ atm, $p_a = 1$ atm, $p_c = 10^{-3}$ atm

What is the limiting factor? Compute $\dot{m}/HD$ for the free-vortex AW and for the expansion AW assuming $\phi = 60$ deg. How realistic is this result?

18.3 Derive the equations for $\dot{m}/DH$ in the compression AW case, assuming that Problem 18.2 parameters are known. What is the limiting factor? Compute $\dot{m}/HD$ for the compression AW. How realistic is this result?

18.4 You are to design a free-vortex AW with the following requirements:

nitrogen gas, $p_a = 1$ atm, $p_c = 0.05$ atm, $p_0 = 7$ atm

$T_0 = 330$ K, $D = 1$ cm, $H = 10$ cm, $\phi = 60$ deg

Assume that the origin is on the left wall. Determine $\dot{m}$, M_1, M_2, $M(y)$, r_c, and r_a. Graph $M(y)$.

18.5 Determine $\dot{m}$ in the limit of $p_c \to 0$ for a free-vortex AW where

$$\text{helium gas}, \quad T_0 = 300 \text{ K}, \quad p_0 = 10 \text{ atm}$$

$$p_a = 1 \text{ atm}, \quad p_c = 10^{-3}, 10^{-4} \text{ atm (2 cases)}$$

Use both the exact and asymptotic formulas when $p_c = 10^{-3}$ atm to determine $(\dot{m} \sin\phi)/DH$. Use just the asymptotic formula when $p_c = 10^{-4}$ atm to determine $(\dot{m} \sin\phi)/DH$.

18.6 You are to design the nozzle for the AW of Problem 18.4. The nozzle is to be designed using a single expansion region. More specifically, you are to determine the (x, y) coordinates (whose origin is point 2 of Fig. 18.10) of points 5, 6, and 8, the Mach number in region III, the nozzle throat half-width w^*, and the angle θ_{III} the centerline has with respect to the x axis. Assume the nozzle upstream of region III is an MLN with a straight sonic line. This problem is to be done analytically.

19. FLOWS WITH SHOCK WAVES

19.1 PRELIMINARY REMARKS

So far, our investigation of flows with shock waves has focused on a few relatively simple flowfields, such as symmetrical flow about a cone. This chapter takes a wider view. Overall, shock waves are a dominant feature of a supersonic flow. They often generate slipstreams or shear layers. For instance, on aerodynamic vehicles, such as the space shuttle, shock interference phenomena can occur and result in surface regions with very high heat-transfer rates.

One difficulty in dealing with oblique shock waves is their lack of mathematical uniqueness. As pointed out for oblique and conical shock waves, strong and weak solutions exist. In Sec. 19.2, we address this issue by discussing the stability of the flowfield and the effect of boundary conditions on stability.

In Sec. 19.3, uniqueness is examined from a different perspective by a detailed analysis of steady two-dimensional flow over a specially shaped compressive ramp. Aside from uniqueness, we discuss an unresolved existence question and the importance of the computational technique.

The structure of an underexpanded supersonic jet is discussed in Sec. 19.4. Even though the jet is underexpanded, shock waves appear in it, located at some distance downstream of the nozzle's exit plane. The primary concerns of this section are the shape of the jet's boundary and the reason for the shock waves.

Shock wave reflection off a wall in a steady flow is the subject of Sec. 19.5. Section 19.6 discusses the pseudo-steady situation, where a normal shock, moving into a quiescent gas, encounters a planar-walled compressive ramp. Finally, in Sec. 19.7, we examine shock wave interference, which occurs when one shock wave impinges on another.

Except for Secs. 19.5 and 19.6, the topics are unrelated, and one may go directly to the section of interest. This chapter thus consists of special shock wave topics. However, a number of shock wave phenomena are not considered. These include:

(1) Nonequilibrium effects that occur only at substantial (hypersonic) freestream Mach numbers.[1,2] In a monatomic gas, electronic excitation first occurs and alters the internal energy. At still higher Mach numbers, ionization must be considered. In a polyatomic gas, vibrational excitation also occurs, followed by dissociation and recombination.

(2) An unsteady normal or oblique shock wave, propagating into a quiescent gas, that encounters a small concave or convex bend in the wall.[3,4] This topic is referred to as diffraction of a blast wave.

348   GASDYNAMICS: THEORY AND APPLICATIONS

(3) Unsteady normal shock wave motion, such as that which occurs in shock tubes. This topic was covered in Chap. 10.

(4) Truly nonstationary flow that occurs when the compressive ramp in Sec. 19.6 is not planar but has curvature.[5-7]

(5) Shock wave refraction that occurs when a normal or oblique shock wave encounters an interface separating two different gases or gas states.[8-10]

(6) Steady or unsteady shock wave/boundary-layer interaction. For instance, when a steady shock wave impinges on a wall, it may cause the boundary layer to separate. As a consequence, the overall flowfield changes, especially the shock wave system predicted by an inviscid analysis.

19.2 STABILITY OF SHOCK WAVES

We follow the analysis of Salas and Morgan.[11] They point out that it is not pertinent to ask whether the weak or strong solutions occur, since both occur in nature. Rather, one should ask under what conditions one or the other occurs.

Consider a flow around a wedge or a cone at zero angle of attack with an attached shock wave. As long as the flow between the shock and body is supersonic, the flowfield is hyperbolic. In this situation, the flowfield is fully determined by freestream conditions and a body half-angle. By "determined" we mean that a velocity tangency condition on the body and the shock wave jump conditions are the only boundary conditions allowed. In particular, a change in body shape, such as a shoulder, has no upstream effect.

If some or all of the flow between the shock and the body becomes subsonic, then part of the flowfield is elliptic. When this happens, an additional downstream boundary condition is required. For instance, this might be the location of a sonic line, as shown in Fig. 19.1. The location of the sonic line is not known a priori, except that it starts at the shoulder. A downstream pressure condition can also be utilized. The essential difference in boundary conditions was not appreciated in the stability analysis that preceded the work of Salas and Morgan.[11] A solution is considered stable when any slight perturbation decays with time.

In both the planar and conical cases, a narrow region exists in which the weak-shock solution is subsonic. For a cone, this region is most discernible in Fig. 4 of Ref. 12. In general, the stability of this narrow region should be

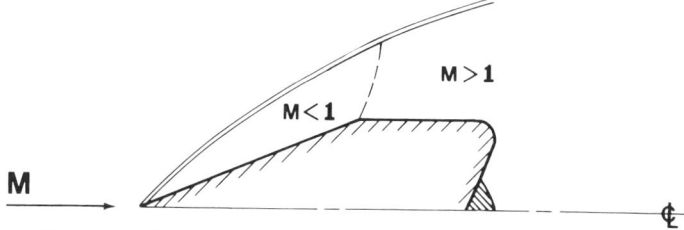

Fig. 19.1 Flow around a cone or wedge with an attached shock and a subsonic flow region.

subject to the same restrictions and conclusions associated with the subsonic strong-shock solution. Most stability studies make a disclaimer that their analysis does not encompass this region.

As previously mentioned, suppose the wedge or cone semivertex angle and freestream conditions are known. The similarity solution of Sec. 15.3 is then correct only when the flow between the shock and body is supersonic. When some or all of the flow is subsonic, the solution can hold only in the vicinity of the apex. The extent to which it holds depends on how closely the downstream pressure matches the pressure near the apex and the angular extent of the subsonic flow.

Salas and Morgan's investigation indicates that a cone or wedge flow is stable when it is hyperbolic. Two types of wall boundary conditions are examined; these are referred to as the fixed-wall and fixed-pressure boundary conditions.

In the planar elliptic case, the flow is unstable with a fixed-wall boundary condition; it converges either to the weak-shock solution or to a detached solution. However, if a fixed-pressure condition is used, the strong-shock solution is stable when the flow is elliptic or hyperbolic.

The conical case is the same as that of the wedge, except for the strong-solution, fixed-pressure condition. In the latter situation, there are two different strong-shock solutions corresponding to the same surface pressure. The stronger solution, i.e., the one with the greater entropy production, is unstable; it either detaches or becomes the weaker but stable strong-shock solution.

19.3 FLOW OVER A COMPRESSIVE RAMP

In a supersonic flow, a compressive turn in a two-dimensional wall can be considered as either sharp or smooth and gradual. Oblique shock theory (Chap. 5) is used for the sharp turn, whereas the flow near the wall is assumed homentropic for a smooth turn (Chap. 6). Both turns result in oblique shock waves. In this section, we discuss two questions: how are these shocks related and, in the smooth turn case, is the disturbance a weak-solution shock wave?

Introductory Remarks

The computation in the smooth case is formidable and requires a large computer code. The shock wave gradually forms as the left-running Mach lines run into each other, starting with the leading-edge Mach line, as shown in Fig. 19.2. The shock wave starts to form at point A and is complete at point B, where point B is the intercept with the Mach line from point C. After point C the wall is presumed to be planar with an angle θ_2. The flow between point A and the wall is homentropic, and the streamlines between A and D remain homentropic downstream. (These streamlines may encounter a shock wave downstream of point C.) The shock wave between A and B increases in strength and thus is curved. The flow downstream of this part of the shock wave is isentropic and vortical. Actually, the shock wave above point B is occasionally slightly curved, and the flow downstream of the curved portion would be slightly vortical. Far from the corner, the smooth turn looks like a sharp one with an angle, θ_2. When the turn is smooth, the foregoing description is incomplete and somewhat simplified, as we shall see.

The flow downstream of the left-running Mach line from D to A must be a simple wave region, since the Mach line borders a uniform flow region. This region terminates at the right-running characteristic that passes through point A, as shown in Fig. 19.3. In region ADE, the left-running Mach lines are straight and each one has a different Riemann invariant; whereas, the right-running Mach lines that originate in the freestream are curved but have the same Riemann invariant. Just downstream of AE, the right-running Mach lines originate behind a curved shock and, therefore, the Riemann invariant changes from Mach line to Mach line. If the bend in the wall continues past point E, the flow is homentropic but nonsimple in the $AFCE$ quadrilateral.

The flow near the wall along the DEC turn is usually treated as a homentropic, simple wave region, such that

$$\nu(M_C) = \nu(M_1) - \theta_2$$

If point C is downstream of point E, then along the EC wall this relation is an approximate one, since the adjacent flow is nonsimple.

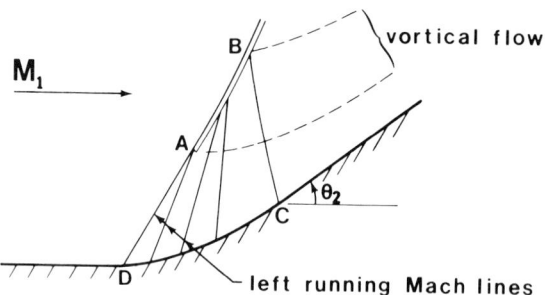

Fig. 19.2 Gradual formation of a shock wave due to the overlap of left-running Mach lines in a two-dimensional supersonic flow.

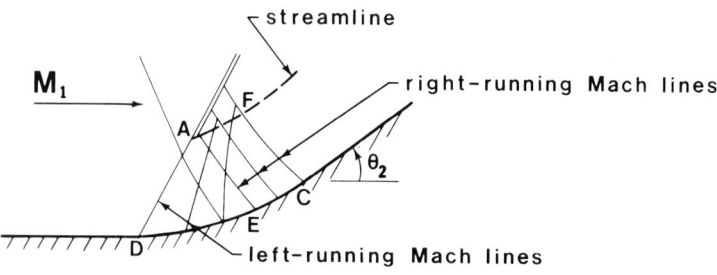

Fig. 19.3 Left- and right-running Mach lines associated with the flow in Fig. 19.2.

FLOWS WITH SHOCK WAVES   351

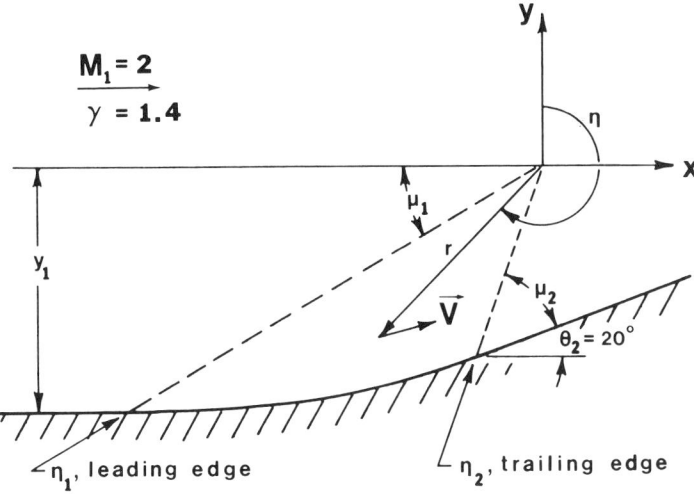

Fig. 19.4 Wall shape and nomenclature.

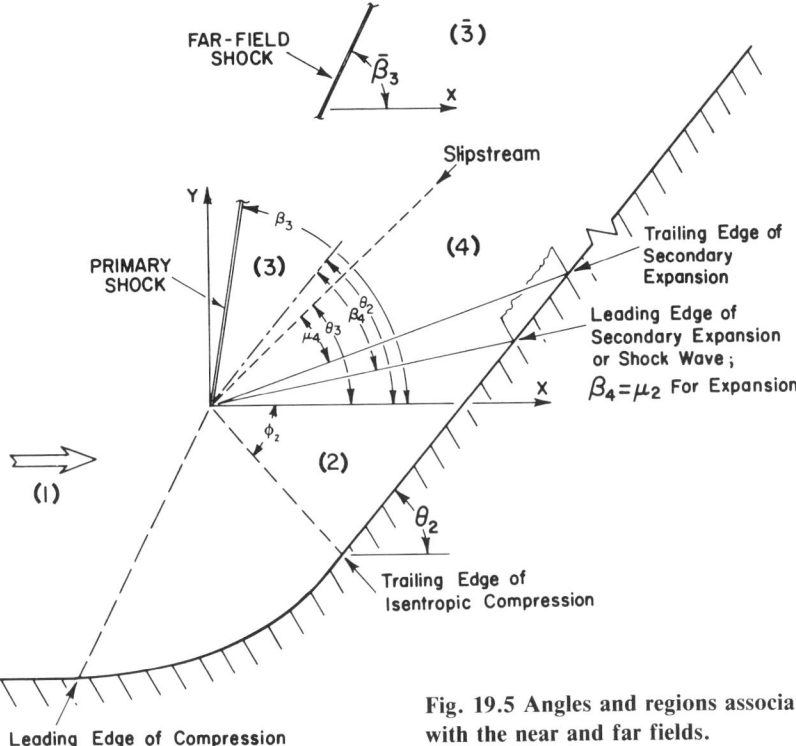

Fig. 19.5 Angles and regions associated with the near and far fields.

We will examine the flowfield of a smooth compressive ramp without benefit of a large-scale computer code. This is possible only when the turn has a special contour. This contour must generate a centered Prandtl–Meyer compression, as shown in Fig. 19.4. This figure is to scale when $\gamma = 1.4$, $\theta_2 = 20$ deg, and $M_1 = 2$. The coordinate x is aligned with the freestream, and the origin is the center of the fan. The angle η is measured from the y axis, and η_1 and η_2 accurately show the locations of the leading and trailing edges of the compression. The flow is isentropically compressed from $M_1 = 2$ to $M_2 = 1.308$ at η_2. Inside the compression we have a simple wave region where the left-running Mach lines are straight, with a focal point at the origin. Thus, the shock wave is at full strength at the origin or focal point.

Many books discuss the sharp and smooth compressive turns that are basic to gasdynamics. For example, Oswatitsch[13] has an extensive discussion of the Prandtl–Meyer compression turn. Unfortunately, the slipstream that emanates from the focal point, Fig. 19.5, is incorrectly assumed in Ref. 13 to be parallel to the downstream wall. Our presentation is based on Refs. 14 and 15.

Wall Shape

We first determine the equations for the wall contour. The upstream straight wall is located a distance y_1 below the origin, and all lengths scale with y_1, whose magnitude is arbitrary. Since the flow inside the compression is a simple wave region, it admits a similarity solution in which the angle η is the independent variable. One can show that the r,η velocity components v_r and v_n satisfy

$$\frac{dv_r}{d\eta} = v_n = -a \tag{19.1}$$

where the speed of sound a is given by the energy equation

$$a^2 = \gamma \frac{p}{\rho} = \frac{\gamma - 1}{2} V_m^2 - \frac{\gamma - 1}{2}(v_r^2 + v_n^2) \tag{19.2}$$

and where the maximum speed V_m is

$$V_m = a_1 \left(\frac{2}{\gamma - 1} X_1\right)^{\frac{1}{2}} \tag{19.3}$$

For a compressive turn, Eq. (19.1) becomes

$$\frac{dv_r}{d\eta} = -\left(\frac{\gamma - 1}{\gamma + 1}\right)^{\frac{1}{2}} (V_m^2 - v_r^2)^{\frac{1}{2}} \tag{19.4}$$

A solution for the velocity components and Mach number is obtained by integrating Eq. (19.4) from the leading edge η_1, to obtain:

$$\frac{v_r}{V_m} = \sin z \tag{19.5a}$$

$$\frac{v_n}{V_m} = -\left(\frac{\gamma-1}{\gamma+1}\right)^{\frac{1}{2}} \cos z \qquad (19.5b)$$

$$M^2 = \frac{[(\gamma-1)/2] + \sin^2 z}{[(\gamma-1)/2]\cos^2 z} \qquad (19.5c)$$

where z is given by

$$z = [(\gamma-1)/(\gamma+1)]^{\frac{1}{2}}(\eta_1 - \eta) + \sin^{-1}\left(\frac{v_{r1}}{V_m}\right) \qquad (19.6a)$$

$$\eta_1 = \tfrac{3}{2}\pi - \mu_1 \qquad (19.6b)$$

$$\mu_i = \sin^{-1}\left(\frac{1}{M_i}\right) \qquad (19.6c)$$

$$\frac{v_{r1}}{V_m} = -\left[\left(\frac{\gamma-1}{2}\right)\frac{M_1^2-1}{X_1}\right]^{\frac{1}{2}} \qquad (19.6d)$$

The Mach number M_2 is determined by the relation

$$\nu_2 = \nu_1 - \theta_2 \qquad (19.7)$$

where $\nu_i = \nu(M_i)$ is the Prandtl–Meyer function. Once M_2 is numerically determined, the trailing-edge angle is given by

$$\eta_2 = \tfrac{3}{2}\pi - \mu_2 - \theta_2 \qquad (19.8)$$

The differential equation for the wall shape is

$$\frac{1}{r}\frac{dr}{d\eta} = \cot\mu = (M^2-1)^{\frac{1}{2}} = -[(\gamma+1)/(\gamma-1)]^{\frac{1}{2}}\tan z \qquad (19.9)$$

where Eq. (19.5c) is used. After integration, this becomes

$$\frac{r(\eta)}{M_1|y_1|} = \left[-\left(\frac{a_1}{V_m}\right)\left(\frac{V_m}{v_n}\right)\right]^{(\gamma+1)/(\gamma-1)} \qquad (19.10)$$

where Eqs. (19.3) and (19.5) are used. In x, y coordinates the wall shape is

$$x_w = r\sin\eta, \qquad y_w = r\cos\eta$$

This shape, which is also a streamline, can now be written as

$$\frac{x_w}{|y_1|} = M_1 \left(\frac{\gamma-1}{2}\right)^{(\gamma+1)/2(\gamma-1)} X_1^{-(\gamma+1)/2(\gamma-1)} \left(-\frac{v_n}{V_m}\right)^{-(\gamma+1)/(\gamma-1)} \sin\eta$$

(19.11)

$$\frac{y_w}{|y_1|} = M_1 \left(\frac{\gamma-1}{2}\right)^{(\gamma+1)/2(\gamma-1)} X_1^{-(\gamma+1)/2(\gamma-1)} \left(-\frac{v_n}{V_m}\right)^{-(\gamma+1)/(\gamma-1)} \cos\eta$$

where v_n/V_m is given by Eq. (19.5b), and η ranges from η_2, Eq. (19.8), to η_1, Eq. (19.6b). In Eq. (19.5b), the variable z is given by Eq. (19.6a), and the constant v_{r1}/V_m is given by Eq. (19.6d).

Flowfield Description

Figure 19.5 defines various angles and regions of the flow. Near the focal point (referred to as the near field), the flow between the primary shock and slipstream is designated as region 3; whereas downstream of the far-field shock, it is designated as region $\bar{3}$. Across the slipstream there is a jump in entropy, velocity, etc.

The flow in region 2 is uniform and parallel to the adjacent wall. Moreover, it is supersonic because v_2 cannot be negative. Since the pressure p_2 does not generally equal p_3, a centered expansion or (secondary) shock wave must emanate from the focal point. Region 2 terminates at the start of this disturbance. When $p_2 > p_3$, there is an expansion between regions 2 and 4. The expansion turns the flow in region 4 away from the wall thus causing the slipstream angle θ_3 to exceed θ_2. Consequently, the shock angle β_3 exceeds $\bar{\beta}_3$, and the near-field primary shock is stronger than its far-field counterpart. The expansion reflects from the downstream wall in a manner that maintains the flow at the wall parallel to it. It then reflects from the primary shock, thereby gradually reducing the strength of the shock. As shown in Fig. 19.5, the leading edge of the expansion has an angle β_4, relative to the downstream wall, which is equal to the Mach angle μ_2. The angle of the trailing edge, relative to the slipstream, is the Mach angle μ_4. In going from region 2 to region 4, the flow makes an expansive turn of angle $\theta_3 - \theta_2$, which is positive. Hence, M_4 is determined by

$$\nu_4 = \nu_2 + (\theta_3 - \theta_2) \qquad (19.12)$$

When $p_2 < p_3$, a shock wave occurs that has an angle β_4 (not equal to μ_2) relative to the downstream wall. In region 4 the flow is turned toward the wall. Thus, θ_3 is less than θ_2, and the near-field primary shock is weaker than its far-field counterpart. The flow next to the wall in region 4 must be parallel to the wall. This is accomplished by either regular or Mach reflection from the downstream wall, as discussed in Sec. 19.5. In either case, the reflected shock impinges on the primary shock, thereby strengthening it. At impingement, a triple point and shock interference occur, as discussed in Sec. 19.7.

The near field can consist of uniform supersonic flow in regions 1–4. In this situation, the near-field solution is exact in these regions and in the centered compression. The downstream extent of validity is limited by the reflected secondary wave or by a Mach reflection, if it occurs. The primary shock wave is straight until a reflected wave intersects it. If the flow is subsonic in regions 3 or 4, it may not be uniform in these regions. If the flow is nonuniform, the near-field solution in these regions then holds only in the vicinity of the focal point.

Along with Eqs. (19.7) and (19.12), there is a slipstream pressure condition, $p_3 = p_4$, and a slipstream tangency condition. One can show that more than one secondary pressure disturbance cannot emanate from the focal point. Thus, an expansion followed by a shock wave, or the reverse, does not occur. This is demonstrated by considering the assumed different angles of the two disturbances.

The foregoing description is for the near-field flow. The far-field solution is the weak- and strong-shock wave solutions wherein the shock wave angle β_3 is determined by the usual oblique shock equations for a sharp wall turn angle of θ_2 deg. This solution is based only on a tangency condition; whereas, the near-field solution also requires a slipstream pressure condition. The near-field solution is more constrained and, therefore, might be single-valued. If this is the case, some clarification of the nonuniqueness of the far-field solution might be expected.

There are several difficulties with this expectation. The first is the nontrivial question of how the primary shock wave evolves into its far-field counterpart. At best, only a partial answer is given in Ref. 14, which is discussed later. We also observe that there are three possibilities for the far-field shock: the weak and strong solutions, and no solution when θ_2 is too large. However, there are seven possibilities in the near field, which are summarized in Table 19.1. This group consists of the weak (W) and strong (S) primary shock wave solutions, the expansion (E), and the weak (W) and strong (S) shock solutions for the secondary disturbance, and no solution.

Table 19.1 Catalog of Near-Field Cases

Case	Primary	Secondary	Bounds	
A	W	E	$\mu_1 \leq \beta_3 \leq \beta_{3\max}$,	$M_2 \leq M_4 \leq M_1$
B	S	E	$\beta_{3\max} \leq \beta_3 \leq \pi/2$,	$M_2 \leq M_4 \leq M_1$
C	W	W	$\mu_1 \leq \beta_3 \leq \beta_{3\max}$,	$\mu_2 \leq \beta_4 \leq \beta_{4\max}$
D	W	S	$\mu_1 \leq \beta_3 \leq \beta_{3\max}$,	$\beta_{4\max} \leq \beta_4 \leq \pi/2$
E	S	W	$\beta_{3\max} \leq \beta_3 \leq \pi/2$,	$\mu_2 \leq \beta_4 \leq \beta_{4\max}$
F	S	S	$\beta_{3\max} \leq \beta_3 \leq \pi/2$,	$\beta_{4\max} \leq \beta_4 \leq \pi/2$
—		No solution	—	

Numerical Formulation

The flowfield associated with a focused Prandtl–Meyer compression was chosen because the near-field solution required only algebraic equations. In an early investigation,[14] only two such equations were required. Once solved, all other near-field variables are easily obtained.

Two algebraic equations in two unknowns should be numerically easily solvable. However, this was not the case. Severe loss of numerical significance occurred frequently. For given values of the ratio of specific heats γ, incident Mach number M_1, and overall turn angle θ_2, difficulty was encountered in determining whether the correct number of near-field solutions was $0, 1, 2, \ldots$. Finally, for M_1 values below 1.88, the algorithm failed to converge.

Reference 15 describes an algorithm that is completely free of the above difficulties. This algorithm, which requires the solution of but a single algebraic equation, is described later. The approach embodied in this algorithm may be of use for other flowfields, such as those that contain one or more triple points.

Let us begin by prescribing γ, M_1, and θ_2. Solution of the wall shape equations is not necessary. In what follows, the functions summarized in Table 19.2 are used. The Mach number functions X, μ, and ν are not listed but have their usual meaning.

First, consider cases A and B in Table 19.1 where the secondary disturbance is an expansion. In going from region 2 to region 4, the flow

Table 19.2 Near-Field Equations

$$(p/p_0)_{sh} = \left[\left(\frac{2\gamma}{\gamma+1}\right) M^2 \sin^2\beta - \frac{\gamma-1}{\gamma+1}\right](X(M))^{-\gamma/(\gamma-1)}$$

$$M_{sh}(M, \beta, \theta) = \frac{1}{\sin(\beta-\theta)} \left[\frac{X(M \sin \beta)}{\gamma M^2 \sin^2\beta - (\gamma-1)/2}\right]^{\frac{1}{2}}$$

$$M_4(M, \beta) = \left(\frac{2}{\gamma-1} \left\{\left[\frac{p}{p_0}(M, \beta)\right]_{sh}^{-(\gamma-1)/\gamma} - 1\right\}\right)^{\frac{1}{2}}$$

$$\theta(M, \beta) = \tan^{-1}\left[\frac{(\cot \beta)(M^2 \sin^2\beta - 1)}{(\gamma+1) M^2/2 - (M^2 \sin^2\beta - 1)}\right]$$

$$\beta_4(M_1, M_2, \beta_3) = \sin^{-1}\left(\frac{1}{M_2}\left\{\frac{\gamma-1}{2\gamma} + \frac{\gamma+1}{2\gamma}[X(M_2)]^{\gamma/(\gamma-1)}\left[\frac{p}{p_0}(M_1, \beta_3)\right]_{sh}\right\}^{\frac{1}{2}}\right)$$

$$\beta_{max}(M) = \sin^{-1}\left(\frac{(\gamma+1)}{4\gamma M^2}\left\{M^2 - \frac{4}{\gamma+1} + \left[M^4 + 8\left(\frac{\gamma-1}{\gamma+1}\right)M^2 + \frac{16}{\gamma+1}\right]^{\frac{1}{2}}\right\}\right)^{\frac{1}{2}}$$

makes an expansive turn of $\theta_3 - \theta_2$. Consequently, Eq. (19.12) holds, where θ_2 and θ_3 are measured relative to the freestream direction. Equation (19.12) constitutes a slipstream tangency condition. There is also a slipstream pressure condition

$$p_3 = p_4 \tag{19.13}$$

These two conditions can be put in the form

$$M_4 = M_4(M_1, \beta_3) \tag{19.14}$$

$$\theta(M_1, \beta_3) - \nu_4 = \theta_2 - \nu_2 \tag{19.15}$$

where the unknowns are β_3 and M_4. Once these are determined, θ_3 is given by $\theta(M_1, \beta_3)$. As shown in Table 19.1 for case A, two conditions must be satisfied if the primary shock is to be weak and the secondary disturbance is an expansion

$$\mu_1 \leq \beta_3 \leq \beta_{3\,\text{max}} \tag{19.16}$$

$$M_2 \leq M_4 \leq M_1 \tag{19.17}$$

where $\beta_{3\,\text{max}} = \beta_{\text{max}}(M_1)$. Case B utilizes Eqs. (19.14), (19.15), and (19.17); only condition (19.16) is altered, as shown.

When the secondary disturbance is a shock, the pressure and tangency conditions have the form

$$\beta_4 = \beta_4(M_1, M_2, \beta_3) \tag{19.18}$$

$$\theta(M_1, \beta_3) + \theta(M_2, \beta_4) = \theta_2 \tag{19.19}$$

where $\theta(M_2, \beta_4) = (\theta_2 - \theta_3)$. This is the angle by which the slipstream turns the flow in region 2. Equations (19.18) and (19.19) provide the solutions for cases C–F. The unknowns β_3 and β_4 are bounded, as shown in Table 19.1, where $\beta_{4\,\text{max}} = \beta_{\text{max}}(M_2)$.

Equations (19.14) and (19.15) or (19.18) and (19.19), along with Eq. (19.7) when M_2 is needed, and the bounds in Table 19.1 are to be solved. As Fig. 19.6 will shortly demonstrate, there is an array of solution possibilities when θ_2 is an independent parameter. However, the numerical algorithm is greatly simplified by interchanging the roles of θ_2 and β_3. For instance, with γ, M_1, and β_3 known, the bounds on β_3 for all cases in Table 19.1 are easily determined. The number of real solutions now is either none or one.

The foregoing interchange also simplifies the solution process. Note that Eq. (19.14) is directly solved for M_4 and that θ_2 can be eliminated from Eqs. (19.7) and (19.15). Thus, we have a single equation with M_2 as the one unknown. After this implicit equation is solved, all other quantities of interest are readily found. When the secondary disturbance is a shock, we eliminate β_4 from Eqs. (19.18) and (19.19) and then eliminate θ_2 by using Eq. (19.7). The result is again a single implicit equation with M_2 as the unknown.

Results and Discussion

Figure 19.6 shows results for incident Mach numbers that range from 1.2 to 16. To the left of its peak, the solid line is the far-field weak-shock solution, while to the right of the peak, the dashed line is the strong-shock solution. For the far-field curve, β_3 here represents $\bar{\beta}_3$. Computation of the far-field curves is straightforward and requires no discussion.

The six near-field cases are shown in the legend. The first letter, W or S, refers to the strength of the primary shock. The second letter describes the secondary disturbance. Whenever the near-field curve is inside the far-field curve, the secondary disturbance is a centered Prandtl–Meyer expansion. Referring to the $M_1 = 1.2$ case, we observe that the near-field WE solution is unique and extends over the limited range 56.4 deg $\leq \beta_3 \leq$ 67.0 deg. Termination occurs at 67 deg, where $M_2 = 1$. For all $M_1 \leq 1.5$ values, termination occurs for the same reason. The extreme closeness of the near-field and far-field curves indicates a very weak secondary disturbance, whether it is an expansion or a shock. This is the regime in the first computational attempt where extreme loss of significance occurred. A gap appears between the maximum near-field value $\theta_{2\,\text{max}}$ of 3.51 deg and the far-field value $\bar{\theta}_{2\,\text{max}}$ of 3.94 deg.

Figure 19.6(b), where M_1 is 1.3, is similar to Fig. 19.6(a), except that the curves cross at $\beta_3 = 56.5$ deg. Below this value the secondary disturbance is a weak-solution shock, while above, it is an expansion.

Two new features are present when $M_1 = 1.5$. For $\beta_3 \leq 62.6$ deg (where the solution terminates because $M_2 = 1$), the near-field solution is entirely WW. A second near-field branch SS occurs in the 88.1 deg $\leq \beta_3 \leq$ 90 deg range. At 88.1 deg, the secondary disturbance is a normal shock; for larger β_3 values, it is oblique. In the 0.630 deg $\leq \theta_2 \leq$ 1.72 deg range, the near-field solution is double-valued relative to θ_2. Outside this range the near-field solution is unique.

As shown in Fig. 19.6(d), where $M_1 = 1.6$, the right branch of the near-field solution is now more extensive. The WS solution appears for the first and only time in this sequence of solutions. Its extent is small, i.e., 63.7 deg < β_3 < 65.8 deg. By the time $M_1 = 1.64$, Fig. 19.6(e), the SS solution replaces the WS solution. The left branch in both Figs. 19.6(d) and 19.6(e) terminates when the secondary shock becomes normal. This shock is also normal when the solution resumes on the right side of the figures.

Gradually, the two near-field branches approach each other. When $M_1 = 2$, Fig. 19.6(f), only a single branch occurs. The near-field solution lies entirely outside the far-field one and consists of three types: WW, SW, and SS. For $\theta_2 \leq 13.1$ deg, the near-field WW solution is unique.

The near-field solutions for $M_1 \geq 4$ are similar, in that they consist of WE, SE, and SW types. For sufficiently small θ_2, the WE solution is unique. A small gap between $\bar{\theta}_{2\,\text{max}}$ and $\theta_{2\,\text{max}}$ is evident when $M_1 = 4$. This gap becomes sizable at the higher Mach numbers. The greater separation between the near-field and far-field curves indicates the increased strength

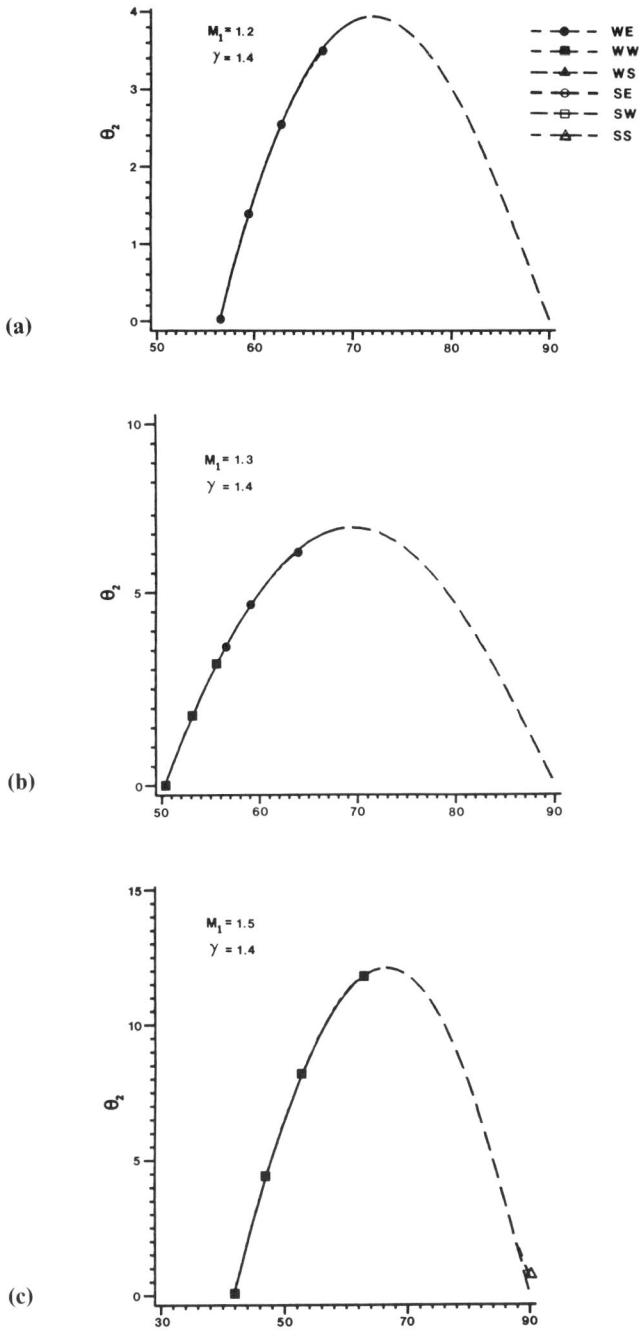

Fig. 19.6 Variation of θ_2 with β_3 when $\gamma = 1.4$ for incident Mach numbers ranging from 1.2 to 16.

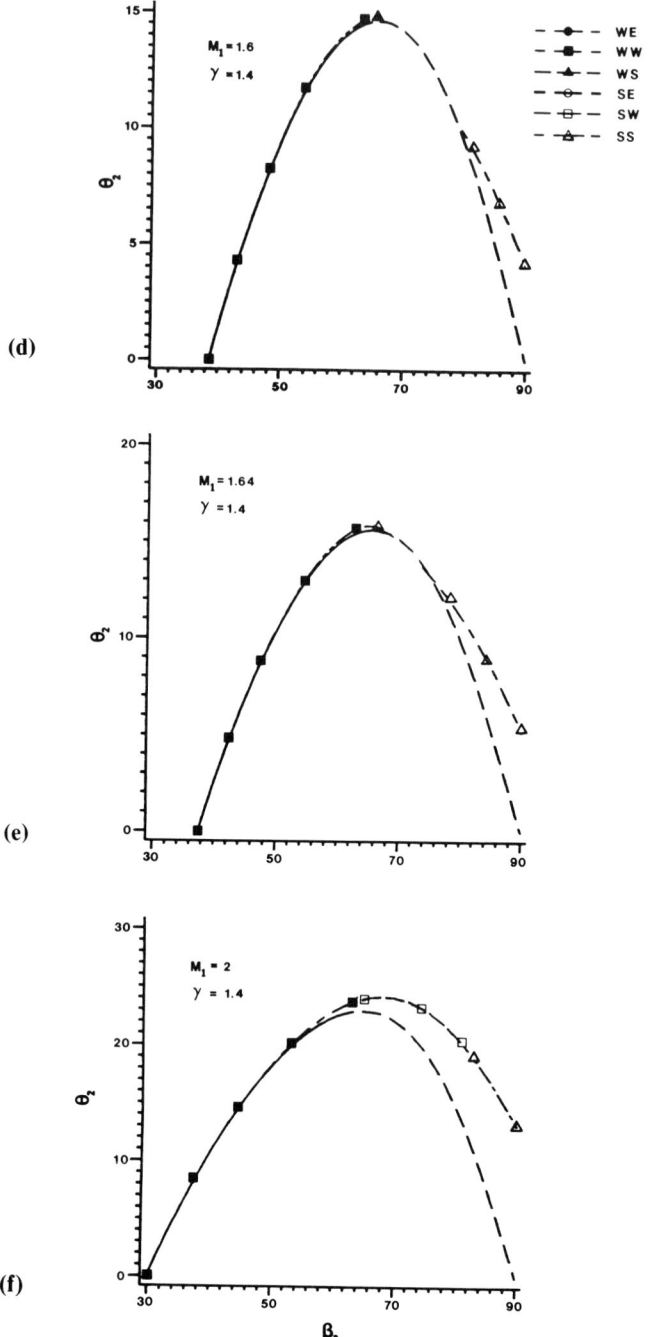

Fig. 19.6 (continued) Variation of θ_2 with β_3 when $\gamma = 1.4$ for incident Mach numbers ranging from 1.2 to 16.

FLOWS WITH SHOCK WAVES 361

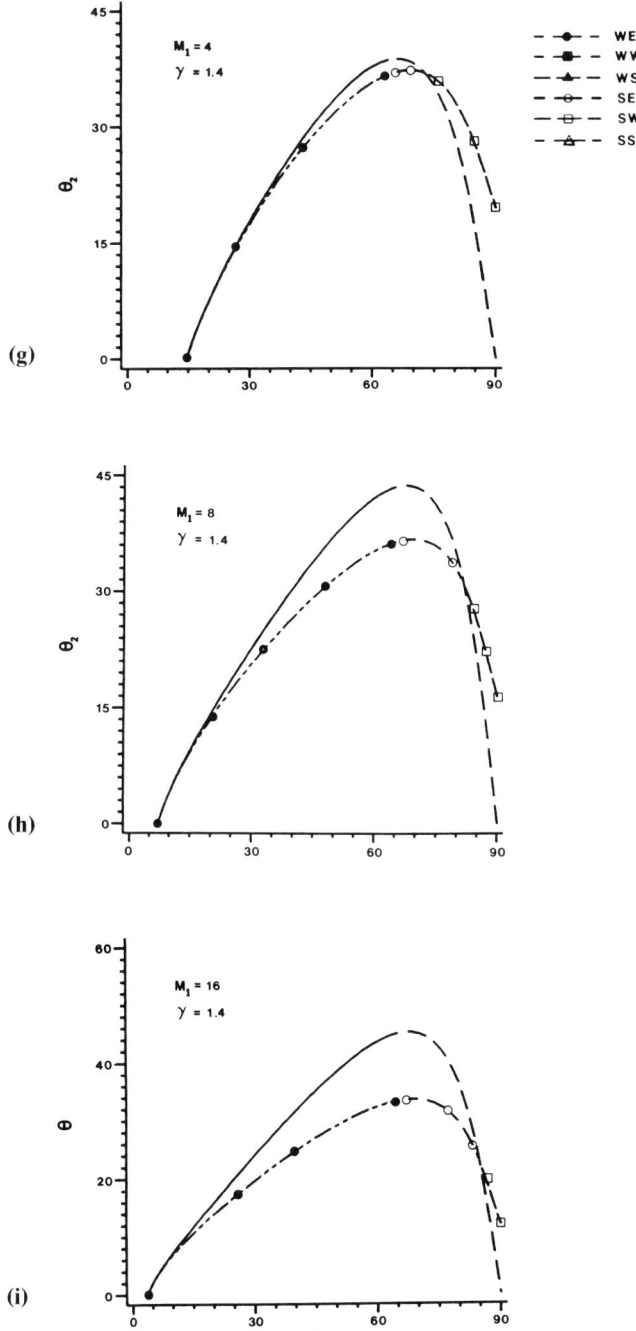

Fig. 19.6 (continued) Variation of θ_2 with β_3 when $\gamma = 1.4$ for incident Mach numbers ranging from 1.2 to 16.

of the secondary disturbance. For instance, when $M_1 = 8$ and $\beta_3 = 60$ deg, the secondary expansion has a pressure ratio of $(p_2/p_4) = 1.96$.

Figure 19.6 shows that the difference in far-field and near-field peaks, $(\bar{\theta}_2 - \theta_2)_{\max}$, is positive at small incident Mach numbers. The difference is negative when $M_1 = 2$ but again becomes positive when $M_1 = 4$. Above a Mach number of 4 the positive difference is substantial. In this discussion, we are comparing two local solutions. The far-field one is by far the more general of the two. It relates the streamline and shock angles immediately downstream of the primary shock at any point on the shock. Most likely, with increasing distance from the corner, the streamline angle is asymptotic to θ_2. On the other hand, the near-field solution is structurally more complex. However, it is limited to describing the flow near the corner when the wall generates a focused compression.

The flow in region 3 and/or region 4 can be subsonic, in which case the near-field solution may be restricted to the immediate vicinity of the focal point. (This proviso does not apply to the focused compression or to region 2.) For example, the restriction is apparent whenever $\beta_3 = 90$ deg. In this instance, the primary shock is normal to the incident flow; and, initially, the slipstream is horizontal. For instance, in Fig. 19.6(i), we have

$$\beta_3 = 90 \text{ deg}, \quad \beta_4 = 16.7 \text{ deg}, \quad \theta_2 = 11.9 \text{ deg}, \quad \bar{\beta}_3 = 87.50 \text{ deg}$$

$$M_2 = 9.45, \quad M_3 = 0.382, \quad M_4 = 5.91, \quad \bar{M}_3 = 0.394$$

Since the subsonic flow in region 3 must turn, the solution in this region is limited to the vicinity of the focal point.

Figure 19.7 shows three angles for the Fig. 19.6(i) case. The slipstream angle θ_3 is shown as $\theta_3 - \theta_2$. In general, the slipstream at the focal point is not parallel to the downstream wall and here deviates from it by as much as ± 11 deg. The angle of the leading edge of the secondary disturbance relative to the x axis is $\beta_4 - \theta_2$. For most β_3 values, the leading edge is below the x axis. Nevertheless, region 2 is substantial in angular extent. This is demonstrated by the 180-ϕ_2 curve, where ϕ_2 is the angle between the x axis and the trailing edge of the Prandtl–Meyer compression (see Fig. 19.5). As shown in Fig. 19.7, ϕ_2 is between 135 and 177 deg. The large value for ϕ_2 is a consequence of the large M_2 value of the preceding paragraph.

If one assumes that the near-field solution evolves into its closest far-field counterpart, then several conclusions follow. (We mean the closest, in terms of β_3, near-field and far-field solutions with the same value for θ_2.) Both solutions should physically exist if an attached shock is to occur. (By attached, we mean that the primary shock originates at the focal point.) Within the constraints of this analysis, an attached shock need not occur when $\theta_{2\max} < \theta_2 \leq \bar{\theta}_{2\max}$, even though predicted by the oblique shock equations.

One difficulty with the above discussion is the assumption that the far-field solution is provided by the oblique shock equations for a sharp turn. Although the secondary pressure disturbance will decay with increasing downstream distance, this is not the case for the slipstream or the

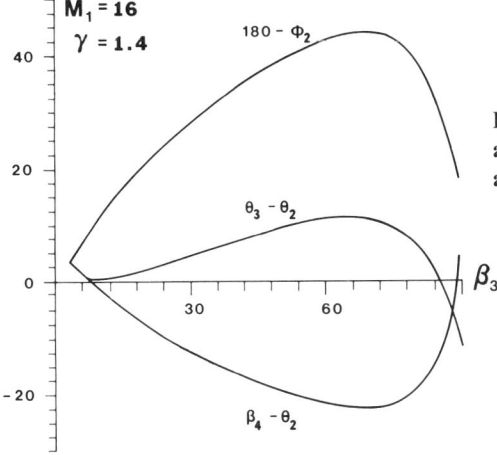

Fig. 19.7 Variation of ϕ_2, $\theta_3 - \theta_2$, and $\beta_4 - \theta_2$ with β_3 when $\gamma = 1.4$ and $M_1 = 16$.

entropy jump across it. This discontinuity persists, and the oblique shock solution is thus globally approximate in the far field.

Suppose two near-field solutions mathematically exist for the same θ_2 value. Based on the discussion in Sec. 19.2, we presume that only one of them is stable. Another conclusion suggests that the solution is globally unique whenever the near-field solution is unique. In this case, the primary shock is invariably given by the weak solution.

One of the more intriguing aspects of the analysis is the large gap between the far-field and near-field peak values for θ_2 that occurs at large M_1 values. Experimental verification, however, appears to be difficult, since avoiding boundary-layer separation along a ramp with θ_2 in excess of 30 deg may not be feasible. A study by the author assuming a laminar boundary layer shows that separation occurs well before $\theta_2 = 10$ deg is reached.

19.4 FORMATION OF SHOCK WAVES IN JETS

Let us consider steady flow in a two-dimensional or axisymmetric nozzle where the flow is uniform in the exit plane of the nozzle. If the exit pressure matches the ambient pressure, the jet is straight and unchanging, except for mixing with the ambient gas. A schlieren photograph of the jet would show a barely visible diamond pattern, due to Mach waves that originate at the lip of the nozzle. If the jet is overexpanded, a shock wave system in the jet is to be expected. The shock waves also begin at the lip of the nozzle, as previously discussed in Chap. 7. When the exit pressure is above the ambient pressure, the jet is underexpanded, and expansion waves originate at the lip of the nozzle. However, shock waves occur further downstream, even when the jet is only slightly underexpanded. As the nozzle exit pressure is increased above the ambient, shock waves appear in the jet, as illustrated in Fig. 19.8.

In Fig. 19.8(a), the nozzle exit pressure p_e equals the ambient pressure p_∞. The Mach waves form a repeating diamond pattern. Within any one cell there is upstream-downstream symmetry about a midplane, which is indicated by the dashed line in the first cell.

As p_∞ decreases or p_e increases, symmetry about a cell's midplane disappears. Weak planar (conical in the axisymmetric case) shock waves appear near the end of the cell as shown in Fig. 19.8(b). As p_∞ further decreases, the shocks coalesce, Fig. 19.8(c). With a still lower value for p_∞, a Mach disk normal shock appears in the central region, Fig. 19.8(d). The flow downstream of this shock is an interior subsonic jet. As a result of the shocks, and especially the subsonic jet, the flow in the second cell differs from that in the first.

In this section, we examine two general features of the flow; the reasons for the oval shape of a cell, particularly the first one, and the shock formation process as shown in Fig. 19.8(b). For convenience, we discuss only a two-dimensional jet, although the mechanisms are the same for an axisymmetric one. Our presentation is based on the MOC analysis by Pack.[16]

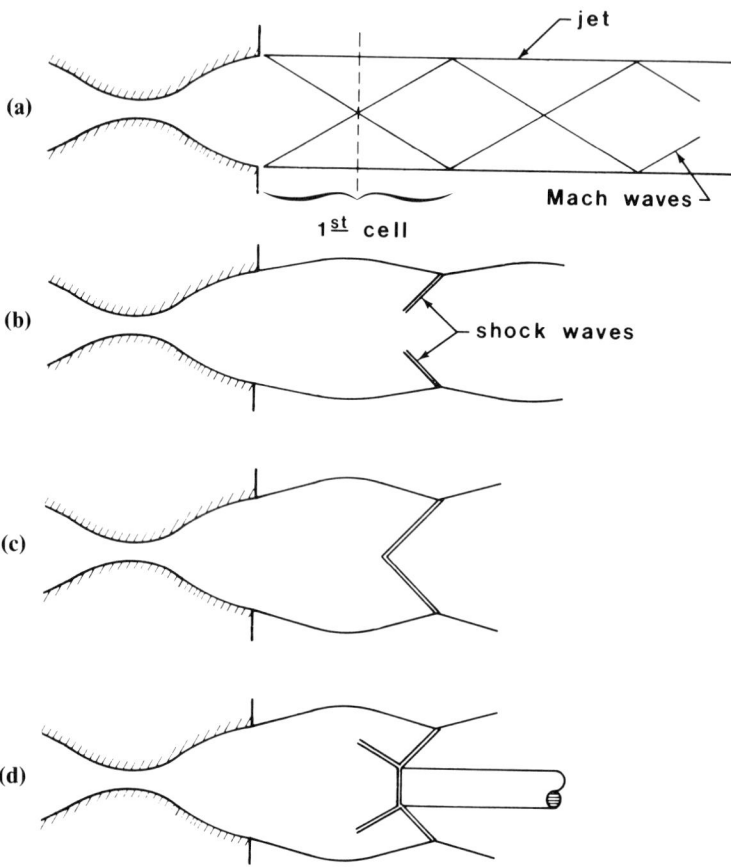

Fig. 19.8 Jet emanating from an underexpanded nozzle.

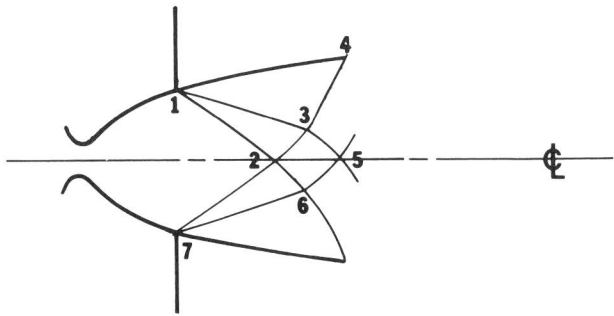

Fig. 19.9 Centered Prandtl–Meyer expansions when the nozzle is underexpanded.

As discussed in Chap. 7, there is a centered Prandtl–Meyer expansion attached to each lip, shown in Fig. 19.9. The lip expansions allow the nozzle exit pressure p_e to adjust to the ambient pressure. The flow, of course, is symmetrical about the centerline and is uniform between the exit plane and the straight Mach line 1–2. Region 1–3–4 is a uniform flow region whose pressure is p_∞. The two Prandtl–Meyer expansions overlap in region 2–3–5–6, which is a nonsimple wave region wherein both sets of Mach lines are curved.

The left-running Mach lines in the expansion that originate at point 7 reflect off the upper boundary of the jet as a compression wave, as shown in Fig. 19.10. (A compression wave is easily spotted. It consists of Mach lines that converge in the flow direction.) The expansion from point 7 is a simple wave in region 3–4–8–5. If we examine the various regions downstream of the 1–2 Mach line, in accord with the discussion near the end of Sec. 16.1, we obtain the classification in Table 19.3. The first cell ends at a plane through point 12 that is perpendicular to the centerline. Within the upper half of the first cell, between Mach lines 1–2 and 12–14, there are 4 uniform flow regions, 4 simple wave regions, and 3 nonsimple wave regions.

The 1–4–9–12 curve is a constant-pressure and constant-entropy streamline which, according to Bernoulli's equation, also has a constant speed. It is straight along 1–4 and 9–12, and regions 1–3–4 and 9–11–12 are uniform flow regions with pressure p_∞. Since arc 4–9 is a streamline, the left-running Mach lines reflect from it with equal incident and reflection angles, as shown in Fig. 19.11. The angles are the local Mach angles. In general, points 9 and 4 are not located the same distance from the centerline, although the difference is small for jets that are only slightly underexpanded. The same statement holds for points 1 and 12.

If the foregoing discussion were complete, shock waves could not form anywhere inside a cell. We need to include the possibility that Mach lines of the same family may overlap. This would result in a nonunique solution which, of course, cannot occur. Rather, overlap is avoided by the insertion of a shock wave.

Certain Mach lines can overlap because they curve more in the nonsimple wave regions than others of the same family. With this in mind, we reconsider Fig. 19.10 in a new form, Fig. 19.12. For purposes of clarity,

366 GASDYNAMICS: THEORY AND APPLICATIONS

Table 19.3 Type of Regions in the First Cell

Region	Type of region
1–2–3	Simple wave
1–3–4	Uniform flow
2–3–5	Nonsimple wave
3–4–8–5	Simple wave
5–8–10	Uniform flow
4–9–8	Nonsimple wave
8–9–11–10	Simple wave
9–12–11	Uniform flow
10–11–13	Nonsimple wave
11–12–13	Simple wave
12–13–14	Uniform flow

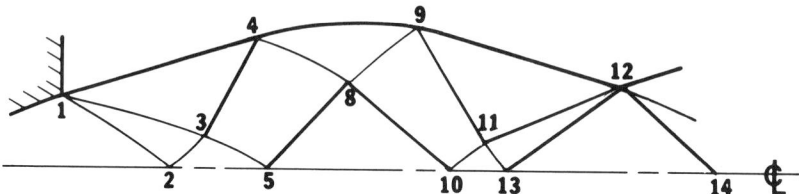

Fig. 19.10 Numbering system for the first cell.

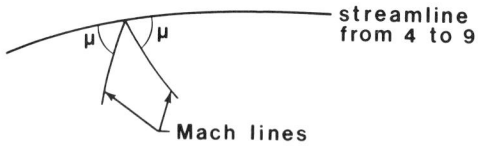

Fig. 19.11 Mach line reflection condition at the edge of the jet.

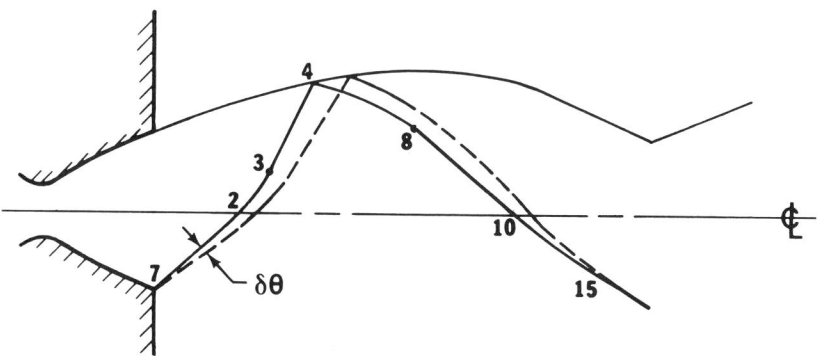

Fig. 19.12 Adjacent Mach lines that overlap at point 15.

most of the earlier details are left out. The leading edge of the Prandtl–Meyer expansion from the lower lip is shown and labeled in accord with the earlier figures. A nearby Mach line at a small angle $\delta\theta$ relative to the first, and within the Prandtl–Meyer expansion, is shown as a dashed line in the simple wave regions and as a solid line in the nonsimple wave regions. After passing through three nonsimple wave regions, this second Mach line runs into the first at point 15. As $\delta\theta$ approaches zero, point 15 moves slightly closer to point 10. A very weak shock begins at the overlap point. The resulting flow appears as in Fig. 19.8(b). MOC calculations[16] indicate that the shock first forms when the Mach lines of the Prandtl–Meyer expansion start to overlap with the leading-edge Mach line from the nozzle lip.

19.5 SHOCK WAVE REFLECTION FROM A WALL IN STEADY FLOW

As discussed in Chap. 12, the flow through a shock wave can be viewed locally as two-dimensional, even though on a global scale the flow is three-dimensional. Thus, we can reduce the general problem of a shock wave reflecting from a wall in a steady flow to the two-dimensional problem of uniform supersonic flow over a finite wedge. In this circumstance, only two patterns exist: regular reflection (RR) or Mach reflection (MR).

Regular Reflection

We first consider the simpler RR case, as illustrated in Fig. 19.13. The top of the wedge is parallel to the freestream, which has a Mach number M_1. An I indicates the incident shock while the streamline and shock wave turn angles are θ_i and β_i, respectively. Both angles, of course, are measured relative to $\mathbf{V}_1$. The incident shock wave is planar and the flow downstream of it, in region 2, is uniform. For a reflected shock wave R to exist, the Mach number M_2 in region 2 must be supersonic. The reflected shock wave turns the flow counterclockwise by an angle θ_r, such that the flow in region 3 is once again parallel to the wall. Since M_2 is supersonic, the back edge of the wedge generates a centered Prandtl–Meyer expansion that weakens and curves R. We will not consider this interaction or the vortical flow downstream of it. Thus, region 2 terminates at the leading edge of the expansion.

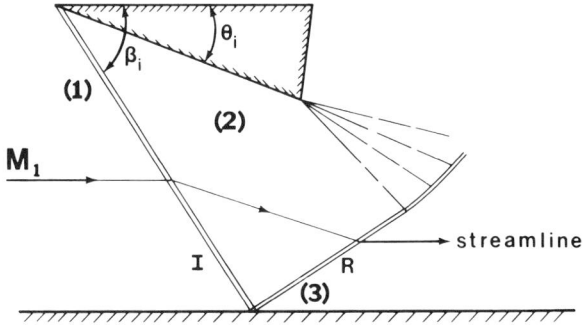

Fig. 19.13 Regular reflection on a planar wall in steady flow.

Since region 2 is supersonic, I is a weak-solution shock wave. Consequently, θ_i and β_i are limited to the heavy line shown in Fig. 19.14, which is part of a β-θ curve for a given value of M_1. Recall that the curve shown in the figure represents Eq. (12.21b). At the sonic point, denoted by an s, $M_2 = 1$, and from Problem 12.1, we have

$$\cos\beta_s = \sin(\beta_s - \theta_s)\left[\cos\beta_s\sin(\beta_s - \theta_s) + \frac{\gamma+1}{2}\sin\theta_s\right] \quad (19.20)$$

The simultaneous solution of this relation with Eq. (12.21b) provides conditions at the sonic point. Later, we show that RR terminates well before the sonic condition in region 2 is reached.

For the reflected shock, θ_r and β_r are shown in Fig. 19.15. For convenience, we have defined both θ_i and θ_r as positive. The RR reflection condition then is

$$\theta_r = \theta_i \quad (19.21)$$

For the latter discussion, it is important to recognize that the flow in this figure has no length scale.

In general, the reflection angle $\beta_r - \theta_r$ does not equal β_i. Hence, the shock wave does not reflect in a specular manner. However, in the limit of $\theta_i \to 0$, both shocks become Mach waves, and specular reflection holds for these waves, as shown in Fig. 19.11.

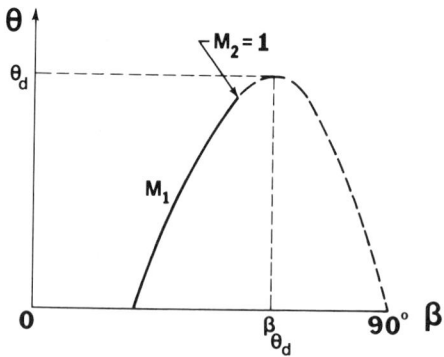

Fig. 19.14 Sketch of a β-θ curve for an oblique shock wave.

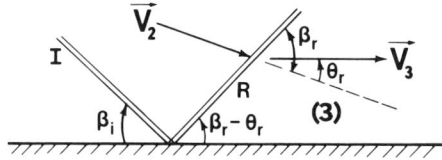

Fig. 19.15 Velocities and angles associated with the reflected shock wave.

FLOWS WITH SHOCK WAVES

For the subsequent analysis, it is necessary to consider the static pressures. All pressures are referenced to p_1; hence, we introduce

$$P_2 = \frac{p_2}{p_1}, \qquad P_3 = \frac{p_3}{p_1}$$

where

$$P_2 = \frac{2\gamma}{\gamma+1} M_1^2 \sin^2\beta_i - \frac{\gamma-1}{\gamma+1} \qquad (19.22a)$$

$$P_3 = P_2 \left(\frac{2\gamma}{\gamma+1} M_2^2 \sin^2\beta_r - \frac{\gamma-1}{\gamma+1} \right) \qquad (19.22b)$$

We also have

$$M_2 = \frac{M_{2n}}{\sin(\beta_i - \theta_i)} = \frac{1}{\sin(\beta_i - \theta_i)} \left\{ \frac{1 + [(\gamma-1)/2] M_1^2 \sin^2\beta_i}{\gamma M_1^2 \sin^2\beta_i - [(\gamma-1)/2]} \right\}^{\frac{1}{2}} \qquad (19.22c)$$

$$\tan\theta_i = \frac{1}{\tan\beta_i} \frac{M_1^2 \sin^2\beta_i - 1}{1 + \{[(\gamma+1)/2] - \sin^2\beta_i\} M_1^2} \qquad (19.22d)$$

$$\tan\theta_r = \frac{1}{\tan\beta_r} \frac{M_2^2 \sin^2\beta_r - 1}{1 + \{[(\gamma+1)/2] - \sin^2\beta_r\} M_2^2} \qquad (19.22e)$$

It is computationally convenient to consider P_2 as the independent variable, such that

$$\theta = \theta_i = \theta_r = \theta(P_2; \gamma, M_1), \qquad \beta_i = \beta_i(P_2; \gamma, M_1),\ldots$$

With the use of Eqs. (19.22), one can show that the following relations of the above form hold:

$$\beta_i = \sin^{-1}\left\{ \frac{1}{M_1} \left[\frac{\gamma+1}{2\gamma} \left(P_2 + \frac{\gamma-1}{\gamma+1} \right) \right]^{\frac{1}{2}} \right\} \qquad (19.23)$$

$$\theta = \tan^{-1}\left[\frac{P_2 - 1}{\gamma M_1^2 - (P_2 - 1)} \right.$$

$$\left. \times \left(\frac{[2\gamma/(\gamma+1)] M_1^2 - \{P_2 + [(\gamma-1)/(\gamma+1)]\}}{P_2 + [(\gamma-1)/(\gamma+1)]} \right) \right] \qquad (19.24)$$

$$M_2 = \frac{1}{\sin(\beta_i - \theta)} \left\{ \frac{[(\gamma+1)/2\gamma] + [(\gamma-1)/2\gamma] P_2}{P_2} \right\}^{\frac{1}{2}} \qquad (19.25)$$

The angle β_r is given implicitly by Eq. (19.22e), which is solved iteratively. It is worth noting that the solution for β_r is single-valued, that β_r increases monotonically with P_2, and that the weak-shock solution is bounded as follows:

$$\sin^{-1}(1/M_2) \le \beta_r \le \beta_{r,\theta_d}$$

where the d subscript on θ stands for detachment. The upper bound is the β_r value when θ is a maximum, θ_d, as shown in Fig. 19.14. By differentiating Eq. (19.22e), we obtain for the upper bound

$$\beta_{r,\theta_d} = \sin^{-1}\left(\frac{\gamma+1}{4\gamma M_2^2}\left\{M_2^2 - \frac{4}{\gamma+1} + \left[M_2^4 + 8\left(\frac{\gamma-1}{\gamma+1}\right)M_2^2 + \frac{16}{\gamma+1}\right]^{\frac{1}{2}}\right\}\right)^{\frac{1}{2}}$$

(19.26)

Once β_r is known, P_3 is determined by Eq. (19.22b). Finally, M_3 is given by an equation similar to Eq. (19.25),

$$M_3 = \frac{1}{\sin(\beta_r - \theta)}\left\{\frac{[(\gamma+1)/2\gamma] + [(\gamma-1)/2\gamma](P_3/P_2)}{(P_3/P_2)}\right\}^{\frac{1}{2}} \quad (19.27)$$

Since $\theta_r = \theta_i$, the value for β_i and for β_r at the two sonic conditions ($M_2 = 1$ and $M_3 = 1$, respectively) is the same and is given by Eq. (19.20).

Shock-Polar Diagrams

The foregoing algorithm is interpreted physically by means of a P-θ shock-polar diagram, Fig. 19.16. Here, M_1 is held fixed, P is p_2/p_1, and P_n is the pressure ratio for a normal shock.

For a given value of M_1, an upper limit on θ occurs when the reflected shock is at the peak of its β-θ curve. This is the detachment condition that applies only to the reflected shock, since M_1 exceeds M_2. It is determined by Eq. (19.26), with M_2 given by Eq. (19.25). The actual value of θ_d is then given by Eq. (19.22e). Slightly before the detachment condition is reached, region 3 becomes subsonic. Once this occurs, pressure disturbances, from downstream and from each side of region 3, can alter the region 3 flowfield so that it is no longer uniform and R is no longer planar. In this situation, the detachment condition becomes a local condition that is valid, at best, only in the immediate vicinity of the reflection point at the wall, where θ_r must still equal θ_i.

As noted earlier, the flow shown in Fig. 19.15 has no length scale when region 3 is supersonic. However, a length scale can be communicated from the sides or the downstream flow to the shock reflection point, once region 3 is subsonic. As we shall see, the sonic condition for region 3 represents an important transition condition.

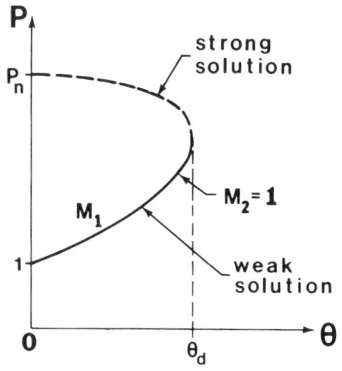
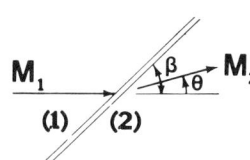

Fig. 19.16 Shock-polar diagram.

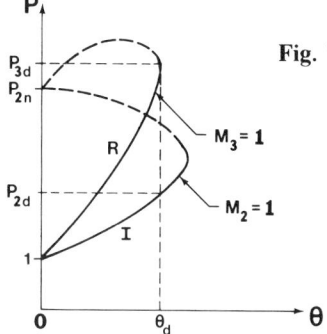

Fig. 19.17 Shock-polar diagrams for regular reflection.

Our discussion is well represented by a polar diagram for both the incident and reflected shocks, as illustrated in Fig. 19.17. The incident shock curve, labeled I, is a conventional shock-polar curve that shows p_2/p_1 vs θ for a fixed value of M_1. In Fig. 19.17 the reflected shock curve R is not a conventional shock polar since M_2 is not fixed but varies in accord with Eq. (19.25). Both weak and strong branches are shown for each shock. However, only that portion of the weak solution of I up to θ_d is now of interest. As will be evident from subsequent numerical results, $\theta \leq \theta_d$ limits M_2 to a supersonic value. At $\theta = \theta_d$, we have detachment for the reflected shock. As θ increases from zero, M_2 decreases. On the weak-solution branch of R, P_3 increases from unity, and this part of the R curve lies above the I curve because $p_3 \geq p_2$. The strong-solution branch of R, however, is also of interest. In this case, the incident shock is still a weak-solution shock. However, P_3 can increase, or decrease, from the normal shock value P_{2n} as θ increases from zero.

RR to MR Transition

As θ increases from a small value, a transition occurs that results in the MR pattern shown in Fig. 19.18. The value of θ at which RR → MR transition occurs can depend only on M_1 and γ. After transition, there is a third shock wave, the Mach stem, indicated by M. This shock is normal to the wall and is of infinitesimal length at the transition condition. A triple point occurs where the three shocks meet. Because of the entropy difference, a slipstream, indicated by an ss, emanates from the triple point. When M is of infinitesimal length, the slipstream is parallel to the wall, and $\theta_r = \theta_i$, as before. Across the slipstream, we have $p_3 = p_4$ with V_3 and V_4 parallel to each other.

Figure 19.19 contains shock-polar diagrams for $M_1 = 3$ and 2 when $\gamma = 1.4$. When M_1 is large, as in Fig. 19.19(a), M_3 is supersonic as long as the wedge angle θ is below θ_s, which equals 21.3 deg. The sonic condition $M_3 = 1$ occurs at the s point on the R curve. At a slightly smaller value than θ_s, a condition occurs that corresponds to the flow pattern in Fig. 19.18. This is called the mechanical equilibrium condition, which is denoted by an m subscript. It occurs when the weak-solution part of the R curve satisfies

$$\frac{p_3}{p_1} = \frac{p_4}{p_1}$$

or

$$P_3 = P_{2n} \qquad (19.28)$$

At this condition, we have

$$\theta = \theta_m, \qquad \frac{p_2}{p_1} = P_{2m}$$

where P_{2m} is associated with the m point on the I curve in Fig. 19.19(a). Note that M_3 is supersonic at the mechanical equilibrium point.

Transition is expected to occur at the smallest possible value for θ that can yield the MR pattern. As shown in Fig. 19.19(a), this occurs at the mechanical equilibrium condition for a sufficiently large M_1 value. In this situation, neither the sonic nor detachment conditions are obtained.

Next, consider a moderate value of 2 for M_1, as shown in Fig. 19.19(b). The difficulty here is that if Eq. (19.28) is to hold, then R is a strong-solution shock and M_3 is subsonic. This situation is not expected to occur. Instead, there are two alternative RR → MR transition conditions. In the

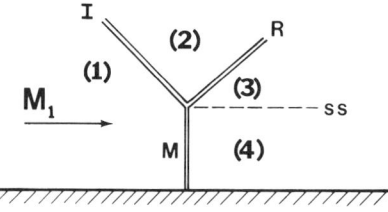

Fig. 19.18 Mach reflection pattern at the mechanical equilibrium condition, when length of the Mach stem M is infinitesimal.

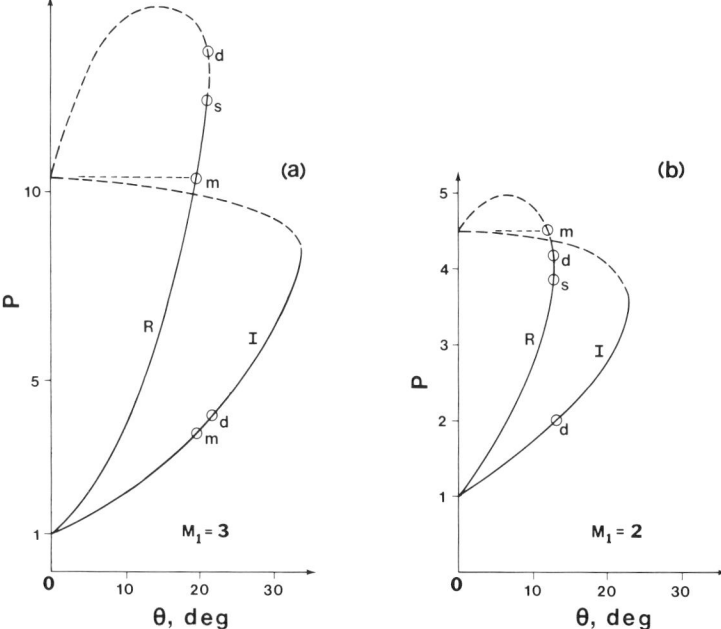

Fig. 19.19 Shock polars when $M_1 = 3, 2$ and $\gamma = 1.4$. The d, m, s points are the detachment, mechanical equilibrium, and sonic points. The sonic point on the I curve is just below the detachment point and within the small circle around the detachment point. The strong-solution part of the R curve when $M_1 = 3$ is only a sketch.

first, transition might occur at detachment when $\theta = \theta_d$, as indicated by the d point. (See also Fig. 19.17.) However, M_3 is already subsonic. A more likely transition condition would occur when $M_3 = 1$. That is, transition occurs at the first opportunity for a Mach stem to develop. At this time, a downstream length scale can communicate with the Mach stem. Further, R is a weak-solution shock. As previously noted, this transition condition is denoted by an s subscript.

At a still smaller M_1 value, the R curve can no longer intersect P_{2n}, in which case mechanical equilibrium is not possible.

We summarize the RR flow and the RR → MR transition situations as follows: Equations (19.24)–(19.27) describe RR. At high M_1 values, transition occurs at the mechanical equilibrium condition given by Eq. (19.28). At smaller M_1 values, transition occurs at the sonic condition when M_3 first becomes unity.

Aside from Fig. 19.19, which accurately shows the d, m, and s points, parametric transition results for $\gamma = 1.4$ are shown in Figs. 19.20 and 19.21. Figure 19.20 shows the M_2 and M_3 values at transition, where a sonic condition is used when M_1 is below 2.405, and mechanical equilibrium applies above this M_1 value. The corresponding wedge angle θ is shown as θ_s in Fig. 19.21 when $M_1 \leq 2.405$ and as θ_m for larger M_1 values. The θ_{2s}

curve shows the θ value required for $M_2 = 1$. Since θ_{2s} appreciably exceeds θ_s, the assertion made below Eq. (19.20) is verified.

Figure 19.21 also shows $\theta_d - \theta_s$, which never exceeds 0.21 deg. Detailed calculations over the range $1.05 \leq \gamma \leq (5/3)$ show that the $\theta_d - \theta_s$ curve invariably has the shape shown in the figure with a peak near $M_1 = 2$. The maximum value for $\theta_d - \theta_s$ never exceeds 0.3 deg.

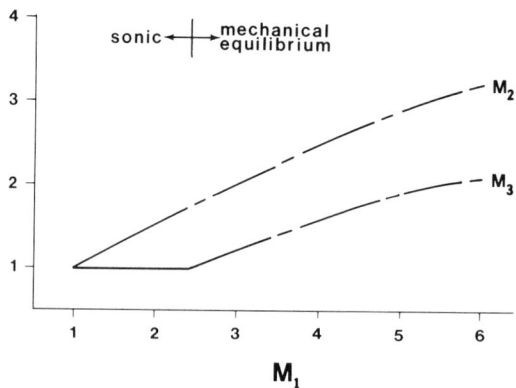

Fig. 19.20 Values of M_2 and M_3 at RR $\rightarrow$ MR transition when $\gamma = 1.4$.

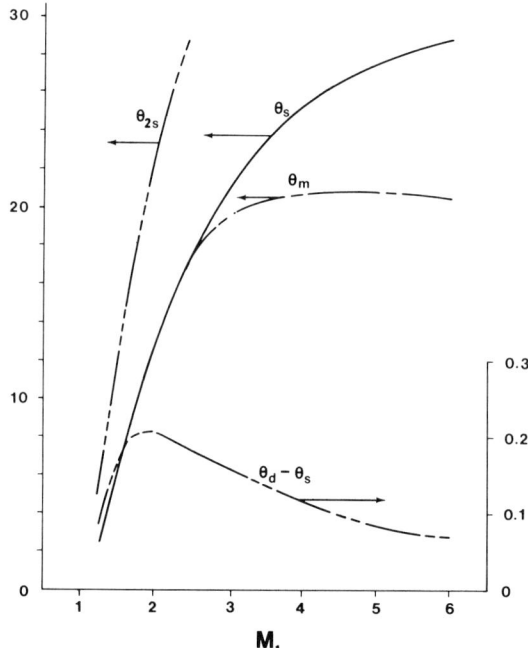

Fig. 19.21 Values of θ_s and θ_m at RR $\rightarrow$ MR transition. θ_{2s} is the wedge angle when $M_2 = 1$. $\theta_d - \theta_s$ is the difference in wedge angles for detachment and for the sonic condition. All curves are for $\gamma = 1.4$ and all angles are in deg.

Until recently, the RR → MR steady-flow transition was thought to occur at the detachment condition. However, experiments[2] at a high Mach number confirm mechanical equilibrium as the transition condition. Reliable data at low or moderate M_1 values, however, are not available. Thus, transition in this Mach number range is still uncertain. The uncertainty may not be resolved soon because the experiments are difficult, as discussed in the next subsection. Indeed, the difference between θ_s and θ_d, shown in Fig. 19.21, may be too small to resolve even in an otherwise sound experiment.

Mach Reflection

When θ increases after the RR → MR transition, the Mach stem increases rapidly in length,[2] thereby resulting in the Mach reflection configuration shown in Fig. 19.22. Both the incident shock and Mach stem are represented by the I curve in the shock-polar diagrams of Fig. 19.19. The Mach stem is a portion of the strong solution part of the I curve. Region 3 is usually supersonic, while region 4 is subsonic. The Mach stem has some curvature, and the slipstream just downstream of the triple point usually has a slightly negative slope. Consequently, $\theta_r \neq \theta_i$ and the R curve in the earlier shock-polar diagrams are no longer applicable.

After a Mach stem appears, its length is determined by some downstream length scale. The length of the stem and its curvature depend on how the flow in region 4 adjusts to downstream disturbances. One of the more important disturbances is the Prandtl–Meyer expansion from the wedge as shown in Fig. 19.22. This disturbance accelerates the flow in region 4 and may cause it to become supersonic. Since a global solution is required for the subsonic flow, there is no general analytical theory for the length and curvature of the Mach stem.

At some large value for θ_i, the MR pattern must change, probably to a fully detached shock system. This second transition must occur before M_2 becomes sonic.

A detailed investigation of the flow in the vicinity of the triple point is not included for two reasons. First, while the analysis might be interesting, it cannot be definitive unless it can be related to the angle of the Mach stem at the triple point. Second, both the RR and MR processes are not of pressing concern in a steady flow. Shock wave-boundary layer interaction

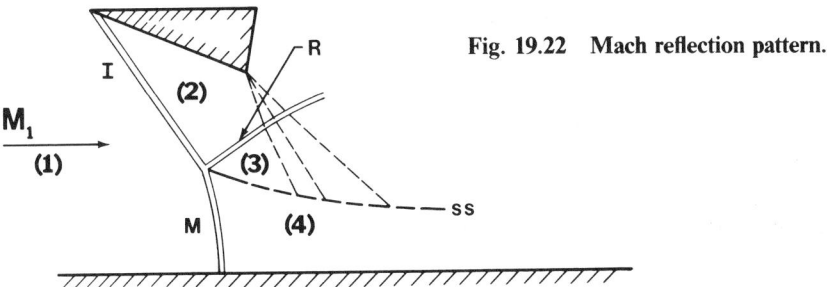

Fig. 19.22 Mach reflection pattern.

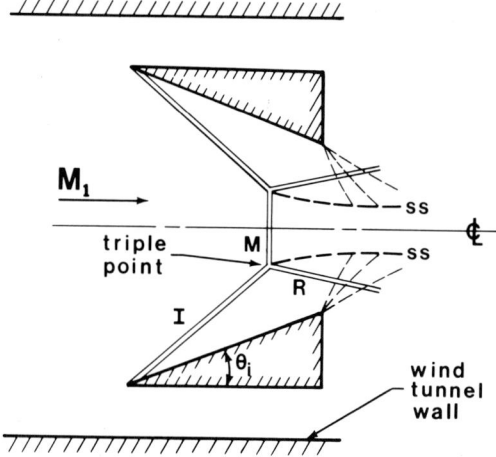

Fig. 19.23 Steady-flow wind-tunnel experiment.[2] The sketch, which shows an MR pattern, could equally well be for an RR pattern.

cannot be avoided. As a consequence of the interaction, the alteration of the flowfield is generally much too large to be neglected.

Reliable, steady-flow experiments can be performed, however, without the above interaction. This has been done at high Mach numbers by using a symmetric double-wedge configuration[2] in a wind tunnel, as illustrated in Fig. 19.23. This figure shows the MR pattern; we could, equally as well, have shown the RR pattern. In this type of experiment, the upstream flow must be uniform, and the wedges should be located outside the tunnel wall boundary layers. The wedges must be identical and accurately positioned. Furthermore, there should be a way of altering θ_i if a range of flows is to be investigated for a given value of M_1. When the foregoing provisos are satisfied, the centerline of the flow, in effect, becomes an inviscid wall. One further difficulty emerges when MR occurs; namely, the length and curvature of the Mach stem depends on the subsonic flow downstream of it. Hence, two MR experiments with the same γ, M_1, and θ_i may yield different results, for instance, owing to different wedge lengths. Any attempt at the computational modeling of an MR flow requires sufficient downstream data for fixing the subsonic part of the flow.

19.6 PSEUDO-STEADY FLOW OVER A PLANAR COMPRESSIVE RAMP

In contrast to the flow in Sec. 19.5, unsteady experiments are more readily performed. These are done by means of a shock tube with a rectangular cross section in which a ramp is placed. A reflection pattern results when the incident shock wave encounters the ramp.

Preliminary Remarks

In Fig. 19.24, we show the incident shock, before reflection, traveling with velocity u_I into region 1, which is quiescent. The flow behind the shock, in region 2, is uniform with the velocity u_2, which is in the positive x direction.

As shown in Chap. 9, we can introduce several Mach numbers; for instance,

$$M_I = \frac{u_I}{a_1} \qquad (19.29a)$$

$$M_2 = \frac{u_2}{a_2} = \frac{M_I^2 - 1}{\{1 + [(\gamma - 1)/2] M_I^2\}^{\frac{1}{2}} \{\gamma M_I^2 - [(\gamma - 1)/2]\}^{\frac{1}{2}}} \qquad (19.29b)$$

Throughout this section we will take γ, M_I, and ω as the principal dimensionless independent parameters, where ω is the angle of the ramp. The notation here differs from the unsteady shock analysis in Chap. 9. In particular, M_2 is associated with a coordinate system fixed to the wall, as indicated in Fig. 19.24. Hereafter, the coordinates x, y and velocity components u, v refer to the unsteady flow. We also note that M_2 can be subsonic or supersonic, as is evident from Table 19.4. Thus, M_2 increases with M_I, becoming sonic when $M_I = 2.068$ and reaches a maximum of 1.890 when $M_I = \infty$, where both values are for $\gamma = 1.4$. (Note that the trend of M_2 increasing with M_I is opposite to the trend for a steady normal shock wave.)

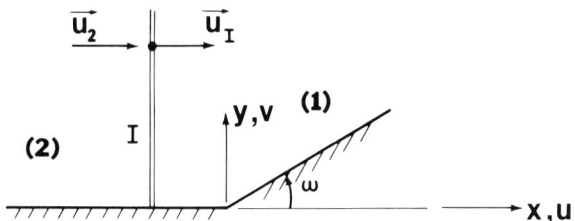

Fig. 19.24 Flow before the incident shock encounters a ramp of angle ω.

Table 19.4 Special M_2 Values

M_I	M_2
1	0
$\left\{ \frac{1}{4(2-\gamma)} \left[7 - \gamma + \left(\gamma^2 + 2\gamma + 17\right)^{\frac{1}{2}} \right] \right\}^{\frac{1}{2}}$	1
∞	$\left[\frac{2}{\gamma(\gamma-1)} \right]^{\frac{1}{2}}$

When ω is small, blast wave diffraction occurs, as mentioned in item (2) of the list in Sec. 19.1. At the other extreme, when ω is 90 deg, the shock reflects as in a shock tube [item (3) in Sec. 19.1]. Our interest is limited to intermediate values of ω for which four different flow patterns are possible. The primary basis of our pattern sketches are the fine shock tube, infinite-fringe interferometric photographs of Ref. 1. In the next subsection, we analyze regular reflection (RR), the simplest of the patterns.

Regular Reflection

The RR pattern tends to occur only at large values of ω and appears as in Fig. 19.25. In the x, y plane, the unsteady velocity is denoted as $\mathbf{q}$ with components u and v, Fig. 19.24. As before, region 1 is quiescent, while region 2 is a uniform flow region. Region 3 is also a uniform flow, and the reflected shock between the wall impingement point w and the sonic line is straight. Between the sonic line and the lower wall, the reflected shock is a curved bow shock wave that is perpendicular to the lower wall. The reflected shock propagates outward into the flow of region 2. Relative to point w, region 3 is supersonic. Region 5, behind the bow shock, is subsonic relative to point w. It is separated from region 3 by a curved sonic line.

The unsteady velocity components in region 2 are (see Appendix L)

$$\frac{u_2}{a_1} = \frac{2}{\gamma+1}\frac{M_I^2-1}{M_I}, \qquad v_2 = 0 \tag{19.30}$$

while the pressure is given by

$$\frac{p_2}{p_1} = \frac{2}{\gamma+1}\left(\gamma M_I^2 - \frac{\gamma-1}{2}\right) \tag{19.31}$$

Point w has the velocity components

$$u_w = u_I, \qquad v_w = u_I \tan \omega$$

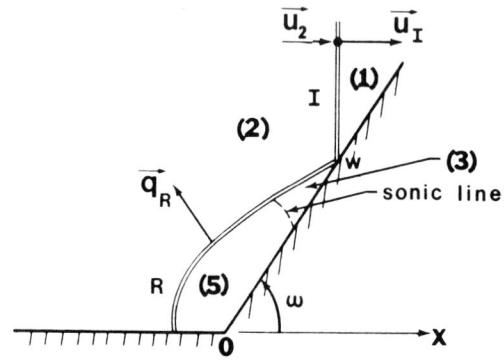

Fig. 19.25 Regular reflection flow pattern.

and a speed of

$$q_w = \left(u_w^2 + v_w^2\right)^{\frac{1}{2}} = \frac{u_I}{\cos \omega} \tag{19.32}$$

The flow in the immediate vicinity of w is investigated using a coordinate system whose origin is at w. The reason for the immediate vicinity limitation is clarified in the last subsection in this section. The flow is shown in Fig. 19.26. For the flow to be at w, we subtract u_w and v_w from all unsteady velocity components. The new, steady components are denoted with a caret,

$$\hat{u}_1 = -u_I, \qquad \hat{v}_1 = -u_I \tan \omega$$
$$\hat{u}_2 = -u_I + u_2, \qquad \hat{v}_2 = -u_I \tan \omega \tag{19.33}$$

For an RR pattern to exist, $\hat{M}_2$ must be supersonic. To impose this condition, the temperature ratio is first determined by means of

$$\frac{T_2}{T_1} = \left(\frac{2}{\gamma+1}\right)^2 \frac{\left\{1 + [(\gamma-1)/2]\hat{M}_1^2 \sin^2\beta_i\right\}\left\{\gamma \hat{M}_1^2 \sin^2\beta_i - [(\gamma-1)/2]\right\}}{\hat{M}_1^2 \sin^2\beta_i}$$

where β_i, the incident oblique shock angle, equals $(\pi/2) - \omega$. However, from Eq. (19.32) we have

$$\hat{M}_1 = \frac{M_I}{\cos \omega} \tag{19.34}$$

so that

$$\left(\frac{a_2}{a_1}\right)^2 = \frac{T_2}{T_1} = \left(\frac{2}{\gamma+1}\right)^2 \frac{\left\{1 + [(\gamma-1)/2]M_I^2\right\}\left\{\gamma M_I^2 - [(\gamma-1)/2]\right\}}{M_I^2}$$

$$\tag{19.35}$$

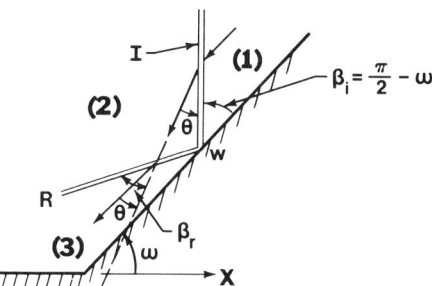

Fig. 19.26 Nomenclature for the flow in the vicinity of the impingement point w in RR.

which could have been obtained directly from Fig. 19.24. Hence, $\hat{M}_2$ is given by

$$\hat{M}_2^2 = \frac{\hat{u}_2^2 + \hat{v}_2^2}{a_2^2} = \frac{\{[(\gamma+1)/2] M_I^2/\cos\omega\}^2 - (\gamma M_I^2 + 1)(M_I^2 - 1)}{\{1 + [(\gamma-1)/2] M_I^2\}\{\gamma M_I^2 - [(\gamma-1)/2]\}} \tag{19.36}$$

where Eqs. (19.33) and (19.35) are utilized. The $\hat{M}_2 \geq 1$ condition for an RR pattern becomes

$$\cos^2\omega_1 = \frac{[(\gamma+1)/2] M_I^4}{\gamma M_I^4 + [(3-\gamma)/2] M_I^2 - 1} \geq \cos^2\omega \tag{19.37}$$

Thus, if the RR pattern is to occur, ω must exceed ω_1. The curve $\omega_1 = \omega_1(M_1)$ is shown in Fig. 19.27. This curve does not correspond to transition to a Mach reflection pattern, which occurs at a substantially larger value for ω. The turn angle θ across the incident and reflected shocks is given by the dot product relation

$$\cos\theta = \frac{\hat{u}_1 \hat{u}_2 + \hat{v}_1 \hat{v}_2}{[(\hat{u}_1^2 + \hat{v}_1^2)(\hat{u}_2^2 + \hat{v}_2^2)]^{\frac{1}{2}}}$$

$$= \frac{[(\gamma+1)/2] M_I^2 - (M_I^2 - 1)\cos^2\omega}{\{[(\gamma+1)/2]^2 M_I^4 - (M_I^2 - 1)(\gamma M_I^2 + 1)\cos^2\omega\}^{\frac{1}{2}}} \tag{19.38}$$

In steady flow, the slipstream is parallel to the wall just after transition. This is not the case here; thus, the mechanical equilibrium condition of Sec. 19.5 does not apply. Transition occurs when ω is decreased below a transition value that depends on γ and M_I. We envision a series of separate

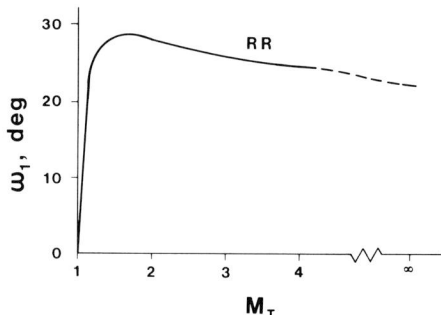

Fig. 19.27 Value of ω_1 when $\gamma = 1.4$. Regular reflection can only occur above the curve, which does not correspond to an RR → MR transition condition.

experiments with M_I fixed and with successively smaller ramp angles ω. As ω decreases, the sonic line in Fig. 19.25 moves closer to point w. Transition is believed to occur when the sonic line and point w first coincide.[2,16] A Mach stem can develop since the subsonic flow in region 5 enables the $\overline{0w}$ distance to function as a length scale for the stem.

Transition to a Mach reflection pattern thus occurs when $\hat{M}_3 = 1$. From Problem 12.1, when $\hat{M}_3 = 1$, we obtain

$$\cos \beta_r = \sin(\beta_r - \theta)\left[\cos\beta_r\sin(\beta_r - \theta) + \frac{\gamma+1}{2}\sin\theta\right] \quad (19.39)$$

where Fig. 19.26 defines the shock wave angle β_r. Consequently, Eqs. (19.36), (19.38), (19.39), and (12.21b) are four equations that can be used to eliminate $\hat{M}_2^2$, θ, and β_r. The result is a functional relation for the transition value ω_s as a function of γ and M_I.

A second possible transition condition would be the value ω_d for detachment of the reflected shock. While $\omega_s > \omega_d$, the difference is too small to resolve experimentally,[2] particularly in view of the boundary-layer interaction we describe shortly. However, the length scale argument is persuasive in both the steady and unsteady flows. Consequently, we consider RR → MR transition to occur at ω_s.

Aside from the transition condition, there are other fundamental differences between the steady and unsteady flows. In a steady flow, the slipstream is parallel to the wall at transition. As previously noted, this is not the case when the flow is unsteady, as discussed in the next subsection.

Another major difference is the manner in which the shock waves interact with the boundary layer. Downstream of the impingement point, there is a growing viscous boundary layer in the unsteady case. In a coordinate frame fixed with point w, the wall is moving faster than the inviscid flow above the viscous layer. Furthermore, the wall is cold relative to the shocked gas. Both effects result in a larger mass flux in the boundary layer than in the external inviscid flow. Consequently, the displacement thickness of the boundary layer is negative, as shown in Fig. 19.28. Thus, the velocity $\mathbf{q}_3$ is inclined toward the wall rather than parallel to it. In turn, this results in a decrease in the value of ω_s that is considerably larger than the difference $\omega_s - \omega_d$. Reference 2 first pointed out this phenomenon in the RR case.

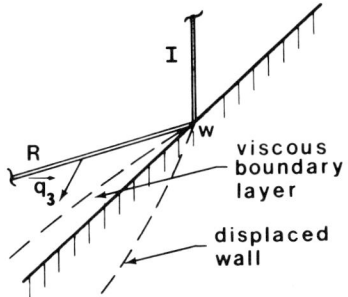

Fig. 19.28 Displaced wall for the RR case.

Their estimate, it should be noted, was based only on the wall-gas speed difference and did not include the wall-gas temperature difference. The effect ought to be appreciable when a Mach stem is present, although this also remains to be evaluated. (Evaluation of the influence of the boundary layer, for regular and Mach reflection patterns, presents an interesting challenge for computational fluid dynamicists.)

Mach Reflection

After transition, three different Mach reflection patterns are possible.[16] The simplest of these is the single Mach reflection (SMR) pattern, shown in Fig. 19.29, which we now discuss in some detail. The figure shows the incident, reflected, and Mach stem shocks, and the slipstream. Examination of published schlieren and interferometric photographs invariably show the Mach stem to be straight and normal to the wall. The slipstream is usually straight or nearly so.[1] (While the slipstream is approximately straight throughout most of its length, it often curls in the immediate vicinity of the wall.) A more important fact is that the slipstream is not nearly parallel to the ramp wall.

Unsteady reflection patterns, including RR, have been experimentally shown to be time-invariant. For example, the SMR pattern does not change with time. All lengths and curvatures simply increase in magnitude in direct proportion to time. Physically, this results from the absence of a length scale available to the flow before the incident shock encounters the ramp. After the encounter, the only new parameter is the angle ω. As a consequence, the unsteady flow is only a function of x/t and y/t; it is referred to as pseudo-steady. The invariance no longer holds if the ramp is not planar, as mentioned in item (4) of the list in Sec. 19.1.

Since u_I is a constant, the Mach stem speed q_M is also constant. This speed is related to u_I by a geometrical derivation based on Fig. 19.30. The path of the triple point is a straight line at an angle χ measured from the ramp's surface. Distance along this surface is denoted as z. Let ℓ_M be

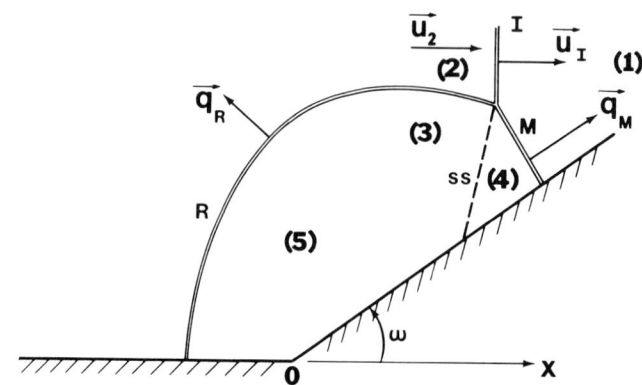

Fig. 19.29 Single Mach reflection (SMR) pattern.

FLOWS WITH SHOCK WAVES 383

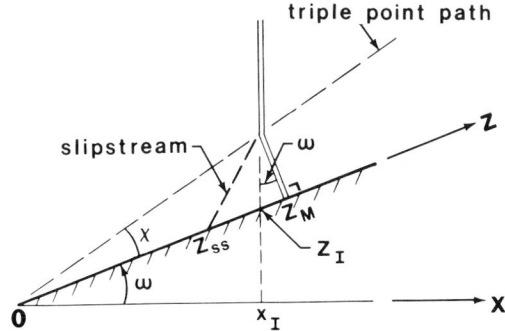

Fig. 19.30 Nomenclature sketch for the triple point and Mach stem.

the length of the Mach stem. From the figure, we obtain

$$\frac{\ell_M}{z_M} = \tan\chi, \qquad \frac{z_M - z_I}{\ell_M} = \tan\omega, \qquad \frac{x_I}{z_I} = \cos\omega$$

Next, we eliminate ℓ_M and z_I to yield

$$z_M = \frac{\cos\chi}{\cos(\omega + \chi)} x_I$$

and, consequently, the two speeds are related by

$$q_M = \frac{\cos\chi}{\cos(\omega + \chi)} u_I \qquad (19.40)$$

Thus, $q_M > u_I$, as one might expect. This is the case even in the $\chi \to 0$ limit, since the Mach stem travels a distance z_I in the same time that the incident shock travels the shorter distance x_I. Hence, q_M/u_I increases with $\omega + \chi$. The unsteady components of $\mathbf{q}_M$ are given by

$$u_M = q_M \cos\omega, \qquad v_M = q_M \sin\omega \qquad (19.41)$$

As already noted, region 2 behind the incident shock is a uniform flow whether it is subsonic or supersonic. The same is true for region 4 behind the Mach stem. Furthermore, since

$$M_M = \frac{q_M}{a_1} = \frac{\cos\chi}{\cos(\omega + \chi)} M_I \qquad (19.42)$$

region 4 is even more likely to be supersonic than is region 2. The velocity components in region 4 are given by

$$\frac{u_4}{a_1} = \frac{2}{\gamma + 1} \frac{M_M^2 - 1}{M_M} \cos\omega, \qquad \frac{v_4}{a_1} = \frac{2}{\gamma + 1} \frac{M_M^2 - 1}{M_M} \sin\omega \qquad (19.43)$$

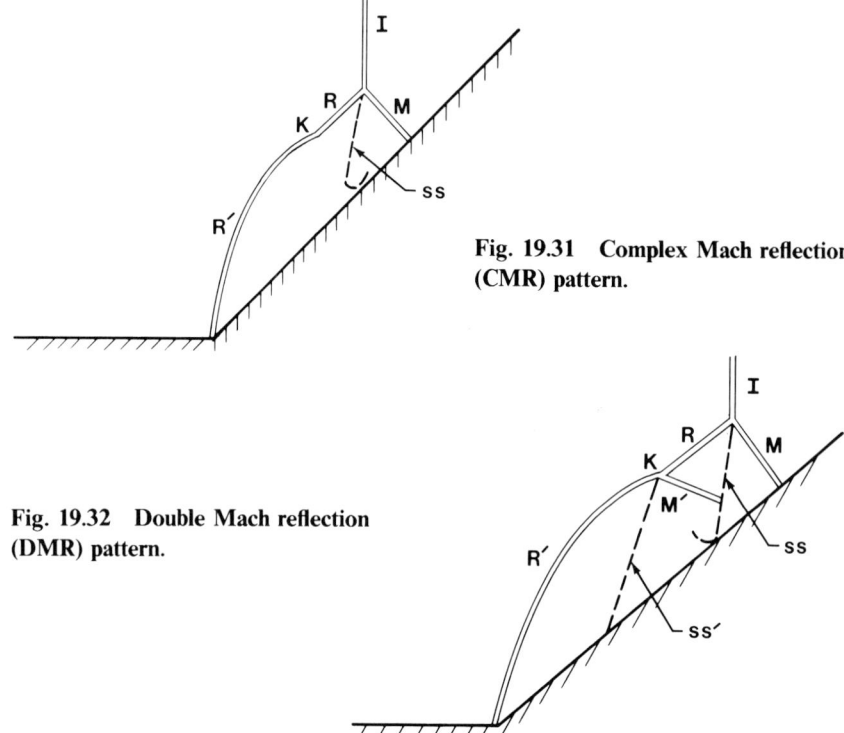

Fig. 19.31 Complex Mach reflection (CMR) pattern.

Fig. 19.32 Double Mach reflection (DMR) pattern.

We note that q_4 is parallel to the ramp and not to the slipstream. Nevertheless, no flow crosses the slipstream, which really functions as a contact surface. The slipstream moves with velocity q_4 and increases in length with time as region 4 accumulates more mass. The location, z_{ss}, where the (straight) slipstream touches the wall (see Fig. 19.30), moves with speed q_4. Consequently, z_{ss} is given by

$$\frac{z_{ss}}{z_M} = \frac{q_4}{q_M} = \frac{\left(u_4^2 + v_4^2\right)^{\frac{1}{2}}}{q_M} = \frac{2}{\gamma + 1} \frac{M_M^2 - 1}{M_M^2} \qquad (19.44)$$

in which Eqs. (19.43) are used. Since z_{ss} cannot be negative, a necessary condition for a Mach stem to exist is $M_M \geq 1$. Note that as $M_M \to \infty$, z_{ss}/z_M has an upper bound of $2/(\gamma + 1)$.

The third pattern is referred to as a complex Mach reflection (CMR) and appears as in Fig. 19.31. The reflected shock R between the incident one and point K, which is called the kink, is planar. Although the kink is not fully understood, it is thought to be the termination point for a sonic line.[17] The flow downstream of R is supersonic.

The final pattern is referred to as double Mach reflection (DMR) and appears as in Fig. 19.32. There are now two Mach stems, two slipstreams, and two triple points.

Pseudo-Steady Theory

Earlier the physical basis for referring to the flow as pseudo-steady was discussed. We also pointed out that important differences exist between a steady and an unsteady reflection process. These aspects are now further examined from a theoretical viewpoint. The pseudo-steady theory to be discussed was first given in Ref. 18. We start with the unsteady two-dimensional equations of motion in the x, y coordinate system of Fig. 19.24:

$$\frac{D\rho}{Dt} + \rho\left(\frac{\partial u}{\partial x} + \frac{\partial v}{\partial y}\right) = 0$$

$$\frac{Du}{Dt} + \frac{1}{\rho}\frac{\partial p}{\partial x} = 0$$

$$\frac{Dv}{Dt} + \frac{1}{\rho}\frac{\partial p}{\partial y} = 0 \quad (19.45)$$

$$\frac{D(p/\rho^\gamma)}{Dt} = 0$$

where

$$\frac{D}{Dt} = \frac{\partial}{\partial t} + u\frac{\partial}{\partial x} + v\frac{\partial}{\partial y}$$

and the energy equation is replaced by isentropic flow along a streamline. Except for time, we introduce the following nondimensional variables:

$$\xi = \frac{x}{a_1 t}, \quad \eta = \frac{y}{a_1 t}, \quad \zeta = t$$

$$U = \frac{u}{a_1} - \xi, \quad V = \frac{v}{a_1} - \eta \quad (19.46)$$

$$P = \frac{p}{\rho_1 a_1^2}, \quad R = \frac{\rho}{\rho_1}$$

Equations (19.45) thus become

$$\zeta\frac{\partial R}{\partial \zeta} + \frac{\partial(RU)}{\partial \xi} + \frac{\partial(RV)}{\partial \eta} = -2R$$

$$\zeta\frac{\partial U}{\partial \zeta} + U\frac{\partial U}{\partial \xi} + V\frac{\partial U}{\partial \eta} + \frac{1}{R}\frac{\partial P}{\partial \xi} = -U$$

$$\zeta\frac{\partial V}{\partial \zeta} + U\frac{\partial V}{\partial \xi} + V\frac{\partial V}{\partial \eta} + \frac{1}{R}\frac{\partial P}{\partial \eta} = -V \quad (19.47)$$

$$\zeta\frac{\partial(P/R^\gamma)}{\partial \zeta} + U\frac{\partial(P/R^\gamma)}{\partial \xi} + V\frac{\partial(P/R^\gamma)}{\partial \eta} = 0$$

Based on the physical reasoning in the preceding subsection, we set the unsteady leftmost terms equal to zero. The remaining terms on the left side are identical to those for a steady two-dimensional flow in the ξ, η plane. Equations (19.47), however, are not appropriate to the steady flow of Sec. 19.5, due to the terms on the right side. These terms correspond to a density sink in the continuity equation and to a variable nonconservative body force in the momentum equations. The sink and body force terms occur in the ξ, η plane, which is not a physical one. Thus, Eqs. (19.47) are not in violation of the conservation laws.

Transformation (19.46) must also be applied to the shock waves, slipstreams, regions 1 and 2, etc. For instance, in region 1, we have

$$U_1 = -\xi, \quad V_1 = -\eta, \quad R_1 = 1, \quad P_1 = 1/\gamma \qquad (19.48)$$

while in region 2

$$U_2 = \frac{2}{\gamma+1} \frac{M_I^2 - 1}{M_I} - \xi$$

$$V_2 = -\eta$$

$$R_2 = \frac{\gamma+1}{2} \frac{M_I^2}{1 + [(\gamma-1)/2] M_I^2} \qquad (19.49)$$

$$P_2 = \frac{2}{\gamma(\gamma+1)} \left(\gamma M_I^2 - \frac{\gamma-1}{2} \right)$$

The location of the incident shock that separates the two regions is given by

$$\xi_I = M_I$$

In the ξ, η plane, regions 1 and 2 are no longer uniform. Furthermore, neither the streamlines nor the Mach number are invariant. Thus, in the unsteady plane, we have

$$M^2 = \frac{u^2 + v^2}{(\gamma p/\rho)} = \frac{R}{\gamma P} (U + \xi)^2 (V + \eta)^2$$

while in the steady ξ, η plane

$$\hat{M}^2 = \frac{R}{\gamma P} (U^2 + V^2)$$

Equations (19.47), with the unsteady terms deleted, hold for RR and all MR patterns. As noted, they must be supplemented by wall, slipstream, and shock wave conditions. Properly supplemented, the equations ought to determine the conditions, i.e., γ, M_I, and ω, for which a given pattern, or patterns, can exist.

The steady version of Eqs. (19.47) can be solved by the method of characteristics for those regions where $\hat{M}$ is supersonic. (In the unsteady plane, the equations are hyperbolic everywhere, regardless of the Mach number. This is not the case in the steady plane.) The terms on the right side of Eqs. (19.47) alter the compatibility equations, but they do not change the equations for the characteristic lines. Further discussion of the characteristic form of these equations can be found in Ref. 19.

Our discussion illustrates the difficulty involved in trying to interpret an unsteady motion as a steady one. The point here is the frequent but sometimes unwarranted comparisons made with the steady flow of Sec. 19.5, for which Eqs. (19.47)–(19.49) do not apply.

19.7 SHOCK WAVE INTERFERENCE

High-speed aerodynamics of complex configurations became important in the 1960s. These configurations involved fins, externally mounted structures, inlet cowls, etc. It was observed in both wind-tunnel and flight testing that small regions with anomalously high heat-transfer rates could occur with an occasional magnitude sufficient to cause structural failure.

In a steady flow, a variety of circumstances, such as boundary-layer transition or the reattachment of a detached boundary layer, can increase the heat-transfer rate. However, a third mechanism results when one shock wave impinges on another, as illustrated in Fig. 19.33. When a relatively weak fuselage bow shock impinges on the upstream part of the fin's bow shock, the result is a very high rate of heat transfer near the fin's leading edge.

This section is based on the brilliant experimental and analytical work of Edney.[20] Much of his analysis stems from interpretations of the flow, utilizing shock-polar diagrams. Because of space limitations, this presentation will focus on a qualitative discussion of a portion of his experimental results. For more details, consult Refs. 20–23.

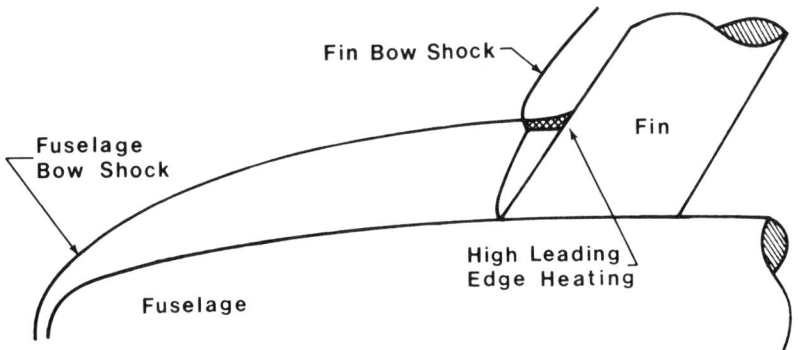

Fig. 19.33 Example of shock impingement heating (after Edney[20]).

One of the configurations Edney examined is a hemisphere-cylinder model, shown in Fig. 19.34. A two-dimensional wedge, whose leading edge is perpendicular to the freestream velocity V_∞ generates a planar shock wave. The strength of the shock depends on the wedge angle θ, which is adjustable. The model is mounted on a movable support such that it can be fully immersed in the uniform flowfield with Mach number M_1. As the support is extended upward, the wedge shock starts to interfere with the upper part of the bow shock of the model. Interference terminates when the model is fully extended into the freestream. Edney made three principal diagnostic measurements while the model was fully immersed in the wedge's shock layer and during interference. These were schlieren photographs, heat-transfer, and static-pressure measurements that were made at various locations on the surface of the model. For a given model shape, the results of these measurements are determined by θ, M_∞, and y. The y distance denotes the location where the wedge shock meets the bow shock, relative to the model's centerline in a plane perpendicular to the wedge. In Fig. 19.34, y has a negative value. It should be noted that all lengths can be normalized by the hemisphere's radius.

As the model progresses upward, six different steady-flow patterns evolve. Figures 19.35–19.40 illustrate the patterns for air with $M_\infty = 4.6$ and $\theta = 10$ deg. They are in a plane perpendicular to the surface of the wedge that contains the model's centerline. In this plane the flow is essentially two-dimensional. Thus, the drawings are based on two-dimensional flow calculations.

Type I interference, Fig. 19.35, occurs when the wedge shock meets the bow shock well below the lower sonic line, i.e., y has its most negative value for an interference pattern. The flow in regions 1–4 is supersonic, and a slipstream or free shear layer separates regions 2 and 4. All shock waves

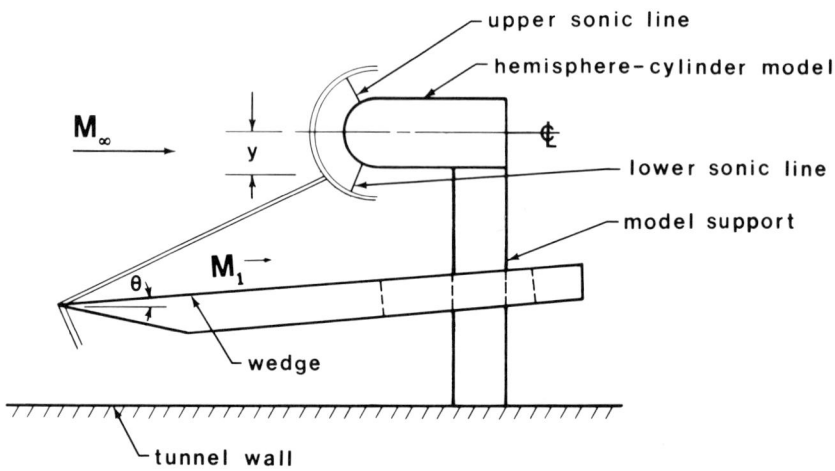

Fig. 19.34 Sketch of Edney's experimental setup for shock interference investigation with a sphere-cylinder model.

FLOWS WITH SHOCK WAVES

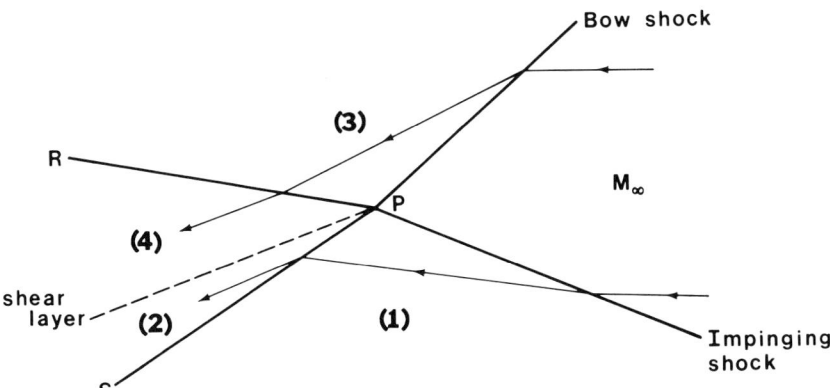

Fig. 19.35 Type I interference pattern (after Edney[20]).

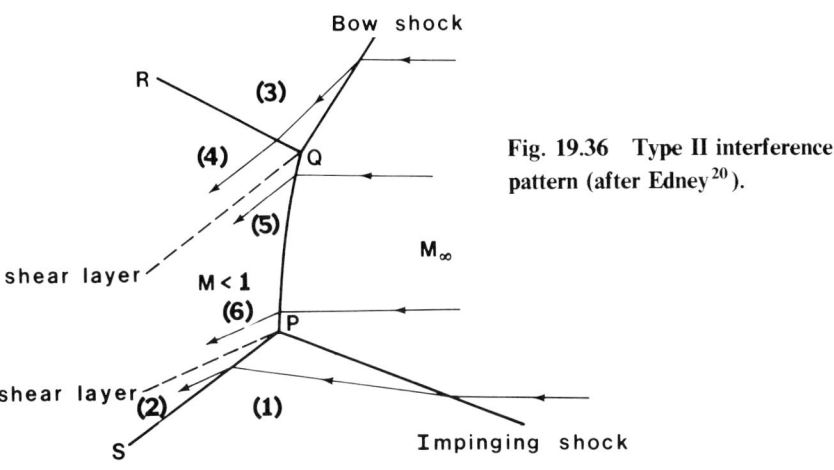

Fig. 19.36 Type II interference pattern (after Edney[20]).

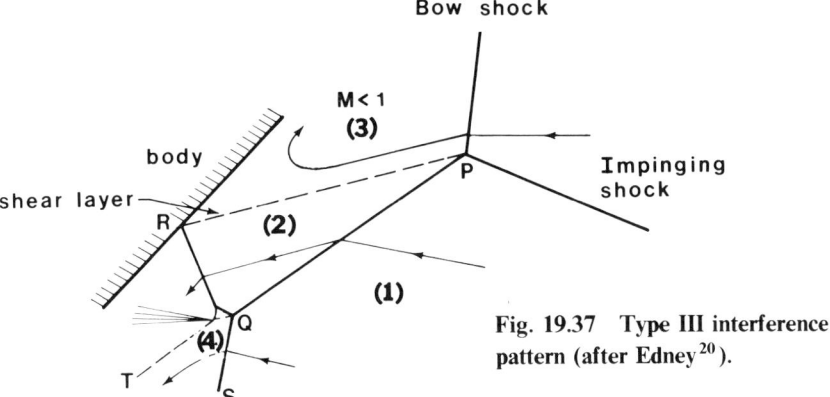

Fig. 19.37 Type III interference pattern (after Edney[20]).

are of the weak-solution variety. Aside from the fact that the shear layer is not aligned with V_∞, the flow corresponds to two steady-flow RR patterns.

The shear layer does not meet the model's surface and does not alter the heat-transfer rate. However, the shock PR can impinge on the surface of the model and will result in shock wave-boundary layer interaction with a consequent increase in the heat-transfer rate in the interaction region. It should be noted that the interaction process can take several forms. In its weakest form, a laminar boundary layer may be tripped into transition. In its strongest form, the boundary layer may separate from the model with subsequent reattachment or impingement and a high heat-transfer rate.

In type II interference, Fig. 19.36, the wedge shock meets the bow shock just below the lower sonic line. Thus, regions 1–4 are supersonic, while from regions 5 to 6 the flow is subsonic. Aside from asymmetry, the flow corresponds to two steady-flow MR patterns. As in Mach reflection, the length of the strong-solution shock between the P and Q triple points depends on the geometry of the body. As in type I interference, the QR

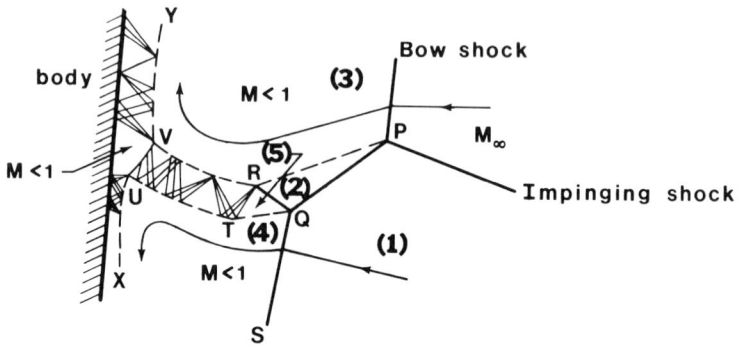

Fig. 19.38 Type IV interference pattern (after Edney[20]).

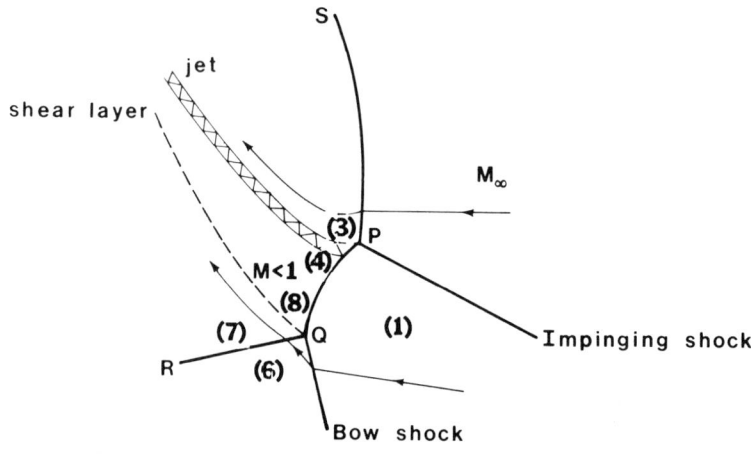

Fig. 19.39 Type V interference pattern (after Edney[20]).

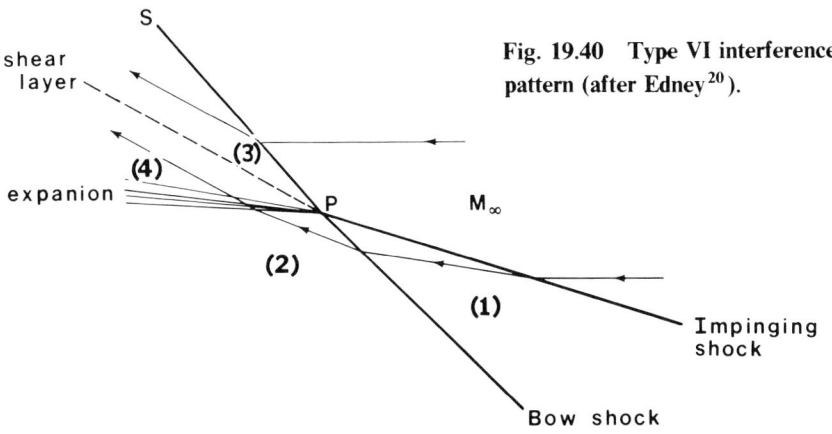

Fig. 19.40 Type VI interference pattern (after Edney[20]).

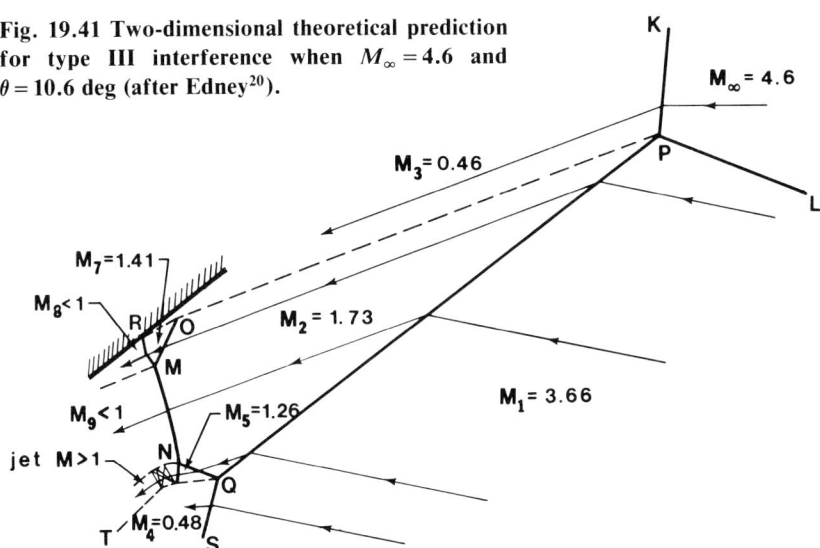

Fig. 19.41 Two-dimensional theoretical prediction for type III interference when $M_\infty = 4.6$ and $\theta = 10.6$ deg (after Edney[20]).

shock causes shock wave-boundary layer interaction. The shear layer from Q may or may not meet the body. If it does, it causes little change in the heat-transfer rate, since it is weak, i.e., the speed change across the shear layer is small.

Type III interference, Fig. 19.37, occurs when the wedge shock meets the bow shock somewhat above the lower sonic line. As a consequence, region 3 is subsonic, while region 2 is supersonic. The shear layer from the triple point P is thus relatively strong. For the supersonic flow in region 2 to become parallel to the body, a strong-solution shock RQ is required. The flow pattern in the vicinity of RQ is somewhat oversimplified. A more realistic pattern is shown in Fig. 19.41, including various Mach numbers.

There is an increase in the wall pressure at point R and an increase in the heat-transfer rate, which occurs over a distance approximately equal to OR. Another interesting difference between Figs. 19.37 and 19.41 is the replacement of the shear layer T by an embedded supersonic jet downstream of the weak-solution shock located between points N and Q. The jet is bounded, on both sides, by shear layers. Neither this jet nor the shear layer from point M meets the body.

As in type III, type IV interference occurs when the wedge shock impinges on the subsonic part of the model's shock layer. However, if the inclination angle between $\mathbf{V}_2$ and the body (Fig. 19.37) is too large, an RQ shock wave is not possible. A dramatic change in the pattern then occurs, as shown in Fig. 19.38. Downstream of the weak-solution shock RQ, a relatively large embedded supersonic jet develops. Regions 3 and 4 are subsonic, and the jet is bounded by a shear layer on both sides. The flow in the jet goes through alternating expansion and compression regions, as shown in Fig. 19.42. This flow is analogous to the multiple cell flow in a jet, as discussed in Sec. 19.4. The reason for the alternating Mach number pattern is the assumption that the pressures in regions 3 and 4 are constant. This assumption is reasonable since both regions are subsonic. Since p_4 is computed to be greater than p_3, $M_6 > M_5$, and the jet must curve upward in a manner analogous to the free-vortex aerowindow of Chap. 18. If p_3 exceeds p_4, the jet would curve downward.

Near the wall, Fig. 19.38, the jet goes through a strong-solution shock UV. As the flow from the jet proceeds around the body, it accelerates to supersonic speeds, but then decelerates via further shock waves, such as the one near point Y.

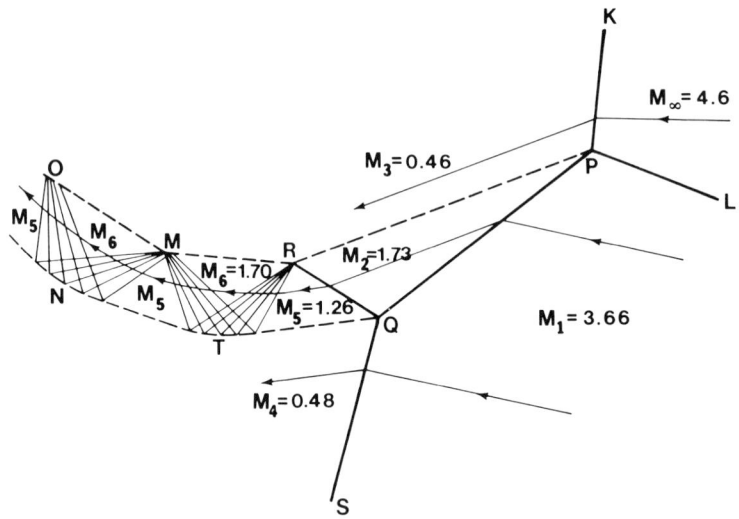

Fig. 19.42 Two-dimensional theoretical prediction for type IV interference when $M_\infty = 4.6$ and $\theta = 10.6$ deg (after Edney[20]).

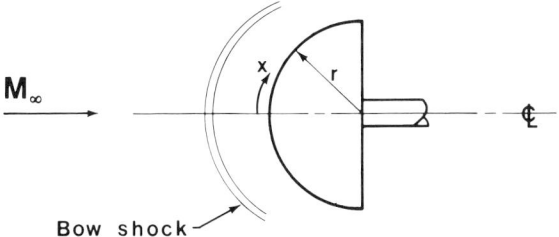

Fig. 19.43 Flow around a hemisphere showing the nomenclature.

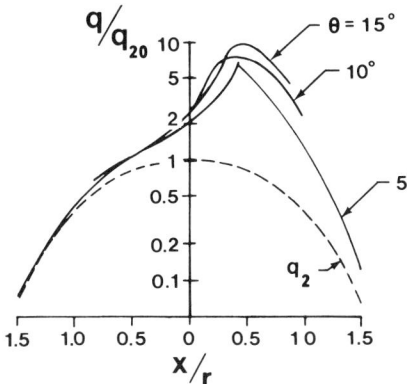

Fig. 19.44 Normalized heat-transfer rate of a hemisphere when $M_\infty = 4.6$ (after Edney[20]).

As the impingement point P moves higher on the bow shock toward the upper sonic line, the jet becomes narrower and tends to curl more. This result stems from an increase in the value of $|p_4 - p_3|$.

Type V interference, Fig. 19.39, occurs when the wedge shock meets the bow shock just above the sonic line. The pattern is similar to type II interference, except that a narrow jet rather than a shear layer separates regions 3 and 4. Neither the shear layer from Q nor the jet impinges on the body. The shock RQ will impinge with a result similar to that in type II interference.

Finally, type VI interference, Fig. 19.40, occurs when the wedge shock impinges on the bow shock well above the upper sonic line. Neither the shear layer nor the expansion fan has any appreciable effect on the heat-transfer rate. In fact, neither disturbance need impinge on the body. The flow about the model corresponds to that in a uniform freestream with a Mach number M_1.

Let us consider the heat-transfer rate q on the hemisphere shown in Fig. 19.43 as a function of position. Experimental data[20] for three wedge angles are shown in Fig. 19.44. The q_2 curve is the heat-transfer rate for a hemisphere with no interference, while q_{20} is the no-interference value at the stagnation point.

Peak heating occurs when the wedge shock is strongest (i.e., $\theta = 15$ deg) and when it impinges on the bow shock at a point slightly below the model's symmetry axis. The peak heat-transfer rate, which exceeds q_{20} by an order of magnitude, is associated with type IV interference. In particular, it is associated with the stagnation point region downstream of the UV shock in Fig. 19.38. Still higher normalized heat-transfer rates are possible at a higher freestream Mach number and at lower γ values.

Figure 19.45 shows static pressure measurements[20] on the lower surface of the hemisphere. The location of the peak pressure coincides well with that for the peak heat-transfer rate and, thus, with type IV interference. The p_1 and p_2 curves refer to noninterference flows with M_1 and M_∞, respectively, as the freestream Mach numbers. The p_{20} value is the stagnation point pressure with no interference when M_∞ is the freestream Mach number. This pressure is also a pitot-tube pressure.

The fact that p_1 and the interference pressures can considerably exceed p_{20} may be surprising, but it is explainable. For this, we compare two pitot pressure measurements, as shown in Fig. 19.46. We assume that each pitot

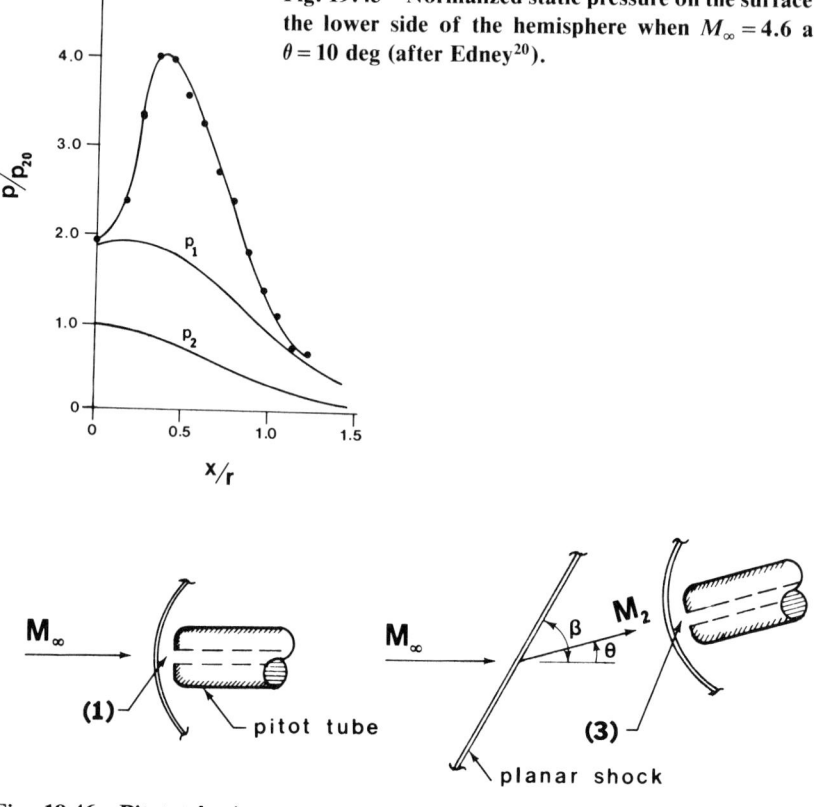

Fig. 19.45 Normalized static pressure on the surface of the lower side of the hemisphere when $M_\infty = 4.6$ and $\theta = 10$ deg (after Edney[20]).

Fig. 19.46 Pitot tube in a supersonic flow with and without an upstream planar shock.

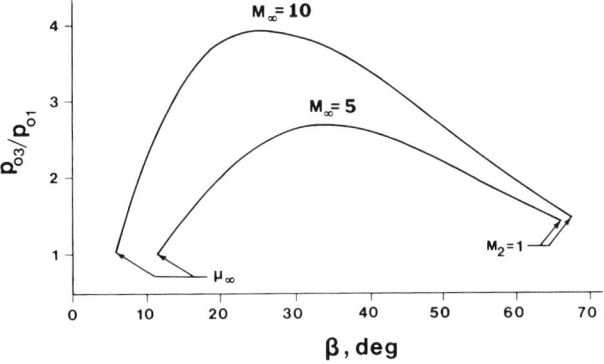

Fig. 19.47 Stagnation pressure ratio vs the planar shock wave angle β for $\gamma = 1.4$.

tube is aligned with the oncoming local flow and that M_2 is supersonic. For the simple flow on the left, the pitot pressure is

$$\frac{p_{01}}{p_\infty} = \left(\frac{\gamma+1}{2}\right)^{(\gamma+1)/(\gamma-1)} \left\{\frac{M_\infty^{2\gamma}}{\gamma M_\infty^2 - [(\gamma-1)/2]}\right\}^{1/(\gamma-1)} \quad (19.50)$$

As in Sec. 19.3, we take β as the independent parameter for the flow on the right. Thus, for region 2, we have

$$\tan\theta = \frac{1}{\tan\beta}\frac{M_\infty^2\sin^2\beta - 1}{1 + \{[(\gamma+1)/2] - \sin^2\beta\}M_\infty^2} \quad (19.51)$$

$$M_2^2 = \frac{\cos\beta}{\sin(\beta-\theta)\{\cos\beta\sin(\beta-\theta) + [(\gamma+1)/2]\sin\theta\}} \quad (19.52)$$

$$\frac{p_2}{p_\infty} = \frac{2}{\gamma+1}\left(\gamma M_\infty^2 \sin^2\beta - \frac{\gamma-1}{2}\right) \quad (19.53)$$

where Eq. (19.51) is Eq. (12.21b) and Eq. (19.52) comes from Problem 12.1. For this flow, the pitot pressure can be written as

$$\frac{p_{03}}{p_\infty} = \frac{p_2}{p_\infty}\frac{p_{03}}{p_2} = \left(\frac{\gamma+1}{2}\right)^{2/(\gamma-1)}\left(\gamma M_\infty^2\sin^2\beta - \frac{\gamma-1}{2}\right)$$

$$\times \left(\frac{M_2^{2\gamma}}{\gamma M_2^2 - [(\gamma-1)/2]}\right)^{1/(\gamma-1)} \quad (19.54)$$

where θ and M_2 are given by Eqs. (19.51) and (19.52). Equations (19.50) and (19.54) provide p_{03}/p_{01}, which appears in Fig. 19.47 for several freestream Mach numbers. Hence, the flow downstream of a weak-solution planar shock has a larger pitot pressure than one without the planar shock.

References

[1] Deschambault, R. L. and Glass, I. I., "An Update on Non-Stationary Oblique Shock-Wave Reflections: Actual Isopycnics and Numerical Experiments," *Journal of Fluid Mechanics*, Vol. 131, June 1983, pp. 27–57.

[2] Hornung, H. G., Oertel, H., and Sandeman, R. J., "Transition to Mach Reflexion of Shock Waves in Steady and Pseudosteady Flow With and Without Relaxation," *Journal of Fluid Mechanics*, Vol. 90, Feb. 1979, pp. 541–560.

[3] Lighthill, M. J., *Proceedings of the Royal Society of London*, Ser. A, Vol. 198, 1949, pp. 454–470.

[4] Srivastava, R. S. and Chopra, M. G., "Diffraction of Blast Wave for the Oblique Case," *Journal of Fluid Mechanics*, Vol. 40, March 1970, pp. 821–831.

[5] Ben-Dor, G., Takayama, K., and Kawauchi, T., "The Transition from Regular to Mach Reflexion and from Mach to Regular Reflexion in Truly Non-Stationary Flows," *Journal of Fluid Mechanics*, Vol. 100, Sept. 1980, pp. 147–160.

[6] Itoh, S., Okazaki, N., and Itaya, M., "On the Transition Between Regular and Mach Reflection in Truly Non-Stationary Flows," *Journal of Fluid Mechanics*, Vol. 108, July 1981, pp. 383–400.

[7] Takayama, K. and Sasaki, M., "Effects of Radius of Curvature and Initial Angle on the Shock Transition over Concave or Convex Walls," *Report of the Institute of High Speed Mechanics*, Tōhoku University, Japan, Vol. 46, 1983, pp. 1–30.

[8] Polachek, H. and Seegar, R. J., "On Shock-Wave Phenomena; Refraction of Shock Waves at a Gaseous Interface," *Physics Review*, Vol. 84, Dec. 1951, pp. 922–929.

[9] Jahn, R. G., "The Refraction of Shock Waves at a Gaseous Interface," *Journal of Fluid Mechanics*, Vol. 1, Nov. 1956, pp. 457–489.

[10] Pack, D. C., "The Reflexion and Diffraction of Shock Waves," *Journal of Fluid Mechanics*, Vol. 18, April 1964, pp. 549–577.

[11] Salas, M. D. and Morgan, B. D., "Stability of Shock Waves Attached to Wedges and Cones," *AIAA Journal*, Vol. 21, Dec. 1983, pp. 1611–1617.

[12] Ames Research Staff, *Equations, Tables, and Charts for Compressible Flow*, NACA Rept. 1135.

[13] Oswatitsch, K., *Gas Dynamics*, Academic Press, Orlando, FL, 1956, pp. 367–372.

[14] Emanuel, G., "Near-Field Analysis of a Compressive Supersonic Ramp," *The Physics of Fluids*, Vol. 25, July 1982, pp. 1127–1133.

[15] Emanuel, G., "Numerical Method and Results for Inviscid Supersonic Flow over a Compressive Ramp," *Computers and Fluids*, Vol. 11, No. 4, 1983, pp. 367–377.

[16] Pack, D. C., "On the Formation of Shock-Waves in Supersonic Gas Jets," *Quarterly Journal of Mechanical and Applied Mathematics*, Vol. 1, 1948, pp. 1–17.

[17] Ben-Dor, G. and Glass, I. I., "Domains and Boundaries of Non-Stationary Oblique Shock-Wave Reflexions. 1. Diatomic Gas," *Journal of Fluid Mechanics*, Vol. 92, June 1979, pp. 459–496.

[18] Jones, D. M., Moira, P., Martin, E., and Thornhill, C. K., "A Note on the Pseudo-Stationary Flow Behind a Strong Shock Diffracted or Reflected at a Corner," *Proceedings of the Royal Society of London*, Ser. A, Vol. 209, 1951, pp. 238–248.

[19] Meyer, R. E., "The Method of Characteristics," *Modern Developments in Fluid Dynamics*, Vol. I, edited by L. Howarth, Oxford University Press, New York, 1953, Chap. III, Sec. 11.

[20] Edney, B., "Anomalous Heat Transfer and Pressure Distributions on Blunt Bodies at Hypersonic Speeds in the Presence of an Impinging Shock," The Aeronautical Research Institute of Sweden, Stockholm, Rept. 115, 1968.

[21] Philpott, D. R. and Zhao, J. Z., "Interference Effects Between Spherically Blunted Cylinders at M = 2.5 and 1.5," *Proceedings of AIAA Atmospheric Flight Mechanics Conference*, Seattle, WA, Aug. 1984, pp. 231–239.

[22] Hains, F. D. and Keyes, J. W., "Shock Interference Heating in Hypersonic Flows," *AIAA Journal*, Vol. 10, Nov. 1972, pp. 1441–1447.

[23] Edney, B. E., "Effects of Shock Impingement on the Heat Transfer Around Blunt Bodies," *AIAA Journal*, Vol. 6, Jan. 1968, pp. 15–21.

Problems

19.1 For single Mach reflection (SMR), determine the angle α between the slipstream and Mach stem as a function of γ, M_I, ω, and χ. Keeping γ, M_I, and ω fixed, sketch how α varies with χ.

19.2 We have a steady flow with $M_1 = 3$, $\gamma = 1.4$, and $\theta_i = 30$ deg (see Fig. 19.13). If the pattern is RR, determine M_2 and M_3. If the pattern is MR, Fig. 19.22, determine M_2, M_3, and M_4. In the MR case, the slipstream is not parallel to the wall, and the M_i are to be evaluated in the immediate vicinity of the triple point.

19.3 Consider RR in a steady flow as shown in Figs. 19.13 and 19.15. Let $M_1 \to \infty$ and $\theta_i \to 0$ in such a way that

$$\theta_i M_1 = K_{\theta_i} = \mathcal{O}(1)$$

Determine an explicit, hypersonic relation of the form

$$K_{\theta_r} = \theta_r M_2 = f(K_{\theta_i}; \gamma)$$

Computational Problems:

19.4 Determine the mechanical equilibrium and sonic transition conditions for steady two-dimensional flow. The parameter space is

$$\gamma = 1.3, 1.4, 5/3$$

$$M_1 = 1.25\,(0.25)\,2.5\,(0.5)\,6$$

In particular, for a given value of γ and M_1, determine P_2, P_3, M_2, M_3, and θ for the sonic and mechanical equilibrium transitions.

19.5 Utilize Eqs. (12.21b), (19.36), (19.38), and (19.39) to determine the transition condition RR $\to$ MR; i.e., for a given value of γ and M_I, determine ω_s. Tabulate ω_s for

$$\gamma = 1.3, 1.4, 5/3$$

$$M_I = 2(1)8$$

along with β, θ, and $\hat{M}_2$ at transition. Compare inequality (19.37) with your results.

19.6 Compute p_{03}/p_{01} vs β, Eq. (19.54), for

$$\gamma = 1.3, 1.4, 5/3$$

$$M_\infty = 2(1)10$$

20. WAVERIDER AERODYNAMICS

20.1 PRELIMINARY REMARKS

The design of supersonic manned vehicles and guided missiles is a complex task. In general, these vehicles are maneuverable, and their shapes and sizes vary widely according to their functions. It is desirable for these vehicles to have high lift and low drag and to provide useful internal storage space. For example, a bullet-shaped body, which is a hemisphere or a cone followed by a circular cylinder, has good volumetric efficiency but no lift at zero angle of attack.

Any supersonic vehicle that is maneuverable and, because of range requirements, has a propulsion system benefits from lift. If the amount of aerodynamic lift in level flight is sufficient to overcome the downward gravitational force, then the thrust of the propulsion system need only overcome the drag.

A class of high lift vehicles, called waveriders, has been studied since the 1960s. This investigation started when it was realized that a conventional space shuttle configuration did not provide a great deal of lift.[1] Any delta-shaped vehicle with a nearly flat bottom, such as that of the space shuttle, allows considerable cross flow, as illustrated in Fig. 20.1(a). The streamlines are curved away from the centerline, where the pressure is a maximum. As a result of the cross flow, the pressure rapidly decreases away from the centerline. If a diverging cross flow can be prevented, as indicated by the parallel streamlines in Fig. 20.1(b), then the lifting pressure remains at a relatively high value for the entire undersurface.

A blunt-nosed vehicle has a bow shock wave whose strength decays rapidly with downstream distance. The shock is detached and, at an angle of attack, a diverging cross flow is unavoidable. The significant feature of a waverider is a body configuration that allows for an attached shock wave that can prevent this type of cross flow.

At supersonic speeds, heat transfer is always large and of major importance. A waverider body requires sharp leading edges that have a high heat-transfer rate relative to the rest of the surface. Nevertheless, the overall heat-transfer rate is decreased[1] as a result of three factors. First, the upstream stagnation point associated with a bow shock, where the heat-transfer rate is very high, is not present on a waverider. Second, the flow speed on the lower surface of a waverider is less than that of a conventional body. (Cross flow increases the speed.) The lower speed reduces the convective heat-transfer rate. Third, because of its greater lift per unit area,

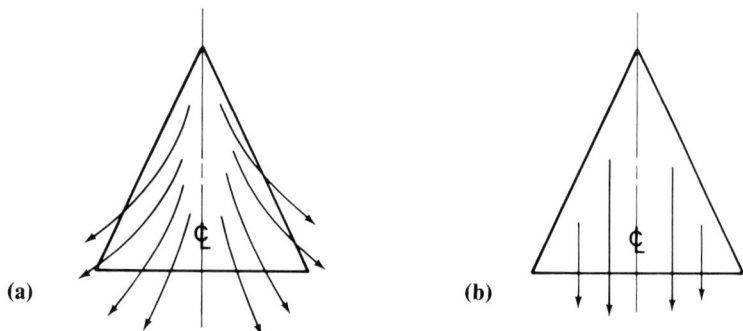

Fig. 20.1 Bottom view of two vehicles. In (a) the surface is flat or convex, while (b) is a waverider. The lines with arrows are streamlines.

the lifting surface area of a waverider can be reduced from that of a conventional body.

Waverider theory is an inverse approach in which the body is found for a known flowfield. It generally involves a much simpler aerodynamic computation than finding the flowfield for a prescribed body. This is true for the two basic types of waveriders. The first is a caret-shaped waverider initially proposed by Nonweiler,[2] which is the subject of the next section. Section 20.3 investigates the lift and drag relations of Kim[3] for an arbitrary waverider shape. These relations are used in Sec. 20.4, where we analyze waveriders that are based on axisymmetric flow about a cone.

Throughout this chapter, the usual gasdynamic assumptions apply, e.g., inviscid, adiabatic flow of a perfect gas. In addition, our interest is often restricted to high supersonic freestream Mach numbers ($M_\infty \geq 4$). Many effects of importance are neglected. These include nonzero angle of attack, ellipticity of a conical body, longitudinal curvature, configuration optimization, etc. These effects have been well analyzed by Rasmussen and his colleagues.[3-7] A comprehensive analysis of potentially attractive aerodynamic configurations has been summarized by Krieger.[8]

In view of the above restrictions, our presentation is of an introductory nature. It is meant to complement the shock-expansion theory of Sec. 6.2. As in that section, emphasis is on analytical techniques for determining lift and drag.

20.2 CARET-SHAPED WAVERIDERS

We first consider two-dimensional flow about an infinite wedge whose cross-sectional shape is a right triangle (Fig. 20.2). The upper surface does not generate a shock or an expansion, since it is aligned with the freestream. The pressure on this surface is therefore p_∞. The lower surface produces an attached planar shock, with a compressive pressure p_c, which is given by

$$\frac{p_c}{p_\infty} = \frac{2}{\gamma + 1}\left(\gamma M_\infty^2 \sin^2\beta - \frac{\gamma - 1}{2}\right) \qquad (20.1)$$

where β is the shock wave angle. The flow in the base region, denoted by a subscript b, is bordered by two Prandtl–Meyer expansions. Consequently, the base pressure p_b is below p_∞.

There is a wake region downstream of the base whose analysis is beyond the scope of this book. Nevertheless, measurements at $M_\infty = 6$ downstream of a symmetric wedge,[9] show that p_b is closer to zero than to p_∞. The value of p_b, however, is Reynolds number and body-configuration dependent. For waveriders, p_b affects only the drag, which is a minimum if p_b is set equal to p_∞. However, at high Mach numbers for a slender body, the magnitude of the base drag is small compared to that of the compression surfaces. Hence, the value chosen for p_b is of little consequence and, following Rasmussen,[4] we set $p_b = p_\infty$. As will be evident, this choice simplifies the analysis.

We develop lift and drag formulas for the wedge of Fig. 20.2, following closely the procedures of shock-expansion theory given in Sec. 6.2. The lift and drag, per unit span, are seen to be

$$L = p_c \ell_c \cos\theta - p_\infty \ell$$

$$D = p_c \ell_c \sin\theta - p_\infty \ell_b$$

where θ is the wedge angle. The compression and base lengths are given by

$$\ell_c = \ell/\cos\theta, \qquad \ell_b = \ell \tan\theta$$

so that

$$L = \left(\frac{p_c}{p_\infty} - 1\right) p_\infty \ell, \qquad D = L \tan\theta$$

The lift and drag coefficients thus become

$$C_L = \frac{2L}{\gamma p_\infty M_\infty^2 \ell} = \frac{2}{\gamma M_\infty^2}\left(\frac{p_c}{p_\infty} - 1\right) \qquad (20.2\text{a})$$

$$C_D = C_L \tan\theta \qquad (20.2\text{b})$$

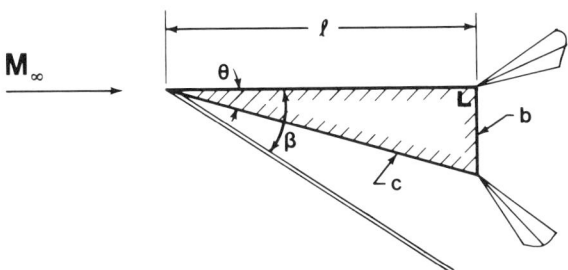

Fig. 20.2 Two-dimensional wedge in a uniform freestream. The upper surface is aligned with the flow.

Equation (20.1) is used to eliminate p_c/p_∞, to obtain

$$C_L = \frac{4}{(\gamma+1)M_\infty^2}(M_\infty^2 \sin^2\beta - 1) \tag{20.3}$$

where C_D is still given by Eq. (20.2b).

The known dimensionless parameters that govern the flow are γ, M_∞, and the wedge angle θ. We would like to use Eq. (5.23) to replace β in Eq. (20.3) with θ. Equation (5.23) is

$$\tan\theta = \cot\beta \frac{M_\infty^2 \sin^2\beta - 1}{1 + \{[(\gamma+1)/2] - \sin^2\beta\}M_\infty^2} \tag{20.4}$$

and cannot be explicitly solved for β. Therefore, we derive an approximate equation, valid for a hypersonic flow, that achieves this purpose.

When $M_\infty^2 \gg 1$, the shock wave approaches the body, and this type of flow is referred to as hypersonic. This is true for either a bow or an attached shock wave. The phenomenon is referred to as the shock wrapping itself around the upstream part of the body. For an attached shock, if θ is relatively large when $M_\infty^2 \gg 1$, the drag and heat transfer become unmanageable. Thus, at hypersonic speeds, θ must be small and the body is slender. Detailed studies show that the correct scaling parameter is

$$K_\theta = M_\infty \sin\theta \tag{20.5}$$

Thus, as $M_\infty \to \infty$, we require $\theta \to 0$ in such a way that K_θ is kept fixed and is of order unity. This parameter is the hypersonic similarity parameter.

In parallel with Eq. (20.5), we introduce for the shock wave angle

$$K_\beta = M_\infty \sin\beta \tag{20.6}$$

so that Eq. (20.4) becomes

$$\tan\beta \tan\theta = \frac{K_\beta^2 - 1}{[(\gamma+1)/2]M_\infty^2 - (K_\beta^2 - 1)}$$

With both Ks of order unity, $M_\infty \to \infty$, and θ and $\beta \to 0$, we obtain

$$[(\gamma+1)/2]K_\beta K_\theta = K_\beta^2 - 1$$

This relation is solved for K_β to yield

$$K_\beta = K + (1 + K^2)^{\frac{1}{2}} \tag{20.7a}$$

or
$$K_\beta^2 - 1 = 2K\left[K + (1 + K^2)^{\frac{1}{2}}\right] \quad (20.7b)$$

where
$$K = \frac{\gamma + 1}{4} K_\theta$$

We use Eq. (20.7b) in conjunction with Eq. (20.3), to obtain for the lift and drag coefficients

$$C_L = \frac{\gamma + 1}{2} \theta^2 \left[1 + (1 + K^{-2})^{\frac{1}{2}}\right] \quad (20.8a)$$

$$C_D = C_L \theta \quad (20.8b)$$

where the approximations $\sin \theta \sim \theta$ and $\tan \theta \sim \theta$ are used. Equations (20.8) are the hypersonic results for these coefficients.

In the limit $\theta \to 0$, the wedge becomes a flat plate aligned with the flow. In this circumstance, we readily obtain from Eqs. (20.8) that $C_L = C_D = 0$. A more interesting limit is $M_\infty \to \infty$, with θ kept fixed at a small finite value. In this case, $K \to \infty$, and Eqs. (20.8) yield

$$C_L \sim (\gamma + 1)\theta^2, \quad C_D \sim (\gamma + 1)\theta^3$$

which can be shown to agree with Eqs. (20.3) and (20.4). Hence, if $\theta = 10$ deg, we have

$$C_L \cong 7.31 \times 10^{-2}, \quad C_D \cong 1.28 \times 10^{-2}$$

when $K \to \infty$ and $\gamma = 1.4$.

Of course, an infinite wedge is not a practical vehicle. However, a wedge waverider is easily constructed by placing vertical fins on the wedge at any two locations, as shown in Fig. 20.3. The fins are of zero thickness and serve to confine the high-pressure shock layer, thereby preventing cross flow. As indicated in the rear-view sketch, where the waverider's base is shaded, the planar shock wave is attached to the forward edge of the fins. This

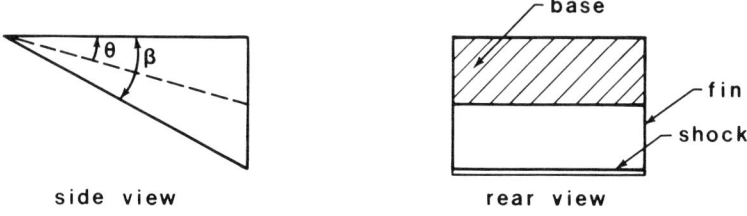

Fig. 20.3 Two-dimensional wedge waverider.

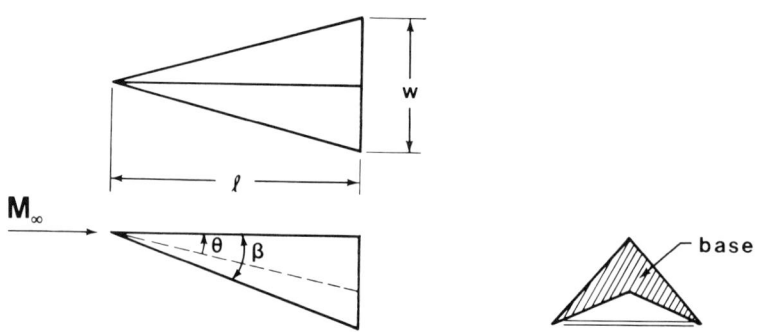

Fig. 20.4 Caret-shaped waverider, which is based on the wedge flow of Fig. 20.2.

configuration explains the waverider's name; the body appears to be riding on a shock wave. The lift and drag coefficients for this vehicle are given by Eqs. (20.8).

A more interesting caret-shaped waverider is shown in Fig. 20.4. The upper two surfaces are canted but are aligned with the freestream. The amount of cant is determined by the base width w. The lower two planar surfaces in Fig. 20.4 are canted stream surfaces located between the body and shock in Fig. 20.2. Thus, the flowfield between the body and shock for a caret waverider is a section of the shock layer in Fig. 20.2. (In an inviscid flow, stream surfaces can be used for walls.) The angles θ and β and the length ℓ are taken as the same in the two figures.

We now determine the lift and drag coefficients for the caret waverider. One can show that the planform area S_p and base area S_b, shown shaded in Fig. 20.4, are given by

$$S_p = \tfrac{1}{2} w\ell, \qquad S_b = \tfrac{1}{2} w\ell \tan \theta$$

Hence, the lift and drag are

$$L = (p_c - p_\infty) S_p = \left(\frac{p_c}{p_\infty} - 1 \right) p_\infty \frac{w\ell}{2}$$

$$D = (p_c - p_\infty) S_b = L \tan \theta$$

and Eqs. (20.2) for C_L and C_D are again obtained, providing $w/2$ is used as the span distance. Consequently, in the hypersonic limit we again obtain Eqs. (20.8). This is expected since the shock layer has the same constant pressure p_c as in the wedge flowfield.

20.3 C_L AND C_D FOR AN ARBITRARY WAVERIDER

We consider a uniform, supersonic freestream flow aligned with the z axis, as shown in Fig. 20.5. Assume the flowfield has a shock wave of known

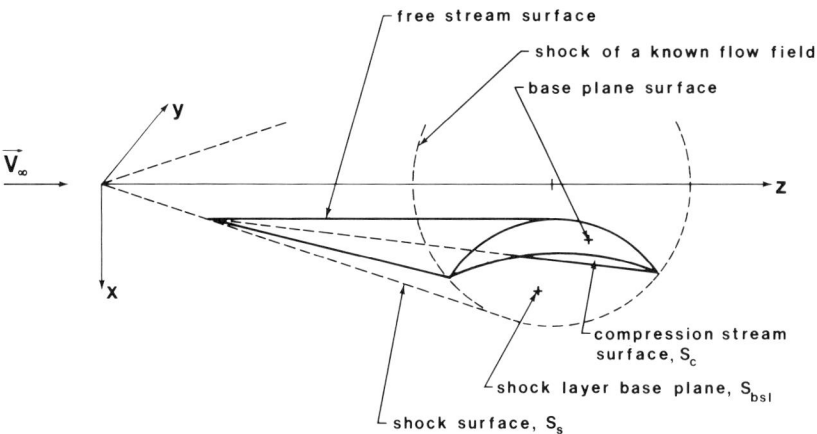

Fig. 20.5 Waverider configuration in a Cartesian coordinate system (after Kim[3]).

shape, as illustrated in the figure. A planar surface perpendicular to z then intersects the shock wave along a closed curve. We draw a second curve in this plane that falls inside the shock wave and intersects it at two points. The streamlines of the known flowfield, downstream of the shock, that pass through this curve constitute one surface of the waverider's body. This is the compression stream surface whose area is S_c. The intersection of this surface with the shock wave is the sharp leading edge of the body.

A second, upper surface of the body is determined by the freestream streamlines that pass through the curve for the leading edge of the body. This upper surface intersects the $z = $ const plane in a curve, as shown in the figure. The curves for the upper and lower body surfaces enclose the base plane surface of the body whose area is S_b.

Thus, the body consists of three surfaces. The upper surface is aligned with the freestream and has a constant pressure p_∞. The base plane is perpendicular to the freestream and has an assumed pressure of p_∞. Finally, the compression surface on the lower side of the body has a pressure that need not be constant.

The shock layer on which the body rides is also enclosed with three surfaces. The first surface would be the part of the shock wave downstream of the leading edge that has an area S_s. This surface is taken on the freestream side of the shock wave and terminates at the base plane. The second surface is the compression surface of area S_c, while the last surface is the shock layer base plane with an area S_{bsl}. This last area is in the same plane as the base of the vehicle but does not include the vehicle's base area S_b.

We apply conservation of mass and momentum to the shock layer flow that is enclosed by the above three surfaces. In integral form, these equations are

$$\int_S \rho \mathbf{V} \cdot \hat{\mathbf{n}} \, \mathrm{d}s = 0 \qquad (20.9a)$$

$$\int_S [\rho \mathbf{V}(\mathbf{V} \cdot \hat{\mathbf{n}}) + p\hat{\mathbf{n}}] \, ds = 0 \tag{20.9b}$$

where the integrals are surface integrals over the area

$$S = S_s + S_c + S_{bs\ell} \tag{20.10}$$

and $\hat{\mathbf{n}}$ is the outward unit normal vector.

Since S is a closed simple surface, we have

$$\int_S \hat{\mathbf{n}} \, ds = 0$$

After multiplication by p_∞, it is convenient to subtract this integral from Eq. (20.9b), to obtain

$$\int_S [\rho \mathbf{V}(\mathbf{V} \cdot \hat{\mathbf{n}}) + (p - p_\infty)\hat{\mathbf{n}}] \, ds = 0 \tag{20.11}$$

On the three surfaces, we have the following conditions:

$$\hat{\mathbf{n}} = \hat{\mathbf{e}}_z \text{ on } S_{bs\ell}$$

$$\mathbf{V} \cdot \hat{\mathbf{n}} = 0, \qquad \hat{\mathbf{n}} = \hat{\mathbf{n}}_c \text{ on } S_c$$

$$\mathbf{V} = \mathbf{V}_\infty, \qquad p = p_\infty, \qquad \rho = \rho_\infty, \qquad \hat{\mathbf{n}} = \hat{\mathbf{n}}_s \text{ on } S_s$$

Note that $\hat{\mathbf{n}}_c$ is into the body and that S_s is on the freestream side of the shock. Consequently, Eqs. (20.9a) and (20.11) become

$$\int_{S_s} \rho_\infty (\mathbf{V}_\infty \cdot \hat{\mathbf{n}}_s) \, ds + \int_{S_{bs\ell}} \rho \mathbf{V} \cdot \hat{\mathbf{e}}_z \, ds = 0$$

$$\int_{S_c} (p - p_\infty)\hat{\mathbf{n}}_c \, ds + \int_{S_{bs\ell}} [\rho \mathbf{V}(\mathbf{V} \cdot \hat{\mathbf{e}}_z) + (p - p_\infty)\hat{\mathbf{e}}_z] \, ds$$

$$+ \int_{S_s} \rho_\infty \mathbf{V}_\infty (\mathbf{V}_\infty \cdot \hat{\mathbf{n}}_s) \, ds = 0$$

The S_s integrals are eliminated from the two equations, to yield

$$\int_{S_c} (p - p_\infty)\hat{\mathbf{n}}_c \, ds = -\int_{S_{bs\ell}} [\rho(\mathbf{V} - \mathbf{V}_\infty)(\mathbf{V} \cdot \hat{\mathbf{e}}_z) + (p - p_\infty)\hat{\mathbf{e}}_z] \, ds \tag{20.12}$$

The left side represents the force on the waverider due to the excess pressure $p - p_\infty$ on the compression surface. We now assume the flow is symmetric about the x-z plane so that the sideways force in the y direction is zero. The resulting force on the waverider is in the x-z plane and can be resolved into lift and drag components.

The lift component is in the negative x direction. It is obtained by dotting Eq. (20.12) with $-\hat{\mathbf{e}}_x$. After noting that $\hat{\mathbf{e}}_x \cdot \hat{\mathbf{e}}_z = 0$ and $\hat{\mathbf{e}}_x \cdot \mathbf{V}_\infty = 0$, we obtain

$$L = -\hat{\mathbf{e}}_x \cdot \int_{S_c} (p - p_\infty)\hat{\mathbf{n}}\, ds = \int_{S_{bs\ell}} \rho(\mathbf{V} \cdot \hat{\mathbf{e}}_x)(\mathbf{V} \cdot \hat{\mathbf{e}}_z)\, ds \quad (20.13a)$$

To obtain the drag, we dot Eq. (20.12) with $\hat{\mathbf{e}}_z$. After noting that $\hat{\mathbf{e}}_z \cdot \hat{\mathbf{e}}_z = 1$ and $\mathbf{V}_\infty \cdot \hat{\mathbf{e}}_z = V_\infty$, we obtain

$$D = \hat{\mathbf{e}}_z \cdot \int_{S_c} (p - p_\infty)\hat{\mathbf{n}}\, ds = \int_{S_{bs\ell}} [\rho(V_\infty - \mathbf{V} \cdot \hat{\mathbf{e}}_z)(\mathbf{V} \cdot \hat{\mathbf{e}}_z) - (p - p_\infty)]\, ds$$

$$(20.13b)$$

If the flow is not symmetric about the x-z plane, then a sideways force equation is readily obtained by dotting Eq. (20.12) with $\hat{\mathbf{e}}_y$.

The right side of Eqs. (20.13) is the desired result. Hence, the lift and drag depend only on the flow in the shock layer base plane. This elegant result, it should be noted, does not invoke any hypersonic assumptions. Finally, the lift and drag coefficients are defined by

$$C_L = \frac{2L}{\gamma p_\infty M_\infty^2 S_p}, \qquad C_D = \frac{2D}{\gamma p_\infty M_\infty^2 S_p} \quad (20.14)$$

where S_p is the planform area.

20.4 WAVERIDERS DERIVED FROM A CONICAL FLOWFIELD

The baseline flowfield is a supersonic flow about a circular cone at zero angle of attack. The shock wave is assumed to be attached and of the weak-solution variety. Thus, the basic flowfield is discussed in Sec. 15.3 and is shown schematically in Figs. 15.2 and 15.3.

A waverider is extracted by truncating the cone and by providing fins, as shown in Fig. 20.6. The body has a base perpendicular to the z axis, which is along the cone's centerline. Two fins are spaced at an angle ϕ_f relative to the x axis, where the angle may be greater or less than 90 deg. The fins are stream surfaces and extend outward only as far as the conical shock wave. They confine the shock layer to the lower part of the body and, for simplicity, are assumed to be of negligible thickness. The upper surface of the cone is truncated at the fin angle. Thus, we have a freestream flow over the entire upper surface.

This configuration is referred to as an idealized conical waverider. Its analysis is especially simple, since all surfaces except the lower fin surface have a constant pressure. As shown in Ref. 7, the configuration is a

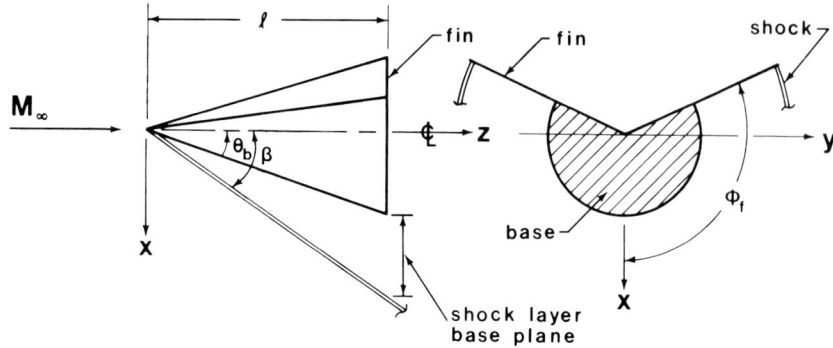

Fig. 20.6 Waverider schematic based on flow about a circular cone at zero angle of attack.

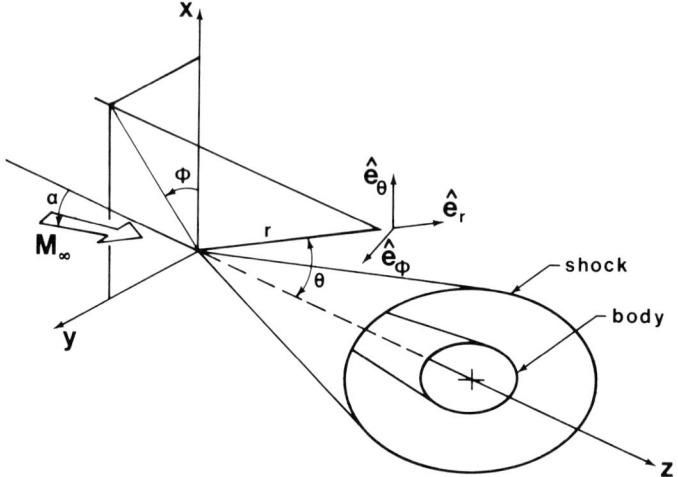

Fig. 20.7 Spherical coordinate system for conical-flow-based waverider (after Rasmussen[4]).

surprisingly good approximation to an optimized (minimum drag for a fixed lift) waverider configuration, when $M_\infty \theta_b > 0.5$. The actual optimized configuration has winglets instead of fins. These winglets are thin, especially when ϕ_f approaches 90 deg. A value of ϕ_f near 90 deg is particularly interesting since it results in a substantial useful interior volume for the vehicle.

Our objective is to develop simple formulas for C_L and C_D for an idealized conical waverider that are functions of γ, M_∞, θ_b, and ϕ_f. As seen from Sec. 20.3, we need only the solution for p, ρ, and $\mathbf{V}$ in the shock layer base plane. As shown in Sec. 15.3, this solution depends only on the angular variable η.

The differential equations for the flow in Sec. 15.3 require a numerical integration. Hence, we follow a different procedure here. An approximate analytical solution for the flowfield is developed that is based on a hypersonic small-disturbance assumption. This solution, in conjunction with Eqs. (20.13) and (20.14), yields the desired expressions for C_L and C_D.

Our analysis is a simplified version of Kim et al.[7] We use a spherical coordinate system, centered at the cone's apex, as shown in Fig. 20.7. The velocity is written as

$$\mathbf{V} = u\hat{\mathbf{e}}_r + v\hat{\mathbf{e}}_\theta + w\hat{\mathbf{e}}_\phi \qquad (20.15)$$

Because the flow is conical, $\mathbf{V}, p, \rho, \ldots$ are constant along a ray and are independent of the azimuthal angle ϕ. Hence, $w = 0$ and u and v depend only on θ. With this in mind, the exact jump conditions across the conical shock are

$$\frac{\rho_2}{\rho_\infty} = \frac{\gamma + 1}{2} \frac{K_\beta^2}{1 + [(\gamma - 1)/2] K_\beta^2}$$

$$\frac{p_2}{p_\infty} = \frac{2}{\gamma + 1}\left(\gamma K_\beta^2 - \frac{\gamma - 1}{2}\right) \qquad (20.16)$$

$$\frac{u_2}{a_\infty} = K_\beta \cot \beta$$

$$\frac{v_2}{a_\infty} = -\frac{2}{\gamma + 1} \frac{1 + [(\gamma - 1)/2] K_\beta^2}{K_\beta}$$

where K_β is defined by Eq. (20.6), and a subscript 2 denotes conditions on the downstream side of the shock. The shock location in terms of the body angle θ_b is given by[10]

$$K_\beta = \left(1 + \frac{\gamma + 1}{2} K_b^2\right)^{\frac{1}{2}} \qquad (20.17)$$

which is an approximate, but accurate, hypersonic relation, where

$$K_b = M_\infty \sin \theta_b$$

Equation (20.17) is the axisymmetric counterpart of Eq. (20.7a).

In the conical-flow shock layer, the density increases smoothly from ρ_2 to its value ρ_b on the body. For a slender body at hypersonic speeds, the magnitude of $(\rho_b - \rho_2)/\rho_2$ is quite small. This result is the basis for the constant-density shock layer approximation. The shock layer flow is homentropic; hence, $p \sim \rho^\gamma$. Therefore, the change in pressure $(p_b - p_2)/p_2$ exceeds that of the density but is still small. In view of the spherical coordinate system, the velocity component v at the shock is normal to it and is zero at the body. However, the magnitude of v_2/a_2 is well below unity since it is the normal component of the Mach number on the downstream side of the shock wave. As a consequence, $|v|$ is relatively small throughout the shock layer. On the other hand, u is not small in the shock layer or zero on the body. With K_β of order unity, we see from Eqs. (20.16) that ρ_2/ρ_∞, p_2/p_∞, and v_2/a_∞ are of order unity but that u_2/a_∞ is of order β^{-1}, which is large compared to unity. This slender body result is used to simplify the subsequent lift and drag derivations.

In Eqs. (20.13) the differential area in spherical coordinates is

$$ds = r^2 \sin\theta \, d\theta \, d\phi = \ell^2 \, d\phi \, \frac{\sin\theta \, d\theta}{\cos^2\theta} \tag{20.18}$$

where $r = \ell/\cos\theta$ in the base plane. Since the flow changes slowly along a streamline in the shock layer, the integrands in Eqs. (20.13) can be approximated by evaluating ρ, p, $\mathbf{V} \cdot \hat{\mathbf{e}}_x$, and $\mathbf{V} \cdot \hat{\mathbf{e}}_z$ just behind the shock, where $\theta = \beta$. With $\mathbf{V}_2$ given by

$$\mathbf{V}_2 = u_2 \hat{\mathbf{e}}_r + v_2 \hat{\mathbf{e}}_\theta$$

we need the basis vector transformation:

$$\hat{\mathbf{e}}_r = \hat{\mathbf{e}}_x \sin\theta \cos\phi + \hat{\mathbf{e}}_y \sin\theta \sin\phi + \hat{\mathbf{e}}_z \cos\theta$$

$$\hat{\mathbf{e}}_\theta = \hat{\mathbf{e}}_x \cos\theta \cos\phi + \hat{\mathbf{e}}_y \cos\theta \sin\phi - \hat{\mathbf{e}}_z \sin\theta$$

$$\hat{\mathbf{e}}_\phi = -\hat{\mathbf{e}}_x \sin\phi + \hat{\mathbf{e}}_y \cos\phi$$

Consequently, the two dot products are given by

$$\mathbf{V}_2 \cdot \hat{\mathbf{e}}_x = (u_2 \sin\beta + v_2 \cos\beta)\cos\phi$$

$$\mathbf{V}_2 \cdot \hat{\mathbf{e}}_z = u_2 \cos\beta - v_2 \sin\beta$$

With the aid of Eqs. (20.16), these become

$$\mathbf{V}_2 \cdot \hat{\mathbf{e}}_x = a_\infty \cos\beta \left\{ M_\infty \sin\beta - \frac{2}{\gamma+1} \frac{1 + [(\gamma-1)/2] K_\beta^2}{K_\beta} \right\} \cos\phi$$

$$= \frac{a_\infty \cos\beta}{K_\beta} \left[K_\beta^2 - \frac{2}{\gamma+1}\left(1 + \frac{\gamma-1}{2} K_\beta^2\right) \right] \cos\phi$$

$$= \frac{a_\infty \cos\beta}{K_\beta} K_b^2 \cos\phi = V_\infty \cot\beta \, \sin^2\theta_b \cos\phi \tag{20.19}$$

$$\mathbf{V}_2 \cdot \mathbf{e}_z = a_\infty \left\{ M_\infty \cos^2\beta + \frac{2}{\gamma+1} \frac{1+[(\gamma-1)/2]K_\beta^2}{M_\infty} \right\}$$

$$= \frac{a_\infty}{M_\infty}\left(M_\infty^2 - K_\beta^2 + \frac{2}{\gamma+1}K_\beta^2\right) = \frac{a_\infty}{M_\infty}\left[M_\infty^2 - \frac{2}{\gamma+1}(K_\beta^2 - 1)\right]$$

$$= \frac{a_\infty}{M_\infty}(M_\infty^2 - K_b^2) = V_\infty \cos^2\theta_b \qquad (20.20)$$

where Eq. (20.17) is used to eliminate K_β^2. With Eqs. (20.18)–(20.20), the lift becomes

$$L = \rho_\infty \frac{\gamma+1}{2} \frac{K_\beta^2}{1+[(\gamma-1)/2]K_\beta^2} \times V_\infty \cot\beta \sin^2\theta_b \times V_\infty \cos^2\theta_b$$

$$\times \frac{\ell^2 \sin\beta}{\cos^2\beta} \int_{-\phi_f}^{\phi_f} \cos\phi \, d\phi \int_{\theta_b}^{\beta} d\theta$$

After simplification, we obtain

$$L = 2\gamma p_\infty \ell^2 K_b^2 \tan\beta \frac{\sin\phi_f}{1 + K_b\{1+[(\gamma+1)/2]K_b^2\}^{-\frac{1}{2}}} \qquad (20.21)$$

where $\cos\theta_b \cong 1$ for a slender body.

A similar procedure for the drag results in

$$D = \{-(p_2 - p_\infty) + \rho_2[V_\infty - (\mathbf{V}_2 \cdot \hat{\mathbf{e}}_z)](\mathbf{V}_2 \cdot \hat{\mathbf{e}}_z)\} S_{bs\ell}$$

where the shock layer base area is

$$S_{bs\ell} = \ell^2(\tan^2\beta - \tan^2\theta_b)\phi_f \cong \ell^2(\beta^2 - \theta_b^2)\phi_f$$

With the aid of Eqs. (20.16), we have

$$p_2 - p_\infty = \gamma p_\infty K_b^2$$

The above results, plus Eq. (20.20), then yield

$$D = \frac{\gamma p_\infty K_b^4}{1+[(\gamma-1)/2]K_b^2} S_{bs\ell} = \gamma p_\infty \ell^2 \phi_f K_b^2 \theta_b^2 \qquad (20.22)$$

The planform area to be used in Eqs. (20.14) is given by

$$S_p = \ell^2 \tan\beta \sin\phi_f \cong \ell^2 \beta \sin\phi_f$$

Hence, the final result for the lift and drag coefficients is

$$C_L = 4\theta_b^2 \frac{\{1 + [(\gamma+1)/2]K_b^2\}^{\frac{1}{2}}}{K_b + \{1 + [(\gamma+1)/2]K_b^2\}^{\frac{1}{2}}} \quad (20.23a)$$

$$C_D = \frac{2K_b\theta_b^3}{\{1 + [(\gamma+1)/2]K_b^2\}^{\frac{1}{2}}} \frac{\phi_f}{\sin\phi_f} \quad (20.23b)$$

We note that Eq. (20.23a) is in exact accord with the precise results of Ref. 7. In the limit of $K_b \to \infty$, we obtain

$$C_L \sim \frac{4}{1 + [2/(\gamma+1)]^{\frac{1}{2}}} \theta_b^2, \quad C_D \sim 2\left(\frac{2}{\gamma+1}\right)^{\frac{1}{2}} \theta_b^3 \frac{\phi_f}{\sin\phi_f}$$

When θ_b is 10 deg and $\gamma = 1.4$, this yields

$$C_L \cong 6.37 \times 10^{-2}, \quad C_D \cong 9.71 \times 10^{-3} \frac{\phi_f}{\sin\phi_f}$$

which is similar to the earlier result for a caret waverider.

We now examine the L/D ratio, which is given by

$$\frac{L}{D} = \frac{C_L}{C_D} = 2M_\infty \frac{1 + [(\gamma+1)/2]K_b^2}{K_b^2\left(K_b + \{1 + [(\gamma+1)/2]K_b^2\}^{\frac{1}{2}}\right)} \frac{\sin\phi_f}{\phi_f}$$

For a fixed value of K_b, this ratio increases with Mach number but slowly decreases as the fin angle ϕ_f increases. Furthermore, with M_∞ and ϕ_f fixed, the ratio has a maximum of infinity when $K_b \to 0$. This result is expected, since it means that the finned body in Fig. 20.6 has a very small drag when the conical part shrinks to a thin sliver. Remember that the fins have no wave drag. This conclusion would be greatly altered if viscous drag were included.

In the hypersonic limit, the caret lift coefficient can be written as

$$C_{L,\text{caret}} = 2\theta\beta$$

while the idealized conical waverider becomes

$$C_{L,\text{cone}} = 2\theta_b\beta_c \frac{2\theta_b}{\beta_c + \theta_b}$$

where β and β_c are the caret and conical shock angles, respectively. We compare lift coefficients, under the assumption that the body angles θ and θ_b are the same. With the aid of Eq. (20.7a) for the caret and Eq. (20.17) for the cone, we obtain

$$C_{L,\text{caret}} = 2\theta^2 \left\{ \frac{\gamma+1}{4} + \left[\left(\frac{\gamma+1}{4}\right)^2 + \frac{1}{(M_\infty \theta)^2} \right]^{\frac{1}{2}} \right\} \quad (20.24a)$$

and

$$C_{L,\text{cone}} = 2\theta^2 \frac{2\{[(\gamma+1)/2] + (M_\infty \theta)^{-2}\}^{\frac{1}{2}}}{1 + \{[(\gamma+1)/2] + (M_\infty \theta)^{-2}\}^{\frac{1}{2}}} \quad (20.24b)$$

By comparing Eqs. (20.24), we see that

$$C_{L,\text{caret}} > C_{L,\text{cone}}$$

for all $M_\infty \theta > 0$, where the difference is largest for small $M_\infty \theta$. A more meaningful comparison is the subject of Problem 20.4, whose results favor the cone-based waverider.

In determining the foregoing results, we have not used a detailed solution for the shock layer flowfield. Indeed, we have not even written the equations of motion in spherical coordinates. Of course, more accurate results would require a flowfield solution, which still can be approximate, as in Ref. 7. A number of factors contribute to the simplicity of the analytical method. One factor is the lift and drag integral formulas, Sec. 20.3, that require a flowfield solution only for the shock layer base plane. Other factors include the use of a hypersonic small-disturbance approximation and simple formulas for the shock jump conditions. These jump condition formulas are simple because we use the flowfield downstream of a constant-strength conical shock wave.

References

[1] Townend, L. H., "Research and Design for Lifting Reentry," *Progress in Aerospace Sciences*, Vol. 18, 1979, pp. 1–80.

[2] Nonweiler, T. R. F., "Delta Wings of Shape Amenable to Exact Shock-Wave Theory," *Journal of the Royal Aeronautical Society*, Vol. 67, 1963, p. 39.

[3] Kim, B. S., "Optimization of Waverider Configurations Generated from Non-Axisymmetric Flows Past a Nearly Circular Cone," Ph.D. Dissertation, University of Oklahoma, Norman, 1983.

[4] Rasmussen, M. L., "Waverider Configurations Derived from Inclined Circular and Elliptic Cones," *Journal of Spacecraft and Rockets*, Vol. 17, Nov.–Dec. 1980, pp. 537–545.

[5] Jischke, M. C., "Supersonic Flow Past Conical Bodies with Nearly Circular Cross Sections," *AIAA Journal*, Vol. 19, Feb. 1981, pp. 242–245.

[6]Rasmussen, M. L., Jischke, M. C., and Daniel, D. C., "Experimental Forces and Moments on Cone-Derived Waveriders for $M_\infty = 3$ to 5," *Journal of Spacecraft and Rockets*, Vol. 19, Nov.–Dec. 1982, pp. 592–598.

[7]Kim, B. S., Rasmussen, M. L., and Jischke, M. C., "Optimization of Waverider Configurations Generated from Axisymmetric Conical Flows," *Journal of Spacecraft and Rockets*, Vol. 20, Sept.–Oct. 1983, pp. 461–469.

[8]Krieger, R. J., "Summary of Design and Performance Characteristics of Aerodynamic Configured Missiles," AIAA Paper 81-0286, Jan. 1981.

[9]Batt, R. G. and Kubota, T., "Experimental Investigation of Laminar Near Wakes behind 20° Wedges at $M_\infty = 6$," *AIAA Journal*, Vol. 6, Nov. 1968, pp. 2077–2083.

[10]Rasmussen, M. L., "On Hypersonic Flow Past an Unyawed Cone," *AIAA Journal*, Vol. 5, Aug. 1967, pp. 1495–1497.

Problems

20.1 Compare the exact and hypersonic values for C_L for a caret waverider when $\gamma = 1.4$, $\theta = 10$ deg, and $M_\infty = 5$.

20.2 Use Eqs. (20.13) to obtain L and D for the wedge in Fig. 20.2. Compare your results with exact values.

20.3 Let $\dot{Q}$ be the rate of heat transfer to the compressive surface of a waverider. Start with an energy equation in integral form that is consistent with Eqs. (20.9). Derive a formula for $\dot{Q}$ that depends only on an integral over the shock layer base plane.

Computational Problems:

20.4 Prepare figures that show L/D vs $V^{2/3}/S_p$ for both the caret and idealized conical waveriders. V is the interior volume of the vehicle. Assume $\gamma = 1.4$ and consider M_∞ as the primary independent parameter.

APPENDIX A. SI UNITS AND NOMENCLATURE

Quantity	Symbol	Units
Length (meter)	ℓ	m
Mass (kilogram)	m	kg
Time (second)	t	s
Temperature (Kelvin)	T	K
Density	ρ	kg/m^3
Speed	V	m/s
Pressure	p	N/m^2 (Pa)
Dynamic viscosity	μ	N-s/m^2
Specific internal energy	e	J/kg
Specific enthalpy	h	J/kg
Specific entropy	s	J/kg-K
Specific heats	c_p, c_v	J/kg-K
Molecular weight	W	kg/kmole
Universal gas constant	$\tilde{R}$	8.314×10^3 J/kmole-K

Frequent Conversions

From	Multiply By	To
Atmosphere	1.013×10^5	N/m^2 (Pa)
Newton (N)	1	kg-m/s^2
Joule (J)	1	N-m
Kilomole	6.023×10^{26}	molecules

NOMENCLATURE

a	= speed of sound
$a_i^{(j)}, b_i^{(j)}, c^{(j)}$	= coefficients in partial differential equations
A	= cross-sectional area; matrix defined by Eq. (13.4)
A, B, C, D	= coefficients of a second-order partial differential equation
A_i, B_i	= defined by Eqs. (16.18)
A_s	= wetted surface area
$\hat{\mathbf{b}}, \hat{\mathbf{n}}, \hat{\mathbf{t}}$	= unit vectors associated with a shock wave
c	= streamtube circumference
c_p, c_v	= specific heats
C_D	= drag coefficient
C_f, C_L	= skin-friction and lift coefficients, respectively
$C_\pm$	= characteristic lines

D	= hydraulic diameter; drag
D, H	= dimensions of the opening of an aerodynamic window
$e, \mathbf{e}, \hat{\mathbf{e}}$	= specific internal energy; vectors tangential to a coordinate system
$E, F, G; e, f, g$	= first and second fundamental forms of a surface, respectively
f	= shock wave shape
F	= force; impulse, function; shock wave equation
G	= mass flux
h	= specific enthalpy
h_f	= film coefficient
h_i	= scale factors
I	= defined by Eq. (18.14)
J	= Jacobian
$J_\pm$	= Riemann invariants
k	= thermal conductivity
K, K_θ, K_β	= hypersonic similarity parameters
ℓ	= length
L	= lift; duct length
m	= mass of gas; degree of homogeneity
$\dot{m}$	= mass flow rate
M	= Mach number
n	= coordinate normal to a streamline
$\hat{\mathbf{n}}$	= unit normal vector
p	= pressure
P	= nondimensional pressure
Pr	= Prandtl number
q	= heat transfer, per unit mass of gas; flow speed
q_j, Q_j	= dependent variables of the equations of motion in conservative form
q_{ij}	= jth element of q_i
Q	= nondimensional speed
r	= position coordinate; radius; radial coordinate
R	= gas constant; nondimensional density; radial coordinate
$\tilde{R}$	= universal gas constant
Re	= Reynolds number
R_g	= gas constant
s	= specific entropy; coordinate along a streamline; surface area
S, S_b	= surface and base areas, respectively
$S_{bs\ell}, S_p$	= shock layer base plane and planform areas, respectively
t	= time
T	= temperature
u, v	= velocity components
$u, v, w; u_i$	= velocity components in a spherical and a Cartesian coordinate system, respectively
U, V	= defined by Eqs. (19.46)
v	= specific volume; flow speed; velocity component
v_j	= velocity components in the ξ_j coordinate system

v_r, v_n	=	velocity components in Sec. 19.3
V	=	volume; flow speed
V_m	=	maximum possible flow speed
w	=	work, per unit mass of gas; width of a waverider
W	=	molecular weight; $1 + \gamma M^2$
x	=	spatial coordinate
x_i	=	Cartesian coordinates
X	=	$1 + (\gamma - 1) M^2/2$; arbitrary independent variable
y	=	spatial coordinate
Y	=	arbitrary independent variable
z	=	length coordinate along a ramp; defined by Eq. (19.6a)
Z	=	$M^2 - 1$
α	=	angle of attack; $[2X/(\gamma + 1)]^{(\gamma+1)/[2(\gamma-1)]}$
β	=	shock wave angle relative to the freestream velocity; $(M^2 - 1)^{\frac{1}{2}}$
γ	=	ratio of specific heats
Γ	=	circulation
δ	=	reflected shock wave angle relative to the wall
δ_{ij}	=	Kronecker delta
η	=	similarity variable
θ	=	angle coordinate; wall turn angle relative to the freestream velocity; streamline angle relative to a wall or a centerline
θ_j	=	variables defined by Eq. (14.33)
κ	=	curvature
κ_a, κ_ℓ	=	principal curvatures of a surface
λ	=	parameter
μ	=	Mach angle; viscosity
ν	=	Prandtl–Meyer function
ξ	=	similarity variable
ξ_i	=	non-Cartesian coordinate system; characteristic coordinates
ρ	=	density
σ	=	0 for two-dimensional flow, 1 for axisymmetric flow
τ	=	shear stress; volume element
ϕ	=	expansion fan angle; wedge half-angle; potential function; velocity angle relative to a conical body; azimuthal angle
ϕ_f	=	fin angle
ϕ_i	=	dependent variables
χ	=	triple point angle relative to a ramp surface
ψ	=	stream function
ω	=	frequency; vorticity; ramp angle

Special Symbols

$\hat{1}_i$	=	unit vectors for a Cartesian coordinate system
$\mathcal{M}$	=	defined by Eq. (18.11b)
$\mathcal{T}$	=	thrust

$(\cdot)$	= vector dot product
$\nabla, \nabla\cdot, \nabla\times$	= spatial gradient, divergence, and curl operators, respectively

Subscripts

a	= ambient
$a, b, \ldots, g$	= nozzle flow cases, see Fig. 7.2
b	= body; base
c	= cavity; compressive surface
cs	= contact surface
d	= detachment
f	= final value
i	= vector component; summation index; incident wave; inlet
I	= unsteady incident flow
j	= vector component; summation index
k	= summation index
LE	= leading edge
m	= choking value; mechanical equilibrium
max	= maximum value
M	= Mach stem
n	= component normal to a shock wave; normal shock wave
p	= piston
P	= point P
r	= reference value; reflected wave
s	= shock wave; sonic condition
ss	= slipstream
t	= component tangential to a shock wave; transposed matrix
TE	= trailing edge
w	= wall
0	= stagnation condition
1	= freestream or preshock wave value; plenum (reservoir) condition
$1, 2, \ldots$	= regions of flow; thermodynamic states; mesh points
$I, II, \ldots$	= regions of flow
$+, -$	= left- and right-running characteristics, respectively
∞	= freestream condition; ambient value

Superscripts

$(\dot{\ })$	= per unit time
$(\ )^*$	= sonic value
$(\ )'$	= perturbation quantity; unsteady flow; wall point
$(\overline{\ })$	= average value; far-field solution in Sec. 19.3
$(\hat{\ })$	= vector of unit magnitude; steady component in Sec. 19.6

APPENDIX B. THERMODYNAMIC SUMMARY

(1) First law: $de = dq + dw$
(2) Second law: $ds = (dq)_{rev}/T \geq 0$ for a closed adiabatic system.
(3) Combined first and second laws, where $dw = -p\,dv$, $v = 1/\rho$, $h = e + pv$:

$$de = T\,ds - p\,dv = T\,ds + \frac{p}{\rho^2}\,d\rho, \qquad dh = T\,ds + v\,dp = T\,ds + \frac{1}{\rho}\,dp$$

(4) Thermal equation of state: $p = p(\rho, T)$

$$p = \rho R T \qquad \text{(th. p.g.)}^*$$

(5) Caloric equation of state: $e = e(\rho, T)$

$$e = c_v T, \qquad c_v = \text{const} \qquad \text{(cal. p.g.)}^*$$

(6) Reciprocity:

$$\frac{\partial e}{\partial v}\bigg|_T = \left.\frac{T\partial p}{\partial T}\right|_v - p$$

(7) Specific heats:

$$c_v = \frac{\partial e}{\partial T}\bigg|_v, \qquad c_p = \frac{\partial h}{\partial T}\bigg|_p$$

$$c_v = \frac{R}{\gamma - 1}, \qquad c_p = \frac{\gamma R}{\gamma - 1}, \qquad \gamma = \frac{c_p}{c_v} \qquad [\text{th. p.g. and } \gamma = \gamma(T)]$$

(8) Entropy for a thermally and calorically perfect gas:

$$\frac{s}{R} = \ell n\left(T^{1/(\gamma-1)}/\rho\right) + \text{const}, \qquad \frac{p_2}{p_1} = \left(\frac{\rho_2}{\rho_1}\right)^\gamma = \left(\frac{T_2}{T_1}\right)^{\gamma/(\gamma-1)} \quad \text{if } ds = 0$$

(9) Speed of sound: $a = \left(\dfrac{\partial p}{\partial \rho}\right)_s^{\frac{1}{2}}$

$$a = \left(\gamma \frac{p}{\rho}\right)^{\frac{1}{2}} = (\gamma R T)^{\frac{1}{2}} \qquad [\text{th. p.g. and } \gamma = \gamma(T)]$$

---

*Denotes thermally or calorically perfect gas, respectively.

APPENDIX C. STREAMTUBE FLOW EQUATIONS

Steady, inviscid, adiabatic quasi-one-dimensional flow of a perfect gas is assumed.

$$M = \frac{V}{a}$$

$$\frac{T}{T_0} = \frac{h}{h_0} = \left(\frac{a}{a_0}\right)^2 = \left(1 + \frac{\gamma - 1}{2} M^2\right)^{-1}$$

$$\frac{\rho}{\rho_0} = \left(1 + \frac{\gamma - 1}{2} M^2\right)^{-1/(\gamma - 1)}$$

$$\frac{p}{p_0} = \left(1 + \frac{\gamma - 1}{2} M^2\right)^{-\gamma/(\gamma - 1)}$$

$$\frac{S}{R} = \frac{S_0}{R} = \ell n\left(T_0^{\gamma/(\gamma - 1)}/p_0\right) + \text{const}$$

$$\frac{A}{A^*} = \left(\frac{2}{\gamma + 1}\right)^{(\gamma + 1)/[2(\gamma - 1)]} \frac{1}{M}\left(1 + \frac{\gamma - 1}{2} M^2\right)^{(\gamma + 1)/[2(\gamma - 1)]}$$

$$\dot{m} = \left(\frac{2}{\gamma + 1}\right)^{(\gamma + 1)/[2(\gamma - 1)]} \left(\frac{\gamma}{RT_0}\right)^{\frac{1}{2}} p_0 A^*$$

$$\dot{m} = \frac{M}{\{1 + [(\gamma - 1)/2] M^2\}^{(\gamma + 1)/[2(\gamma - 1)]}} \left(\frac{\gamma}{RT_0}\right)^{\frac{1}{2}} p_0 A$$

APPENDIX D. NORMAL AND OBLIQUE SHOCK SUMMARY

Steady two-dimensional flow of a perfect gas is assumed with M_1 and M_{1n} supersonic.

$$M_2^2 = \frac{1 + [(\gamma - 1)/2] M_1^2}{\gamma M_1^2 - [(\gamma - 1)/2]}$$

$$\frac{p_2}{p_1} = \frac{2}{\gamma + 1} \left(\gamma M_1^2 - \frac{\gamma - 1}{2} \right)$$

$$\frac{T_2}{T_1} = \left(\frac{2}{\gamma + 1} \right)^2 \frac{\{1 + [(\gamma - 1)/2] M_1^2\}\{\gamma M_1^2 - [(\gamma - 1)/2]\}}{M_1^2}$$

$$\frac{\rho_2}{\rho_1} = \frac{V_1}{V_2} = \left(\frac{\gamma + 1}{2} \right) \frac{M_1^2}{1 + [(\gamma - 1)/2] M_1^2}$$

$$s_2 - s_1 = R \ln(p_{01}/p_{02})$$

$$\frac{p_{02}}{p_{01}} = \left(\frac{\gamma + 1}{2} \right)^{(\gamma + 1)/(\gamma - 1)}$$

$$\times \frac{M_1^{2\gamma/(\gamma - 1)}}{\{1 + [(\gamma - 1)/2] M_1^2\}^{\gamma/(\gamma - 1)} \{\gamma M_1^2 - [(\gamma - 1)/2]\}^{1/(\gamma - 1)}}$$

$$\frac{T_{02}}{T_{01}} = 1$$

$$\frac{p_{02}}{p_1} = \left(\frac{\gamma + 1}{2} \right)^{(\gamma + 1)/(\gamma - 1)} \frac{M_1^{2\gamma/(\gamma - 1)}}{\left(\gamma M_1^2 - \frac{\gamma - 1}{2} \right)^{1/(\gamma - 1)}}$$

For an oblique shock, as shown in Fig. 5.5, only the weak solution occurs when the shock is attached. In conjunction with the above equations, the equations for an oblique shock wave are:

$$M_{1n} = M_1 \sin \beta$$

$$M_{2n} = M_2 \sin(\beta - \theta)$$

$$\tan \theta = \cot \beta \frac{M_1^2 \sin^2 \beta - 1}{1 + \{[(\gamma + 1)/2] - \sin^2 \beta\} M_1^2}$$

This transformation, however, cannot be used with the p_{02}/p_1 or the V_1/V_2 equations.

APPENDIX E. SHOCK WAVE ANGLE β VS FLOW DEFLECTION ANGLE θ

γ = 1.4

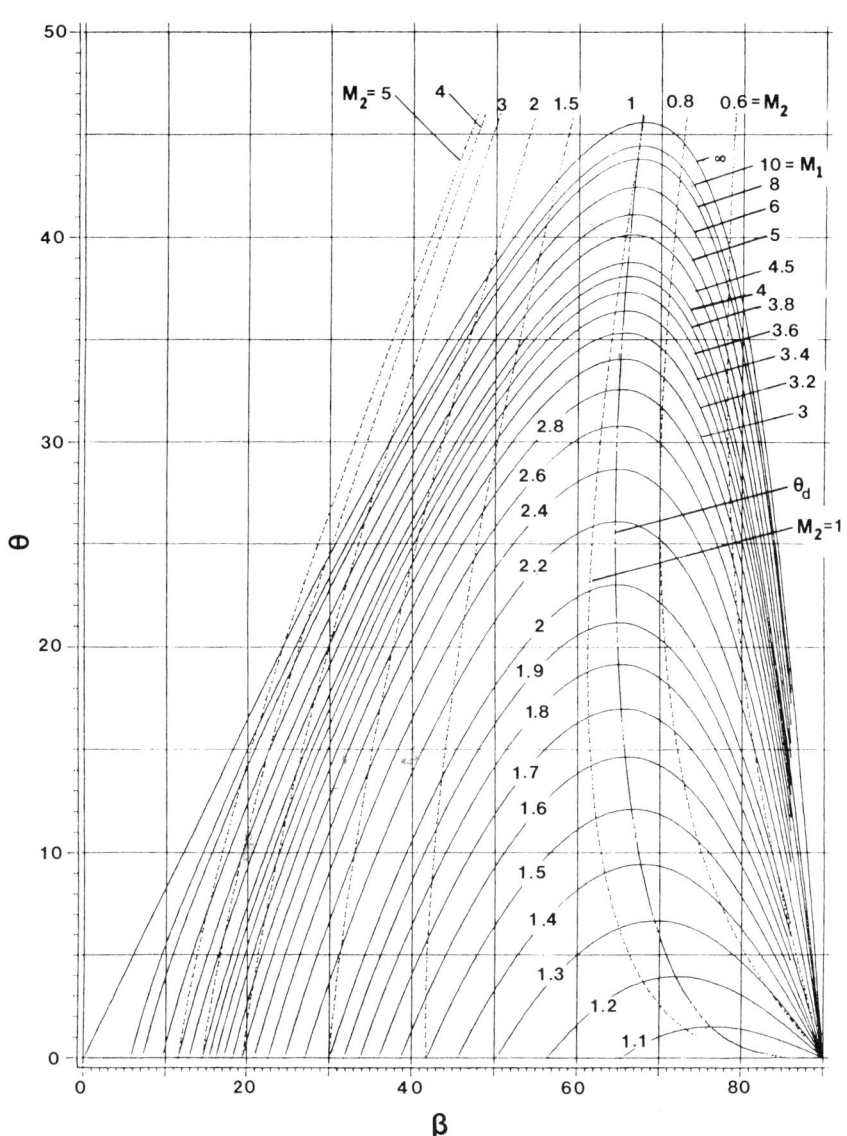

$\gamma = 5/3$

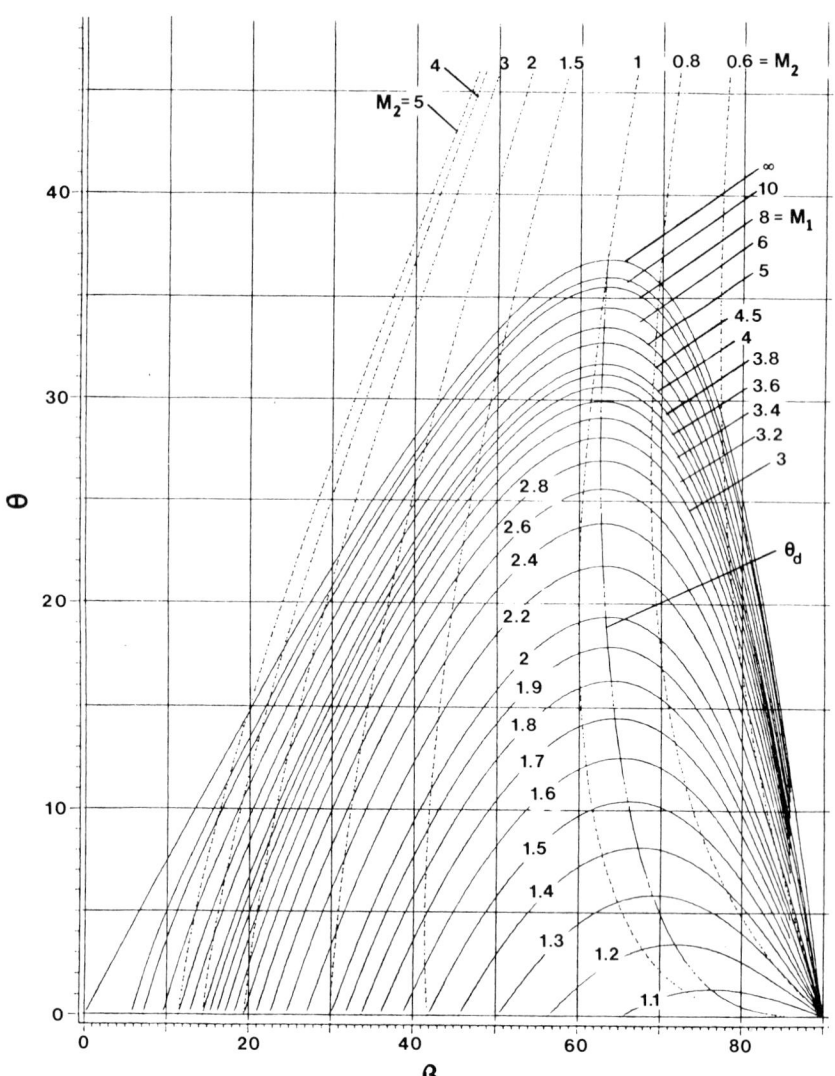

APPENDIX F. PRANDTL–MEYER FLOW SUMMARY

Steady, inviscid, adiabatic, two-dimensional supersonic flow of a perfect gas is assumed.

$$\mu = \sin^{-1}\left(\frac{1}{M}\right)$$

$$\nu = [(\gamma+1)/(\gamma-1)]^{\frac{1}{2}} \tan^{-1}\left\{[(\gamma-1)/(\gamma+1)]^{\frac{1}{2}}(M^2-1)\right\}$$

$$- \tan^{-1}(M^2-1)^{\frac{1}{2}}$$

$$\nu(M_2) = \nu(M_1) + \begin{cases} |\theta|, & \text{expansion} \\ -|\theta|, & \text{compression} \end{cases}$$

Isentropic relations can be used to connect conditions on any two Mach lines.

(1) Sharp expansion turn:

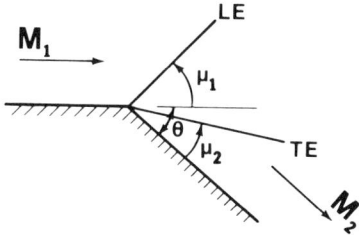

(2) Gradual expansion turn:

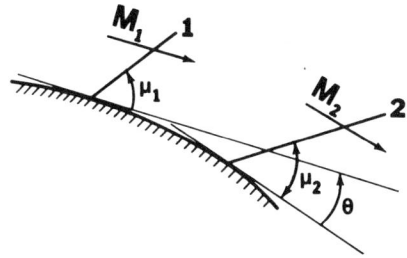

(3) Vacuum expansion turn:

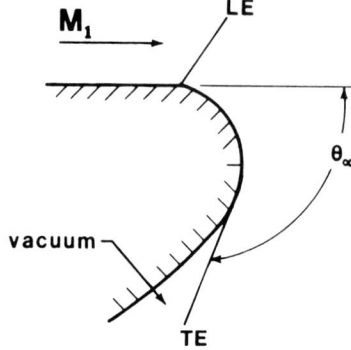

(4) Gradual compression turn for the flow near the wall:

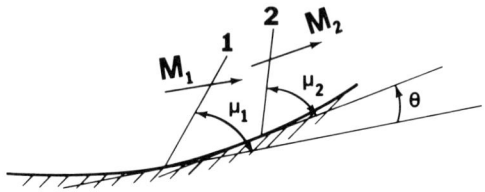

APPENDIX G. SHOCK-EXPANSION SUMMARY

Steady, inviscid, adiabatic, two-dimensional supersonic flow of a perfect gas is assumed.

$$C_L = \frac{2L}{(\rho V^2)_\infty \ell}, \quad C_D = \frac{2D}{(\rho V^2)_\infty \ell}, \quad (\rho V^2)_\infty = \gamma p_\infty M_\infty^2$$

C_L and C_D are developed for a quadrilateral airfoil at an angle of attack α as shown in Fig. G.1. The following items are assumed to be known:

$$\gamma, M_\infty, \ell, \alpha, \phi_1, \phi_2, \phi_3$$

By simple geometry, we have for ϕ_4

$$\phi_4 = 2\pi - (\phi_1 + \phi_2 + \phi_3)$$

and for the lengths of the four sides,

$$\ell_1 = \frac{\sin \phi_2}{\sin(\phi_1 + \phi_2)} \ell, \quad \ell_2 = \frac{\sin \phi_1}{\sin(\phi_1 + \phi_2)} \ell$$

$$\ell_3 = \frac{\sin \phi_4}{\sin(\phi_3 + \phi_4)} \ell, \quad \ell_4 = \frac{\sin \phi_3}{\sin(\phi_3 + \phi_4)} \ell$$

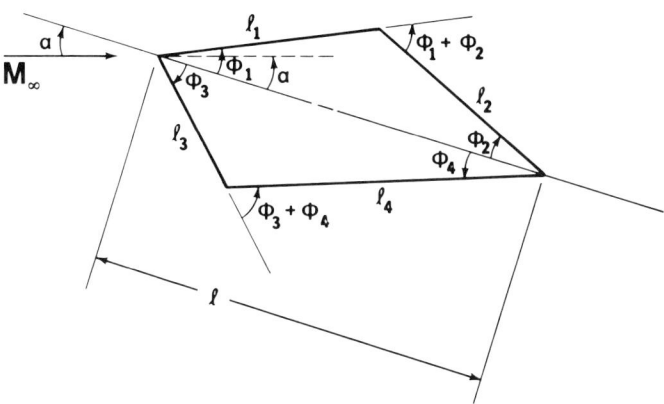

Fig. G.1 Nomenclature for a quadrilateral airfoil.

For surface 1, the lift and drag are given by

$$L_1 = -p_1 \ell_1 \cos(\phi_1 - \alpha), \qquad D_1 = p_1 \ell_1 \sin(\phi_1 - \alpha)$$

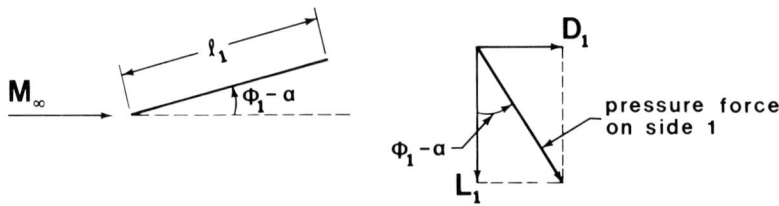

Similarly, for the other surfaces, we have the following:

$$L_2 = -p_2 \ell_2 \cos(\phi_2 + \alpha), \qquad D_2 = -p_2 \ell_2 \sin(\phi_2 + \alpha)$$
$$L_3 = p_3 \ell_3 \cos(\phi_3 + \alpha), \qquad D_3 = p_3 \ell_3 \sin(\phi_3 + \alpha)$$
$$L_4 = p_4 \ell_4 \cos(\phi_4 - \alpha), \qquad D_4 = -p_4 \ell_4 \sin(\phi_4 - \alpha)$$

The lift and drag coefficients are then given by

$$C_L = \frac{2}{\gamma M_\infty^2} \left\{ -\frac{1}{\sin(\phi_1 + \phi_2)} \left[\sin\phi_2 \cos(\phi_1 - \alpha) \frac{p_1}{p_\infty} + \sin\phi_1 \cos(\phi_2 + \alpha) \frac{p_2}{p_\infty} \right] \right.$$
$$\left. + \frac{1}{\sin(\phi_3 + \phi_4)} \left[\sin\phi_4 \cos(\phi_3 + \alpha) \frac{p_3}{p_\infty} + \sin\phi_3 \cos(\phi_4 - \alpha) \frac{p_4}{p_\infty} \right] \right\}$$

$$C_D = \frac{2}{\gamma M_\infty^2} \left\{ \frac{1}{\sin(\phi_1 + \phi_2)} \left[\sin\phi_2 \sin(\phi_1 - \alpha) \frac{p_1}{p_\infty} - \sin\phi_1 \sin(\phi_2 + \alpha) \frac{p_2}{p_\infty} \right] \right.$$
$$\left. + \frac{1}{\sin(\phi_3 + \phi_4)} \left[\sin\phi_4 \sin(\phi_3 + \alpha) \frac{p_3}{p_\infty} - \sin\phi_3 \sin(\phi_4 - \alpha) \frac{p_4}{p_\infty} \right] \right\}$$

When the chord is defined as passing through the leading and trailing edges of the airfoil as in Fig. G.1, the lift coefficient may not be zero when $\alpha = 0$. It may, in fact, be positive or negative. With $\alpha \geq 0$, there is a shock ahead of surface 1 when $\phi_1 > \alpha$; otherwise, there is a Prandtl–Meyer expansion. The only remaining computation is to determine the p_i/p_∞, $i = 1, \ldots, 4$.

APPENDIX H. NOZZLE FLOW SUMMARY

Steady, inviscid, adiabatic quasi-one-dimensional flow of a perfect gas is assumed.
Limiting conditions:
(1) Case (b) in Fig. 7.2:

$$\left(\frac{p_b}{p_0}\right)^{2/\gamma} - \left(\frac{p_b}{p_0}\right)^{(\gamma+1)/\gamma} = \frac{\gamma-1}{2}\left(\frac{2}{\gamma+1}\right)^{(\gamma+1)/(\gamma-1)}\left(\frac{A_2}{A_3}\right)^2$$

Of the two roots use the larger.
(2) Case (f) in Fig. 7.2:
Use the same equation as in item (1), but replace p_b with p_f. Of the two roots, use the smaller.
(3) Case (d) in Fig. 7.2:
Determine p_f/p_0 as in item (2), and p_d/p_3, Fig. 7.4, from the normal shock value for M_3. Then p_d/p_0 is given by

$$\frac{p_d}{p_0} = \frac{p_d}{p_3}\frac{p_f}{p_0}$$

For an internal normal shock wave, use

$$\frac{p}{p_0}\frac{A}{A^*} = \left(\frac{2}{\gamma+1}\right)^{(\gamma+1)/2(\gamma-1)} \frac{1}{M\{1+[(\gamma-1)/2]M^2\}^{\frac{1}{2}}}$$

APPENDIX I. SUMMARY OF EQUATIONS FOR DUCTS WITH AREA CHANGE, HEAT TRANSFER, AND FRICTION

$$\frac{dM^2}{M^2} = \frac{2X}{Z}\frac{dA}{A} - \frac{WX}{Z}\frac{dT_0}{T_0} - \frac{\gamma M^2 X}{Z}\left(4C_f\frac{dx}{D}\right)$$

$$\frac{dp}{p} = -\frac{\gamma M^2}{Z}\frac{dA}{A} + \frac{\gamma M^2 X}{Z}\frac{dT_0}{T_0} + \frac{\gamma M^2(W-M^2)}{2Z}\left(4C_f\frac{dx}{D}\right)$$

$$\frac{dT}{T} = -\frac{(\gamma-1)M^2}{Z}\frac{dA}{A} - \frac{(1-\gamma M^2)X}{Z}\frac{dT_0}{T_0} + \frac{\gamma(\gamma-1)M^4}{2Z}\left(4C_f\frac{dx}{D}\right)$$

$$\frac{d\rho}{\rho} = -\frac{M^2}{Z}\frac{dA}{A} + \frac{X}{Z}\frac{dT_0}{T_0} + \frac{\gamma M^2}{2Z}\left(4C_f\frac{dx}{D}\right)$$

$$\frac{dp_0}{p_0} = -\frac{\gamma M^2}{2}\frac{dT_0}{T_0} - \frac{\gamma M^2}{2}\left(4C_f\frac{dx}{D}\right)$$

$$\frac{dV}{V} = \frac{1}{Z}\frac{dA}{A} - \frac{X}{Z}\frac{dT_0}{T_0} - \frac{\gamma M^2}{2Z}\left(4C_f\frac{dx}{D}\right)$$

$$\frac{ds}{R} = \frac{\gamma X}{\gamma-1}\frac{dT_0}{T_0} + \frac{\gamma M^2}{2}\left(4C_f\frac{dx}{D}\right)$$

The equations assume steady quasi-one-dimensional flow of a perfect gas. The variables W, X, and Z are defined by

$$W = 1 + \gamma M^2$$

$$X = 1 + \frac{\gamma-1}{2}M^2$$

$$Z = M^2 - 1$$

APPENDIX J. RAYLEIGH FLOW SUMMARY

$$\frac{T_0}{T_0^*} = 2(\gamma + 1)\frac{M^2 X}{W^2}$$

$$\frac{p}{p^*} = \frac{\gamma + 1}{W}$$

$$\frac{T}{T^*} = (\gamma + 1)^2 \frac{M^2}{W^2}$$

$$\frac{T^*}{T_0^*} = \frac{2}{\gamma + 1}$$

$$\frac{\rho}{\rho^*} = \frac{W}{(\gamma + 1) M^2}$$

$$\frac{p_0}{p_0^*} = \frac{\gamma + 1}{W}\left(\frac{2X}{\gamma + 1}\right)^{\gamma/(\gamma - 1)}$$

$$\frac{V}{V^*} = (\gamma + 1)\frac{M^2}{W}$$

$$\frac{s - s^*}{R} = \frac{\gamma}{\gamma - 1} \ln\left[M^2\left(\frac{\gamma + 1}{W}\right)^{(\gamma + 1)/\gamma}\right]$$

$$q = h_{02} - h_{01} = \frac{\gamma R}{\gamma - 1}(T_{02} - T_{01})$$

$$\frac{q}{RT_0^*} = \left(\frac{q_m}{RT_0^*}\right)_1 - \left(\frac{q_m}{RT_0^*}\right)_2$$

$$\frac{q_m}{RT_0^*} = \frac{\gamma}{\gamma - 1}\left(\frac{Z}{W}\right)^2$$

See Appendix I for the definitions of W, X, and Z. Equations assume steady flow of a perfect gas in a constant-area frictionless duct. Subscripts 1 and 2 denote inlet and outlet conditions, respectively. If M_1 is supersonic, a normal shock may occur in the duct. The location of the shock does not alter subsonic outlet conditions.

APPENDIX K. FANNO FLOW SUMMARY

$$\frac{p}{p^*} = \frac{1}{M}\left(\frac{\gamma+1}{2X}\right)^{\frac{1}{2}}$$

$$\frac{T}{T^*} = \frac{\gamma+1}{2X}$$

$$\frac{\rho}{\rho^*} = \frac{1}{M}\left(\frac{2X}{\gamma+1}\right)^{\frac{1}{2}}$$

$$\frac{p_0}{p_0^*} = \frac{1}{M}\left(\frac{2X}{\gamma+1}\right)^{[(\gamma+1)/2(\gamma-1)]}$$

$$\frac{V}{V^*} = M\left(\frac{\gamma+1}{2X}\right)^{\frac{1}{2}}$$

$$\frac{s-s^*}{R} = -\ell n\left(\frac{p_0}{p_0^*}\right)$$

$$\frac{F}{F^*} = \frac{W}{[2(\gamma+1)M^2 X]^{\frac{1}{2}}}$$

$$4\bar{C}_f \frac{L_m}{D} = -\frac{Z}{\gamma M^2} + \frac{\gamma+1}{2\gamma}\ell n\left[\frac{(\gamma+1)M^2}{2X}\right]$$

$$4\bar{C}_f \frac{(x_2-x_1)}{D} = \left(4\bar{C}_f \frac{L_m}{D}\right)_1 - \left(4\bar{C}_f \frac{L_m}{D}\right)_2$$

$$\mathcal{T} = F_2 - F_1$$

See Appendix I for the definitions of X and Z. Equations assume steady flow of a perfect gas in a constant-area duct with no heat transfer. Subscripts 1 and 2 denote inlet and outlet conditions, respectively. If M_1 is supersonic, a normal shock may occur as shown in Fig. 8.10. In this case,

$$4\bar{C}_f \frac{(x_s-x_1)}{D} = \left(4\bar{C}_f \frac{L_m}{D}\right)_1 - \left(4\bar{C}_f \frac{L_m}{D}\right)_2$$

$$4\bar{C}_f \frac{(x_4-x_s)}{D} = \left(4\bar{C}_f \frac{L_m}{D}\right)_3 - \left(4\bar{C}_f \frac{L_m}{D}\right)_4$$

where M_2 and M_3 are related by the normal shock equation in Appendix D, and x_s is the location of the shock.

APPENDIX L. UNSTEADY, NORMAL SHOCK SUMMARY

A constant-speed shock wave moving into a quiescent perfect gas is assumed. A prime denotes the unsteady flow variables, and the numbering refers to Fig. 9.3.

$$V_1 = V_s', \qquad V_2 = V_s' - V_2'$$

$$p_1 = p_1', \qquad p_2 = p_2', \qquad a_1 = a_1', \ldots$$

$$M_s = \frac{V_s'}{a_1'} = M_1$$

$$\frac{V_p'}{a_1} = \frac{2}{\gamma+1} \frac{M_s^2 - 1}{M_s}$$

$$M_2' = \frac{M_s^2 - 1}{\{1 + [(\gamma-1)/2] M_s^2\}^{\frac{1}{2}} \{\gamma M_s^2 - [(\gamma-1)/2]\}^{\frac{1}{2}}}$$

$$\frac{p_{01}'}{p_1} = 1, \qquad \frac{T_{01}'}{T_1} = 1$$

$$\frac{p_{02}'}{p_{01}'} = \left\{ \frac{(\gamma+1)/2}{\gamma M_s^2 - [(\gamma-1)/2]} \right\}^{1/(\gamma-1)} \left\{ M_s^2 \frac{(\gamma-1) M_s^2 + [(3-\gamma)/2]}{1 + [(\gamma-1)/2] M_s^2} \right\}^{\gamma/(\gamma-1)}$$

$$\frac{T_{02}'}{T_{01}'} = \frac{2}{\gamma+1} \{(\gamma-1) M_s^2 + [(3-\gamma)/2]\}$$

APPENDIX M. REFLECTED SHOCK WAVE SUMMARY

Normal reflection from a planar wall is assumed for the incident shock wave of Appendix L. A prime denotes the unsteady flow variables, and the numbering refers to Figs. 9.4–9.6.

$$\hat{V}_2 = V_2' - V_r', \qquad V_3 = -V_r'$$

$$p_{03}' = p_3' = p_3, \qquad T_{03}' = T_3' = T_3, \ldots$$

$$M_r = \frac{\hat{V}_2}{a_2} = \left\{ \frac{\gamma M_s^2 - [(\gamma-1)/2]}{1 + [(\gamma-1)/2] M_s^2} \right\}^{\frac{1}{2}}$$

$$M_3' = \frac{V_3'}{a_3} = 0$$

$$\frac{p_3}{p_1} = \frac{p_{03}'}{p_1} = \frac{2}{\gamma+1} \frac{\gamma M_s^2 - [(\gamma-1)/2]}{1 + [(\gamma-1)/2] M_s^2} \left[\frac{3\gamma-1}{2} M_s^2 - (\gamma-1) \right]$$

$$\frac{T_3}{T_1} = \frac{T_{03}'}{T_1} = \left(\frac{2}{\gamma+1}\right)^2 \frac{1}{M_s^2} \left[(\gamma-1) M_s^2 + \frac{3-\gamma}{2} \right] \left[\frac{3\gamma-1}{2} M_s^2 - (\gamma-1) \right]$$

APPENDIX N. JACOBIAN THEORY

A frequency practice in fluid dynamics and thermodynamics is to change either the independent or the dependent variables, or both. Such changes are required for a wide variety of reasons, from finding an analytical solution to recasting the equations in an appropriate form for numerical computation. Any change of variables can be regarded as a transformation. To expedite this type of manipulation, Jacobian theory is useful.

Most calculus texts discuss Jacobians in an elementary fashion. The reason for this interest occurs when changing variables inside an integral. For instance, suppose we have the integral

$$I = \int f(x)\,dx$$

which we hope to evaluate by changing variables. We use the transformation

$$x = g(y)$$

$$dx = g'\,dy$$

to obtain

$$I = \int f[g(y)]g'\,dy$$

where g' is the Jacobian of the transformation. The foregoing process generalizes for multiple integrals of the form

$$I = \int \ldots \int f(x_1,\ldots,x_n)\,dx_1\ldots dx_n$$

as follows. If we change variables by means of the transformation

$$x_1 = x_1(\xi_1,\ldots,\xi_n),\ldots, x_n = x_n(\xi_1,\ldots,\xi_n) \tag{N.1}$$

then I becomes

$$I = \int \ldots \int f\bigl(x_1(\xi_1,\ldots,\xi_n),\ldots\bigr)|J|\,d\xi_1\ldots d\xi_n$$

The Jacobian J is given by the determinant

$$J(x_1,\ldots,x_n) = \frac{\partial(x_1,\ldots,x_n)}{\partial(\xi_1,\ldots,\xi_n)} = \begin{vmatrix} \dfrac{\partial x_1}{\partial \xi_1} & \dfrac{\partial x_1}{\partial \xi_2} & \cdots & \dfrac{\partial x_1}{\partial \xi_n} \\ \dfrac{\partial x_2}{\partial \xi_1} & & \cdots & \vdots \\ \dfrac{\partial x_n}{\partial \xi_1} & & \cdots & \dfrac{\partial x_n}{\partial \xi_n} \end{vmatrix} \tag{N.2}$$

One can show that $J \neq 0$, if the inverse transformation

$$\xi_1 = \xi_1(x, \ldots, x_n), \ldots$$

is to exist. If it does exist, then its Jacobian

$$J^{-1} = \frac{\partial(\xi_1, \ldots, \xi_n)}{\partial(x_1, \ldots, x_n)}$$

is related to J by

$$JJ^{-1} = \frac{\partial(x_1, \ldots, x_n)}{\partial(\xi_1, \ldots, \xi_n)} \frac{\partial(\xi_1, \ldots, \xi_n)}{\partial(x_1, \ldots, x_n)} = 1 \qquad \text{(N.3)}$$

For simplicity, let us consider the transformation

$$x_1 = x_1(\xi_1, \xi_2)$$
$$x_2 = x_2(\xi_1, \xi_2) \qquad \text{(N.4)}$$

which has the Jacobian

$$J(x_1, x_2) = \frac{\partial(x_1, x_2)}{\partial(\xi_1, \xi_2)} = \begin{vmatrix} \dfrac{\partial x_1}{\partial \xi_1} & \dfrac{\partial x_1}{\partial \xi_2} \\ \dfrac{\partial x_2}{\partial \xi_1} & \dfrac{\partial x_2}{\partial \xi_2} \end{vmatrix} = \begin{vmatrix} \dfrac{\partial x_1}{\partial \xi_1} & \dfrac{\partial x_2}{\partial \xi_1} \\ \dfrac{\partial x_1}{\partial \xi_2} & \dfrac{\partial x_2}{\partial \xi_2} \end{vmatrix}$$

$$= \frac{\partial x_1}{\partial \xi_1} \frac{\partial x_2}{\partial \xi_2} - \frac{\partial x_1}{\partial \xi_2} \frac{\partial x_2}{\partial \xi_1} \qquad \text{(N.5)}$$

One can also show that

$$\frac{\partial x_1}{\partial \xi_1} = J \frac{\partial \xi_2}{\partial x_2}$$

$$\frac{\partial x_1}{\partial \xi_2} = -J \frac{\partial \xi_1}{\partial x_2}$$

$$\frac{\partial x_2}{\partial \xi_1} = -J \frac{\partial \xi_2}{\partial x_1} \qquad \text{(N.6)}$$

$$\frac{\partial x_2}{\partial \xi_2} = J \frac{\partial \xi_1}{\partial x_1}$$

whose generalization is given in the "Invariance of the Conservative Form" subsection of Sec. 13.1. Hence, if Eqs. (N.4) are explicitly known, then Eq. (N.5) yields J, and Eqs. (N.6) produce the $\partial \xi_j / \partial x_i$. Note that in Eqs. (N.6), we have

$$\frac{\partial x_1}{\partial \xi_1} = \frac{\partial x_1}{\partial \xi_1}\bigg|_{\xi_2}, \qquad \frac{\partial \xi_1}{\partial x_1} = \frac{\partial \xi_1}{\partial x_1}\bigg|_{x_2}$$

Consequently, the relation

$$\frac{\partial x_1}{\partial \xi_1}\bigg|_{\xi_2} = 1 \bigg/ \frac{\partial \xi_1}{\partial x_1}\bigg|_{x_2}$$

is incorrect.

Suppose we have two consecutive transformations, given by Eqs. (N.4) and

$$\xi_1 = \xi_1(\eta_1, \eta_2)$$
$$\xi_2 = \xi_2(\eta_1, \eta_2)$$

The Jacobian for the transformation

$$x_1 = \phi_1(\eta_1, \eta_2)$$
$$x_2 = \phi_2(\eta_1, \eta_2)$$

is given by the product

$$J(x_1, x_2) = \frac{\partial(x_1, x_2)}{\partial(\eta_1, \eta_2)} = \frac{\partial(x_1, x_2)}{\partial(\xi_1, \xi_2)} \frac{\partial(\xi_1, \xi_2)}{\partial(\eta_1, \eta_2)}$$

From Eq. (N.5), one can also readily show that

$$J(x_1, x_2) = -J(x_2, x_1)$$

All the above discussion can be generalized for the Jacobian, Eq. (N.2), of the transformation given by Eqs. (N.1). The derivatives inside the determinant in Eq. (N.2) are understood as

$$\frac{\partial x_1}{\partial \xi_1} = \frac{\partial x_1}{\partial \xi_1}\bigg|_{\xi_2, \ldots, \xi_n}, \ldots$$

We have the following (self-evident) rules:

(1) The sign of J is changed whenever any pair of xs or ξs are interchanged, provided the order of the rest is preserved. This property

stems directly from the sign change when two columns or two rows of a determinant are interchanged.

(2) Whenever a common variable occurs among the xs and ξs, a reduction in order takes place. We thus have

$$\frac{\partial(\xi_1, x_2, \ldots, x_n)}{\partial(\xi_1, \xi_2, \ldots, \xi_n)} = \frac{\partial(x_2, \ldots, x_n)}{\partial(\xi_2, \ldots, \xi_n)}\bigg|_{\xi_1}$$

where the subscript ξ_1 on the right is held constant throughout. If not written, the ξ_1 subscript is understood. Similarly,

$$\frac{\partial(\xi_1, \ldots, \xi_m, x_{m+1}, \ldots, x_n)}{\partial(\xi_1, \ldots, \xi_m, \xi_{m+1}, \ldots, \xi_n)} = \frac{\partial(x_{m+1}, \ldots, x_n)}{\partial(\xi_{m+1}, \ldots, \xi_n)}$$

and if $m = n - 1$, then

$$\frac{\partial(\xi_1, \ldots, \xi_{n-1}, x_n)}{\partial(\xi_1, \ldots, \xi_n)} = \frac{\partial x_n}{\partial \xi_n}$$

A first-order Jacobian is merely an ordinary first-order partial derivative. Of equal importance is the converse; i.e., any partial derivative can be written as an nth-order Jacobian with $n - 1$ common x and ξ variables.

(3) A necessary and sufficient condition that Eqs. (N.1) have a unique inverse is that $J \neq 0$.

(4) The general transformation property

$$\frac{\partial(x_1, \ldots, x_n)}{\partial(\xi_1, \ldots, \xi_n)} = \frac{\partial(x_1, \ldots, x_n)/\partial(\eta_1, \ldots, \eta_n)}{\partial(\xi_1, \ldots, \xi_n)/\partial(\eta_1, \ldots, \eta_n)} = \frac{J(x_1, \ldots, x_n)}{J(\xi_1, \ldots, \xi_n)}$$

holds. In the Jacobians on the right, the (understood) independent variables are $\eta_1, \ldots, \eta_n$.

An immediate but useful consequence of rules (2) and (4) is

$$\frac{\partial x_r}{\partial \xi_m}\bigg|_{\xi_1, \ldots, \xi_{m-1}, \xi_{m+1}, \ldots, \xi_n} = 1 \bigg/ \frac{\partial \xi_m}{\partial x_r}\bigg|_{\xi_1, \ldots, \xi_{m-1}, \xi_{m+1}, \ldots, \xi_n}$$

This brief exposition of Jacobian theory has been included because of its use in Chap. 13. Reference 1 develops this approach for classical thermodynamics and is the basis of the presentation in this appendix.

Reference

[1] Crawford, F. H., "Thermodynamic Relations in n-Variable Systems in Jacobian Form: Part I, General Theory and Application to Unrestricted Systems," *Proceedings of the American Academy of Arts and Sciences*, Vol. 78, April 1950, pp. 165–184.

Index

Adiabatic:
    flow, 26
    process, 7
Aerodynamic window (AW):
    compression, 330-331
    expansion, 330-331
    infinite pressure ratio, 336
    jet design, 337-340
    nozzle design, 340-344
    vortex, 331-336
Airfoil:
    diamond or double wedge, 80
    flat plate, 78
    quadrilateral, 82
    triangular, 82
Angle of attack, 77, 400, 429

Ballistics, internal, 175
Bernoulli, 249:
    equation, 40, 365
Blast wave diffraction, 347, 378
Boundary conditions, 202, 349
Boundary layer (*see also* shock wave boundary-layer interaction), 2, 78, 82, 94, 101-103, 343
Busemann, A., 276

Cartesian coordinates, 197
Characteristics, 69, 283-303:
    applications, 169-175, 294-303, 312, 322, 364, 387
    existence, 286, 292
    lines, 162, 293, 295, 298-301, 387
    method (MOC), 147, 157-163
    right and left running, 298
    rotational flow, 296-297
    termination, 288
    theory, 286-290
    uniqueness, 287, 292
Choked flow:
    area change, 94
    frictional, 137
    thermal, 122-123, 125, 130, 133, 141
Chord, 77
Circulation, 250
Clausius, 17
Cluster (condensation):
    evaporation, 190
    formation, 190
Compatibility equations, 163, 287, 294-302, 387
Complex Mach reflection (CMR), 384
Conical flow, see Taylor-Maccoll flow
Conservation equations:
    conservative form invariance, 217-221
    general, 199-200
    hodograph, 242
    influence coefficient form, 115, 433
    natural coordinates, 239
    transformation properties, 216-217
    two-dimensional or axisymmetric, 225
    unsteady one-dimensional, 158
Conservative form, 200-202
Contact surface, 181
Continuity equation, 28
Control volume derivation, 27
Coordinates:
    Cartesian, 197
    characteristic, 290-294
    natural, 238-240
    orthogonality condition, 229
Courant, R. and Friedrichs, K.O., 283
Crocco's equation, 249, 253-255
Curvature, 227:
    azimuthal, 236
    condition, 229
    Gaussian, 236
    longitudinal, 236
    principal, 235

Detonation wave, 133, 181
Diffuser, 89, 101-107, 329, 331, 337:
    blockage, 104
    frictional losses, 104
    ideal flow, 104, 106
    starting process, 104
    variable geometry, 106
Diffusion cloud chamber, 190
Dissipation, 6
Double Mach reflection (DMR), 384
Drag (*see also* waverider):
    coefficient, 77, 429-430
    definition, 77
    viscous, 78, 412
    wave, 78, 412

Edney, B.E., 387-388
Energy:
    conservation of, 4, 29
    internal, 3
    kinetic, 29
    total, 30
Engine:
    inlet, 89, 101, 276
    jet, 1, 118
    ramjet, 142, 144
    rocket, 103
    scramjet, 144
Enthalpy, 6:
    stagnation, 30
Entropy, 7
Euler, 249:
    equation, 262

theorem, 236
Eulerian derivative, 23
Expansion:
   Prandtl-Meyer, 72-77
   unsteady, 163-175

Fanno flow, 113, 116, 132-140, 437:
   equation, 133
   validity, 140
Far field, 354
Film coefficient, 14, 131
First law of thermodynamics, 33, 419
Flat plate airfoil, 78-80
Forcing functions, 116

Gas:
   diatomic, 12
   monatomic, 12
   perfect, 2, 9, 11
Gas constant, 9:
   universal, 10
Governing equations, see Conservation equations
Grain silo, 154

Helmholtz vortex theorems, 249
Homentropic, 31, 221, 240, 249, 253
Homogeneity, 262-267:
   degree, 262
   utility, 266-267
Hydraulic diameter, 114, 131
Hypersonic flow, 402, 409:
   similarity parameter, 402
   small disturbance theory, 409

Impulse function, 118, 121, 137
Incompressible flow, 1-2, 249
Influence coefficient method, 116, 118, 132
Initial conditions, 202
Inviscid flow, 2, 26
Irrotational flow, 221, 226, 249, 251, 253
Isentropic:
   equations, 41
   flow, 221, 226, 249, 253
   point relations, 151, 168
Isoenergetic flow, 221, 226, 240, 249, 253

Jacobian theory, 443-446:
   contraction, 217
   coordinates, 214
   zero value, 217
Jet, 127, 337:
   free, 242
   overexpanded, 93, 363
   theory, 97-101
   underexpanded, 93, 363-367

Kármán-Nikuradse equation, 140
Kelvin's equation, 250

Kim, B.S., 400, 409
Kink, 384
Krieger, R.J., 400
Kronecker delta, 201

Laplace's equation, 158
Laser:
   beam, 329, 331
   cavity, 329
   chemical, 133, 308, 336
   excimer, 308
   free-electron, 336
   gas, 1, 89
   gasdynamic, 133, 308
   induced chemistry, 133, 308
   isotope separation, 133, 308
   Raman shifting, 308
Leading edge, 72, 163
Legendre transformation, 261
Length scale, 72, 368, 373, 375, 381
Lift (see also waverider):
   coefficient, 77, 429-430
   definition, 77
Limit line, 275

Mach angle, 69, 427
Mach disk, 100, 364
Mach lines, waves, or cones, 59, 69, 286
Mach number, 41, 421:
   freestream, 77
   reflected shock, 154
   shock wave, 51, 147
   sonic, 41
   subsonic, 41
   supersonic, 41
Mach reflection (MR), 64, 354, 367, 375-376, 390
Mach stem, 64, 372, 382-384
Mass flow rate, 44, 421
Matrix:
   A, 214, 222, 230
   identity, 266
   inverse, 215
   transpose, 215
Maxwell equation, 9
Meridian plane, 235
Meyer, R.E., 226
Minimum length nozzle (MLN), 307-326:
   applications, 307-308, 330, 342
   comparison, 324-326
   curved sonic line, 308-321
   kernel, 321
   rapid expansion, 308
   straight sonic line, 321-324
Missing solutions, 242
Molecular weight, 9:
   table, 194
Mollier diagram, 121, 123, 129, 133-134, 137, 139, 142
Moody chart, 135, 140

Near field, 354
Newton's second law, 28, 113-114, 199
Nomenclature, 2, 415
Nonequilibrium effects, 347
Nonweiler, T.R.F., 400
Nozzle (*see also* minimum length nozzle):
    area ratio, 90
    converging/diverging, 44
    isentropic, 44
    rocket, 1, 23, 89, 103-104, 118
    theory, 39-46, 89-97
    throat, 44
Nusselt number, 131

Oswatitsch, K., 352
Owczarek, J.A., 249

Pack, D.C., 364
Partial differential equations (PDE) (*see also* quasi-linear equation):
    elliptic, 157
    hyperbolic, 157
    linear, 158
    parabolic, 157
Perfect gas, 2, 11:
    calorically, 11
    thermally, 9
Pipe flow, 78, 119, 131, 140
Piston/cylinder, 4, 6, 147
Piston expansion tube (PET), 181, 189-194:
    carrier gas, 190, 192
    utility, 190
    vapor, 190, 192
Planform area, 78
Plenum, 90
Prandtl-Meyer:
    derivative, 299
    expansion, 69
    function, 72, 280, 300, 427
Prandtl number, 132
Pressure, 5:
    ambient, 89
    back, 89
    dynamic, 43, 78
    pitot, 55, 394-396
    stagnation, 44
    static, 44
Pressure gradient:
    adverse, 102, 311, 341
    favorable, 29, 102, 311
Pressure recovery, 104-105
Process:
    irreversible, 6
    isentropic, 17
    reversible, 6
Pseudo-steady reflection, 382:
    theory, 385-387

Quasi-linear equation, 157, 260, 290
Quasi-one-dimensional flow, 23, 113, 431, 433

Radial flow, 273
Ramp:
    compressive, 58, 61, 102, 349-352, 358-363
    contour, 352-354
    expansive, 71
Rankine-Hugoniot equations, 205
Rarefaction wave, 181
Rasmussen, M.L., 280, 400-401
Rayleigh flow, 113, 116, 121-133, 435:
    convective heat transfer, 131-132
    equation, 121
    validity, 132
Real-gas effects, 157
Reciprocity equation, 8, 419
Reducible equations, 289-290
Region:
    dependence, 288
    influence, 288
    uniform flow, 289
Regular reflection (RR), 64, 354, 367-375, 378-382, 390
Reservoir, 90
Reynolds number, 131-132, 135, 140, 308, 401
Riemann invariants, 163, 173, 296, 299, 350

Saadat, A., 322
Salas, M.D. and Morgan, B.D., 348-349
Scale factors, 215, 223, 229
Second law of thermodynamics, 17, 122, 137, 419
Shapiro, A.H., 113, 307
Shock-expansion theory, 77-82, 429-430:
    validity, 82
Shock polar diagrams, 370-371, 387
Shock tube, 154, 376:
    experiments, 185, 190
    flow, 181-189
    validity, 188-189
Shock wave:
    attached, 60, 280
    boundary-layer interaction, 348, 375, 381-382
    bow, 55, 60-61, 378
    condensation, 190-191
    detached, 60-61, 280
    formation, 349, 365
    interference, 355, 387-396
    nonplanar, 203
    normal, 49-56
    oblique, 56-64
    primary, 354
    reflected, 62-64, 100-101, 147, 153-157

refraction, 348
secondary disturbance, 354
strong solution, 59, 98, 280
theory of surfaces, 234-238
three-dimensional, 206-211
transformation to a steady frame, 148-149, 154, 193, 205
two-dimensional or axisymmetric, 234
unsteady, 147-153, 203-205, 439, 441
vector form, 207
weak solution, 59, 98, 280
Shock wave transition, steady flow:
detachment condition, 372-375
mechanical equilibrium, 372-375
RR to MR, 371-375
sonic condition, 372-374
Shock wave transition, unsteady flow:
RR to MR, 380-381
Similarity method, 189, 191, 242, 273, 349, 352
Single Mach reflection (SMR), 382
SI units, 2, 9, 415
Skin friction coefficient, 78, 114, 134
Slipstream, 64, 100
Small disturbance, 31:
propagation, 31
Sonic line, 378:
curved, 307
straight, 307
Sonic point, 368
Specific, 4
Specific heat, 419:
at constant pressure, 12
at constant volume, 12
ratio, 12
Speed, maximum, 257
Speed of sound, 18, 419:
perfect gas, 18, 419
Spherical source or sink flow, 309:
into uniform flow, 311-312
Spiral flow, 273-276, 331
Stable solution, 348
Stagnation, 30
State equations:
caloric, 8, 419
thermal, 8, 419
Steady flow, 25
Stream function:
conventional, 255
Crocco, 256-259
three-dimensional, 257
Streamline, 23
Streamtube, 23
Substantial derivative, 23, 197
Superposition of solutions, 158, 242
Surface:
elliptic, parabolic, 236
fundamental differential forms, 235

Taylor-Maccoll flow, 276-280:
applications, 276
equations, 277
limiting cases, 277-280
validity, 280
Temperature, 6:
film, 132
stagnation, 41
Thermal conductivity, 132
Thermal state equation, 8:
Clausius-II, 13
Dieterici, 21
perfect gas, 9
Redlich-Kwong, 21
van der Waals, 21
virial, 21
Thermodynamic system, 3:
closed, 3
simple, 3
Thermodynamic variable:
equilibrium, 7
intensive, 5
state, 5
Thrust, 117, 437
Time invariance, 382
Trailing edge, 72, 163
Transformation:
hodograph, 240-243, 260, 289
Legendre, 261
orthogonal, 215
theory, 213, 217
Transonic flow, 242
Triple point, 64, 100, 355-356, 372, 382, 384, 390
Truly nonstationary flow, 348
Two-dimensional or axisymmetric flow:
Cartesian or polar coordinates, 221-226
orthogonal curvilinear coordinates, 226-234

Uniform flow, 89:
into source or sink flow, 311-312

Vacuum:
expansion, 75
unsteady flow, 168
Velocity, 23:
component transformation, 213-216
divergence, 198, 232
potential, 251
Venturi meter, 92
Volume dilation, 197-198
Vortex flow, 273
Vorticity, 250, 259-260

Wall:
compressive turn, 69
convex turn, 74, 302
expansive turn, 69, 72

reflection condition, 312
roughness, 135
smooth, 73, 76

Wave drag, 78:
camber, 82
lift, 79
thickness, 81

Wave equation, 33

Wave region:
nonsimple, 166, 172-174, 289
simple, 166, 289
uniform flow, 166

Waverider:
aerodynamics, 400-407
caret-shaped, 400-404
cone-derived, 407-413
drag coefficient, 401, 403, 407, 412-413
heat transfer, 399-400
idealized conical, 407
lift coefficient, 401-403, 407, 412-413
planform area, 404, 407, 411

Waves, 34:
acoustic, 35
left-running, 34
right-running, 34

Wetted surface area, 131
Wilson cloud chamber, 190
Wind tunnel, 1, 89, 102-103:
blowdown, 108
condensation, 190
cryogenic, 11
hypersonic, 307, 325

Work, 3

Xenon vapor pressure, 194